AN INTRODUCTION TO THE MATHEMATICS OF FINANCIAL DERIVATIVES

AN INTRODUCTION TO THE MATHEMATICS OF FINANCIAL DERIVATIVES

Edited by

ALI HIRSA

SALIH N. NEFTCI†

ELSEVIER

AMSTERDAM • BOSTON • HEIDELBERG • LONDON
NEW YORK • OXFORD • PARIS • SAN DIEGO
SAN FRANCISCO • SINGAPORE • SYDNEY • TOKYO
Academic Press is an imprint of Elsevier

An Introduction to the Mathematics of Financial Derivatives, 3rd Edition

Ali Hirsa

Salih N. Neftci

Original English language edition published by Elsevier, Academic Press

ISBN : 978-0-12-384682-2

Reprint ISBN : 978-89-94961-98-9

Elsevier Korea L.L.C

4FL, Chunwoo Bldg, 534 Itaewon-dong,

Yongsan-gu, Seoul 140-861, S. Korea

First Published March 2014

Print in Taiwan

Contents

List of Symbols and Acronyms

$\mathbb{R}$	set of real numbers
$\mathbb{R}^+$	set of positive real numbers
$\mathbb{Z}$	set of integers
$\mathbb{N}$	set of natural numbers
$p(t, T)$	price of a zero-coupon at time t maturing at T
r_t	instantaneous short rate at calendar time t
B_t	money market account at time t starting with \$1 at time 0 and rolling at the instantaneous short rate
$f(t, T)$	instantaneous forward rate at calendar time t for the forward period $[T, T+dt]$
$\mathbb{P}$	real-world (physical) measure
$\mathbb{Q}$	Risk neutral measure
$\mathbb{E}(x)$	expectation of x under some measure
$\mathbb{E}_t(x)$	expectation of x under some measure conditional on knowing all information up to t

CHAPTER

1

Financial Derivatives—A Brief Introduction

OUTLINE

1.1 INTRODUCTION

This book is an introduction to quantitative tools used in pricing financial derivatives. Hence, it is mainly about mathematics. It is a simple and heuristic introduction to mathematical concepts that have practical use in financial markets.

Such an introduction requires a discussion of the logic behind asset pricing. In addition, at various points we provide examples that also require an understanding of formal asset pricing methods. All these necessitate a brief discussion of the securities under consideration. This introductory chapter has that aim. Readers can consult other books to obtain more background on derivatives. Hull (2009) is an excellent source for derivatives. Jarrow and Turnbull (1996) gives another approach. The more advanced books by

Financial Derivatives, Third Edition.
http://dx.doi.org/10.1016/B978-0-12-384682-2.00001-3

Ingersoll (1987) and Duffie (1996) provide strong links to the underlying theory. The manual by Das (1994) provides a summary of the practical issues associated with derivative contracts. A comprehensive new source is Wilmott (1998).

This chapter first deals with the two basic building blocks of financial derivatives: options and forwards (futures). Next, we introduce the more complicated class of derivatives known as swaps. The chapter concludes by showing that a complicated swap can be decomposed into a number of forwards and options. This decomposition is very practical. If one succeeds in pricing forwards and options, one can then reconstitute any swap and obtain its price. This chapter also introduces some formal notation that will be used throughout the book.

1.2 DEFINITIONS

In the words of practitioners, "Derivative securities are financial contracts that 'derive' their value from *cash market* instruments such as stocks, bonds, currencies and commodities."[1]

The academic definition of a "derivative instrument" is more precise.

Definition 1 (Ingersoll, 1987). A financial contract is a derivative security, or a contingent claim, if its value at expiration date T is determined exactly by the market price of the underlying cash instrument at time T.

Hence, at the time of the expiration of the derivative contract, denoted by T, the price $F(T)$ of a derivative asset is completely determined by S_T, the value of the "underlying asset." After that date, the security ceases to exist. This simple characteristic of derivative assets plays a very important role in their valuation.

In the rest of this book, the symbols $F(t)$ and $F(S_t, t)$ will be used alternately to denote the price of a derivative product written on the underlying asset S_t at time t. The financial derivative is sometimes assumed to yield a payout dt. At other times, the payout is zero. T will always denote the expiration date.

1.3 TYPES OF DERIVATIVES

We can group derivative securities under three general headings:

1. Futures and forwards.
2. Options.
3. Swaps.

Forwards and options are considered basic building blocks. Swaps and some other complicated structures are considered hybrid securities, which can eventually be decomposed into sets of basic forwards and options.

We let S_t denote the price of the relevant cash instrument, which we call the underlying security.

We can list five main groups of underlying assets:

1. *Stocks:* These are claims to "real" returns generated in the production sector for goods and services.
2. *Currencies:* These are liabilities of governments or, sometimes, banks. They are not direct claims on real assets.
3. *Interest rates:* In fact, interest rates are not assets. Hence, a *notional asset* needs to be devised so that one can take a position on the direction of future interest rates. Futures on Eurodollars is one example. In this category, we can also include derivatives on bonds, notes, and T-bills, which are government debt instruments. They are promises by governments to make certain payments on set dates. By dealing with derivatives on bonds, notes, and T-bills, one takes positions on the direction of various interest rates. In most cases,[2]

[1] See pages 2–3, Klein and Lederman (1994).

[2] There is a significant amount of trading on "notional" French government bonds in Paris.

these derivative instruments are not *notionals* and can result in actual delivery of the underlying asset.

4. *Indexes:* The S&P 500 and the FTSE 100 are two examples of stock indexes. The CRB commodity index is an index of commodity prices. Again, these are not "assets" themselves. But derivative contracts can be written on notional amounts and a position taken with respect to the direction of the underlying index.
5. *Commodities:* The main classes are
   - *Soft commodities:* cocoa, coffee, and sugar.
   - *Grains and oilseeds:* barley, corn, cotton, oats, palm oil, potato, soybean, winter wheat, spring wheat, and others.
   - *Metals:* copper, nickel, tin, and others.
   - *Precious metals:* gold, platinum, and silver.
   - *Livestock:* cattle, hogs, pork bellies, and others.
   - *Energy:* crude oil, fuel oil, and others.

These underlying commodities are not *financial* assets. They are goods in kind. Hence, in most cases, they can be physically purchased and stored.

There is another method of classifying the underlying asset, which is important for our purposes.

1.3.1 Cash-and-Carry Markets

Some derivative instruments are written on products of *cash-and-carry* markets. Gold, silver, currencies, and T-bonds are some examples of cash-and-carry products.

In these markets, one can *borrow* at risk-free rates (by collateralizing the underlying physical asset), *buy* and *store* the product, and *insure* it until the expiration date of any derivative contract. One can therefore easily build an alternative to holding a forward or futures contract on these commodities.

For example, one can borrow at risk-free rates, buy a T-bond, and hold it until the delivery date of a futures contract on T-bonds. This is equivalent to buying a futures contract and accepting the delivery of the underlying instrument at expiration. One can construct similar examples with currencies, gold, silver, crude oil, etc.[3]

Pure cash-and-carry markets have one more property. Information about future demand and supplies of the underlying instrument should not influence the "spread" between cash and futures (forward) prices. After all, this spread will depend mostly on the level of risk-free interest rates, storage, and insurance costs. Any relevant information concerning future supplies and demands of the underlying instrument is expected to make the cash price and the future price change by the same amount.

1.3.2 Price-Discovery Markets

The second type of underlying asset comes from *price-discovery* markets. Here, it is physically impossible to buy the underlying instrument for cash and store it until some future expiration date. Such goods either are too *perishable* to be stored or may not have a cash market at the time the derivative is trading. One example is a contract on spring wheat. When the futures contract for this commodity is traded in the exchange, the corresponding cash market may not yet exist.

The strategy of borrowing, buying, and storing the asset until some later expiration date is not applicable to price-discovery markets. Under these conditions, any information about the *future* supply and demand of the underlying commodity cannot influence the corresponding cash price. Such information can be *discovered* in the futures market, hence the terminology.

1.3.3 Expiration Date

The relationship between $F(t)$, the price of the derivative, and $S(t)$, the value of the underlying

[3]However, as in the case of crude oil, the storage process may end up being very costly. Environmental and other effects make it very expensive to store crude oil.

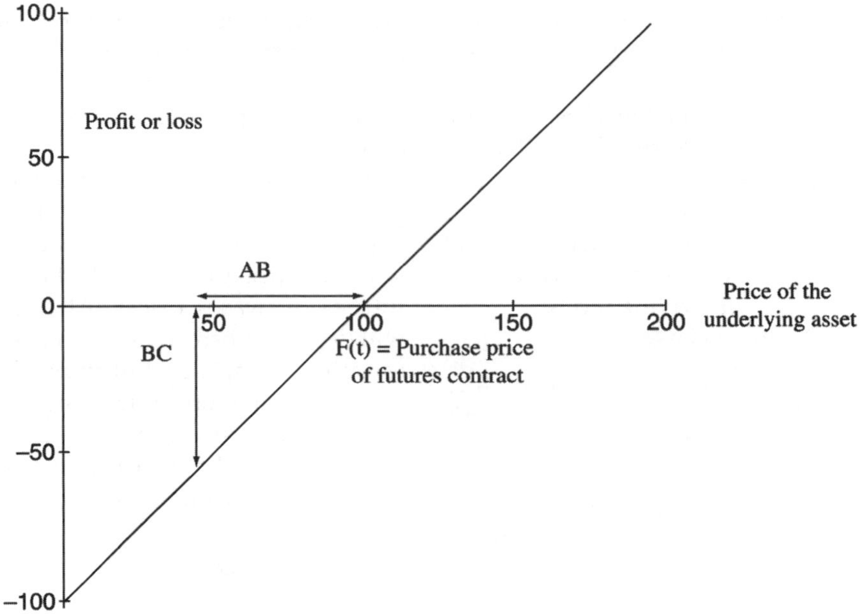

FIGURE 1.1 Payoff diagram for a long position.

asset, is known exactly (or deterministically), only at the expiration date T. In the case of forwards or futures, we naturally expect

$$F(T) = S_T \tag{1.1}$$

that is, at expiration the value of the futures contract should be equal to its cash equivalent.

For example, the (exchange-traded) futures contract promising the delivery of 100 troy ounces of gold cannot have a value different from the actual market value of 100 troy ounces of gold on the expiration date of the contract. They both represent the same thing at time T. So, in the case of gold futures, we can indeed say that the equality in (1.1) holds at expiration.

At $t < T, F(t)$ may not equal S_t. Yet we can determine a function that ties S_t to $F(t)$.

1.4 FORWARDS AND FUTURES

Futures and forwards are linear instruments. This section will discuss forwards; their differences from futures will be briefly indicated at the end.

1.4.1 Forwards

Definition 2. A forward contract is an obligation to buy (sell) an underlying asset at a specified forward price on a known date.

The expiration date of the contract and the forward price are written when the contract is entered into. If a forward purchase is made, the holder of such a contract is said to be long in the underlying asset. If at expiration the cash price is higher than the forward price, the long position makes a profit; otherwise there is a loss.

The payoff diagram for a simplified long position is shown in Figure 1.1. The contract is purchased for $F(t)$ at time t. It is assumed that the contract expires at time $t + 1$. The upward-sloping line indicates the profit or loss of the purchaser at expiration. The slope of the line is one.

If S_{t+1} exceeds $F(t)$, then the long position ends up with a profit.[4] Given that the line has unitary slope, the segment AB equals the vertical line BC. At time $t+1$ the gain or loss can be read directly as being the vertical distance between this "payoff" line and the horizontal axis.

Figure 1.2 displays the payoff diagram of a *short position* under similar circumstances.

Such payoff diagrams are useful in understanding the mechanics of derivative products. In this book we treat them briefly. The reader can consult Hull (2009) for an extensive discussion.

1.4.2 Futures

Futures and forwards are similar instruments. The major differences can be stated briefly as follows.

Futures are traded in formalized *exchanges*. The exchange designs a standard contract and sets some specific expiration dates. Forwards are custom-made and are traded *over-the-counter*.

Futures exchanges are cleared through exchange clearing houses, and there is an intricate mechanism designed to reduce the default risk.

Finally, futures contracts are *marked to market*. That is, every day the contract is settled and simultaneously a new contract is written. Any profit or loss during the day is recorded accordingly in the account of the contract holder.

1.4.3 Repos, Reverse Repos, and Flexible Repos

A repurchase agreement, also known as a repo, is a transaction in which one party sells securities to another party in return for cash, with an agreement to repurchase equivalent securities at an agreed price and on an agreed future date. The difference between the repurchase price and the original sale price is effectively interest which is referred to as the repo rate. The buyer of the security acts as a lender and the seller acts as a borrower. The security is used as collateral for a secured cash loan at a fixed rate of interest. A repo can be seen as a spot sale combined with a forward contract. The spot sale results in transfer of money to the borrower in exchange for legal transfer of the security to the lender, while the forward contract ensures repayment of the loan to the lender and return of the collateral of the borrower. The difference between the forward price and the spot price is effectively the interest on the loan, while the settlement date of the forward contract is the maturity date of the loan.

There are three types of repos: (a) overnight, (b) term, and (c) open repo. An overnight repo is a one-day maturity transaction. A term repo is a repo with a specified end date. An open repo has no end date. Repo transactions occur in three forms: specified delivery, tri-party, and hold-in-custody (wherein the selling party holds the security during the term of the repo). Hold-in-custody is used primarily when there is a risk that the seller will become insolvent prior to maturity of the repo and the buyer will be unable to recover the securities that were posted as collateral to secure the transaction.

For the party selling the security and agreeing to repurchase it in the future, it is a repo. For the party on the other end of the transaction, buying the security and agreeing to sell in the future, it is a *reverse repo*.

Repos are usually used to raise short-term capital. They are classified as a money market instrument.

A flexible repo is a repo with a (flexible) withdrawal schedule. That means the party who bought the security could sell it partially at times before and at maturity as opposed to just at maturity (in case of a repo). The amounts and the time of withdrawals are flexible.

There are four major differences with a traditional repo, which are:

- Convexity due to cash withdrawals.
- Formal written auction like trade.

[4]Note that because the contract expires at $t+1$, S_{t+1} will equal $F(t+1)$.

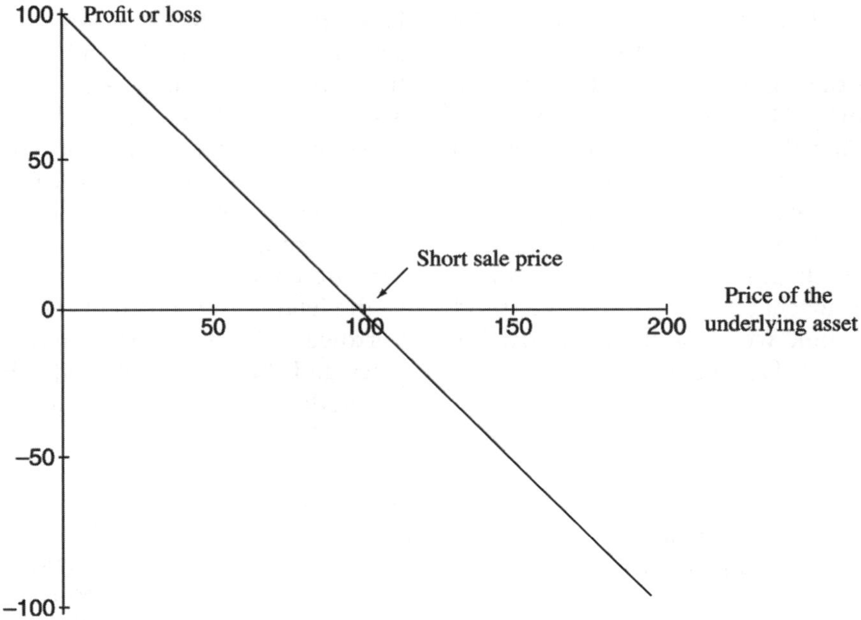

FIGURE 1.2 Payoff diagram of a short position.

- Enhanced documentation (additional documentation is provided for credit support and bankruptcy court protection).
- In the flexible repo market, counterparties are municipal bond issuers.

There are two types of flexible repos:
Secured.
Unsecured.

In a secured flexible repo, the municipality receives collateral. Municipalities trustees would monitor and clear the collateral for the municipality. Collateral could be treasury bonds, GNMA bonds, agency MBS/CMO, or the like. The type of security to be used as collateral is explained in detail in the documentation and request for proposal (RFP). Collateral in most cases comes from the reverse repo market, i.e. a bank buys it from someone else with an agreement to sell it back.

In an unsecured flexible repo, the customer does not receive collateral. In that case, a customer may receive a higher spread. The average size of a single unsecured flexible repo is 10–20 million. The size of a single unsecured flexible repo is usually less than a single secured one.

1.5 OPTIONS

Options constitute the second basic building block of asset pricing. In later chapters we often use pricing models for standard call options as a major example to introduce concepts of stochastic calculus.

Forwards and futures *obligate* the contract holder to deliver or accept the delivery of the underlying instrument at expiration. Options, on the other hand, give the owner the right, but not the obligation, to purchase or sell an asset.

There are two types of options.

Definition 3. A European-type call option on a security S_t is the right to buy the security at a present strike price K. This right may be exercised at the expiration date T of the option. The call option can be purchased for a price of C_t dollars, called the premium, at time $t < T$.

A European *put option* is similar, but gives the owner the right to *sell* an asset at a specified price at expiration.

In contrast to European options, *American* options can be exercised *any* time between the writing and the expiration of the contract.

There are several reasons that traders and investors may want to calculate the arbitrage-free price, C_t, of a call option. Before the option is first *written* at time t, C_t is not known. A trader may want to obtain some estimate of what this price will be if the option is written. If the option is an exchange-traded security, it will start trading and a market price will emerge. If the option trades over-the-counter, it may also trade heavily and a price can be observed.

However, the option may be traded infrequently. Then a trader may want to know the daily value of C_t in order to evaluate its risks. Another trader may think that the market is mispricing the call option, and the extent of this mispricing may be of interest. Again, the arbitrage-free value of C_t needs to be determined.

1.5.1 Some Notation

The most desirable way of pricing a call option is to find a *closed-form* formula for C_t that expresses the latter as a function of the underlying asset's price and the relevant parameters.

At time t, the only known "formula" concerning C_t is the one that determines its value at the time of expiration denoted by T. In fact,

- if there are no commissions and/or fees, and
- if the bid-ask spreads on St and Ct are zero,

then at expiration, C_T can assume only two possible values.

If the option is expiring *out-of-money*, that is, if at expiration the option holder faces

$$S_T < K \tag{1.2}$$

then the option will have no value. The underlying asset can be purchased in the market for S_T, and this is less than the strike price K. No option holder will exercise his or her right to buy the underlying asset at K. Thus,

$$S_T < K \Rightarrow C_T = 0 \tag{1.3}$$

But, if the option expires *in-the-money*, that is, if at time T,

$$S_T > K \tag{1.4}$$

the option will have some value. One should clearly exercise the option. One can buy the underlying security at price K and sell it at a higher price S_T. Since there are no commissions or bid-ask spreads, the net profit will be $S_T - K$. Market participants, being aware of this, will place a value of $S_T - K$ on the option, and we have

$$S_T > K \Rightarrow C_T = S_T - K \tag{1.5}$$

We can use a shorthand notation to express both of these possibilities by writing,

$$C_T = \max[S_T - K, 0] \tag{1.6}$$

This means that the C_T will equal the greater of the two values inside the brackets. In later chapters, this notation will be used frequently.

Equation (1.6), which gives the relation between S_T and C_T, can be graphed easily. Figure 1.3 shows this relationship. Note that for $S_T \leq K$, the C_T is zero. For values of S_T such that $K < S_T$, the C_T increases at the same rate as S_T. Hence, for this range of values, the graph of Eq. (1.6) is a straight line with unitary slope. Options are *nonlinear* instruments.

Figure 1.4 displays the value of a call option at various times before expiration. Note that for $t < T$ the value of the function can be represented by a smooth continuous curve. Only at expiration does the option value become a piecewise linear function with a kink at the strike price.

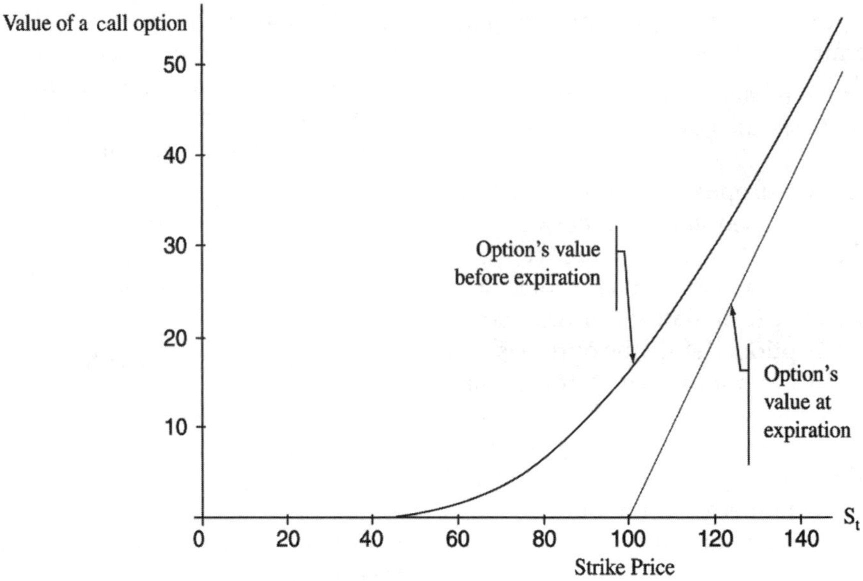

FIGURE 1.3 The relationship between the stock value and the option value before expiration and at expiration.

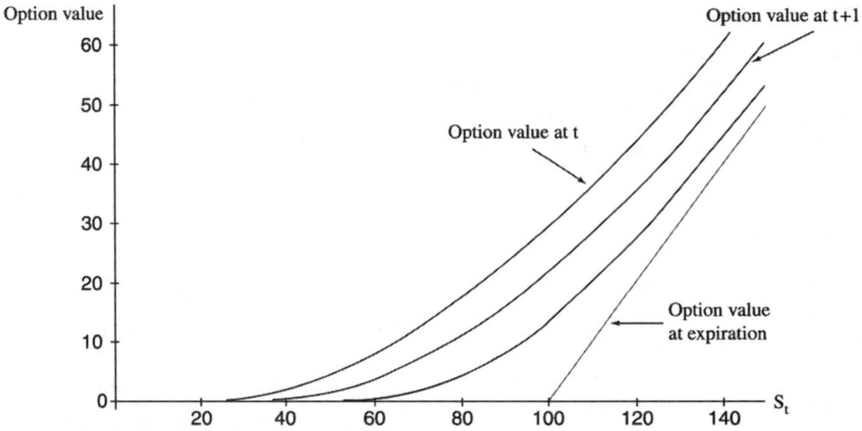

FIGURE 1.4 The value of a call option at various times before its expiration.

1.6 SWAPS

Swaps and swoptions are among some of the most common types of derivatives. But this is not why we are interested in them. It turns out that one method for pricing swaps and swoptions is to *decompose* them into forwards and options. This illustrates the special role played by forwards and options as basic building blocks and justifies the special emphasis put on them in following chapters.

Definition 4. A swap is the simultaneous selling and purchasing of cash flows involving various currencies, interest rates, and a number of other financial assets.

Even a brief summary of swap instruments is outside the scope of this book. As mentioned earlier, our intention is to provide a heuristic introduction of the mathematics behind derivative asset pricing, and not to discuss the derivative products themselves. We limit our discussion to a typical example that illustrates the main points.

1.6.1 A Simple Interest Rate Swap

Decomposing a swap into its constituent components is a potent example of financial engineering and derivative asset pricing. It also illustrates the special role played by simple forwards and options. We discuss an interest rate swap in detail. Das (1994) can be consulted for more advanced swap structures.

In its simplest form, an interest rate swap between two *counterparties* A and B is created as a result of the following steps:

1. Counterparty A needs a \$1 million floating-rate loan. Counterparty B needs a \$1 million fixed-rate loan. But because of market conditions and their relationships with various banks, B has a comparative advantage in borrowing at a floating rate.
2. A and B decide to exploit this comparative advantage. Each counterparty borrows at the market where he had a comparative advantage, and then decides to exchange the interest payments.
3. Counterparty A borrows \$1 million at a fixed rate. The interest payments will be received from counterparty B and paid back to the lending bank.
4. Counterparty B borrows \$1 million at the floating rate. Interest payments will be received from counterparty A and will be repaid to the lending bank.
5. Note that the initial sums, each being \$1 million, are identical. Hence, they do not have to be exchanged. They are called *notional principals*. The interest payments are also in the same currency. Hence, the counterparties exchange only the interest *differentials*. This concludes the interest rate swap.

This very basic interest rate swap consists of exchanges of interest payments. The counterparties borrow in sectors where they have an advantage and then exchange the interest payments. At the end both counterparties will secure lower rates and the *swap dealer* will earn a fee.

It is always possible to decompose simple swap deals into a basket of simpler forward contracts. The basket will replicate the swap. The forwards can then be priced separately, and the corresponding value of the swap can be determined from these numbers. This decomposition into building blocks of forwards will significantly facilitate the valuation of the swap contract.

1.6.2 Cancelable Swaps

A cancelable swap is a swap where one or both parties has the right not obligation to cancel the swap before its maturity. Those dates are typically prespecified in the contract. There are two types of cancelable swaps: callable swaps and puttable swaps.

In a callable swap, payer of the fixed rate in swap has the option to cancel the swap at any of those specified days before the maturity date and extinguish the obligation to pay the present value of future payments for an exchange for a premium. In a puttable swap, conversely, receiver of fixed rate in swap has the option to terminate the swap at any of those specified days before the maturity date. A counterparty in a plain vanilla swap may be able to close out a swap before maturity, but only by paying the net present value of future payments.

Cancelable swaps are popular among institutions with an obligation in which they can repay principal before the maturity date on the obligation, such as callable bonds. Cancelable swaps can be used as a hedge. They allow institutions to avoid maturity mismatches between their assets and liabilities with prepayment options and the swaps put in place.

1.7 CONCLUSION

In this chapter, we have reviewed some basic derivative instruments. Our purpose was twofold: first, to give a brief treatment of the basic derivative securities so we can use them in examples; and second, to discuss some notation in derivative asset pricing, where one first develops pricing formulas for simple building blocks, such as options and forwards, and then decomposes more complicated structures into baskets of forwards and options. This way, pricing formulas for simpler structures can be used to value more complicated structured products.

1.8 REFERENCES

Hull (2009) is an excellent source on derivatives that is unique in many ways. Practitioners use it as a manual; beginning graduate students utilize it as a textbook. It has a practical approach and is meticulously written. Jarrow and Turnbull (1996) is a welcome addition to books on derivatives. Duffie (1996) is an excellent source on dynamic asset pricing theory. However, it is not a source on the details of actual instruments traded in the markets. Yet, practitioners with a very strong math background may find it useful. Das (1994) is a useful reference on the practical aspects of derivative instruments.

1.9 EXERCISES

1. Consider the following investments:

- An investor short sells a stock at a price S, and writes an at-the-money call option on the same stock with a strike price of K.
- An investor buys one put with a strike price of K_1 and one call option at a strike price of K_2 with $K_1 \leq K_2$.
- An investor buys one put and writes one call with strike price K_1, and buys one call and writes one put with strike price K_2 $(K_1 \leq K_2)$.

(a) Plot the expiration payoff diagrams in each case.
(b) How would these diagrams look some time before expiration?

2. Consider a fixed-payer, plain vanilla, interest rate swap paid in arrears with the following characteristics:

- The start date is in 12 months, the maturity is 24 months.
- Floating rate is 6 month USD Libor.
- The swap rate is $\kappa = 5\%$.

(a) Represent the cash flows generated by this swap on a graph.
(b) Create a synthetic equivalent of this swap using two Forward Rate Agreements (FRA) contracts. Describe the parameters of the selected FRAs in detail.
(c) Could you generate a synthetic swap using appropriate interest rate options?

3. Let the arbitrage-free 3-month futures price for wheat be denoted by F_t. Suppose it costs c\$ to store 1 ton of wheat for 12 months and s\$ per year to insure the same quantity. The (simple) interest rate applicable to traders of spot wheat is $r\%$. Finally assume that the wheat has no convenience yield.

(a) Obtain a formula for F_t.
(b) Let the $F_t = 1500, r = 5\%, s = \$100, c = \$150$ and the spot price of wheat be $S_t = 1470$. Is this F_t arbitrage-free? How would you form an arbitrage portfolio?
(c) Assuming that all the parameters of the problem remain the same, what would be

the profit or loss of an arbitrage portfolio at expiration?

4. An at-the-money call written on a stock with current price $S_t = 100$ trades at 3. The corresponding at-the-money put trades at 3.5. There are no transaction costs and the stock does not pay any dividends. Traders can borrow and lend at a rate of 5% per year and all markets are liquid.

   (a) A trader writes a forward contract on the delivery of this stock. The delivery will be within 12 months and the price is F_t. What is the value of F_t?
   (b) Suppose the market starts quoting a price $F_t = 101$ for this contract. Form two arbitrage portfolios.

5. Using a combination of call and put options, explain what kind of portfolio an investor may use to express the following views on the future state of the asset:

   (a) High upcoming volatility.
   (b) Low upcoming volatility.
   (c) Bet on impending rare event, such as default, which would negatively impact asset price.

   For each portfolio chosen draw the corresponding payoff diagram.

6. Plot the payoff function of Bull call spread under different stock price at expiration.

CHAPTER

2

A Primer on the Arbitrage Theorem

OUTLINE

2.1 INTRODUCTION

All current methods of pricing derivative assets utilize the notion of *arbitrage*. In *arbitrage pricing methods* this utilization is direct. Asset prices are obtained from conditions that preclude arbitrage opportunities. In *equilibrium pricing methods* lack of arbitrage

Financial Derivatives, Third Edition.
http://dx.doi.org/10.1016/B978-0-12-384682-2.00002-5

opportunities is part of general equilibrium conditions.

In its simplest form, arbitrage means taking simultaneous positions in different assets so that one guarantees a riskless profit higher than the riskless return given by US Treasury bills. If such profits exist, we say that there is an *arbitrage opportunity*.

Arbitrage opportunities can arise in two different fashions. In the first way, one can make a series of investments with no current net commitment, yet expect to make a positive profit. For example, one can short sell a stock and use the proceeds to buy call options written on the same security. In this portfolio, one finances a long position in call options with short positions in the underlying stock. If this is done properly, unpredictable movements in the short and long positions will cancel out, and the portfolio will be riskless. Once commissions and fees are deducted, such investment opportunities should not yield any excess profits. Otherwise, we say that there are *arbitrage opportunities* of the first kind.

In arbitrage opportunities of the *second kind*, a portfolio can ensure a negative net commitment today, while yielding nonnegative profits in the future.

We use these concepts to obtain a practical definition of a "fair price" for a financial asset. We say that the price of a security is at a "fair" level, or that the security is *correctly priced*, if there are no-arbitrage opportunities of the first or second kind at those prices. Such *arbitrage-free* asset prices will be utilized as benchmarks. Deviations from these indicate opportunities for excess profits.

In practice, arbitrage opportunities may exist. This, however, would not reduce our interest in "arbitrage-free" prices. In fact, determining arbitrage-free prices is at the center of valuing derivative assets. We can imagine at least four possible utilizations of arbitrage-free prices.

One case may be when a derivatives house decides to engineer a *new* financial product. Because the product is new, the price at which it should be sold cannot be obtained by observing actual trading in financial markets. Under these conditions, calculating the arbitrage-free price will be very helpful in determining a market price for this product.

A second example is from *risk management*. Often, risk managers would like to measure the risks associated with their portfolios by running some "worst case" scenarios. These simulations are repeated periodically. Each time some benchmark price needs to be utilized, given that what is in question is a hypothetical event that has not been observed.[1]

A third example is *marking to market* of assets held in portfolios. A treasurer may want to know the current market value of a nonliquid asset for which no trades have been observed lately. Calculating the corresponding arbitrage-free price may provide a solution.

Finally, arbitrage-free benchmark prices can be *compared* with prices observed in actual trading. Significant differences between observed and arbitrage-free values might indicate excess profit opportunities. This way arbitrage-free prices can be used to detect mispricings that may occur during short intervals. If the arbitrage-free price is *above* the observed price, the derivative is *cheap*. A long position may be called for. When the opposite occurs, the derivative instrument is *overvalued*.

The mathematical environment provided by the no-arbitrage theorem is the major tool used to calculate such benchmark prices.

2.2 NOTATION

We begin with some formalism and start developing the notation that is an integral part of every mathematical approach. A correct understanding of the notation is sometimes as impor-

[1]Note that devising such scenarios is not at all straightforward. For example, it is not clear that markets will have the necessary liquidity to secure no-arbitrage conditions if they are hit by some extreme shock.

tant as an understanding of the underlying mathematical logic.

2.2.1 Asset Prices

The index t will represent time. Securities such as options, futures, forwards, and stocks will be represented by a *vector* of asset prices denoted by S_t. This array groups all securities in financial markets under one symbol:

$$S_t = \begin{bmatrix} S_1(t) \\ \vdots \\ S_N(t) \end{bmatrix} \tag{2.1}$$

Here, $S_1(t)$ may be riskless borrowing or lending, $S_2(t)$ may denote a particular stock, $S_3(t)$ may be a call option written on this stock, $S_4(t)$ may represent the corresponding put option, and so on. The t subscript in S_t means that prices belong to time represented by the value of t. In *discrete* time, securities prices can be expressed as $S_0, S_1, \ldots, S_t, S_{t+1}, \ldots$. However, in continuous time, the t subscript can assume any value between zero and infinity. We formally write this as

$$t \in [0, \infty) \tag{2.2}$$

In general, 0 denotes the *initial point*, and t represents the *present*. If we write

$$t < s \tag{2.3}$$

then s is meant to be a *future* date.

2.2.2 States of the World

To proceed with the rest of this chapter, we need one more concept—a concept that, at the outset, may appear to be very abstract, yet has significant practical relevance.

We let the vector W denote all possible *states of the world*,

$$W = \begin{bmatrix} \omega_1 \\ \vdots \\ \omega_N \end{bmatrix} \tag{2.4}$$

where each ω_i represents a distinct outcome that may occur. These states are *mutually exclusive*, and at least one of them is guaranteed to occur.

In general, financial assets will have different values and give different payouts at different states of the world ω_i. It is assumed that there are a finite number K of such possible states.

It is not very difficult to visualize this concept. Suppose that from a trader's point of view, the only time of interest is the "next" instant. Clearly, securities prices may change, and we do not necessarily know how. Yet, in a small time interval, securities prices may have an "uptick" or a "downtick," or may not show any movement at all. Hence, we may act as if there are a total of three possible states of the world.

2.2.3 Returns and Payoffs

The states of the world ω_i matter because in different states of the world returns to securities would be different. We let d_{ij} denote the number of units of account paid by one unit of security i in state j. These payoffs will have two components.

The first component is capital gains or losses. Asset values appreciate or depreciate. For an investor who is "long" in the asset, an appreciation leads to a capital gain and a depreciation leads to a capital loss. For somebody who is "short" in the asset, capital gains and losses will be reversed.[2]

The second component of the d_{ij} is *payouts*, such as dividends or coupon interest payments.[3]

[2] To realize a capital gain, one must unwind the position.

[3] Another example, besides dividend-paying stocks and coupon bonds, is investment in futures. The practice of "marking to market" leads to daily payouts to a contract holder. However, in the case of futures, these payouts may be negative or positive.

Some assets, though, do not have such payouts, call and put options, and discount bonds among these.

The existence of several assets, along with the assumption of many states of the world, means that for each asset there are several possible d_{ij}. *Matrices* are used to represent such arrays.

Thus, for the N assets under consideration, the payoffs d_{ij} can be grouped in a matrix D:

$$D = \begin{bmatrix} d_{11} & \cdots & d_{1K} \\ \vdots & \vdots & \vdots \\ d_{N1} & \cdots & d_{NK} \end{bmatrix} \tag{2.5}$$

There are two different ways one can visualize such a matrix. One can look at the matrix D as if each row represents payoffs to one unit of a given security in different states of the world. Conversely, one can look at D columnwise. Each column of D represents payoffs to different assets in a given state of the world.

If current prices of all assets are nonzero, then one can divide the ith row of D by the corresponding $S_i(t)$ and obtain the gross returns in different states of the world. The D will have a t subscript in the general case when payoffs depend on time.

2.2.4 Portfolio

A portfolio is a particular combination of assets in question. To form a portfolio, one needs to know the positions taken in each asset under consideration. The symbol θ_i represents the commitment with respect to the ith asset. Identifying all $\{\theta_i, i = 1...N\}$ specifies the portfolio.

A positive θ_i implies a long position in that asset, while a negative θ_i implies a short position. If an asset is not included in the portfolio, its corresponding θ_i is zero.

If a portfolio delivers the same payoff in all states of the world, then its value is known exactly and the portfolio is riskless.

2.2.5 A Basic Example of Asset Pricing

We use a simple model to explain most of the important results in pricing derivative assets. With this example, we first intend to illustrate the *logic* used in derivative asset pricing. Second, we hope to introduce the *mathematical tools* needed to carry out this logic in practical applications. The model is kept simple on purpose. A more general case is discussed at the end of the chapter.

We assume that time consists of "now"and a "next period"and that these two periods are separated by an interval of length Δ. Throughout this book Δ will represent a "small" but noninfinitesimal interval.

We consider a case where the market participant is interested only in three assets:

1. A risk-free asset such as a Treasury bill, whose gross return until next period is $(1 + r\Delta)$.[4] This return is "risk-free,"in that it is constant regardless of the realized state of the world.
2. An *underlying asset*, for example, a stock $S(t)$. We assume that during the small interval Δ, $S(t)$ can assume one of only *two* possible values. This means a minimum of two states of the world. $S(t)$ is risky because its payoff is different in each of the two states.
3. A derivative asset, a call option with premium $C(t)$ and a strike price C_o. The option expires "next"period. Given that the underlying asset has two possible values, the call option will assume two possible values as well.

This setup is fairly simple. There are three assets ($N = 3$), and two states of the world ($K = 2$). The first asset is risk-free borrowing and lending, the second is the underlying security, and the third is the option.

The example is not altogether unrealistic. A trader operating in real (continuous) time may contemplate taking a (covered) position in a particular option. If the time interval under con-

[4]We must multiply the risk-free rate, r, by the time that elapses, Δ, to get the proper return.

sideration is "small," prices of these assets may not change by more than an up- or a downtick. Hence, the assumption of two states of the world may be a reasonable approximation.[5]

We summarize this information in terms of the formal notation discussed earlier. Asset prices will form a vector S_t of only three elements,

$$S_t = \begin{bmatrix} B(t) \\ S(t) \\ C(t) \end{bmatrix} \tag{2.6}$$

where $B(t)$ is riskless borrowing or lending, $S(t)$ is a stock, and $C(t)$ is the value of a call option written on this stock. The t indicates the time for which these prices apply.

Payoffs will be grouped in a matrix D_t, as discussed earlier. There are three assets, which means that matrix D_t will have three rows. Also, there are two states of the world; the D_t matrix will thus have two columns. The $B(t)$ is riskless borrowing or lending. Its payoff will be the same, regardless of the state of the world that applies in the "next instant." The $S(t)$ is risky and its value may go either up to $S_1(t+\Delta)$ or down to $S_2(t+\Delta)$. Finally, the market value of the call option $C(t)$ will change in line with movements in the underlying asset price $S(t)$. Thus, D_t will be given by:

$$D_t = \begin{bmatrix} (1+r\Delta)B(t) & (1+r\Delta)B(t) \\ S_1(t+\Delta) & S_2(t+\Delta) \\ C_1(t+\Delta) & C_2(t+\Delta) \end{bmatrix} \tag{2.7}$$

where r is the annual riskless rate of return.

2.2.6 A First Glance at the Arbitrage Theorem

We are now ready to introduce a fundamental result in financial theory that can be used in calculating fair market values of derivative assets. But first we will simplify the notation even further. The amount of risk-free borrowing and lending is selected by the investor. Hence, we can always let

$$B(t) = 1 \tag{2.8}$$

Earlier, the time that elapses was called Δ. In this particular example we let

$$\Delta = 1 \tag{2.9}$$

The arbitrage theorem can now be stated:

Theorem 1. *Given the S_t, D_t defined in* (2.6) *and* (2.7), *and given that the two states have positive probabilities of occurrence,*

1. *if positive constants ψ_1, ψ_2 can be found such that asset prices satisfy*

$$\begin{bmatrix} 1 \\ S(t) \\ C(t) \end{bmatrix} = \begin{bmatrix} (1+r\Delta) & (1+r\Delta) \\ S_1(t+\Delta) & S_2(t+\Delta) \\ C_1(t+\Delta) & C_2(t+\Delta) \end{bmatrix} \begin{bmatrix} \psi_1 \\ \psi_2 \end{bmatrix} \tag{2.10}$$

then there are no-arbitrage possibilities[6]; *and*

2. *if there are no-arbitrage opportunities, then positive constants ψ_1, ψ_2 satisfying* (2.10) *can be found.*

The relationship in (2.10) is called a *representation*. It is not a relation that can be observed in reality. In fact, $S_1(t+1)$ and $S_2(t+1)$ are "possible"future values of the underlying asset. Only one of them—namely, the one that belongs to the state that is realized—will be observed.

What do the constants ψ_1, ψ_2 represent? According to the second row of the representation implied by the arbitrage theorem, if a security pays 1 in state 1, and 0 in state 2, then

$$S(t) = (1)\psi_1 \tag{2.11}$$

Thus, investors are willing to pay ψ_1 (current) units for an "insurance policy"that offers one unit of account in state 1 and nothing in state 2. Similarly, ψ_2 indicates how much investors

[5]In fact, we show later that a continuous-time Wiener process, or Brownian motion, can be approximated arbitrarily well by such two-state processes, as we let the Δ go toward zero.

[6]Note that if $1+r>1$, we need to have $\psi_1+\psi_2<1$ as well. This is obtained from the first row of the matrix equation.

would like to pay for an "insurance policy" that pays 1 in state 2 and nothing in state 1. Clearly, by spending $\psi_1 + \psi_2$, one can guarantee 1 unit of account in the future, regardless of which state is realized. This is confirmed by the first row of representation (2.10). Consistent with this interpretation, $\psi_i, i = 1, 2$ are called *state prices*.[7]

At this point there are several other issues that may not be clear. One can, in fact, ask the following questions:

- How does one obtain this theorem?
- What does the existence of ψ_1, ψ_2 have to do with no-arbitrage?
- Why is this result relevant for asset pricing?

For the moment, let us put the first two questions aside and answer the third question: What types of practical results (if any) does one obtain from the existence of ψ_1, ψ_2? It turns out that the representation given by the arbitrage theorem is very important for practical asset pricing.

2.2.7 Relevance of the Arbitrage Theorem

The arbitrage theorem provides a very elegant and general method for pricing derivative assets.

Consider again the representation:

$$\begin{bmatrix} 1 \\ S(t) \\ C(t) \end{bmatrix} = \begin{bmatrix} (1+r\Delta) & (1+r\Delta) \\ S_1(t+\Delta) & S_2(t+\Delta) \\ C_1(t+\Delta) & C_2(t+\Delta) \end{bmatrix} \begin{bmatrix} \psi_1 \\ \psi_2 \end{bmatrix} \tag{2.12}$$

Multiplying the first row of the dividend matrix D_t by the vector of ψ_1, ψ_2, we get

$$1 = (1+r)\psi_1 + (1+r)\psi_2 \tag{2.13}$$

Define:

$$\begin{aligned} \mathbb{Q}_1 &= (1+r)\psi_1 \\ \mathbb{Q}_2 &= (1+r)\psi_2 \end{aligned} \tag{2.14}$$

[7]Note that, in general, state prices will be time-dependent; hence, they should carry t subscripts. This is omitted here for notational simplicity.

Because of the positivity of state prices, and because of (2.13),

$$0 < \mathbb{Q}_1 \leq 1$$
$$\mathbb{Q}_2 + \mathbb{Q}_1 = 1$$

Hence, $\mathbb{Q}_i$'s are positive numbers, and they sum to one. As such, they can be interpreted as two *probabilities* associated with the two states under consideration. We say "interpreted" because the true probabilities that govern the occurrence of the two states of the world will in general be different from the $\mathbb{Q}_1$ and $\mathbb{Q}_2$. These are defined by Eq. (2.14) and provide no direct information concerning the true probabilities associated with the two states of the world. For this reason, $\{\mathbb{Q}_1, \mathbb{Q}_2\}$ are called risk-adjusted synthetic probabilities.

2.2.8 The Use of Synthetic Probabilities

Risk-adjusted probabilities exist if there are no-arbitrage opportunities. In other words, if there are no "mispriced assets," we are guaranteed to find positive constants $\{\psi_1, \psi_2\}$. Multiplying these by the riskless gross return $1 + r$ guarantees the existence of $\{\mathbb{Q}_1, \mathbb{Q}_2\}$.[8]

The importance of risk-adjusted probabilities for asset pricing stems from the following: Expectations calculated with them, once discounted by the risk-free rate r, equal the current value of the asset.

Consider the equality implied by the arbitrage theorem again. Note that the representation (2.10) implies three separate equalities:

$$1 = (1+r)\psi_1 + (1+r)\psi_2 \tag{2.15}$$

$$S(t) = \psi_1 S_1(t+1) + \psi_2 S_2(t+1) \tag{2.16}$$

$$C(t) = \psi_1 C_1(t+1) + \psi_2 S_2(t+2) \tag{2.17}$$

Now multiply the right-hand side of the last two equations by

[8]This is the case with finite states of the world. With uncountably many states one needs further conditions for the existence of risk-adjusted probabilities.

$$\frac{1+r}{1+r} \tag{2.18}$$

to obtain[9]

$$S(t) = \frac{1}{1+r}$$
$$\left[(1+r)\psi_1 S_1(t+1) + (1+r)\psi_2 S_2(t+1)\right] \tag{2.19}$$

$$C(t) = \frac{1}{1+r}$$
$$\left[(1+r)\psi_1 C_1(t+1) + (1+r)\psi_2 C_2(t+1)\right] \tag{2.20}$$

But, we can replace $(1+r)\psi_i, i = 1,2$, with the corresponding $\widetilde{P}_i$. This means that the two equations become

$$S(t) = \frac{1}{1+r}\left[\mathbb{Q}_1 S_1(t+1) + \mathbb{Q}_2 S_2(t+1)\right] \tag{2.21}$$

$$C(t) = \frac{1}{1+r}\left[\mathbb{Q}_1 C_1(t+1) + \mathbb{Q}_2 C_2(t+1)\right] \tag{2.22}$$

Now consider how these expressions can be interpreted. The expression on the right-hand side multiplies the term in the brackets by $1/(1+r)$, which is a riskless one-period discount factor. On the other hand, the term inside the brackets can be interpreted as some sort of *expected value*. It is the sum of possible future values of $S(t)$ or $C(t)$ weighted by the "probabilities" $\mathbb{Q}_1, \mathbb{Q}_2$. Hence, the terms in the brackets are expectations calculated using the risk-adjusted probabilities.

As such, the equalities in (2.21) and (2.22) do not represent "true"expected values. Yet as long as there is no-arbitrage, these equalities are valid, and they can be used in practical calculations. We can use them in asset pricing, as long as the underlying probabilities are explicitly specified.

With this interpretation of $\mathbb{Q}_1, \mathbb{Q}_2$, the *current prices of all assets under consideration become equal to their discounted expected payoffs*. Further, the discounting is done using the risk-free rate, although the assets themselves are risky.

[9]As long as r is not equal to -1, we can always do this.

In order to emphasize the important role played by risk-adjusted probabilities, consider what happens when one uses the "true"probabilities dictated by their nature.

First, we obtain the "true"expected values by using the true probabilities denoted by $\mathbb{P}_1, \mathbb{P}_2$:

$$\mathbb{E}^{true}[S(t+1)] = \left[\mathbb{P}_1 S_1(t+1) + \mathbb{P}_2 S_2(t+1)\right] \tag{2.23}$$

$$\mathbb{E}^{true}[C(t+1)] = \left[\mathbb{P}_1 C_1(t+1) + \mathbb{P}_2 C_2(t+1)\right] \tag{2.24}$$

Because these are "risky" assets, when discounted by the risk-free rate, these expectations will in general[10] satisfy

$$S(t) < \frac{1}{1+r}\mathbb{E}^{\mathbb{P}}[S(t+1)] \tag{2.25}$$

$$C(t) < \frac{1}{1+r}\mathbb{E}^{\mathbb{P}}[C(t+1)] \tag{2.26}$$

To see why one obtains such inequalities, assume otherwise:

$$S(t) = \frac{1}{1+r}\mathbb{E}^{\mathbb{P}}[S(t+1)] \tag{2.27}$$

$$C(t) = \frac{1}{1+r}\mathbb{E}^{\mathbb{P}}[C(t+1)] \tag{2.28}$$

Rearranging, and assuming that asset prices are nonzero,

$$1+r = \frac{\mathbb{E}^{\mathbb{P}}[S(t+1)]}{S(t)} \tag{2.29}$$

$$1+r = \frac{\mathbb{E}^{\mathbb{P}}[C(t+1)]}{C(t)} \tag{2.30}$$

But this means that (true) expected returns from the risky assets equal riskless return. This, however, is a contradiction, because in general risky assets will command a positive risk premium. If there is no such compensation for risk,

[10]We say "in general"because one can imagine risky assets that are negatively correlated with the "market."Such assets may have negative risk premiums and are called "negative beta"assets.

no investor would hold them. Thus, for risky assets we generally have

$$(1 + r + \text{ risk premium for } S(t)) = \frac{\mathbb{E}^{\mathbb{P}}[S(t+1)]}{S(t)} \tag{2.31}$$

$$(1 + r + \text{ risk premium for } C(t)) = \frac{\mathbb{E}^{\mathbb{P}}[C(t+1)]}{C(t)} \tag{2.32}$$

This implies, in general, the following inequalities for risky assets[11]:

$$S(t) < \frac{1}{1+r}\mathbb{E}^{\mathbb{P}}[S(t+1)] \tag{2.33}$$

$$C(t) < \frac{1}{1+r}\mathbb{E}^{\mathbb{P}}[C(t+1)] \tag{2.34}$$

The importance of the no-arbitrage assumption in asset pricing should become clear at this point. If no-arbitrage implies the existence of positive constants such as ψ_1, ψ_2, then we can always obtain from these constants the risk-adjusted probabilities $\mathbb{Q}_1, \mathbb{Q}_2$ and work with "synthetic" expectations that satisfy

$$1 + r = \frac{\mathbb{E}^{\mathbb{Q}}[S(t+1)]}{S(t)} \tag{2.35}$$

$$1 + r = \frac{\mathbb{E}^{\mathbb{Q}}[C(t+1)]}{C(t)} \tag{2.36}$$

These equations are very convenient to use, and they internalize any risk premiums. Indeed, one does not need to calculate the risk premiums if one uses synthetic expectations. The corresponding discounting is done using the risk-free rate, which is easily observable.

2.2.9 Martingales and Submartingales

This is the right time to introduce a concept that is at the foundation of pricing financial assets. We give a simple definition of the terms and leave technicalities for later chapters.

[11] For negative beta assets the inequalities are reversed.

Suppose at time t one has information summarized by I_t. A random variable X_t that for all $s > 0$ satisfies the equality

$$\mathbb{E}^{\mathbb{P}}[X_{t+s}|\, I_t] = X_t \tag{2.37}$$

is called a *martingale with respect to the probability* $\mathbb{P}$.[12]

If instead we have for all $s > 0$

$$\mathbb{E}^{\mathbb{Q}}[X_{t+s}|\, I_t] \geq X_t \tag{2.38}$$

is called a *submartingale with respect to the probability* $\mathbb{Q}$.

Here is why these concepts are fundamental. According to the discussion in the previous section, asset prices discounted by the risk-free rate will be submartingales under the true probabilities, but become martingales under the risk-adjusted probabilities. Thus, as long as we utilize the latter, the tools available to martingale theory become applicable, and "fair market values" of the assets under consideration can be obtained by exploiting the martingale equality

$$X_t = \mathbb{E}^{\mathbb{Q}}[X_{t+s}|\, I_t] \tag{2.39}$$

where $s > 0$, and where X_{t+s} is defined by

$$X_{t+s} = \frac{S_{t+s}}{(1+r)^s} \tag{2.40}$$

Here S_{t+s} and r are the security price and risk-free return, respectively. $\mathbb{Q}$ is the risk-adjusted probability. According to this, utilization of risk-adjusted probabilities will convert all (discounted) asset prices into martingales.

2.2.10 Normalization

It is important to realize that, in finance, the notion of martingale is always associated

[12] There are other conditions that a martingale must satisfy. In later chapters, we discuss them in detail. In the meantime, we assume implicitly that these conditional expectations exist—that is, they are finite.

with two concepts. First, a martingale is always defined with respect to a certain probability. Hence, in Section 2.2.9 the discounted stock price, was a martingale with respect to the risk-adjusted probability $\mathbb{Q}$. Second, note that it is not the $\mathbb{S}_t$ that is a martingale, but rather the S_t divided, or normalized, by the $(1+r)^s$. The latter is the earnings of 1$ over s periods if invested and rolled-over in the risk-free investment. What is a martingale is the ratio

$$X_{t+s} = \frac{S_{t+s}}{(1+r)^s} \tag{2.41}$$

An interesting question that we investigate in the second half of this book is then the following. Suppose we divide the S_t by some other asset's price, say C_t; would the new ratio,

$$X^*_{t+s} = \frac{S_{t+s}}{C_{t+s}} \tag{2.42}$$

be a martingale with respect to some other probability, say $\mathbb{P}^*$? The answer to this question is positive and is quite useful in pricing interest sensitive derivative instruments. Essentially, it gives us the flexibility to work with a more convenient probability by normalizing with an asset of our choice. But these issues have to wait until Chapter 17.

2.2.11 Equalization of Rates of Return

By using risk-adjusted probabilities, we can derive another important result useful in asset pricing.

In the arbitrage-free representation given in (2.10), divide both sides of the equality by the current price of the asset and multiply both sides by $(1+r)$, the gross rate of riskless return. Assuming nonzero asset prices, we obtain

$$\mathbb{Q}_1 \frac{S_1(t+1)}{S(t)} + \mathbb{Q}_2 \frac{S_2(t+2)}{S(t)} = 1+r \tag{2.43}$$

$$\mathbb{Q}_1 \frac{C_1(t+1)}{C(t)} + \mathbb{Q}_2 \frac{C_2(t+2)}{C(t)} = 1+r \tag{2.44}$$

First note that ratios such as

$$\frac{S_1(t+1)}{S(t)}, \quad \frac{S_2(t+1)}{S(t)} \tag{2.45}$$

are the gross rates of return of $S(t)$ in states 1 and 2, respectively. The equalities (2.43) and (2.44) imply that if one uses $\mathbb{Q}_1$ and $\mathbb{Q}_2$ in calculating the expected values, all assets would have the same expected return. According to this new result, "under $\mathbb{Q}_1$ and $\mathbb{Q}_2$", all expected returns equal the risk-free return r.[13] This is another widely used result in pricing financial assets.

2.2.12 The No-Arbitrage Condition

Within this simple setup we can also see explicitly the connection between the no-arbitrage condition and the existence of ψ_1 and ψ_2. Let the gross returns in states 1 and 2 be given by $R_1(t+1)$ and $R_2(t+1)$ respectively:

$$R_1(t+1) = \frac{S_1(t+1)}{S(t)} \tag{2.46}$$

$$R_2(t+1) = \frac{S_2(t+1)}{S(t)} \tag{2.47}$$

Now write the first two rows of (2.12) using these new symbols:

$$\begin{aligned} 1 &= (1+r)\psi_1 + (1+r)\psi_2 \\ 1 &= R_1\psi_1 + R_2\psi_2 \end{aligned}$$

Subtract the first equation from the second to obtain:

$$0 = \big((1+r) - R_1\big)\,\psi_1 + \big((1+r) - R_2\big)\,\psi_2 \tag{2.48}$$

where we want ψ_1, ψ_2 to be positive. This will be the case and, at the same time, the above equation will be satisfied if and only if:

$$R_1 < (1+r) < R_2$$

[13] In probability theory, the phrase "under $\mathbb{Q}_1, \mathbb{Q}_2$" means "if one uses the probabilities $\mathbb{Q}_1$ and $\mathbb{Q}_2$."

For example, suppose we have

$$(1+r) < R_1 < R_2$$

This means that by borrowing infinite sums at rate r, and going long in $S(t)$, we can guarantee positive returns. So there is an arbitrage opportunity. But then, the right-hand side of (2.48) will be negative and the equality will not be satisfied with positive. Hence no $0 < \psi_1, 0 < \psi_2$ will exist. A similar argument can be made if we have

$$R_1 < R_2 < (1+r)$$

If this was the case, then one could *short* the $S(t)$ and invest the proceeds in the risk-free investment to realize infinite gains. Again Eq. (2.48) will not be satisfied with positive, because the right-hand side will always be positive under these conditions.

Thus, we see that the existence of positive is closely tied to the condition

$$R_1 < (1+r) < R_2$$

which implies, in this simple setting, that there are no-arbitrage possibilities.

2.3 A NUMERICAL EXAMPLE

A simple example needs to be discussed. Let the current value of a stock be given by

$$S_t = 100 \tag{2.49}$$

The stock can assume only two possible values in the next instant:

$$S_1(t+1) = 100 \tag{2.50}$$

and

$$S_2(t+1) = 150 \tag{2.51}$$

Hence, there are only *two* states of the world.

There exists a call option with premium C, and strike price 100. The option expires next period.

Finally, it is assumed that 1 unit of account is invested in the risk-free asset with a return of 10%.

We obtain the following representation under no-arbitrage:

$$\begin{bmatrix} 1 \\ 100 \\ C \end{bmatrix} = \begin{bmatrix} 1.1 & 1.1 \\ 100 & 150 \\ 0 & 50 \end{bmatrix} \begin{bmatrix} \psi_1 \\ \psi_2 \end{bmatrix} \tag{2.52}$$

Note that the numerical value of the call premium C is left unspecified. Using this as a variable, we intend to show the role played by the arbitrage theorem.

2.3.1 Case 1: Arbitrage Possibilities

Multiplying the dividend matrix with the vector of 's yields three equations:

$$1 = 1.1\psi_1 + 1.1\psi_2 \tag{2.53}$$

$$100 = 100\psi_1 + 150\psi_2 \tag{2.54}$$

$$C = 0\psi_1 + 50\psi_2 \tag{2.55}$$

Now suppose a premium $C = 25$ is observed in financial markets. Then the last equation yields

$$50\psi_2 = 25 \tag{2.56}$$

or

$$\psi_2 = \frac{1}{2} \tag{2.57}$$

Substituting this in (2.54) gives

$$\psi_1 = 0.25 \tag{2.58}$$

But at these values of ψ_1 and ψ_2, the first equation is not satisfied:

$$1.1 \times 0.25 + 1.1 \times 1.5 \neq 1 \tag{2.59}$$

Clearly, at the observed value for the call premium, $C = 25$, it is impossible to find ψ_1 and ψ_2 that satisfies all three equations given by the arbitrage-free representation. Arbitrage opportunities therefore exist.

2.3.2 Case 2: Arbitrage-Free Prices

Consider the same system as before

$$\begin{bmatrix} 1 \\ 100 \\ C \end{bmatrix} = \begin{bmatrix} 1.1 & 1.1 \\ 100 & 150 \\ 0 & 50 \end{bmatrix} \begin{bmatrix} \psi_1 \\ \psi_2 \end{bmatrix} \tag{2.60}$$

But now, instead of starting with an observed value of C, solve the first two equations for ψ_1, ψ_2. These form a system of two equations in two unknowns. The unique solution gives

$$\psi_1 = 0.7273, \quad \psi_2 = 0.1818 \tag{2.61}$$

Now use the third equation to calculate a value of C consistent with this solution:

$$C = 9.09 \tag{2.62}$$

At this price, arbitrage profits do not exist.

Note that, using the constants ψ_1, ψ_2, we *derived* the arbitrage-free price $C = 9.09$. In this sense, we used the arbitrage theorem as an asset pricing tool.

It turns out that in this particular case, the representation given by the arbitrage theorem is satisfied with positive and unique ψ_i. This may not always be true.

2.3.3 An Indeterminacy

The same method of determining the unique arbitrage-free value of the call option would not work if there were more than two states of the world. For example, consider the system

$$\begin{bmatrix} 1 \\ 100 \\ C \end{bmatrix} = \begin{bmatrix} 1.1 & 1.1 & 1.1 \\ 100 & 50 & 150 \\ 0 & 0 & 50 \end{bmatrix} \begin{bmatrix} \psi_1 \\ \psi_2 \\ \psi_3 \end{bmatrix} \tag{2.63}$$

Here, the first two equations cannot be used to determine a unique set of $\psi_i > 0$ that can be plugged into the third equation to obtain a C. There are many such sets of ψ_i's.

In order to determine the arbitrage-free value of the call premium C, one would need to select the "correct" ψ_i. In principle, this can be done using the underlying economic equilibrium.

2.4 AN APPLICATION: LATTICE MODELS

Simple as it is, the example just discussed gives the logic behind one of the most common asset pricing methods, namely, the so-called.[14] The binomial model is the simplest example.

We briefly show how this pricing methodology uses the results of the arbitrage theorem.

Consider a call option C_t written on the underlying asset S_t. The call option has strike price C_0 and expires at time $T, t < T$. It is known that at expiration the value of the option is given by

$$C_T = \max\left[S_T - C_0, 0\right] \tag{2.64}$$

We first divide the time interval $(T - t)$ into n smaller intervals, each of size Δ. We choose a "small" Δ, in the sense that the variations of S_t during Δ can be approximated reasonably well by an up or down movement only. According to this, we hope that for small enough Δ the underlying asset price S_t cannot wander too far from the currently observed price S_t.

Thus, we assume that during Δ the only possible changes in S_t are an *up* movement by $\sigma\sqrt{\Delta}$ or a *down* movement by $-\sigma\sqrt{\Delta}$:

$$S_{t+\Delta} = \begin{cases} S_t + \sigma\sqrt{\Delta} \\ S_t - \sigma\sqrt{\Delta} \end{cases} \tag{2.65}$$

Clearly, the size of the parameter σ determines how far $S_{t+\Delta}$ can wander during a time interval of length Δ. For that reason it is called the *volatility* parameter. The σ is known. Note that regardless of σ, in smaller intervals, S_t will change less.

[14] Also called tree models.

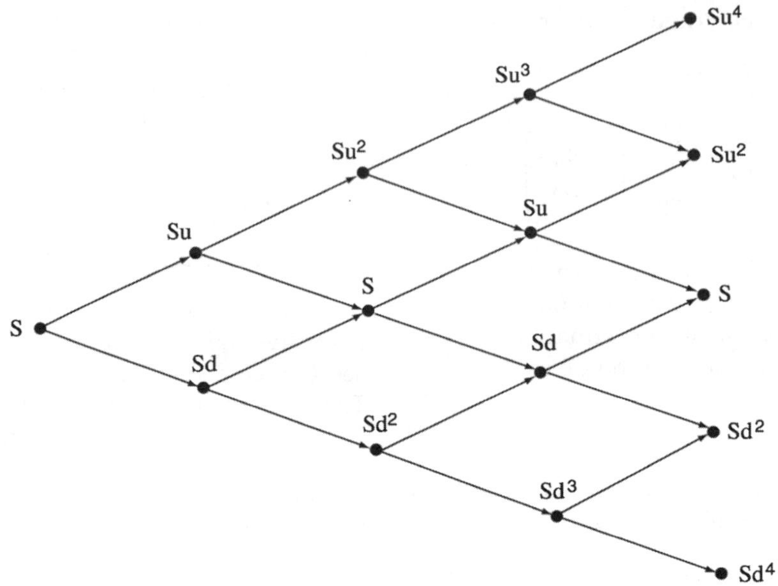

FIGURE 2.1 The multiplicative lattice with up and down movements.

The dynamics described by Eq. (2.65) represent a *lattice* or a *binomial tree*. Figure 2.1 displays these dynamics in the case of *multiplicative* up and down movements.

Suppose now that we are given the (constant) risk-free rate r for the period Δ. Can we determine the risk-adjusted probabilities?[15]

We know from the arbitrage theorem that the risk-adjusted probabilities $\mathbb{Q}_{up}$ and $\mathbb{Q}_{down}$ must satisfy

$$S_t = \frac{1}{1+r}\left[\mathbb{Q}_{up}\left(S_t + \sigma\sqrt{\Delta}\right) + \mathbb{Q}_{down}\left(S_t - \sigma\sqrt{\Delta}\right)\right] \tag{2.66}$$

In this equation, r, S_t, σ, and Δ are known. The first three are observed in the markets, while Δ is selected by us. Thus, the only unknown is the $\mathbb{Q}_{up}$, which can be determined easily.[16]

[15]In the second half of the book, we will relax the assumption that r is constant. But for now we maintain this assumption.

[16]Remember that $\mathbb{Q}_{down} = 1 - \mathbb{Q}_{up}$.

Once this is done, the $\mathbb{Q}_{up}$ can be used to calculate the current arbitrage-free value of the call option. In fact, the equation

$$C_t = \frac{1}{1+r}\left[\mathbb{Q}_{up}C^{up}_{t+\Delta} + \mathbb{Q}_{down}C^{down}_{t+\Delta}\right] \tag{2.67}$$

"ties" two (arbitrage-free) values of the call option at any time $t + \Delta$ to the (arbitrage-free) value of the option as of time t. The $\mathbb{Q}_{up}$ is known at this point. In order to make the equation usable, we need the two values and $C^{up}_{t+\Delta}$ and $C^{down}_{t+\Delta}$. Given these, we can calculate the value of the call option C_t at time t.

Figure 2.2 shows the multiplicative lattice for the option price C_t. The arbitrage-free values of C_t are at this point indeterminate, except for the expiration "nodes."In fact, given the lattice for S_t, we can determine the values of C_t at the expiration using the *boundary condition*

$$C_T = \max\left[S_T - C_0, 0\right] \tag{2.68}$$

Once this is done, one can go backward using

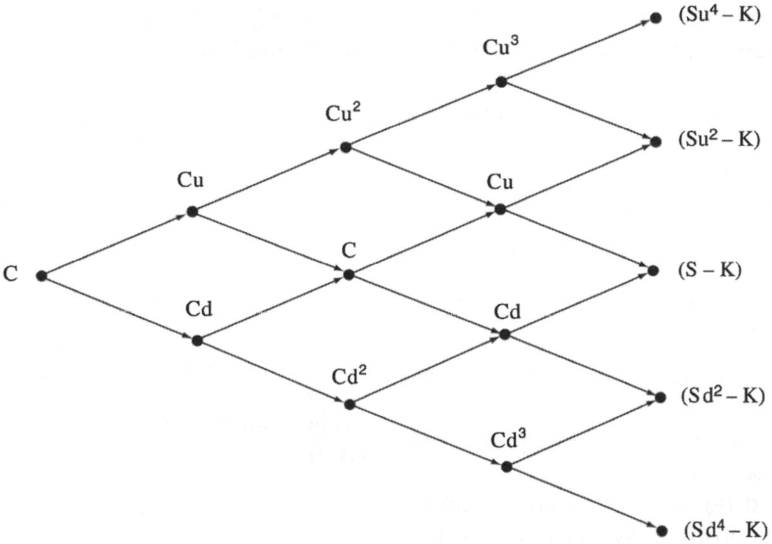

FIGURE 2.2 The multiplicative lattice for the option price C_t.

$$C_t = \frac{1}{1+r}\left[\mathbb{Q}_{up}C^{up}_{t+\Delta} + \mathbb{Q}_{down}C^{down}_{t+\Delta}\right] \qquad (2.69)$$

Repeating this several times, one eventually reaches the initial node that gives the current value of the option.

Hence, the procedure is to use the dynamics of S_t to go *forward* and determine the expiration date values of the call option. Then, using the risk-adjusted probabilities and the boundary condition, one works *backward* with the lattice for the call option to determine the current value C_t.

It is the arbitrage theorem and the implied martingale equalities that make it possible to calculate the risk-adjusted probabilities $\mathbb{Q}_{up}$ and $\mathbb{Q}_{down}$.

In this procedure, Figure 2.1 gives an approximation of all the possible paths that S_t may take during the period $T - t$. The tree in Figure 2.2 gives an approximation of all possible paths that can be taken by the price of the call option written on S_t. If Δ is small, then the lattices will be close approximations to the true paths that can be followed by S_t and C_t.

2.5 PAYOUTS AND FOREIGN CURRENCIES

In this section we modify the simple two-state model introduced in this chapter to introduce two complications that are more often the case in practical situations. The first is the payment of interim payouts such as dividends and coupons. Many securities make such payments *before* the expiration date of the derivative under consideration. These payouts do change the pricing formulas in a simple yet, at first sight, counterintuitive fashion. The second complication is the case of foreign currency denominated assets. Here also the pricing formulas change slightly.

2.5.1 The Case with Dividends

The setup of Section 2.3 is first modified by adding a dividend equal to d_t percent of $S_{t+\Delta}$. Note two points. First, the dividends are not lump-sum, but are paid as a percentage of the price at time $t+\Delta$. Second, the dividend payment rate has subscript t instead of $t + \Delta$. According to this, the d_t is known as of time t. Hence, it is not a random variable given the information set at I_t.

The simple model in (2.10) now becomes:

$$\begin{bmatrix} 1 \\ S_t \\ C_t \end{bmatrix} = \begin{bmatrix} B^u_{t+\Delta} & B^d_{t+\Delta} \\ S^u_{t+\Delta} + d_t S^u_{t+\Delta} & S^d_{t+\Delta} + d_t S^d_{t+\Delta} \\ C^u_{t+\Delta} & C^d_{t+\Delta} \end{bmatrix} \times \begin{bmatrix} \psi_1 \\ \psi_2 \end{bmatrix}$$

where B, S, C denote the savings account, the stock, and a call option, as usual. Note that the notation has now changed slightly to reflect the discussion of Section 2.5.

Can we proceed the same way as in Section 2.3? The answer is positive. With minor modifications, we can apply the same steps and obtain two equations:

$$S = \frac{1+d}{1+r}\left[S^u \mathbb{Q}_{up} + S^d \mathbb{Q}_{down}\right] \tag{2.70}$$

$$C = \frac{1}{1+r}\left[C^u \mathbb{Q}_{up} + C^d \mathbb{Q}_{down}\right] \tag{2.71}$$

where $\mathbb{Q}$ is the risk-neutral probability, and where we ignore the time subscripts. Note that the first equation is now different from the case with no dividends, but that the second equation is the same. According to this, each time an asset has some known percentage payout d during the period Δ, the risk-neutral discounting of the dividend-paying asset has to be done using the factor $(1 + d)/(1 + r)$ instead of multiplying by $1/(1 + r)$ only. It is also worth emphasizing that the discounting of the derivative itself did not change.

Now consider the following transformation:

$$\frac{1+r}{1+d} = \left[\frac{S^u \mathbb{Q}_{up} + S^d \mathbb{Q}_{down}}{S}\right]$$

which means that the expected return under the risk-free measure is now given by:

$$\mathbb{E}^{\mathbb{Q}}\left[\frac{S_{t+\Delta}}{S}\right] = \frac{1+r\Delta}{1+d\Delta}$$

Clearly, as a first-order approximation, if d, r are defined over, say, a year, and are small:

$$\frac{1+r\Delta}{1+q\Delta} \approx 1 + (r-d)\Delta$$

Using this in the previous equation:

$$\mathbb{E}^{\mathbb{Q}}\left[\frac{S_{t+\Delta}}{S_t}\right] \approx 1 + (r-d)\Delta$$

or

$$\mathbb{E}^{\mathbb{Q}}\left[S_{t+\Delta}\right] \approx S_t + (r-d)S_t\Delta$$

or, again, after adding a random, unpredictable component, $\sigma S_t \Delta W_{t+\Delta}$:

$$S_{t+\Delta} \approx S_t + (r-d)S_t\Delta + \sigma S_t \Delta W_{t+\Delta}$$

According to this last equation, we can state the following.

If we were to let Δ go to zero and switch to continuous time, the drift term for dS_t, which represents expected change in the underlying asset's price, will be given by $(r - d)S_t dt$ and the corresponding dynamics can be written as[17]:

$$dS_t = (r-d)S_t dt + \sigma S_t dW_t$$

where dt represents an infinitesimal time period.

There is a second interesting point to be made with the introduction of payouts.

Suppose now we try to go over similar steps using, this time, the equation for C_t shown in (2.71):

$$C = \frac{1}{1+r}\left[C^u \mathbb{Q}_{up} + C^d \mathbb{Q}_{down}\right]$$

We would obtain

$$\mathbb{E}^{\mathbb{Q}}\left[\frac{C_{t+\Delta}}{C_t}\right] 1 + r\Delta$$

Thus, we see that even though there is a divided payout made by the underlying stock,

[17]This stochastic differential equation will be studied with more detail in later chapters.

the risk-neutral expected return and the risk-free discounting remains the same for the call option written on this stock. Hence, in a risk-neutral world future returns to C_t have to be discounted exactly by the same factor as in the case of no dividends.

In other words:

- The expected rate of returns of the S_t and C_t during a period Δ are now *different* under the risk-free probability $\mathbb{Q}$:

$$\begin{aligned}\mathbb{E}^{\mathbb{Q}}\left[\frac{S_{t+\Delta}}{S_t}\right] &= \frac{1+r\Delta}{1+d\Delta} \approx 1+(r-d)\Delta \\ \mathbb{E}^{\mathbb{Q}}\left[\frac{C_{t+\Delta}}{C_t}\right] &= 1+r\Delta \approx 1+r\Delta\end{aligned}$$

These are slight modifications in the formulas, but in practice they may make a significant difference in pricing calculations. The case of foreign currencies below yields similar results.

2.5.2 The Case with Foreign Currencies

The standard setup is now modified by adding an investment opportunity in a foreign currency savings account.

In particular, suppose we spend e_t units of *domestic* currency to buy one unit of *foreign* currency. Thus the e_t is the exchange rate at time t. Assume US dollars (USD) is the domestic currency.

Suppose also that the foreign savings interest rate is known and is given by r^f.

The opportunities in investment and the yields of these investments over Δ can now be summarized using the following setup:

$$\begin{bmatrix} 1 \\ 1 \\ C_t \end{bmatrix} = \begin{bmatrix} 1+r & 1+r \\ \frac{e^u_{t+\Delta}}{e_t}\left(1+r^f\right) & \frac{e^d_{t+\Delta}}{e_t}\left(1+r^f\right) \\ C^u_{t+\Delta} & C^d_{t+\Delta} \end{bmatrix} \begin{bmatrix} \psi_1 \\ \psi_2 \end{bmatrix}$$

where the C_t denotes a call option on price e_t of one unit of foreign currency. The strike price is K.[18]

We proceed in a similar fashion to the case of dividends and obtain the following pricing equations[19]:

$$\begin{aligned} e &= \frac{1}{1+r}\left[e^u \mathbb{Q}_{up} + e^d \mathbb{Q}_{down}\right] \\ C &= \frac{1}{1+r}\left[C^u \mathbb{Q}_{up} + C^d \mathbb{Q}_{down}\right]\end{aligned}$$

Again, note that the first equation is different but the second equation is the same. Thus, each time we deal with a foreign currency denominated asset that has payout r^f during Δ, the risk-neutral discounting of the foreign asset has to be done using the factor $(1+r)/(1+r^f)$.

Note the first-order approximation if r^f is small:

$$\frac{(1+r\Delta)}{(1+r^f\Delta)} \approx 1+(r-r^f)\Delta$$

We again obtained a different result.

- The expected rate of return of the e_t and C are different under the probability $\mathbb{Q}$:

$$\begin{aligned}\mathbb{E}^{\mathbb{Q}}\left[\frac{e_{t+\Delta}}{e_t}\right] &\approx 1+\left(r-r^f\right)\Delta \\ \mathbb{E}^{\mathbb{Q}}\left[\frac{C_{t+\Delta}}{C_t}\right] &\approx 1+r\Delta\end{aligned}$$

According to the last remark, if we were to let Δ go to zero and switch to SDE's, the drift terms for dC_t will be given by $rC_t dt$. But the drift term for the foreign currency denominated asset, de_t, will now have to be $(r-r^f)e_t dt$.

[18] Here the K is a strike price on the exchange rate e_t. If the exchange rate exceeds the K at time $t+\Delta$, the buyer of the call will receive the difference $e_{t+\Delta} - K$ times a notional amount N.

[19] As usual, we omit the time subscripts for convenience.

2.6 SOME GENERALIZATIONS

Up to this point, the setup has been very simple. In general, such simple examples cannot be used to price real-life financial assets. Let us briefly consider some generalizations that are needed to do so.

2.6.1 Time Index

Up to this point we considered discrete time with $t = 1, 2, 3, \ldots$. In continuous-time asset pricing models, this will change. We have to assume that t is continuous:

$$t \in [0, \infty) \tag{2.72}$$

This way, in addition to the "small" time interval Δ dealt with in this chapter, we can consider infinitesimal intervals denoted by the symbol dt.

2.6.2 States of the World

In continuous time, the values that an asset can assume are not limited to two. There may be uncountably many possibilities and a continuum of states of the world.

To capture such generalizations, we need to introduce stochastic differential equations. For example, as mentioned above, increments in security prices S_t may be modeled using

$$dS_t = \mu_t S_t dt + \sigma_t S_t dW_t \tag{2.73}$$

where the symbol dS_t represents an infinitesimal change in the price of the security, the $\mu_t S_t dt$ is the predicted movement during an infinitesimal interval dt, and $\sigma_t S_t dW_t$ is an unpredictable, infinitesimal random shock.

It is obvious that most of the concepts used in defining stochastic differential equations need to be developed step by step.

2.6.3 Discounting

Using continuous-time models leads to a change in the way discounting is done. In fact, if t is continuous, then the discount factor for an interval of length Δ will be given by the exponential function

$$e^{-r\Delta} \tag{2.74}$$

The r becomes the continuously compounded interest rate. If there exist dividends or foreign currencies, the r needs to be modified as explained in Section 2.6.

2.7 CONCLUSIONS: A METHODOLOGY FOR PRICING ASSETS

The arbitrage theorem provides a powerful methodology for determining fair market values of financial assets in practice. The major steps of this methodology as applied to financial derivatives can be summarized as follows:

1. Obtain a model (approximate) to track the dynamics of the underlying asset's price.
2. Calculate how the derivative asset price relates to the price of the underlying asset at *expiration* or at other *boundaries*.
3. Obtain risk-adjusted probabilities.
4. Calculate expected payoffs of derivatives at *expiration* using these risk-adjusted probabilities.
5. Discount this expectation using the risk-free return.

In order to be able to apply this pricing methodology, one needs familiarity with the following types of mathematical tools.

First, the notion of time needs to be defined carefully. Tools for handling changes in asset prices during "infinitesimal" time periods must be developed. This requires *continuous-time analysis*.

Second, we need to handle the notion of "randomness"during such infinitesimal periods. Concepts such as probability, expectation, average value, and volatility during infinitesimal periods need to be carefully defined. This requires the study of the so-called *stochastic calculus*. We try to discuss the intuition behind the assumptions that lead to major results in stochastic calculus.

Third, we need to understand how to obtain risk-adjusted probabilities and how to determine the correct discounting factor. The *Girsanov theorem* states the conditions under which such risk-adjusted probabilities can be used. The theorem also gives the form of these probability distributions.

Further, the notion of *martingales* is essential to Girsanov theorem, and, consequently, to the understanding of the "risk-neutral" world.

Finally, there is the question of how to relate the movements of various quantities to one another over time. In standard calculus, this is done using differential equations. In a random environment, the equivalent concept is a *stochastic differential equation* (SDE).

Needless to say, in order to attack these topics in turn, one must have some notion of the well-known concepts and results of "standard" calculus. There are basically three: (1) the notion of derivative, (2) the notion of integral, and (3) the Taylor series expansion.

2.8 REFERENCES

In this chapter, arbitrage theorem was treated in a simple way. Ingersoll (1987) provides a much more detailed treatment that is quite accessible, even to a beginner. Readers with a strong quantitative background may prefer Duffie (1996). The original article by Harrison and Kreps (1979) may also be consulted. Other related material can be found in Harrison and Pliska (1981). The first chapter in Musiela and Rutkowski (1997) is excellent and very easy to read after this chapter.

2.9 APPENDIX: GENERALIZATION OF THE ARBITRAGE THEOREM

According to the arbitrage theorem, if there are no-arbitrage possibilities, then there are "supporting"state prices $\{\psi_i\}$, such that each asset's price today equals a linear combination of possible future values. The theorem is also true in reverse. If there are such (supporting) state prices, then there are no-arbitrage opportunities.

In this section, we state the general form of the arbitrage theorem. First, we briefly define the underlying symbols.

- Define a matrix of payoffs, D:

$$D_t = \begin{bmatrix} d_{11} & \cdots & d_{1K} \\ \vdots & \vdots & \vdots \\ d_{N1} & \cdots & d_{NK} \end{bmatrix} \tag{2.75}$$

  N is the total number of securities and K is the total number of states of the world.
- Now define a portfolio, θ, as the vector of commitments to each asset:

$$\theta = \begin{bmatrix} \theta_1 \\ \vdots \\ \theta_N \end{bmatrix} \tag{2.76}$$

  In dealer's terminology, θ gives the *positions* taken at a certain time. Multiplying the θ by S_t, we obtain the value of portfolio θ:

$$S_t'\theta = \sum_{i=1}^{N} S_i(t)\theta_i \tag{2.77}$$

  This is total investment in portfolio θ at time t.
- Payoff to portfolio θ in state j is $\sum_{i=1}^{N} d_{ij}\theta_i$. In matrix form, this is expressed as

$$D'\theta = \begin{bmatrix} d_{11} & \cdots & d_{N1} \\ \vdots & \vdots & \vdots \\ d_{1K} & \cdots & d_{NK} \end{bmatrix} \begin{bmatrix} \theta_1 \\ \vdots \\ \theta_N \end{bmatrix} \tag{2.78}$$

- We can now define an arbitrage portfolio:

Definition 5. θ is an arbitrage portfolio, or simply an arbitrage, if either one of the following conditions is satisfied:

1. $S'\theta \leq 0$ and $D'\theta > 0$.
2. $S'\theta < 0$ and $D'\theta \geq 0$.

According to this, the portfolio θ guarantees some positive return in all states, yet it costs nothing to purchase. Or it guarantees a nonnegative return while having a negative cost today.

The following theorem is the generalization of the arbitrage conditions discussed earlier.

Theorem 2. *If there are no-arbitrage opportunities, then there exists a > 0 such that*

$$S = D\psi \tag{2.79}$$

If the condition in (2.77) *is true, then there are no-arbitrage opportunities.*

This means that in an arbitrage-free world there exist ψ_i such that

$$\begin{bmatrix} S_1 \\ \vdots \\ S_N \end{bmatrix} = \begin{bmatrix} d_{11} & \cdots & d_{1K} \\ \vdots & \vdots & \vdots \\ d_{N1} & \cdots & d_{NK} \end{bmatrix} \begin{bmatrix} \psi_1 \\ \vdots \\ \psi_K \end{bmatrix} \tag{2.80}$$

Note that according to the theorem we must have

$$\psi_i > 0 \text{ for all } i$$

if each state under consideration has a nonzero probability of occurrence.

Now suppose we consider a special type of return matrix where

$$D = \begin{bmatrix} 1 & \cdots & 1 \\ d_{21} & \cdots & d_{2K} \\ \vdots & \vdots & \vdots \\ d_{N1} & \cdots & d_{NK} \end{bmatrix} \tag{2.81}$$

In this matrix D, the first row is constant and equals 1. This implies that the return for the first asset is the same no matter which state of the world is realized. So, the first security is riskless.

Using the arbitrage theorem, and multiplying the first row of D with the state-price vector, we obtain

$$S = \psi_1 + \cdots + \psi_K \tag{2.82}$$

and define

$$\sum_{i=1}^{K} \psi_i = \psi_0 \tag{2.83}$$

The ψ_0 is the *discount in riskless borrowing*.

2.10 EXERCISES

1. 1. You are given the price of a nondividend-paying stock S_t and a European call option C_t in a world where there are only two possible states:

$$S_t = \begin{cases} 320 \text{ if } u \text{ occurs} \\ 290 \text{ if } d \text{ occurs} \end{cases}$$

The *true* probabilities of the two states are given by $\{\mathbb{P}_{up} = 0.5, \mathbb{P}_{down} = 0.5\}$. The current price is $S_t = 280$. The annual interest rate is constant at $r = 5\%$. The time is discrete, with $\Delta = 3$ months. The option has a strike price of $K = 280$ and expires at time $t + \Delta$.

(a) Find the risk-neutral martingale measure $\mathbb{Q}$ using the normalization by risk-free borrowing and lending.

(b) Calculate the value of the option under the risk-neutral martingale measure using

$$C_t = \frac{1}{1 + r\Delta} \mathbb{E}^{\mathbb{Q}} [C_{t+\Delta}]$$

(c) Now use the normalization by S_t and find a new measure $\widetilde{\mathbb{P}}$ under which the normalized variable is a martingale.

(d) What is the martingale equality that corresponds to normalization by S_t?

(e) Calculate the option's fair market value using the $\widetilde{\mathbb{P}}$.

(f) Can we state that the option's fair market value is independent of the choice of martingale measure?

(g) How can it be that we obtain the same arbitrage-free price although we are using two different probability measures?

(h) Finally, what is the *risk premium* incorporated in the option's price? Can we calculate this value in the real world? Why not?

2. In an economy there are two states of the world and four assets. You are given the following prices for three of these securities in different states of the world:

	Price		Dividends	
	State 1	State 2	State 1	State 2
Security A	120	70	4	1
Security B	80	60	3	1
Security C	90	150	2	10

"Current" prices for A, B, C are 100, 70, and 180, respectively.

(a) Are the "current" prices of the three securities arbitrage-free?

(b) If not, what type of arbitrage portfolio should one form?

(c) Determine a set of arbitrage-free prices for securities A, B, and C.

(d) Suppose we introduce a fourth security, which is a one-period futures contract written on B. What is its price?

(e) Suppose a put option with strike price $K = 125$ is written on C. The option expires in period 2. What is its arbitrage-free price?

3. Consider a stock S_t and a plain vanilla, at-the-money put option written on this stock. The option expires at time $t + \Delta$, where Δ denotes a small interval. At time t, there are only two possible ways the S_t can move. It can either go *up* to $S^u_{t+\Delta}$, or go *down* to $S^d_{t+\Delta}$. Also available to traders is risk-free borrowing and lending at annual rate r.

(a) Using the arbitrage theorem, write down a three-equation system with two states that gives the arbitrage-free values of S_t and C_t.

(b) Now plot a two-step binomial tree for S_t. Suppose at every node of the tree the markets are arbitrage-free. How many three-equation systems similar to the preceding case could then be written for the entire tree?

(c) Can you find a three-equation system with four states that corresponds to the same tree?

(d) How do we know that all the implied state prices are internally consistent?

4. A four-step binomial tree for the price of a stock S_t is to be calculated using the up- and downticks given as follows:

$$u = 1.15$$
$$d = \frac{1}{u}$$

These up and down movements apply to one-month periods denoted by $\Delta = 1$. We have the following dynamics for S_t,

$$S^u_{t+\Delta} = S_t u$$
$$S^d_{t+\Delta} = S_t d$$

where up and down describe the two states of the world at each node. Assume that time is measured in months and that $t = 4$ is the expiration date for a European call option C_t written on S_t. The stock does not pay any dividends and its price is expected (by "market participants") to grow at an annual rate of 15%. The risk-free interest rate r is known to be constant at 5%.

(a) According to the data given above, what is the (approximate) annual volatility of S_t if this process is known to have a log-normal distribution?

(b) Calculate the four-step binomial trees for the S_t and the C_t.

(c) Calculate the arbitrage-free price C_0 of the option at time $t = 0$.

5. You are given the following information concerning a stock denoted by S_t.

- Current value $= 102$.
- Annual volatility $= 30\%$.
- You are also given the spot rate $r = 5\%$, which is known to be constant during the next 3 months.

It is hoped that the dynamic behavior of S_t can be approximated reasonably well by a binomial process if one assumes observation intervals of length 1 month.

(a) Consider a European call option written on S_t. The call has a strike price $K = 120$ and an expiration of 3 months. Using the S_t and the risk-free borrowing and lending B_t, construct a portfolio that replicates the option.
(b) Using the replicating portfolio, price this call.
(c) Suppose you sell, over-the-counter, 100 such calls to your customers. How would you hedge this position? Be precise.
(d) Suppose the market price of this call is 5. How would you form an arbitrage portfolio?

6. Suppose you are given the following data:

- Risk-free yearly interest rate is $r = 6\%$.
- The stock price follows:

$$S_t - S_{t-1} = \mu S_t + \sigma S_t \varepsilon_t$$

where the ε_t is a serially uncorrelated binomial process assuming the following values:

$$\varepsilon_t = \begin{cases} +1 \text{ with probability } p \\ -1 \text{ with probability } 1 - p \end{cases}$$

The $0 < p < 1$ is a parameter.
- Volatility is 12% a year.
- The stock pays no dividends and the current stock price is 100.

Now consider the following questions.

(a) Suppose μ is equal to the risk-free interest rate:

$$\mu = r$$

and that the S_t is arbitrage-free. What is the value of p?
(b) Would a $p = 1/3$ be consistent with arbitrage-free S_t?
Now suppose μ is given by:

$$\mu = r + \text{ risk premium}$$

(c) What do the p and ε_t represent under these conditions?
(d) Is it possible to determine the value of p?

7. Using the data in the previous question, you are now asked to approximate the current value of a European call option on the stock S_t. The option has a strike price of 100, and a maturity of 200 days.

(a) Determine an appropriate time interval Δ, such that the binomial tree has 5 steps.
(b) What would be the implied u and d?
(c) What is the implied "up"probability?
(d) Determine the tree for the stock price S_t.
(e) Determine the tree for the call premium C_t.

8. Assume that the annual interest rate, r, is equal to 1%. Also assume that the annualized volatility on the asset is 10%. Using a binomial tree, price a digital option expiring in 3 years that pays \$1 if the asset price is above 55. The current stock price is \$50. Use a three period tree, with each period spanning a year.
9. Write a Matlab program to do the above problem.

CHAPTER

3

Review of Deterministic Calculus

OUTLINE

3.1 INTRODUCTION

The mathematics of derivative assets assumes that time passes continuously. As a result, new information is revealed continuously, and decision makers may face instantaneous changes in random news. Hence, technical tools for pricing derivative products require ways of handling random variables over infinitesimal time intervals. The mathematics of such random variables is known as *stochastic calculus*.

Stochastic calculus is an internally consistent set of operational rules that are different from the tools of "standard" calculus in some fundamental ways.

At the outset, stochastic calculus may appear too abstract to be of any use to a practitioner. This first impression is not correct. Continuous time finance is both *simpler* and *richer*. Once a market participant gets some practice, it is easier to work with continuous-time tools than their discrete-time equivalents.

Financial Derivatives, Third Edition.
http://dx.doi.org/10.1016/B978-0-12-384682-2.00003-7

In fact, sometimes there are no equivalent results in discrete time. In this sense stochastic calculus offers a wider variety of tools to the financial analyst. For example, continuous time permits infinitesimal adjustments in portfolio weights. This way, replicating "nonlinear" assets with "simple" portfolios becomes possible. In order to *replicate* an option, the underlying asset and risk-free borrowing may be used. Such an *exact* replication will be impossible in discrete time.[1]

3.1.1 Information Flows

It may be argued that the manner in which information flows in financial markets is more consistent with stochastic calculus than with "standard calculus."

For example, the relevant "time interval" may be different on different trading days. During some days an analyst may face more volatile markets, on others less. Changing volatility may require changing the basic "observation period," i.e., the Δ of the previous chapter.

Also, numerical methods used in pricing securities are costly in terms of computer time. Hence, the pace of activity may make the analyst choose coarser or finer time intervals depending on the level of volatility. Such approximations can best be accomplished using random variables defined over continuous time. The tools of stochastic calculus will be needed to define these models.

3.1.2 Modeling Random Behavior

A more technical advantage of stochastic calculus is that a complicated random variable can have a very simple structure in continuous time, once the attention is focused on infinitesimal intervals. For example, if the time period under consideration is denoted by dt, and if dt is "infinitesimal," then asset prices may safely be assumed to have two likely movements: uptick or downtick.

Under some conditions, such a "binomial" structure may be a good approximation to reality during an infinitesimal interval dt, but not necessarily in a large "discrete time" interval denoted by Δ.[2]

Finally, the main tool of stochastic calculus—namely, the Ito integral—may be more appropriate to use in financial markets than the Riemann integral used in standard calculus.

These are some reasons behind developing a new calculus. Before doing this, however, a review of standard calculus will be helpful. After all, although the rules of stochastic calculus are different, the reasons for developing such rules are the same as in standard calculus:

- We would like to calculate the response of one variable to a (random) change in another variable. That is, we would like to be able to *differentiate* various functions of interest.
- We would like to calculate sums of random increments that are of interest to us. This leads to the notion of (stochastic) *integral*.
- We would like to *approximate* an arbitrary function by using simpler functions. This leads us to (stochastic) Taylor series approximations.
- Finally, we would like to model the dynamic behavior of continuous-time random variables. This leads to *stochastic differential equations*.

3.2 SOME TOOLS OF STANDARD CALCULUS

In this section we review the major concepts of *standard* (deterministic) calculus. Even if the

[1]Unless, of course, the underlying state space is itself discrete. This would be the case when the underlying asset price can assume only a finite number of possible values in the future.

[2]A binomial random variable can assume one of the two possible values, and it may be significantly easier to work with than, say, a random variable that may assume any one of an uncountable number of possible values.

reader is familiar with elementary concepts of standard calculus discussed here, it may still be worthwhile to go over the examples in this section. The examples are devised to highlight exactly those points at which standard calculus will fail to be a good approximation when underlying variables are stochastic.

3.3 FUNCTIONS

Suppose A and B are two sets, and let f be a rule which associates to every element x of A, exactly one element y in B.[3] Such a rule is called a *function* or a *mapping*. In mathematical analysis, functions are denoted by

$$f: A \to B \tag{3.1}$$

or by

$$y = f(x), \quad x \in A \tag{3.2}$$

If the set B is made of real numbers, then we say that f is a *real-valued function* and write

$$f: A \to \mathbb{R} \tag{3.3}$$

If the sets A and B are themselves collections of functions, then f transforms a function into another function, and is called an *operator*.

Most readers will be familiar with the standard notion of functions. Fewer readers may have had exposure to *random* functions.

3.3.1 Random Functions

In the function

$$y = f(x), \quad x \in A \tag{3.4}$$

once the value of x is given, we get the element y. Often y is assumed to be a real number. Now consider the following significant alteration.

[3]The set A is called the domain, and the set B is called the range of f.

There is a set Ω, where $\omega \in \Omega$ denotes a state of the world. The function f depends on $x \in \mathbb{R}$ *and* on $\omega \in \Omega$:

$$f: \mathbb{R} \times \Omega \to \mathbb{R} \tag{3.5}$$

or

$$y = f(x, \omega), \quad x \in \mathbb{R}, \quad \omega \in \Omega \tag{3.6}$$

where the notation $\mathbb{R} \times \Omega$ implies that one has to "plug in" to $f(\cdot)$ two variables, one from the set Ω, and the other from $\mathbb{R}$.

The function $f(x, \omega)$ has the following property: Given a $\omega \in \Omega$, the $f(\cdot, \omega)$ becomes a function of x only. Thus, for different values of $\omega \in \Omega$ we get different functions of x. Two such cases are shown in Figure 3.1. $f(x, \omega_1)$ and $f(x, \omega_2)$ are two functions of x that differ because the second element ω is different.

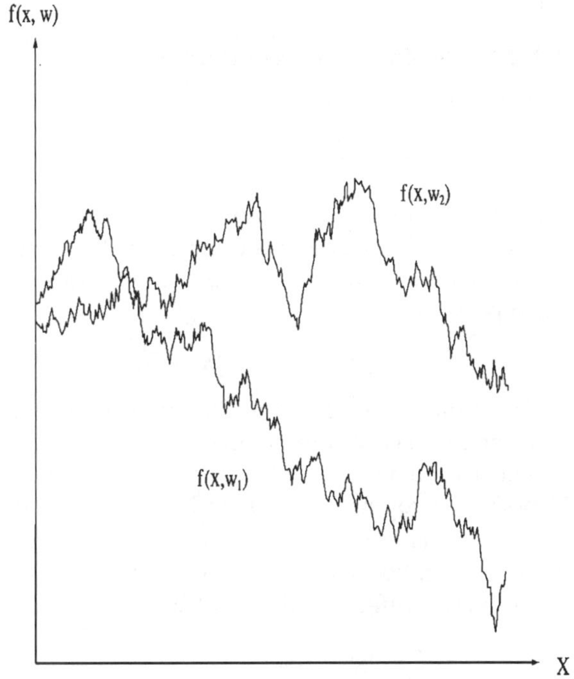

FIGURE 3.1 Plot of the function $f(x, \omega)$ for two different values of ω.

When x represents time, we can interpret $f(x, \omega_1)$ and $f(x, \omega_2)$ as two different *trajectories* that depend on different states of the world.

Hence, if ω represents the underlying randomness, the function $f(x, \omega)$ can be called a *random function*. Another name for random functions is *stochastic processes*. With stochastic processes, x will represent time, and we often limit our attention to the set $x \geq 0$.

Note this fundamental point. Randomness of a stochastic process is in terms of the trajectory as a whole, rather than a particular value at a specific point in time. In other words, the random drawing is done from a collection of trajectories. Choosing the state of the world, ω, determines the complete trajectory.

3.3.2 Examples of Functions

There are some important functions that play special roles in our discussion. We will briefly review them.

3.3.2.1 *The Exponential Function*

The infinite sum

$$1 + 1 + \frac{1}{2!} + \frac{1}{3!} + \cdots + \frac{1}{n!} + \cdots \tag{3.7}$$

converges to an irrational number between 2 and 3 as $n \to \infty$. This number is denoted by the letter e. The *exponential function* is obtained by raising e to a power of x:

$$y = e^x, \quad x \in \mathbb{R} \tag{3.8}$$

This function is generally used in discounting asset prices in continuous time.

The exponential function has a number of important properties. It is infinitely differentiable. That is, beginning with $y = e^{f(x)}$, the following operation can be repeated infinitely by recursively letting y be the right-hand side in:

$$\frac{dy}{dx} = e^{f(x)} \frac{df(x)}{dx} \tag{3.9}$$

The exponential function also has the interesting multiplicative property:

$$e^x e^z = e^{x+z} \tag{3.10}$$

Finally, if x is a random variable, then $y = e^x$ will be random as well.

3.3.2.2 *The Logarithmic Function*

The *logarithmic function* is defined as the inverse of the exponential function. Given

$$y = e^x, \quad x \in \mathbb{R} \tag{3.11}$$

the natural logarithm of y is given by

$$\ln(y) = x, \quad y > 0 \tag{3.12}$$

A practitioner may sometimes work with the logarithm of asset prices. Note that while y is always positive, there is no such restriction on x. Hence, the logarithm of an asset price may extend from minus to plus infinity.

3.3.2.3 *Functions of Bounded Variation*

The following construction will be used several times in later chapters.

Suppose a time interval is given by $[0, T]$. We *partition* this interval into n subintervals by selecting the $t_i, i = 1, \ldots, n$, as

$$0 = t_0, \ldots, t_n \tag{3.13}$$

The $[t_i - t_{i-1}]$ represents the length of the ith subinterval.

Now consider a function of time $f(t)$, defined on the interval $[0, T]$:

$$f\colon [0.T] \to \mathbb{R} \tag{3.14}$$

We form the sum

$$\sum_{i=1}^{n} \left| f(t_i) - f(t_{i-1}) \right| \tag{3.15}$$

This is the sum of the absolute values of all changes in $f(\cdot)$ from one t_i to the next.

Clearly, for each partition of the interval $[0, T]$, we can form such a sum. Given that uncountably many partitions are possible, the sum can assume uncountably many values. If these sums are bounded from above, the function $f(\cdot)$ is said to be of bounded variation. Thus, bounded variation implies

$$V_0 = \max \sum_{i=1}^{n} \left| f\left(t_i\right) - f\left(t_{i-1}\right) \right| < \infty \qquad (3.16)$$

where the maximum is taken over all possible partitions of the interval $[0, T]$. In this sense, V_0 is the maximum of all possible variations in $f(\cdot)$, and it is finite. V_0 is the total variation of f on $[0, T]$. Roughly speaking, V_0 measures the length of the trajectory followed by $f(\cdot)$ as t goes from 0 to T.

Thus, functions of bounded variation are not excessively "irregular." In fact, any "smooth" function will be of bounded variation.[4]

3.3.2.4 *An Example*

Consider the function

$$f(t) = \begin{cases} t \sin\left(\frac{\pi}{t}\right), \text{when } 0 < t \leq 1 \\ 0, \text{when } t = 0 \end{cases} \qquad (3.17)$$

It can be shown that $f(t)$ is not of bounded variation.[5]

That this is the case is shown in Figure 3.2. Note that as $t \to 0, f$ becomes excessively "irregular."

The concept of bounded variation will play an important role in our discussions later. One reason is the following: asset prices in continuous time will have some unpredictable part. No matter how finely we slice the time interval, they will still be partially unpredictable. But this means that trajectories of asset prices will have to be very irregular.

[4]It can be shown that if a function has a derivative everywhere on $[0, T]$, then the function is of bounded variation.

[5]To show this formally, choose the partition

$$0 < \frac{2}{2n+1} < \frac{2}{2n-1} < \cdots < \frac{2}{5} < \frac{2}{3} < 1 \qquad (3.18)$$

Then the variation over this partition is

$$\sum_{i=1}^{n} \left| f\left(t_i\right) - f\left(t_{i-1}\right) \right| = 4\frac{1}{3} + \frac{1}{5} + \frac{1}{7} + \cdots + \frac{1}{2n+1} \qquad (3.19)$$

The right-hand side of this quantity becomes arbitrarily large as $n \to \infty$.

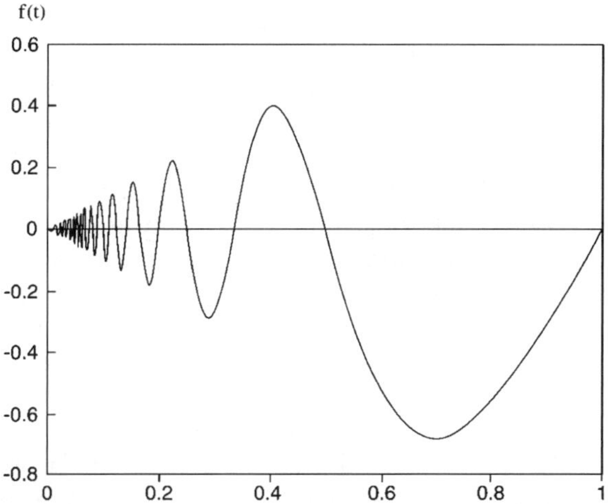

FIGURE 3.2 A plot of a function that is not of bounded variation.

As will be seen later, continuous-time processes that we use to represent asset prices have trajectories with unbounded variation.

3.4 CONVERGENCE AND LIMIT

Suppose we are given a *sequence*

$$x_0, x_1, \ldots, x_n, \ldots \qquad (3.20)$$

where x_n represents an object that changes as n is increased. This "object" can be a sequence of numbers, a sequence of functions, or a sequence of operations. The essential point is that we are observing successive versions of x_n.

The notion of *convergence* of a sequence has to do with the "eventual" value of x_n as $n \to \infty$. In the case where x_n represents real numbers, we can state this more formally:

Definition 6. We say that a sequence of real numbers x_n converges to $x^* < \infty$ if for arbitrary $\varepsilon > 0$, there exists a $N < \infty$ such that

$$\left| x_n - x^* \right| < \varepsilon \text{ for all } n > N \qquad (3.21)$$

We call x^* the limit of x_n.

In words, x_n converges to x^* if x_n stays arbitrarily close to the point x^* after a finite number of steps. Two important questions can be asked.

First can we deal with convergence of x_n if these were *random* variables instead of deterministic numbers? This question is relevant, since a random number x_n could conceivably assume an extreme value and suddenly may fall very far from any x^*, even if $n > N$.

Second since one can define different measures of "closeness," we should in principle be able to define convergence in different ways as well. Are these definitions all equivalent?

We will answer these questions later. However, convergence is clearly a very important concept in approximating a quantity that does not easily lend itself to direct calculation. For example, we may want to define the notion of integral as the limit of a sequence.

3.4.1 The Derivative

The notion of the derivative[6] can be looked at in (at least) two different ways. First, the derivative is a way of dealing with the "smoothness" of functions. It is a way of defining rates of change of variables under consideration. In particular, if trajectories of asset prices are "too irregular," then their derivative with respect to time may not exist.

Second, the derivative is a way of calculating how one variable responds to a change in another variable. For example, given a change in the price of the underlying asset, we may want to know how the market value of an option written on it may move. These types of derivatives are usually taken using the *chain rule*.

The derivative is a *rate* of change. But it is a rate of change for infinitesimal movements. We give a formal definition first.

Definition 7. Let

$$y = f(x) \tag{3.22}$$

be a function of $x \in \mathbb{R}$. Then the derivative of $f(x)$ with respect to x, if it exists, is formally denoted by the symbol f_x and is given by

$$f_x = \lim_{\Delta \to 0} \frac{f(x + \Delta) - f(x)}{\Delta} \tag{3.23}$$

where Δ is an increment in x.

The variable x can represent any real-life phenomenon. Suppose it represents *time*.[7] Then Δ would correspond to a finite time interval. The $f(x)$ would be the value of y at time x, and the $f(x+\Delta)$ would represent the value of y at time $x + \Delta$. Hence, the numerator in (3.23) is the change in y during a time interval Δ. The ratio itself becomes the *rate of change* in y during the same interval. For example, if y is the price of a certain asset at time x, the ratio in (3.23) would represent the rate at which the price changes during an interval Δ.

Why is a limit being taken in (3.23)? In defining the derivative, the limit has a practical use. It is taken to make the ratio in (3.23) independent of the size of Δ, the time interval that passes.

For making the ratio independent of the size of Δ, one pays a price. The derivative is defined for infinitesimal intervals. For larger intervals, the derivative becomes an approximation that deteriorates as Δ gets larger and larger.

3.4.1.1 *Example: The Exponential Function*

As an example of derivatives, consider the exponential function:

$$f(x) = Ae^{rx}, \quad x \in \mathbb{R} \tag{3.24}$$

A graph of this function with $r > 0$ is shown in Figure 3.3. Taking the derivative with respect to x formally:

$$\begin{aligned} f_x &= \frac{df(x)}{dx} = r\left[Ae^{rx}\right] \\ &= rf(x) \end{aligned} \tag{3.25}$$

[6]The reader should not confuse the mathematical operation of differentiation or taking a derivative with the term "derivative securities" used in finance.

[7]Time is one of the few deterministic variables one can imagine.

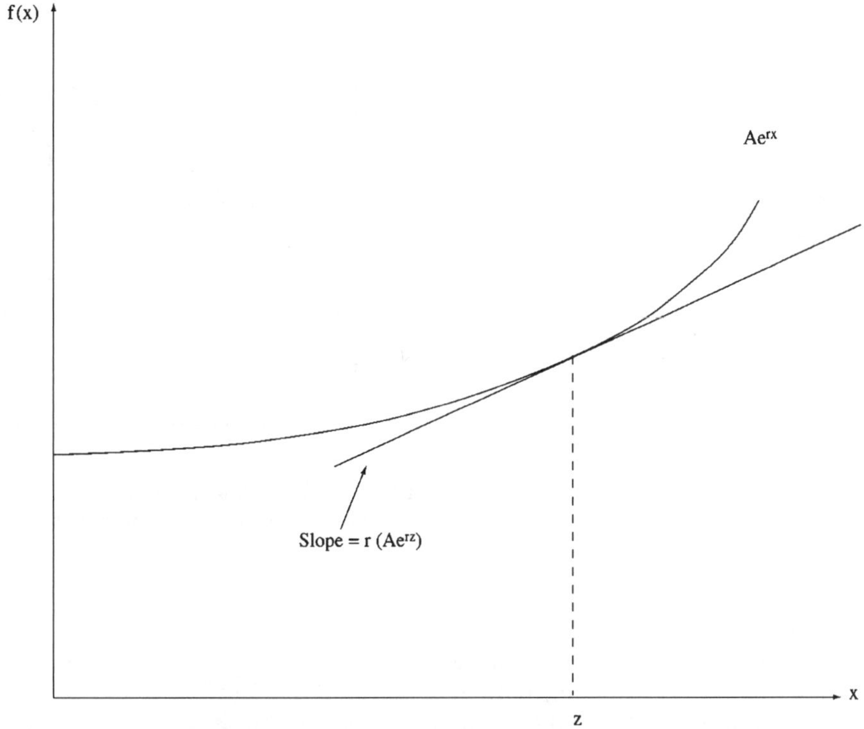

FIGURE 3.3 Graphical illustration of the slope of an exponential function.

The quantity f_x is the rate of change of $f(x)$ at point x. Note that as x gets larger, the term e^{rx} increases. This can be seen in Figure 3.3 from the increasing growth the $f(\cdot)$ exhibits. The ratio

$$\frac{f_x}{f(x)} = r \tag{3.26}$$

is the *percentage* rate of change. In particular, we see that an exponential function has a constant percentage rate of change with respect to x.

3.4.1.2 Example: The Derivative as an Approximation

To see an example of how derivatives can be used in approximations, consider the following argument.

Let Δ be a finite interval. Then, using the definition of derivative in (3.23) and if Δ is "small," we can write approximately

$$f(x + \Delta) \approx f(x) + f_x \Delta \tag{3.27}$$

This equality means that the value assumed by $f(\cdot)$ at point $x + \Delta$, can be approximated by the value of $f(\cdot)$ at point x, plus the derivative f_x multiplied by Δ. Note that when one does not know the *exact* value of $f(x + \Delta)$, the knowledge of $f(x), f_x$, and Δ is sufficient to obtain an approximation.[8]

This result is shown in Figure 3.4, where the ratio

$$\frac{f(x + \Delta) - f(x)}{\Delta} \tag{3.28}$$

[8]If x represents time, and if x is the "present," then $f(x + \Delta)$ will belong to the "future." However, $f(x), f_x$, and Δ are all quantities that relate to the "present." In this sense, they can be used for obtaining a crude "prediction" of $f(x + \Delta)$ in real time. This prediction requires having a numerical value for f_x, the value of the derivative at the point x.

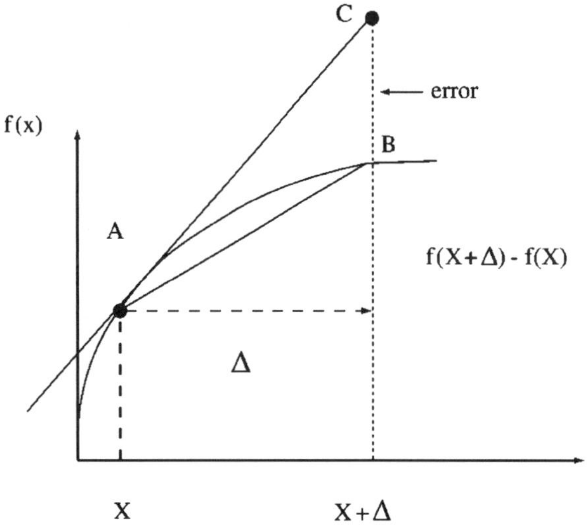

FIGURE 3.4 Illustration of derivative of f at x is the slope of the tangent line at x.

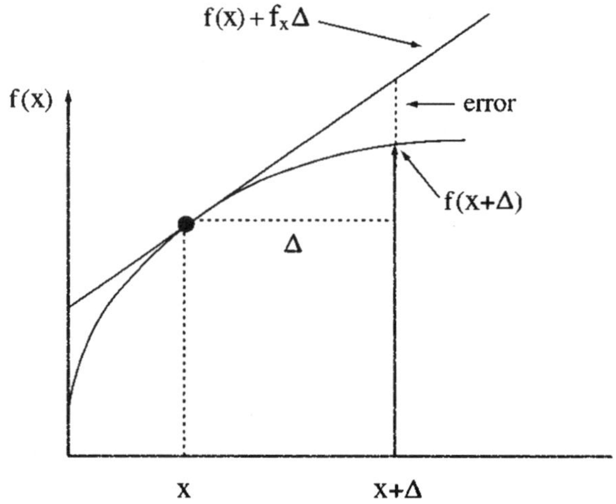

FIGURE 3.5 Illustration of approximating $f(x+\Delta)$ using $f(x)$ plus the derivative at x multiplied by Δ.

represents the slope of the segment denoted by AB. As Δ becomes smaller and smaller, with A fixed, the segment AB converges toward the tangent at the point A. Hence, the derivative f_x is the slope of this tangent.

When we add the product $f_x\Delta$ to $f(x)$, we obtain the point C. This point can be taken as an approximation of B. Whether this will be a "good" or a "bad" approximation depends on the size of Δ and on the shape of the function $f(\cdot)$

Two simple examples will illustrate these points. First, consider Figure 3.5. Here, Δ is large. As expected, the approximation $f(x)+f_x\Delta$ is not very near $f(x+\Delta)$.

Figure 3.6 illustrates a more relevant example. We consider a function $f(\cdot)$ that is not very smooth. The approximating $\widehat{f}(x+\Delta)$ obtained from

$$\widehat{f}(x+\Delta) = f(x) + f_x\Delta \tag{3.29}$$

may end up being a very unsatisfactory approximation to the true $f(x+\Delta)$. Clearly, the more "irregular" the function $f(\cdot)$ becomes, the more such approximations are likely to fail.

Consider an extreme case in the next example.

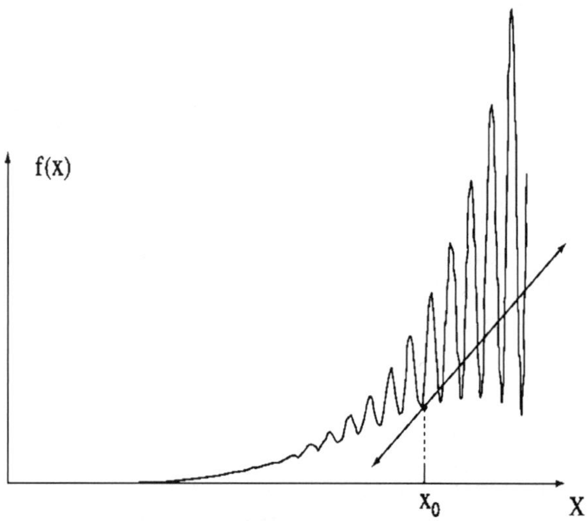

FIGURE 3.6 Slope of a function that is not very smooth.

3.4.1.3 *Example: High Variation*

Consider Figure 3.7, where the function $f(x)$ is continuous, but exhibits extreme variations even in small intervals Δ. Here, not only is the prediction

$$f(x+\Delta) \approx f(x) + f_x\Delta \tag{3.30}$$

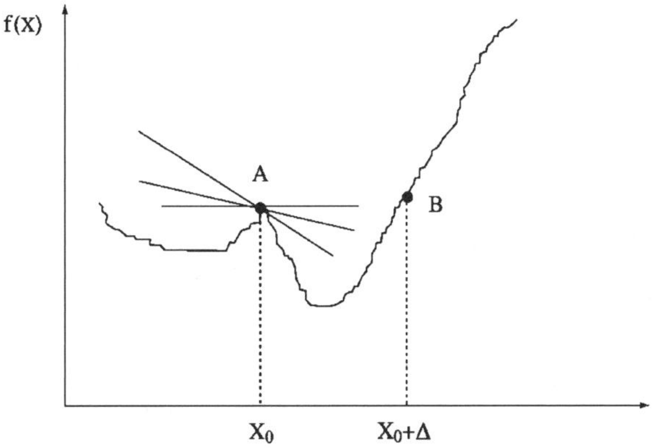

FIGURE 3.7 A plot of a continuous function with extreme variations in small intervals.

likely to fail, but even a satisfactory definition of f_x may not be obtained. Take, for example, the point x_0. What is the rate of change of the function $f(x)$ at the point x_0? It is difficult to answer. Indeed, one can draw many tangents with differing slopes to $f(x)$ at that particular point. It appears that the function $f(x)$ is not differentiable.

3.4.2 The Chain Rule

The second use of the derivative is the "chain rule". In the examples discussed earlier, $f(x)$ was a function of x, and x was assumed to represent time. The derivative was introduced as the response of a variable to a variation in time.

In pricing derivative securities, we face a somewhat different problem. The price of a derivative asset, e.g., a call option, will depend on the price of the underlying asset, and the price of the underlying asset depends on time.[9]

Hence, there is a chain effect. Time passes, new (small) events occur, the price of the underlying asset changes, and this affects the derivative asset's price. In standard calculus, the tool used to analyze these sorts of chain effects is known as the chain rule.

[9]As time passes, the expiration date of a contract comes closer, and even if the underlying asset's price remains constant, the price of the call option will fall.

Suppose in the example just given x was not itself the time, but a deterministic function of time, denoted by the symbol $t \geq 0$:

$$x_t = g(t) \tag{3.31}$$

Then the function $f(\cdot)$ is called a composite function and is expressed as

$$y_t = f(g(t)) \tag{3.32}$$

The question is how to obtain a formula that gives the ultimate effect of a change in t on the y_t.

In standard calculus the chain rule is defined as follows:

Definition 8. For f and g defined as above, we have

$$\frac{dy}{dt} = \frac{df(g(t))}{dg(t)}\frac{dg(t)}{dt} \tag{3.33}$$

According to this, the chain rule is the product of two derivatives. First, the derivative of $f(g(t))$ is taken with respect to $g(t)$. Second, the derivative of $g(t)$ is taken with respect to t. The final effect of t on y_t is then equal to the product of these two expressions.

The chain rule is a useful tool in approximating the responses of one variable to changes in other variables.

Take the case of derivative asset prices. A trader observes the price of the underlying asset continuously and wants to know how the valuation of the complex derivative products written on this asset would change. If the derivative is an exchange-traded product, these changes can be observed from the markets directly.[10] However, if the derivative is a "structured" product, its valuation needs to be calculated in-house, using theoretical pricing models. These pricing models will use some tool such as the chain rule shown in (3.33).

In the example just given, $f(x)$ was a function of x_t, and x_t was a deterministic variable. There was no randomness associated with x_t. What happens if x_t is random, or if the function $f(\cdot)$

[10]Of course, there is always the question of whether the markets are correctly pricing the security at that instant.

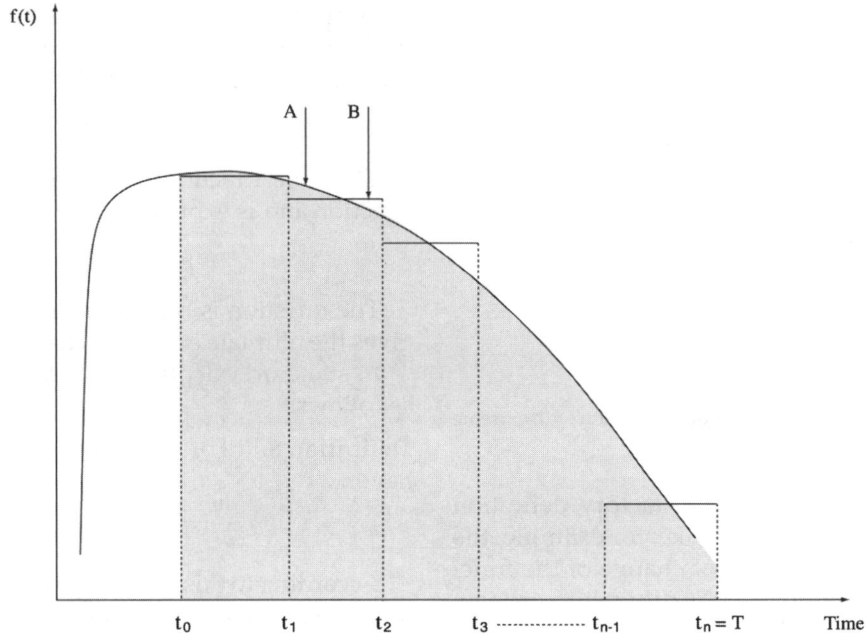

FIGURE 3.8 The construction steps to find the area under the graph via the Riemann integral.

depends on some random variable z_t as well? In other words,

1. Can we still use the *same* chain rule formula?
2. How does the chain rule formula change in stochastic environments?

The answer to the first question is No. The chain rule formula given in (3.33) cannot be used in a continuous-time stochastic environment. In fact, by "stochastic calculus," we mean a set of methods that yield the formulas equivalent to the chain rule and that approximate the laws of motion of random variables in continuous time.

The *purpose* of stochastic calculus is the same as that of standard calculus. The rules, though, are different.

3.4.3 The Integral

The integral is the mathematical tool used for calculating sums. In contrast to the $\sum$ operator, which is used for sums of a countable number of objects, integrals denote sums of *uncountably infinite* objects. Since it is not clear how one could "sum" objects that are not even countable, a formal definition of integral has to be derived.

The general approach in defining integrals is, in a sense, obvious. One would begin with an approximation involving a countable number of objects, and then take some limit and move into uncountable objects. Given that different types of limits may be taken, the integral can be defined in various ways. In standard calculus the most common form is the Riemann integral. A somewhat more general integral defined similarly is the Riemann–Stieltjes integral. In this section we will review these definitions.

3.4.3.1 The Riemann Integral

We are given a deterministic function $f(t)$ of time $t \in [0, T]$. Suppose we are interested in integrating this function over an interval $[0, T]$

$$\int_0^T f(s)\, ds \tag{3.34}$$

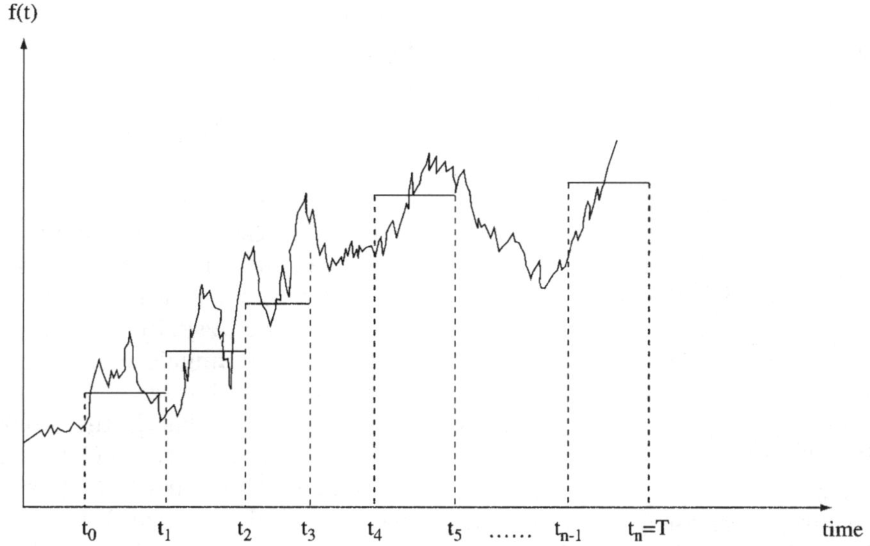

FIGURE 3.9 A example of a function with steep variations that is not Riemann-integrable.

which corresponds to the area shown in Figure 3.8.

In order to calculate the Riemann integral, we partition the interval $[0, T]$ into n disjoint subintervals

$$0 = t_0 < t_1 < \cdots < t_n = T \tag{3.35}$$

then consider the approximating sum

$$\sum_{i=1}^{n} f\left(\frac{t_i + t_{i-1}}{2}\right)(t_i - t_{i-1}) \tag{3.36}$$

Definition 9. Given that $\max |t_i - t_{i-1}| \to 0$, the Riemann integral will be defined by the limit

$$\sum_{i=1}^{n} f\left(\frac{t_i + t_{i-1}}{2}\right)(t_i - t_{i-1}) \to \int_0^T f(s)ds \tag{3.37}$$

where the limit is taken in a standard fashion.

The term on the left-hand side of (3.37) involves adding the areas of n rectangles constructed using $|t_i - t_{i-1}|$ as the base and $f((t_i - t_{i-1})/2)$ as the height. Figure 3.8 displays this construction. Note that the small area A is approximately equal to the area B. This is especially true if the base of the rectangles is small *and* if the function $f(t)$ is smooth—that is, does not vary heavily in small intervals.

In case the sum of the rectangles fails to approximate the area under the curve, we may be able to correct this by considering a *finer* partition. As the $|t_i - t_{i-1}|$'s get smaller, the base of the rectangles will get smaller. More rectangles will be available, and the area can be better approximated.

Obviously, the condition that $f(t)$ should be smooth plays an important role during this process. In fact, a very "irregular" path followed by $f(t)$ may be much more difficult to approximate by this method. Using the terminology discussed before, in order for this method to work, the function $f(t)$ must be Riemann-integrable.

A counterexample is shown in Figure 3.9. Here, the function $f(t)$ shows steep variations. If such variations do not smooth out as the base of the rectangles gets smaller, the approximation by rectangles may fail.

We have one more comment that will be important in dealing with the Ito integral later in the text. The rectangles used to approximate the

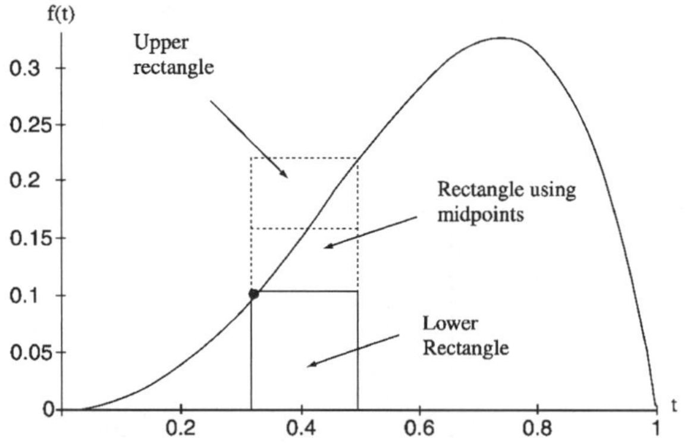

FIGURE 3.10 Illustration of approximating the area under the curve using upper rectangles, lower rectangles, and rectangles using midpoints.

area under the curve were constructed in a particular way. To do this, we used the value of $f(t)$ evaluated at the midpoint of the intervals $t_i - t_{i-1}$. Would the same approximation be valid if the rectangles were defined in a different fashion? For example, if one defined the rectangles either by

$$f(t_i)(t_i - t_{i-1}) \tag{3.38}$$

or by

$$f\left(t_{i-1}\right)\left(t_i - t_{i-1}\right) \tag{3.39}$$

would the integral be different? To answer this question, consider Figure 3.10. Note that as the partitions get finer and finer, rectangles defined either way would eventually approximate the same area. Hence, at the limit, the approximation by rectangles would not give a different integral even when one uses different heights for defining the rectangles.

It turns out that a similar conclusion cannot be reached in stochastic environments. Suppose $f(W_t)$ is a function of a random variable W_t and that we are interested in calculating

$$\int_0^T f(W_s)\, dW_s \tag{3.40}$$

Unlike the deterministic case, the choice of rectangles defined by

$$f\left(W_{t_i}\right)\left(W_{t_i} - W_{t_{i-1}}\right) \tag{3.41}$$

will in general result in a different expression from the rectangles:

$$f\left(W_{t_{i-1}}\right)\left(W_{t_i} - W_{t_{i-1}}\right) \tag{3.42}$$

To see the reason behind this fundamental point, consider the case where W_t is a martingale. Then the expectation of the term in (3.43), conditional on information at time t_{i-1}, will vanish. This will be the case because, by definition, future increments of a martingale will be unrelated to the current information set.

On the other hand, the same conditional expectation of the term in (3.42) will, in general, be nonzero.[11] Clearly, in stochastic calculus, expressions that utilize different definitions of approximating rectangles may lead to different results.

Finally, we would like to emphasize an important result. Note that when $f(\cdot)$ depends on a random variable, the resulting integral itself will be a random variable. In this sense, we will be dealing with *random integrals*.

3.4.3.2 The Stieltjes Integral

The Stieltjes integral is a different definition of the integral. Define the differential df as a small variation in the function $f(x)$ due to an infinitesimal variation in x:

$$df(x) = f(x + dx) - f(x) \tag{3.43}$$

We have already discussed the equality

$$df(x) = f_x(x)\, dx \tag{3.44}$$

(Note that according to the notation used here, the derivative $f_x(x)$ is a function of x as well.) Now suppose we want to integrate a function $h(x)$ with respect to x:

[11]Note that $\left(W_{t_i} - W_{t_{i-1}}\right)$ and W_{t_i} are correlated.

$$\int_{x_0}^{x_n} h(x)\, dx \tag{3.45}$$

where the function $h(x)$ is given by

$$h(x) = g(x)f_x(x) \tag{3.46}$$

Then the Stieltjes integral is defined as

$$\int_{x_0}^{x_n} g(x)\, df(x) \tag{3.47}$$

with

$$df(x) = f_x(x)\, dx \tag{3.48}$$

This definition is not very different from that of the Riemann integral. In fact, similar approximating sums are used in both cases.

If x represents time t, the Stieltjes integral over a partitioned interval, $[0, T]$, is given by

$$\int_0^T g(s)\, df(s) \approx \sum_{i=1}^{n} g\left(\frac{t_i + t_{i-1}}{2}\right)\left(f\left(t_i\right) - f\left(t_{i-1}\right)\right) \tag{3.49}$$

Because of these similarities, the limit as $\max_i \left|t_i - t_{i-1}\right|$ of the right-hand side is known as the Riemann–Stieltjes integral.

The Riemann–Stieltjes integral is useful when the integration is with respect to increments in $f(x)$ rather than the x itself. Clearly, in dealing with financial derivatives, this is often the case. The price of the derivative asset depends on the underlying asset's price, which in turn depends on time. Hence, it may appear that the Riemann–Stieltjes integral is a more appropriate tool for dealing with derivative asset prices.

However, before coming to such a conclusion, note that all the discussion thus far involved deterministic functions of time. Would the same definitions be valid in a stochastic environment? Can we use the same rectangles to approximate integrals in random environments? Would the choice of the rectangle make a difference?

The answer to these questions is, in general, no. It turns out that in stochastic environments the functions to be integrated may vary too much for a straightforward extension of the Riemann integral to the stochastic case. A new definition of integral will be needed.

3.4.3.3 *Example*

In this section, we would like to discuss an example of a Riemann–Stieltjes integral. We do this by using a simple function. We let

$$g\left(S_t\right) = aS_t \tag{3.50}$$

where a is a constant. This makes $g(\cdot)$ a *linear* function of S_t.[12] What is the value of the integral

$$\int_0^T aS_t dS(t) \tag{3.51}$$

if the Riemann–Stieltjes definition is used?

Directly "taking" the integral gives

$$\int_0^T aS_t dS(t) = a\left[\frac{1}{2}S_t^2\right]_0^T \tag{3.52}$$

or

$$\int_0^T aS_t dS_t = a\left[\frac{1}{2}S_T^2 - \frac{1}{2}S_0^2\right] \tag{3.53}$$

Now, let us see if we can get the same result using approximation by rectangles.

Because $g(\cdot)$ is linear, in this particular case the approximation by rectangles works well. This is especially true if we evaluate the height of the rectangle at the midpoint of the base. Figure 3.11 shows this setup, with $a = 4$.

Due to the linearity of $g(\cdot)$, a single rectangle whose height is measured at the midpoint of the interval $S_0 - S_T$ is sufficient to replicate the shaded area. In fact, the area of the rectangle S_0ABS_T is

$$a\left[\frac{S_T + S_0}{2}\right]\left[S_T - S_0\right] = a\left[\frac{1}{2}S_T^2 - \frac{1}{2}S_0^2\right] \tag{3.54}$$

The Riemann–Stieltjes approximating sums measure the area under the rectangle exactly, with no need to augment the number of approximating rectangles.

[12] S_t is a function of time.

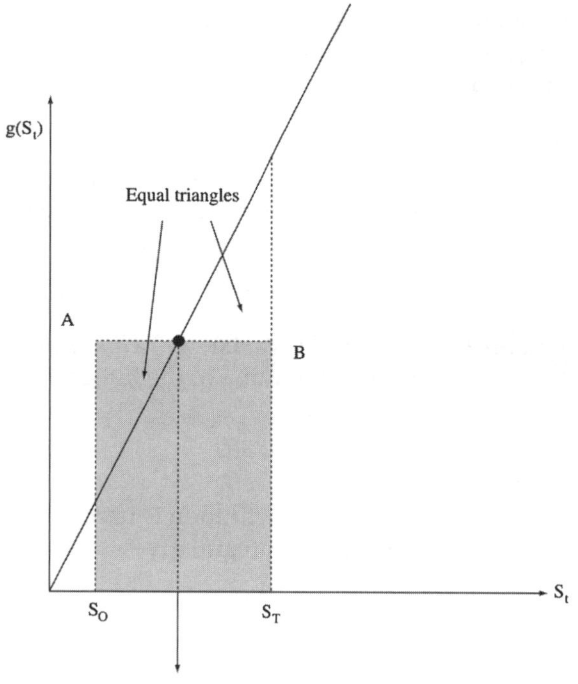

FIGURE 3.11 An example that approximation by rectangles works well.

3.4.4 Integration by Parts

In standard calculus there is a useful result known as integration by parts. It can be used to transform some integrals into a form more convenient to deal with. A similar result is also very useful in stochastic calculus, even though the resulting formula is different.

Consider two differentiable functions $f(t)$ and h(t), where $t \in [0, T]$ represents time. Then it can be shown that

$$\int_0^T f_t(t)\, h(t)\, dt = [f(T)\, h(T) - f(0)\, h(0)] - \int_0^T f(t)\, h_t(t)\, dt \tag{3.55}$$

where $h_t(t)$ and $f_t(t)$ are the derivatives of the corresponding functions with respect to time. They are themselves functions of time t.

In the notation of the Stieltjes integral, this transformation means that an expression that involves an integral

$$\int_0^T h(t)\, df(t) \tag{3.56}$$

can now be transformed so that it ends up containing the integral

$$\int_0^T f(t)\, dh(t) \tag{3.57}$$

The stochastic version of this transformation is very useful in evaluating Ito integrals. In fact, imagine that $f(\cdot)$ is *random* while $f(\cdot)$ is (conditionally) a deterministic function of time. Then, using integration by parts, we can express *stochastic integrals* as a function of integrals with respect to a deterministic variable. In stochastic calculus, this important role will be played by Ito's formula.

3.5 PARTIAL DERIVATIVES

Consider a call option. Time to expiration affects the price (premium) of the call in two different ways. First, as time passes, the expiration date will approach, and the remaining life of the option gets shorter. This lowers the premium. But at the same time, as time passes, the price of the underlying asset will change. This will also affect the premium. Hence, the price of a call is a function of two variables. It is more appropriate to write

$$C_t = F(S_t, t) \tag{3.58}$$

where C_t is the call premium, S_t is the price of the underlying asset, and t is time.

Now suppose we "fix" the time variable t and differentiate $F(S_t, t)$ with respect to S_t. The resulting *partial* derivative,

$$\frac{\partial F(S_t, t)}{\partial S_t} = F_s \tag{3.59}$$

would represent the (theoretical) effect of a change in the price of the underlying asset when time is kept fixed. This effect is anabstraction, because in practice one needs *some* time to pass before S_t can change.

The partial derivative with respect to time variable can be defined similarly as

$$\frac{\partial F(S_t, t)}{\partial t} = F_t \tag{3.60}$$

Note that even though S_t is a function of time, we are acting as if it does not change. Again, this shows the abstract character of the partial derivative. As t changes, S_t will change as well. But in taking partial derivatives, we behave as if it is a constant.

Because of this abstract nature of partial derivatives, this type of differentiation cannot be used directly in representing actual changes of asset price in financial markets. However, partial derivatives are very useful as intermediary tools. They are useful in taking a *total* change and then splitting it into components that come from different sources, and they are useful in *total differentiations*.

Before dealing with total differentiation, we have one last comment on partial derivatives. Because the latter do not represent "observed" changes, there is no difference between their use in stochastic or deterministic environments. We do not have to develop a new theory of partial differentiation in stochastic environments.

To make this clearer, consider the following example.

3.5.1 Example

Consider a function of two variables

$$F(S_t, t) = .3S_t + t^2 \tag{3.61}$$

where S_t is the (random) price of a financial asset and t is time.

Taking the partial with respect to S_t involves simply differentiating $F(\cdot)$ with respect to S_t:

$$\frac{\partial F(S_t, t)}{\partial S_t} = .3 \tag{3.62}$$

Here ∂S_t is an abstract increment in S_t and does not imply a similar actual change inreality. In fact, the partial derivative F_s is simply how much the function $F(\cdot)$ would have changed if we changed the S_t by one unit. The F_s is just a multiplier.

3.5.2 Total Differentials

Suppose we observe a small change in the price of a call option at time t. Let this total change be denoted by the differential dC_t. How much of this variation is due to a change in the underlying asset's price? How much of the variation is the result of the expiration date getting nearer as time passes? Total differentiation is used to answer such questions.

Let $f(S_t, t)$ be a function of the two variables. Then the total differential is defined as

$$df = \left[\frac{\partial f(S_t, t)}{\partial S_t}\right] dS_t + \left[\frac{\partial f(S_t, t)}{\partial t}\right] dt \tag{3.63}$$

In other words, we take the total change in S_t and multiply it by the *partial* derivative f_s. We take the total change in time dt and multiply it by the partial derivative f_t. The total change in $f(\cdot)$ is the sum of these two products. According to this, total differentiation is calculated by splitting an observed change into different abstract components.

3.5.3 Taylor Series Expansion

3.5.3.1 *One-Dimensional Taylor Expansion*

Let $f(x)$ be an infinitely differentiable function of $x \in \mathbb{R}$, and pick an arbitrary value of x; call this x_0.

Definition 10. The Taylor series expansion of $f(x)$ around $x_0 \in \mathbb{R}$ is defined as

$$\begin{aligned} f(x) &= f(x_0) + f_x(x_0)(x - x_0) \\ &\quad + \frac{1}{2} f_{xx}(x_0)(x - x_0)^2 \\ &\quad + \frac{1}{3!} f_{xxx}(x_0)(x - x_0)^3 + \cdots \\ &= \sum_{i=0}^{\infty} \frac{1}{i!} f^i(x_0)(x - x_0)^i \end{aligned} \tag{3.64}$$

where $f^i(x)$ is the ith order derivative of $f(x)$, with respect to x evaluated at the point x_0.[13]

We are not going to elaborate on why the expansion in (3.65) is valid if $f(x)$ is continuous and smooth enough. Taylor series expansion is taken for granted. We will, however, discuss some of its implications.

First, note that at this point the expression in (3.65) is *not* an approximation. The right-hand side involves an *infinite* series. Each element involves "simple" powers of x only, but there are an infinite number of such elements. Because of this, Taylor series *expansion* is not very useful in practice.

Yet, the expansion in (3.65) can be used to obtain useful approximations. Suppose we consider Eq. (3.65) and only look at those x near x_0. That is, suppose

$$(x - x_0) \approx \text{"small"} \tag{3.65}$$

Then, we surely have

$$|x_1 - x_0| > |x_1 - x_0|^2 > |x_1 - x_0|^3 > \cdots \tag{3.66}$$

(Each time we raise $|x_1 - x_0|$ to a higher power, we multiply it by a small number and make the result even smaller.)

Under these conditions, we may want to drop some of the terms on the right-hand side of (3.65) *if* we can argue that they are negligible. To do this, we must adopt a "convention" for smallness and then eliminate all terms that are "negligible" according to this criterion. But when is a term small enough to be negligible?

The convention in calculus is that, in general, terms of order $(dx)^2$ or higher are assumed to be negligible if x is a deterministic variable.[14] Thus, if we assume that x is deterministic, and let $(x - x_0)$ be small, then we could use the first-order Taylor series approximation:

$$f(x) \approx f(x_0) + f_x(x_0)(x - x_0) \tag{3.67}$$

This becomes an equality if the $f(x)$ has a derivative at x_0 and if we let

$$(x - x_0) \to 0 \tag{3.68}$$

Under these conditions, the infinitesimal variation $(x - x_0)$ is denoted by

$$dx \approx (x - x_0) \tag{3.69}$$

and the one in $f(\cdot)$ by

$$df(x) \approx f(x) - f(x_0) \tag{3.70}$$

As a result we obtain the familiar notation in terms of the differentials dx and df:

$$df(x) = f_x(x)\, dx \tag{3.71}$$

Here, the $f_x(x)$ is written as a function of x instead of the usual $f_x(x_0)$, because we are considering the limit when x approaches x_0.

3.5.3.2 Second-Order Approximations

The equation

$$f(x) \approx f(x_0) + f_x(x_0)(x - x_0) \tag{3.72}$$

is called a first-order Taylor series approximation. Often, a better approximation can be obtained by including the second-order term:

$$f(x) \approx f(x_0) + f_x(x_0)(x - x_0) + \frac{1}{2} f_{xx}(x_0)(x - x_0)^2 \tag{3.73}$$

This point is quite relevant for the later discussion of stochastic calculus. In fact, in order to prepare the groundwork for Ito's Lemma, we would like to consider a specific example.

3.5.3.3 Example: Duration and Convexity

Consider the exponential function where t denotes time, T is fixed, $r > 0$, and $t \in [0, T]$:

$$B_t = 100e^{-r(T-t)} \tag{3.74}$$

This function begins at $t = 0$ with a value of $B_0 = 100e^{-rT}$. Then it increases at a constant

[13] This last point implies that once x_0 is plugged in $f^i(\cdot)$, the latter become constants, independent of x.

[14] If so, the terms $(dx)^3, (dx)^4, \ldots$, will be smaller than $(dx)^2$.

percentage rate r. As $t \to T$, the value of B_t approaches 100. Hence, B_t could be visualized as the value, as of time t, of 100 to be paid at time T. It is the present value of a default-free zero-coupon bond that matures at time T, and r is the corresponding continuously compounding yield to maturity.

We are interested in the Taylor series approximation of B_t with respect to t, assuming that r, T remain constant. A first-order Taylor series expansion around $t = t_0$ will be given by

$$B_t \approx 100e^{-r(T-t_0)} + (r)100e^{r(T-t_0)}\,(t - t_0) \qquad t \in [0, T] \qquad (3.75)$$

where the first term on the right-hand side is B_t evaluated at $t = t_0$. The second term on the right-hand side is the first derivative of B_t with respect to t, evaluated at t_0, times the increment $t - t_0$.

Figure 3.12 displays this approximation. The equation is represented by a convex curve that increases as $t \to T$. The first-order Taylor series approximation is shown as a straight line tangent to the curve at point A. Note that as we go away from t_0 in either direction, the line becomes a worse approximation of the exponential curve. At t near t_0, on the other hand, the approximation is quite close.

Figure 3.13 plots the exponential curve with the second-order Taylor series approximation:

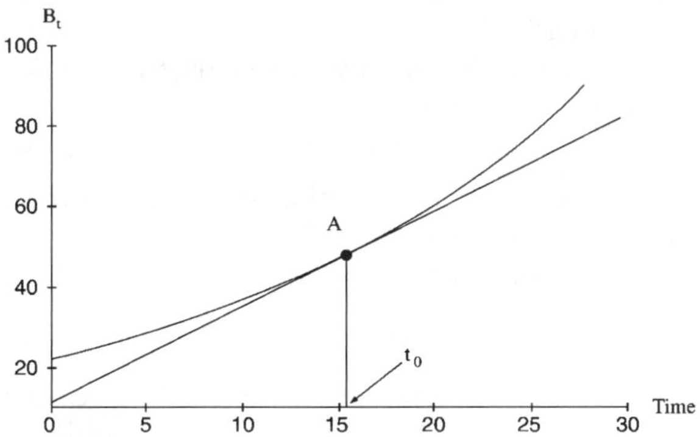

FIGURE 3.12 Plot of an exponential function and its first order approximation at an arbitrary point.

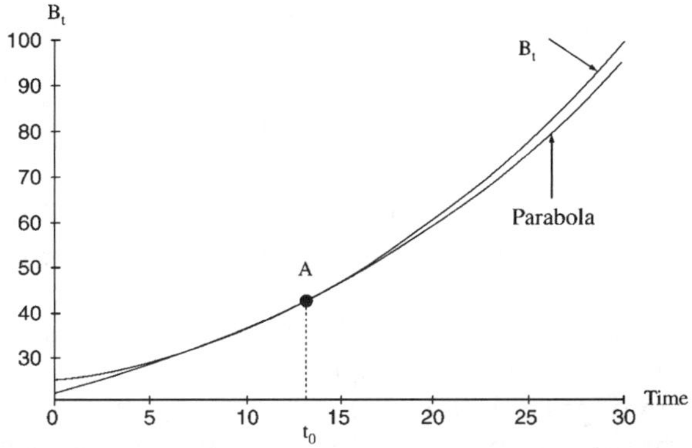

FIGURE 3.13 Plot of an exponential function and its second order approximation at an arbitrary point.

$$B_t \approx 100e^{-r(T-t_0)} + (r)100e^{-r(T-t_0)}\,(t - t_0) + \frac{1}{2}\left(r^2\right)100e^{-r(T-t_0)}\,(t - t_0)^2, \quad t \in [0, T] \qquad (3.76)$$

The right-hand side of this equation is a *parabola* that touches the exponential curve at point A. Because of the curvature of the parabola near t_0, we expect this curve to be nearer the exponential function.

Note that the difference between the first-order and second-order Taylor series approximations hinges on the size of the term $(t - t_0)^2$. As t nears t_0, this terms becomes smaller. More importantly, it becomes smaller faster than the term $(t - t_0)$.

These Taylor series approximations show how the valuation of a discount bond changes as the maturity date approaches.

A second set of Taylor series approximations can be obtained by expanding B_t with respect to r, keeping t, T fixed. Consider a second-order approximation around the rate r_0:

$$B_t \approx \left[100e^{-r_0(T-t)}\right]\left[1 - (T - t)\,(r - r_0) + \frac{1}{2}(T - t)^2\,(r - r_0)^2\right], \qquad t \in [0, T], \quad r > 0$$

or, dividing by $100e^{-r_0(T-t)}$,

$$\frac{dB_t}{B_t} \approx -(T-t)(r-r_0) + \frac{1}{2}(T-t)^2(r-r_0)^2$$

This expression provides a second-order Taylor series expansion for the percentage rate of change in the value of a zero-coupon bond as r changes infinitesimally. The right-hand side measures the percentage rate of change in the bond price as r changes by $r - r_0$, where r_0 can be interpreted as the current rate. We see two terms containing $r - r_0$ on the right-hand side. In financial markets the coefficient of the first term is called the modified duration. The second term is positive and has a coefficient of $(T-t)^2/2$. It represents the so-called convexity of the bond. Overall, the second-order Taylor series expansion of B_t with respect to r shows that, as interest rates increase (decrease), the value of the bond decreases (increases). The "convexity" of the bond implies that the bigger these changes, the smaller their *relative* effects.

3.5.3.4 Multi-Dimensional Taylor Series Expansion

Let $f(x)$ be an infinitely differentiable function of $x \in \mathbb{R}^d$, where $x = (x_1, \ldots, x_d)$ and pick an arbitrary vector of x; call this a.

Definition 11. The Taylor series expansion of $f(x)$ around $a \in \mathbb{R}^d$ is defined as

$$f(x) = f(x_1, \ldots, x_d) \tag{3.77}$$

$$= f(a_1, \ldots, a_d) + \sum_{i=1}^{d} \frac{\partial f(x)}{\partial x_i}(x_i - a_i) \tag{3.78}$$

$$+\frac{1}{2}\sum_{i=1}^{d}\sum_{j=1}^{d} \frac{\partial^2 f(x)}{\partial x_i \partial x_j}(x_i - a_i)(x_j - a_j) \tag{3.79}$$

$$+\frac{1}{3!}\frac{1}{2}\sum_{i=1}^{d}\sum_{j=1}^{d}\sum_{k=1}^{d} \frac{\partial^3 f(x)}{\partial x_i \partial x_j} (x_i - a_i)(x_j - a_j)(x_k - a_k) + \cdots \tag{3.80}$$

In order to find an approximation for mixed derivatives in partial differential equations which arise in case of stochastic volatility models and the like we utilize Taylor series expansion in higher dimension.

3.5.4 Ordinary Differential Equations

The third major notion from standard calculus that we would like to review is the concept of an ordinary differential equation (ODE). For example, consider the expression

$$dB_t = -r_t B_t dt \tag{3.81}$$

with known $B_0, r_t > 0$. This expression states that B_t is a quantity that varies with t—i.e., changes in B_t are a function of t and of B_t. The equation is called an *ordinary differential equation*. Here, the percentage variation in B_t is proportional to some factor r_t times dt:

$$\frac{dB_t}{B_t} = -r_t dt \tag{3.82}$$

Now, we say that the function B_t, defined by

$$B_t = e^{-\int_0^t r_u du} \tag{3.83}$$

solves the ODE in (3.78) in that plugging it into (3.80) satisfies the equality (3.78).

Thus, an ordinary differential equation is first of all an *equation*. That is, it is an equality where there exist one or more *unknowns* that need to be determined.

A very simple analogy may be useful. In a *simple equation*,

$$3x + 1 = x \tag{3.84}$$

the unknown is x, a number to be determined. Here the solution is $x = -1/2$.

In a *matrix equation*,

$$Ax - b = 0 \tag{3.85}$$

the unknown element is a vector. Under appropriate conditions, the solution would be $x = A^{-1}b$—i.e., the inverse of A multiplied by the vector b.

In an *ordinary differential equation,*

$$\frac{dx_t}{dt} = ax_t + b \tag{3.86}$$

where the unknown is x_t, a function. More precisely, it is a function of t:

$$x_t = f(t)$$

In the case of the ODE,

$$dB_t = -rB_t dt \tag{3.87}$$

the solution, with the condition $B_T = 1$, was

$$B_t = e^{-\int_0^t r_u du} \tag{3.88}$$

Readers will recognize this as the valuation function for a zero-coupon bond. This example shows that the pricing functions for fixed income securities can be characterized as solutions of some appropriate differential equations. In stochastic settings, we will obtain more complex versions of this result.

Finally, we need to define the integral equation

$$\int_0^t \left(ax_s + b\right) ds = x_t \tag{3.89}$$

where the unknown x_t is again a function of t.

3.6 CONCLUSIONS

This chapter reviewed basic notions in calculus. Most of these concepts were elementary. While the notions of derivative, integral, and Taylor series may all be well known, it is important to review them for later purposes.

Stochastic calculus is an attempt to perform similar operations when the underlying phenomena are continuous-time random processes. It turns out that in such an environment, the usual definitions of derivative, integral, and Taylor series approximations do not apply. In order to understand stochastic versions of such concepts, one first has to understand their deterministic equivalents.

The other important concept of the chapter was the notion of "smallness." In particular, we need a convention to decide when an increment is small enough to be ignored.

3.7 REFERENCES

The reader may at this point prefer to skim through an elementary calculus textbook. A review of basic differentiation and integration rules may especially help, along with solving some practice exercises.

3.8 EXERCISES

1. Write the sequences $\{X_n\}$ for $n = 1, 2, 3,$ where

   (a) $X_n = a_n$
   (b) $X_n = \left(1 + \frac{1}{n}\right)^n$
   (c) $X_n = (-1)^{n-1}/n$
   (d) Are the sequences $\{X_n\}$, given above, convergent?
   (e) Suppose the yearly interest rate is 5%. Let Δ be a time interval that repeats n times during 1 year, such that we have:

$$n\Delta = 1$$

      i. What is the gross return to 1$ invested during Δ?
      ii. Now suppose 5% is the annual yield on a T-bill with maturity Δ. What is the compound return during one year?

2. If it exists, find the limit of the following sequences for $n = 1, 2, 3 \ldots$:

   (a) $x_n = (-1)^n$
   (b) $x_n = \sin\left(\frac{n\pi}{3}\right)$
   (c) $x_n = n(-1)^n$

(d) $x_n = \sin\left(\frac{n\pi}{3}\right) + (-1)^n/n$. Is this sequence bounded?

3. Determine the following limits:

$$\lim_{n\to\infty} \left(3 + \sqrt{n}\right)/\sqrt{n}$$
$$\lim_{n\to\infty} n^{1/n}$$

4. Show that the partial sum

$$S_n = \sum_{k=1}^{n} \frac{1}{k!}$$

is convergent.

5. Show that the partial sum S_n defined by the recursion formula:

$$S_{n+1} = \sqrt{3S_n}$$

with $S_1 = 1$, converges to 3. Use mathematical induction.

6. Does the series

$$\sum_{n=1}^{N} \frac{1}{n}$$

converge as $N \to \infty$?

7. Suppose

$$X_n = aX_n + 1$$

with X_0 given. Write X_n as a partial sum. When does this partial sum converge?

8. Consider the function:

$$f(x) = x^3$$

(a) Take the integral and calculate

$$\int_0^1 f(x)\, dx$$

(b) Now consider splitting the interval $[0, 1]$ into 4 pieces, where you choose the x_i. They may or may not be equally spaced. Calculate the following sums numerically:

$$\sum_{i=1}^{4} f(x_i)\left(x_i - x_{i-1}\right)$$
$$\sum_{i=1}^{4} f\left(x_{i-1}\right)\left(x_i - x_{i-1}\right)$$

(c) What are the differences between these two sums and how well do they approximate the true value of the integral?

9. Now consider the function $f(x)$ discussed in this chapter:

$$f(x) = \begin{cases} x \sin\left(\frac{\pi}{x}\right) & 0 < x < 1 \\ 0 & x = 0 \end{cases}$$

(a) Take the integral and calculate

$$\int_0^1 f(x)\, dx$$

(b) Again, split the interval $[0, 1]$ into 4 pieces,

$$0 = x_0 < x_1 < x_2 < x_3 < x_4 = 1$$

by choosing the x_i numerically. Calculate the following sums:

$$\sum_{i=1}^{4} f(x_i)\left(x_i - x_{i-1}\right)$$
$$\sum_{i=1}^{4} f\left(x_{i-1}\right)\left(x_i - x_{i-1}\right)$$

(c) How do these sums approximate the true integral?

(d) Why?

10. Consider the following functions:

$$f(x, y, z) = \frac{x + y + z}{(1 + x)(1 + y)(1 + z)}$$
$$f(x, y, z) = \frac{x + y + z}{(1 + x)(1 + y)(1 + z)}$$

Take the partials with respect to x, y, z, respectively.

11. Does the series

$$\sum_{n=1}^{N} \frac{1}{n^2} \tag{3.90}$$

converge as $N \to \infty$?

12. Write a program that calculates a numerical approximation to the following integral:

$$\int_1^2 x^4 e^{2x} dx \tag{3.91}$$

CHAPTER

4

Pricing Derivatives: Models and Notation

OUTLINE

4.1 INTRODUCTION

There are some aspects of pricing-derivative instruments that set them apart from the general theory of asset valuation. Under simplifying assumptions, one can express the arbitrage-free price of a derivative as a function of some "basic" securities, and then obtain a set of *formulas* that can be used to price the asset without having to consider any linkages to other financial markets or to the real side of the economy.

There exist specific ways to obtain such formulas. One method was discussed in Chapter 2. The notion of *arbitrage* can be used to determine a probability measure under which financial assets behave as *martingales*, once discounted properly. The tools of martingale arithmetic become available, and one can easily calculate arbitrage-free prices, by evaluating the implied expectations. This approach to pricing derivatives is called the *method of equivalent martingale measures*.

The second pricing method that utilizes arbitrage takes a somewhat more direct approach. One first constructs a risk-free portfolio, and then obtains a *partial differential equation* (PDE) that is implied by the lack of arbitrage opportunities. This PDE is either solved analytically or evaluated numerically.

Financial Derivatives, Third Edition.
http://dx.doi.org/10.1016/B978-0-12-384682-2.00004-9

In either case, the problem of pricing derivatives is to find a function $F(S_t, t)$ that relates the price of the derivative product to S_t, the price of the underlying asset, and possibly to some other market risk factors. When a *closed-form* formula is impossible to determine, one finds numerical ways to describe the dynamics of $F(S_t, t)$.

This chapter provides examples of how to determine such pricing functions $F(S_t, t)$ for linear and *nonlinear* derivatives. These concepts are clarified and an example of partial differential equation methods is given. This discussion provides some motivation for the fundamental tools of stochastic calculus that we introduce later.

4.2 PRICING FUNCTIONS

The unknown of a derivative pricing problem is a *function* $F(S_t, t)$, where S_t is the price of the underlying asset and t is time. Ideally, the financial analyst will try to obtain a *closed-form* formula for $F(S_t, t)$. The Black–Scholes formula that gives the price of a call option in terms of the underlying asset and some other relevant parameters is perhaps the best-known case. There are, however, many other examples, some considerably simpler.

In cases in which a closed-form formula does not exist, the analyst tries to obtain an equation that governs the *dynamics* of $F(S_t, t)$.[1]

In this section, we show examples of how to determine such $F(S_t, t)$. The discussion is intended to introduce new mathematical tools and concepts that have common use in pricing derivative products.

4.2.1 Forwards

Consider the class of cash-and-carry goods.[2] Here we show how a pricing function $F(S_t, t)$, where S_t is the underlying asset, can be obtained for forward contracts. In particular, we consider a forward contract with the following provisions:

- At some future date T, where
$$t < T \tag{4.1}$$
F dollars will be paid for one unit of gold.
- The contract is signed at time t, but no payment changes hands until time T.

Hence, we have a contract that imposes an *obligation* on both counterparties—the one that delivers the gold, and the one that accepts the delivery. How can one determine a *function* $F(S_t, t)$ that gives the fair market value of such a contract at time t in terms of the underlying parameters?[3] We use an *arbitrage* argument.

Suppose one buys one unit of physical gold at time t for S_t dollars using funds borrowed at the continuously compounding risk-free rate r_t. The r_t is assumed to be fixed during the contract period. Let the insurance and storage costs per time unit be c dollars and let them be paid at time T. The total cost of *holding* this gold during a period of length $T - t$ will be given by

$$e^{-r_t(T-t)}S_t + (T - t)\,c \tag{4.2}$$

where the first term is the principal and interest to be returned to the bank at time T, and the second represents *total* storage and insurance costs paid at time T.

This is one method of securing one unit of physical gold at time T. One borrows the nec-

[1] The nonexistence of a *closed-form* formula does not necessarily imply the nonexistence of a pricing function. It may simply mean that we are not able to *express* the pricing function in terms of a simple formula. For example, all continuous and "smooth" functions can be expanded as an infinite Taylor series expansion. At the same time, truncating Taylor series in order to obtain a closed-form formula would in general lead to an approximation error.

[2] See Chapter 1 for definition.

[3] Note the sense in which this is a *derivative* contract. Once the contract is signed, it becomes a separate security and can be traded on its own. To trade the forward contract, one need not have in possession any physical gold. In fact, such instruments can be derived from "notional" underlying assets that do not even exist concretely. Derivatives written on equity indices are one such class.

essary funds, buys the underlying commodity, and stores it until time T.

The forward contract is another way of obtaining a unit of gold at time T. One signs a contract now for delivery of one unit of gold at time T, with the understanding that all payments will be made at expiration.

Hence, the outcomes of the two sets of transactions are identical.[4] This means that they must cost the same; otherwise, there will be arbitrage opportunities. An astute player will enter two separate contracts, buying the cheaper gold and selling the expensive one simultaneously. Mathematically, this gives the equality

$$F(S_t, t) = e^{-r_t(T-t)}S_t + (T-t)\,c \tag{4.3}$$

Thus we used the possibility of exploiting any arbitrage opportunities and obtained an equality that expresses the price of a forward contract $F(S_t, t)$ as a *function* of S_t, t, and other *parameters*. In fact, we determined a function $F(S_t, t)$ that gives the value of the forward contract at any time t.

Of the arguments in $F(S_t, t)$ and t are *variables*. They may change during the life of the contract. On the other hand, c, r_t, and T are *parameters*. It is assumed that they will remain constant during $T-t$.

The function $F(S_t, t)$ in (4.3) is *linear* in S_t. Thus, forward contracts are called *linear products*. Later we will derive the Black–Scholes formula, which provides a pricing function $F(S_t, t)$ for call options. This formula will be *nonlinear* in S_t. Instruments that have option-like characteristics are called *nonlinear* products.

4.2.1.1 *Boundary Conditions*

Here we have to mention briefly what a *boundary condition* is. Suppose we want to express formally the notion that the "expiration date gets nearer." To do this, we use the concept of limits. We let

$$t \to T \tag{4.4}$$

Note that as this happens,

$$\lim_{T\to t} e^{r_t(T-t)} = 1 \tag{4.5}$$

One question here is the presence of r_t. In reality, this and S_t are *random* variables, and one may ask if the use of a standard *limit* concept is valid. Ignoring this and applying the limit to the left-hand side of the expression in (4.3), we obtain

$$S_T = F(S_T, T) \tag{4.6}$$

According to this, at expiration, the cash price of the underlying asset and the price of the forward contract will be equal.

This is an example of a boundary condition. At the expiration date—i.e., at the boundary for time variable t—the pricing function $F(S_t, t)$ assumes a special value, S_T. The boundary condition is known at time t, although the value that S_t will assume at T is unknown.

4.2.2 Option

Determining the pricing function $F(S_t, t)$ for nonlinear assets is not as easy as in the case of forward contracts. This will be done in later chapters. At this point, we only introduce an important property that the $F(S_t, t)$ should satisfy in the case of nonlinear products. This will prepare the groundwork for further mathematical tools.

Suppose C_t is a call option written on the stock S_t. Let r be the constant risk-free rate. K is the strike price, and $T, t < T$, is the expiration date. Then the price of the call option can be expressed as[5]

$$C_t = F(S_t, t) \tag{4.7}$$

The pricing function $F(S_t, t)$ for options will have a fundamental property. Under simplifying conditions, the S_t will be the only source of

[4]Behind this statement there are assumptions, such as zero default risk of the forward contract.

[5]The interest rate r is constant and, hence, is dropped as an argument of $F(\cdot)$.

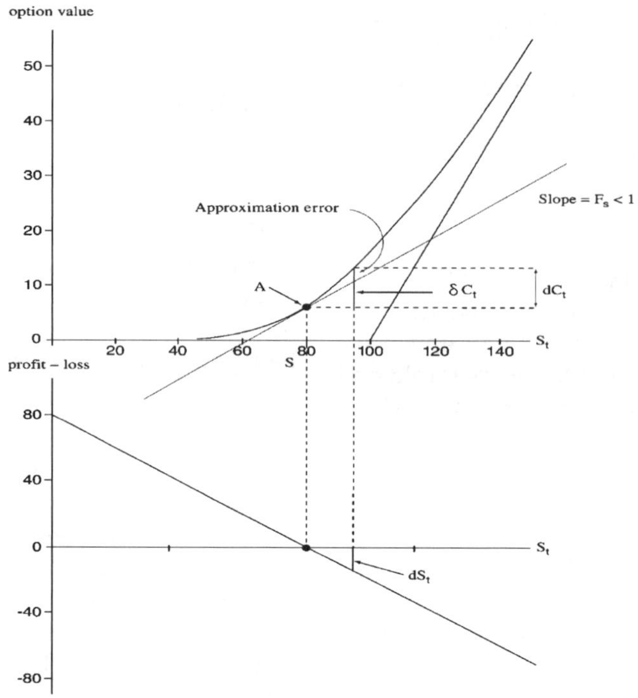

FIGURE 4.1 Change in stock price versus the change in the call option price.

randomness affecting the option's price. Hence, unpredictable movements in S_t can be offset by opposite positions taken simultaneously in C_t. This property imposes some conditions on the way $F(S_t, t)$ can change over time once the time path of S_t is given.

To see how this property can be made more explicit, consider Figure 4.1. The lower part of this figure displays a payoff diagram for a short position in S_t. A unit of the underlying asset, S_t, is borrowed and sold at price S.

The first panel of Figure 4.1 displays the price $F(S_t, t)$ of a call option written on S_t. At this point, we leave aside how the formula for $F(S_t, t)$ is obtained and graphed.[6]

Suppose, originally, the underlying asset's price is S. That is, initially we are at point A on the $F(S_t, t)$ curve. If the stock price increases by dS_t, the short position will lose exactly the amount dS_t. But the option position gains.

[6]It comes from the Black–Scholes formula that we prove later.

However, we see a critical point. According to Figure 4.1, when S_t increases by dS_t, the price of the call option will increase only by dC_t; this latter change is smaller because the slope of the curve is less than one, i.e.,

$$dC_t < dS_t \tag{4.8}$$

Hence, if we owned one call option and sold one stock, a price increase equal to dS_t would lead to a net loss.

But this reasoning suggests that with careful adjustments of positions, such losses could be eliminated. Consider the slope of the tangent to $F(S_t, t)$ at point A. This slope is given by

$$\frac{\partial F(S_t, t)}{\partial S_t} = F_s \tag{4.9}$$

Now, suppose we are short by not *one*, but by F_s units of the underlying stock. Then, as S_t increases by dS_t, the total loss on the short position will be $F_s dS_t$. But according to Figure 4.1, this amount is very close to dC_t. It is indicated by ∂C_t.

Clearly, if dS_t is a small incremental change, then the ∂C_t will be a very good approximation of the actual change dC_t. As a result, the gain in the option position will (approximately) offset the loss in the short position. Such a portfolio will not move unpredictably.

Thus, incremental movements in $F(S_t, t)$ and S_t should be related by some equation such as

$$d[F_S S_t] + d[F(S_t, t)] = g(t)$$

where $g(t)$ is a completely predictable function of time t.[7]

If we learn how to calculate such differentials, the equation above can be used in finding a closed-form formula for $F(S_t, t)$. When such closed-form formulas do not exist, numerical methods can be used to trace the trajectories followed by $F(S_t, t)$.

The following definition formalizes some of the concepts discussed in this section.

Definition 12. Offsetting changes in C_t by taking the opposite position in F_s units of the underlying asset is called delta hedging. Such a portfolio is delta neutral, and the parameter F_s is called the delta.

It is important to realize that when dS_t is "large," the approximation

$$\partial C_t \approx dC_t \tag{4.10}$$

will fail. With an extreme movement, the "hedge" may be less satisfactory. This can be seen in Figure 4.2. If the change in S_t is equal to dS_t, then the corresponding dC_t would far exceed the loss $-F_s dS_t$.

Clearly, the assumption of *continuous time* plays implicitly a fundamental role in asset pricing. In fact, we were able to replicate the movements in the option position by infinitesimally adjusting our short position in the underlying asset. The ability to make such infinitesimal adjustments in the portfolio clearly hinges on the assumption of continuous time and the absence of transaction costs. As shown earlier, with "large" increments, such approximations will deteriorate quickly.

[7]And of other possible parameters of the problem.

4.3 APPLICATION: ANOTHER PRICING MODEL

This book deals with the *mathematics* of derivative asset pricing. It is not a text on asset pricing per se. However, a discussion of general methods of pricing derivative assets is unavoidable. This is needed to illustrate the type of mathematics that we intend to discuss and to provide examples.

We use the discussion of the previous section to summarize the pricing method that uses partial differential equations (PDEs).

1. Assume that an analyst observes the current price of a derivative product $F(S_t, t)$ and the underlying asset price S_t in real time. Suppose the analyst would like to calculate the change in the derivative asset's price $dF(S_t, t)$, given a change in the price of the underlying asset dS_t.
2. Here the notions that we introduced in Chapter 3 start to become useful. Remember that the concept of differentiation is a tool that one can use to approximate small *changes* in a function. In this particular case, we indeed have a function $F(\cdot)$ that depends on S_t, t. Thus, *if* we can use the standard calculus, we could write

$$dF(S_t, t) = F_s dS_t + F_t\, dt \tag{4.11}$$

where the F_i are partial derivatives,[8]

$$F_s = \frac{\partial F}{\partial S}, \quad F_t = \frac{\partial F}{\partial t} \tag{4.12}$$

and where $dF(S_t, t)$ denotes the total change.

[8]Note the important difference between $F(S_t, t)$, which denotes the price of the derivative at time t, and F_t, which denotes the partial derivative of $F(S_t, t)$ with respect to t.

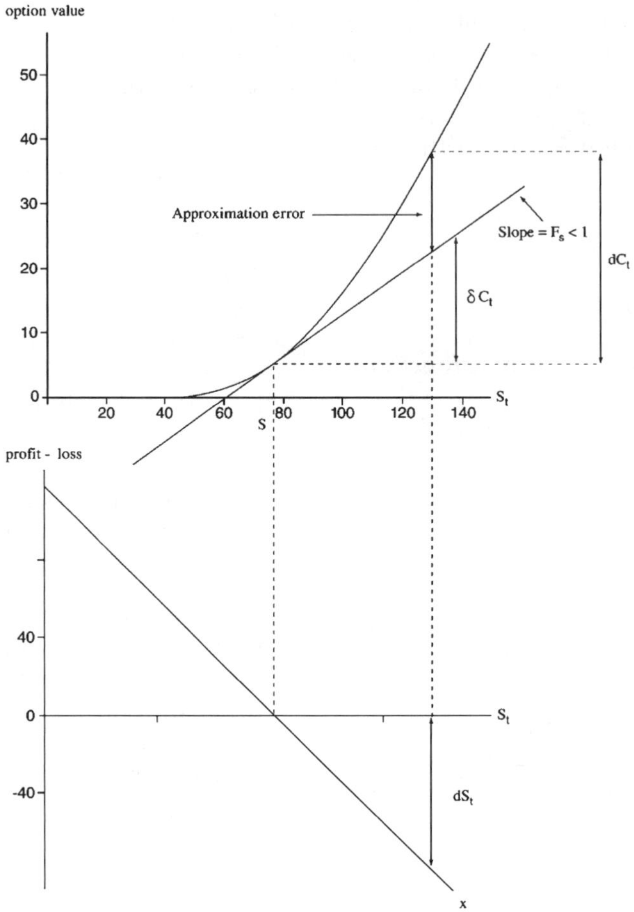

FIGURE 4.2 Change in stock price versus the change in the call option price for the case that the move in stock price is large.

3. Equation (4.11), called the total differential of $F(\cdot)$, gives the change in a derivative product's price in terms of changes in its determinants. Hence, one might think of an analyst who first obtains estimates of dS_t and then uses the equation for the total differential to evaluate the $dF(S_t, t)$. Equation (4.11) can be used once the partial derivatives F_s, F_t are evaluated numerically. This, on the other hand, requires that the functional form of $F(S_t, t)$ be known. However, all these depend on our ability to take total differentials as in (4.11). Can this be done in a straightforward fashion if underlying variables are continuous-time *stochastic processes*? The answer is no. Yet, with the new tools of *stochastic calculus*, it can be done.
4. Once the stochastic version of Eq. (4.11) is determined, one can complete the "program" for valuing a derivative asset in the following way. Using delta-hedging and risk-free portfolios, one can obtain additional relationships among $dF(S_t, t)$, dS_t, and dt. These can be used to eliminate all differentials from (4.11).
5. One would then obtain a relationship that ties only the partial derivatives of $F(\cdot)$ to each other. Such equations are called partial differential equations and can be solved for $F(S_t, t)$ if one has enough boundary conditions, and if

a closed-form solution exists. Thus, we are led to a problem where the *unknown* is a *function*. This argument shows that partial differential equations and their solutions are topics that need to be studied.

An example might be helpful at this point.

4.3.1 Example

Suppose you know that the partial derivative of $F(x)$ with respect to $x \in [0, X]$ is a known constant, b:

$$F_x = b \tag{4.13}$$

This equation is a trivial PDE. It is an expression involving a partial derivative of $F(x)$, a term with unknown functional form.

Using this PDE, can we tell the *form* of the function $F(x)$? The answer is yes. Only linear relationships have a property such as (4.13). Thus, $F(x)$ must be given by

$$F(x) = a + bx \tag{4.14}$$

The *form* of $F(x)$ is pinned down. However, the parameter a is still unknown. It is found by using the so-called "boundary conditions."

For example, if we knew that at the *boundary* $x = X$,

$$F(X) = 10 \tag{4.15}$$

then a can be determined by

$$a = 10 - bX \tag{4.16}$$

Remember that in the case of derivative products, one generally has some information about the form of $F(\cdot)$ at the expiration date. Such information can sometimes be used to determine the function $F(\cdot)$ explicitly, given a PDE.

4.4 THE PROBLEM

The program discussed earlier may appear quite technical at the outset, but in fact is a straightforward approach. However, there is a fundamental problem.[9]

Financial market data are not *deterministic*. In fact, all the variables under consideration, with the exception of the time variable t, are likely to be *random*. Since time is continuous, we observe uncountably many random variables as time passes. Hence, $F(S_t, t)$, S_t, and possibly the risk-free rate r_t are all *continuous-time stochastic processes*.

Can we then apply the same reasoning and use the same tools as in standard calculus to write

$$dF(t) = F_s dS_t + F_r dr_t + F_t\, dt \tag{4.18}$$

The answer to this question is no. It turns out that one needs a "new" calculus and a different formula when the variables under consideration are random processes. The following is a first look at some of these difficulties.

4.4.1 A First Look at Ito's Lemma

In standard calculus, variables under consideration are deterministic. Hence, to get a relation such as

$$dF(t) = F_s dS_t + F_r dr_t + F_t\, dt \tag{4.19}$$

one uses total differentiation. The change in $F(\cdot)$ is given by the relation on the right-hand side of (4.19). But according to the rules of calculus, this equation holds exactly only during infinitesimal intervals. In *finite* time intervals, Eq. (4.19) will hold only as an approximation.

Consider again the univariate Taylor series expansion. Let $f(x)$ be an infinitely differentiable function of $x \in \mathbb{R}$. One can then write the Taylor

[9] In fact, at this point, there are *two* problems. For one, given the equation

$$dF(t) = F_s dS_t + F_r dr_t + F_t\, dt \tag{4.17}$$

we still do not know how arbitrage can be used to eliminate terms such as $dt, dF(t), dS_t$, and dr_t. We leave this aside for the time being.

series expansion of $f(x)$ around $x_0 \in \mathbb{R}$ as

$$f(x) = f(x_0) + f_x(x_0)(x - x_0) + \frac{1}{2} f_{xx}(x_0)(x - x_0)^2 + \frac{1}{3!} f_{xxx}(x_0)(x - x_0)^3 + \cdots = \sum_{i=0}^{\infty} \frac{1}{i!} f^i(x_0)(x - x_0)^i \quad (4.20)$$

where $f^i(x_0)$ is the i th-order partial derivative of $f(x)$ with respect to x, evaluated at x_0.

We can reinterpret $df(x_0)$ using the approximation

$$df(x) \approx f(x) - f(x_0) \quad (4.21)$$

and dx as

$$dx \approx x - x_0 \quad (4.22)$$

Thus, an expression such as

$$dF(t) = F_s\, dS_t + F_r\, dr_t + F_t\, dt \quad (4.23)$$

depends on the assumption that the terms (dt^2), $(dS_t)^2$, and $(dr_t)^2$, and those of higher order, are "small" enough that they can be omitted from a multivariate Taylor series expansion.[10] Because of such an approximation, higher powers of the differentials dS_t, dt, or dr_t do not show up on the right-hand side of (4.23).

Now, dt is a small deterministic change in t. So to say that $(dt)^2, (dt)^3, \ldots$ are "small" with respect to dt is an internally consistent statement. However, the same argument cannot be used for $(dS_t)^2$ and, possibly, for $(dr_t)^2$.[11]

First, it is maintained that $(dS_t)^2$ and $(dr_t)^2$ are random during small intervals.[12] Thus, they have nonzero variances during dt.

This poses a problem. On one hand, we want to use continuous-time random processes with nonzero variances during dt. So, we use positive numbers for the average values of $(dS_t)^2$ and $(dr_t)^2$. But under these conditions, it would be inconsistent to call $(dS_t)^2$ and $(dr_t)^2$ "small" with respect to dt, and then equate them to zero, a step that can be taken if the variables in question are deterministic, as in the case of standard calculus.

Hence, in a stochastic environment with a continuous flow of randomness, we have to write the relevant total differentials as:

$$dF(t) = F_s\, dS_t + F_r\, dr_t + F_t\, dt + \frac{1}{2} F_{ss}\, dS_t^2 + \frac{1}{2} F_{rr}\, dr_t^2 + F_{sr}\, dS_t\, dr_t \quad (4.24)$$

This is an example of why we need to study stochastic calculus. We want to learn how to exploit the chain rule in a stochastic environment and understand what a *differential* means in such a setting. The example above shows that the resulting expressions would be different from the ones obtained in deterministic calculus.

If the notion of differential needs to be changed, then that notion of the integral should also be reformulated. In fact, in such a stochastic environment, we define *differentials* by using a new definition of integral. Otherwise, in continuous-time stochastic environments, a formal definition of *derivative* does not exist.

4.5 CONCLUSIONS

One approach used to find the "fair market value" of derivative securities may at this point be summarized informally.

Using arbitrage, determine an equation that ties various partial derivatives of an (unknown) function $F(S_t, t)$ to each other. Then, solve this (partial differential) equation for the form of $F(\cdot)$. Using the boundary conditions, determine the parameters of this function.

[10]This would make the expression a Taylor series *approximation*.

[11]For that matter, it may not be true for the cross-product term $(dS_t dr_t)$ either.

[12]In *infinitesimal* intervals, we will see that the mean square limits of these terms are deterministic and proportional to dt.

This chapter also introduced the fundamental mathematical problem faced in continuous-time finance. Standard formulas from calculus are not applicable when the variables under consideration are continuous-time stochastic processes. Increments of these processes have nonzero variances. This will make the average "size" of the second-order terms such as $(dS_t)^2$ nonnegligible.

4.6 REFERENCES

Duffie (1996) is an excellent source on dynamic asset valuation. Ingersoll (1987) also provides a very good treatment. There are, however, several less complicated books to consider for an understanding of simple asset valuation formulas. Cox and Rubinstein (1985) is a very good example. Finally, most of the valuation theory can be found in the excellent collection of papers in Merton (1990). There are also some recent sources that give a broad summary of valuation theory. Björk (1999), Nielsen (1999), and Kwok (1998) are three such books.

4.7 EXERCISES

1. Suppose you can bet on an American presidential election in which one of the candidates is an incumbent. The market offers you the following payoffs R:

$$R = \begin{cases} \$1000 \text{ If incumbent wins} \\ -\$1500 \text{ If incumbent loses} \end{cases}$$

You can take either side of the bet. Let the true probability of the incumbent winning be denoted by $p, 0 < p < 1$.

(a) What is the expected gain if $p = .6$?

(b) Is the value of p important for you to make a decision on this bet?

(c) Would two people taking this bet agree on their assessment of p? Which one would be correct? Can you tell?

(d) Would statistical or econometric theory help in determining the p?

(e) What weight would you put on the word of a statistician in making your decision about this bet?

(f) How much would you pay for this bet?

2. Now place yourself exactly in the same setting as before, where the market quotes the above R. It just happens that you have a close friend who offers you the following separate bet, R^*:

$$R = \begin{cases} \$1500 \text{ If incumbent wins} \\ -\$1000 \text{ If incumbent loses} \end{cases} \tag{4.25}$$

Note that the random event behind this bet is the same as in R. Now consider the following:

(a) Using the R and the R^*, construct a portfolio of bets such that you get a guaranteed risk-free return (assuming that your friend or the market does not default).

(b) Is the value of the probability p important in selecting this portfolio? Do you care what the p is? Suppose you are given the R, but the payoff of R^* when the incumbent wins is an unknown to be determined. Can the above portfolio help you determine this unknown value?

(c) What role would a statistician or econometrician play in making all these decisions? Why?

3. Consider a simplified game of roulette with two outcomes: red and black. A bet of any fraction may be placed on red or black for any individual draw. If red or black is chosen correctly, \$1 is paid for each unit bet, while if chosen incorrectly the payoff is \$0. Assuming the events red and black are equally likely, what is the expected payoff of the game? How much should an investor be willing to pay for a chance to play this game once? Are risk preferences involved in the calculation? Does the answer change if the investor is allowed to play the game infinitely many times?

4. Consider flipping a fair coin, what is the expected number of flips that is needed to have three heads. Write a simulation program to confirm the result.

CHAPTER

5

Tools in Probability Theory

OUTLINE

5.1 INTRODUCTION

In this chapter, we review some basic notions in probability theory. Our first purpose here is to prepare the groundwork for a discussion of martingales and martingale-related tools. In doing this, we discuss properties of random variables and stochastic processes. A reader with a good background in probability theory may want to skip these sections.

Financial Derivatives, Third Edition.
http://dx.doi.org/10.1016/B978-0-12-384682-2.00005-0

The second purpose of this chapter is to introduce the binomial process, which plays an important role in derivative asset valuation. Pricing models for derivative assets are formulated in continuous time, but will be applied in discrete, "small" time intervals. Practical methods of asset pricing using "finite difference methods" or lattice methods fall within this category. Prices of underlying assets are assumed to be observed at time periods separated by small finite intervals of length Δ.[1] In such small intervals, it is further assumed, prices can have only a limited number of possible movements. These methods all rely on the idea that a continuous-time stochastic process representing the price of the underlying asset can be approximated arbitrarily well by a binomial process. This chapter introduces the mechanics of justifying such approximations.

5.2 PROBABILITY

Derivative products are contracts written on underlying assets whose prices fluctuate randomly. A mathematical model of randomness is thus needed.

Some elementary models of probability theory are especially well suited to pricing derivative assets.

This can be a bit surprising, given that many investors appear to be driven by "intuitive" notions of probabilities rather than by an axiomatic and formal probabilistic model. However, the discussion in Chapter 2 indicated that no matter what the "true" probabilities are, if there are no arbitrage opportunities, one can *represent* the fair market value of financial assets using *probability measures* constructed "synthetically." Hence, regardless of any subjective chances perceived by market participants, mathematical probability models have a natural use in pricing derivative products.

[1] For example, prices can move up and down by some preset amounts.

In working with random variables, one first defines a *probability space*. That is, one explicitly lays out the framework where the notion of chance and the resulting probability can be defined without falling into some inconsistencies.

To define probability models formally, one needs a set of basic states of the world. A particular state of the world is denoted by the symbol ω. The symbol Ω represents all possible states of the world. The outcome of an experiment is determined by the choice of an ω.

The intuitive notion of an *event* corresponds to a set of elementary ω's. The set of all possible events is represented by the symbol $\mathcal{I}$. To each event $A \in \mathcal{I}$, one assigns a probability $P(A)$.

These probabilities must be consistently defined. Two conditions of consistency are the following:

$$P(A) \geq 0, \quad \text{any} \quad A \in \mathcal{I} \tag{5.1}$$

$$\int_{A \in \mathcal{I}} dP(A) = 1 \tag{5.2}$$

The first of these conditions implies that probabilities of events are either zero or positive. The second says that the probabilities should sum to one. Here, note the notation $dP(A)$. This is a measure theoretic notation and may be read as the incremental probability associated with an event A.

The triplet $\{\Omega, \mathcal{I}, P\}$ is called a *probability space*. According to this, a point ω of Ω is chosen randomly. $P(A)$, where $A \in \mathcal{I}$, represents the probability that the chosen point belongs to the set A.

5.2.1 Example

Suppose the price of an exchange-traded commodity future during a given day depends only on a harvest report the U.S. Department of Agriculture (USDA) will make public during that day.

The specifics of the report written by the USDA are equivalent to an ω.

Depending on what is in the report, we can call it either favorable or unfavorable. This constitutes an example of an *event*. Note that there may be *several* ω's that may lead us to call the harvest report "favorable." It is in this sense that events are collections of ω's.

Hence, we may want to know the probability of a "favorable report." This is given by

$$P(\text{harvest report} = \text{favorable}) \tag{5.3}$$

Finally, note that in this particular example the Ω is the set of all possible reports that the USDA may make public.

5.2.2 Random Variable

In general, there is no reason for a probability to be representable by a simple mathematical formula. However, some convenient and simple mathematical models are found to be acceptable approximations for representing probabilities associated with financial data.[2]

A random variable X is a function, a mapping, defined on the set $\mathcal{I}$. Given an event $A \in \mathcal{I}$, a random variable will assume a particular numerical value. Thus, we have

$$X: \mathcal{I} \rightarrow B \tag{5.4}$$

where B is the set made of all possible subsets of the real numbers $\mathbb{R}$.

In terms of the example just discussed, note that a "favorable harvest report" may contain several judgmental statements besides some accompanying numbers. Let X be the value of the numerical estimate provided by the USDA and let 100 be some minimum desirable harvest. Then mappings such as

$$\text{favorable reports} \Longrightarrow 100 < X \tag{5.5}$$

define the random variable X. Clearly, the values assumed by X are real numbers.

[2]The sense in which a formula becomes a good approximation to a probability is an important question that we will discuss below.

A mathematical model for the probabilities associated with a random variable X is given by the distribution function $G(x)$:

$$G(x) = P(X \leq x) \tag{5.6}$$

Note that $G(\cdot)$ is a function of x.[3]

When the function $G(x)$ is smooth and has a derivative, we can define the density function of X. This function is denoted by $g(x)$ and is obtained by

$$g(x) = \frac{dG(x)}{dx} \tag{5.7}$$

It can be shown that under some technical conditions there always exists a distribution function $G(x)$. However, whether this function $G(x)$ can be written as a convenient formula is a different question. It turns out that there are some well-known models where this is possible. We review three basic probability models that are frequently used in pricing derivative products.

These examples are specially constructed so as to facilitate understanding of more complicated asset pricing methods to be discussed later. But first we need to review the notions of expectations and conditional expectations.

5.3 MOMENTS

There are different ways one can classify models of distribution functions. One classification uses the notion of "moments." Some random variables can be fully characterized by their first *two moments*. Others need *higher-order moments* for a full characterization.

5.3.1 First Two Moments

The expected value $\mathbb{E}[X]$ of a random variable X, with density $f(x)$, is called the *first* moment. It

[3]Here, X represents a random variable, whereas the lowercase x represents a certain "threshold."

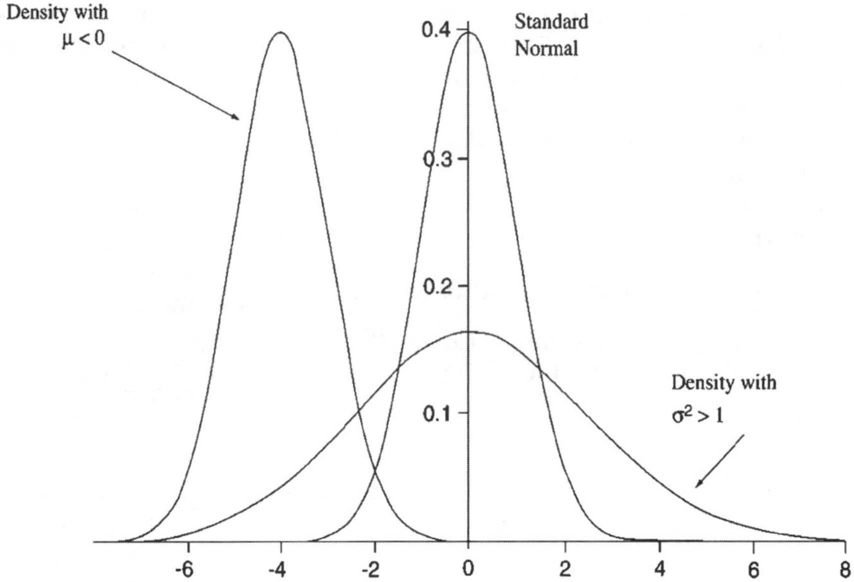

FIGURE 5.1 Examples of normal distributions.

is defined by

$$\mathbb{E}[X] = \int_{-\infty}^{\infty} xf(x)\,dx$$

where $f(x)$ is the corresponding probability density function.[4] The variance $\mathbb{E}[X - \mathbb{E}[X]]^2$ is the *second* moment around the mean. The first moment of a random variable is the "center of gravity" of the distribution, while the second moment gives information about the way the distribution is spread out. The square root of the second moment is the standard deviation. It is a measure of the *average deviation of observations from the mean*. In financial markets, the standard deviation of a price change is called the *volatility*.

For example, in the case of a normally distributed random variable X, the density function is given by the well-known formula

$$f(x) = \frac{1}{\sqrt{2\pi\sigma^2}} e^{-\frac{(x-\mu)^2}{2\sigma^2}} \tag{5.8}$$

[4]If the density does not exists, we replace $f(x)\,dx$ by $dF(x)$.

where the variance parameter σ^2 is the second moment around the mean and the parameter μ is the first moment. Figure 5.1 shows examples of normal distributions.

Integrals of this formula determine the probabilities associated with various values that the random variable x can assume. Note that $f(x)$ depends on only two parameters, σ^2 and μ. Hence, the probabilities associated with a normally distributed random variable can be inferred if one has the sample estimates of these two moments.

A normally distributed random variable X would also have higher-order moments. For example, the centered third moment of any normally distributed random variable X will be given by

$$\mathbb{E}[X - \mathbb{E}[X]]^3 = 0$$

In fact, all higher-order moments of normally distributed random variables can be expressed as functions of μ and σ^2. In other words, given the

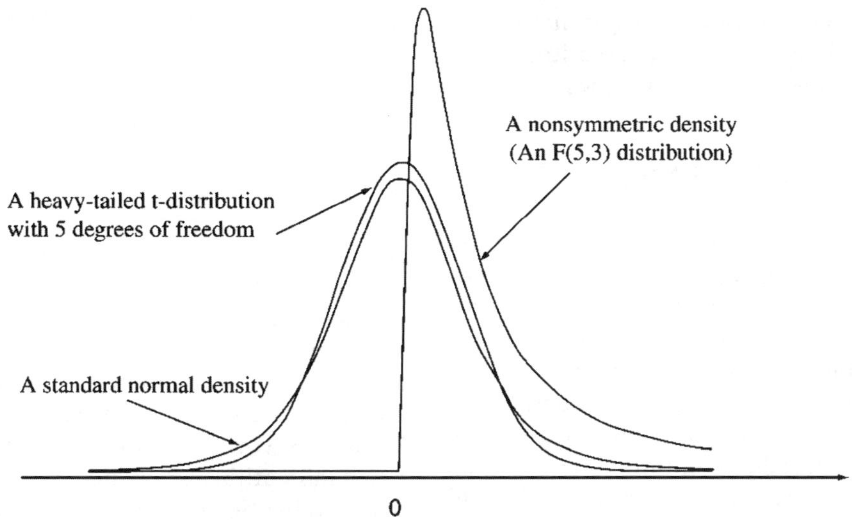

FIGURE 5.2 Examples of symmetric and nonsymmetric distributions.

first two moments, higher-order moments of normally distributed variables do not provide any additional information.

5.3.2 Higher-Order Moments

Consider the nonsymmetric density shown in Figure 5.2. If the mean is the center of gravity and standard deviation is a rough measure of the width of the distribution, then one would need another parameter to characterize the skewness of the distribution. *Third moments* are indeed informative about such asymmetries.

In financial markets, a more important notion is the phenomenon of *heavy tails*. Figure 5.2 displays a symmetric density which has another characteristic that differentiates it from normal distributions. The tails of this distribution are heavier *relative* to the middle part of the tails. Such densities are called heavy-tailed and are fairly common with financial data. Again, one would need a parameter other than variance and the mean to characterize the heavy-tailed distributions. *Fourth moments* are used for that end.

5.3.2.1 Heavy Tails

What is the meaning of heavy tails?

A distribution that has heavier tails than the normal curve means a higher probability of extreme observations. But this point should be carefully made. Note that the normal density also has tails that extend to plus and minus infinities. Thus, a normally distributed random variable could also assume extreme values from time to time. However, in the case of a heavy-tailed distribution, these extreme observations have, relatively speaking, a higher frequency.

But there is more to heavy-tailed distributions than that. In a normal distribution, most of the observations would naturally be occurring around the center. More importantly, the occurrence of extremes is gradual, in that the passage from ordinary, to large, and then to extreme observations occurs in a gradual fashion. In the case of a heavy-tailed distribution, on the other hand, the passage from "ordinary" to extreme observations is more sudden. The middle tail region of the distribution contains relatively less weight than in the normal density. Compared to the normal density, one is likely to get "too many extreme observations."

In other words, a casual observer is more likely to be "surprised" by extreme observations in the case of heavy-tailed random variables.

5.4 CONDITIONAL EXPECTATIONS

The operation of taking expectations of random variables is the formal equivalent of the heuristic notion of "forecasting." To forecast a random variable, one utilizes some information denoted by the symbol I_t. Expectations calculated using such information are called *conditional* expectations. The corresponding mathematical operation is the "conditional expectation operator."[5] Since the information utilized could be, and in general is, different from one time to another, the conditional expectation operator is itself indexed by the time index.

In general, the information used by decision makers will increase as time passes. If we also assume that the decision maker never forgets past data, the information sets must be increasing over time:

$$I_{t_0} \subseteq I_{t_1} \subseteq \cdots \subseteq I_{t_k} \subseteq I_{t_{k+1}} \subseteq \cdots \tag{5.9}$$

where $t_i, i = 0, 1, \ldots$ are times when the information set becomes available.

In the mathematical analysis, such information sets are called an increasing sequence of *sigma fields*. When such information sets become available continuously, a different term is used, and the family I_t satisfying (5.9) is called a *filtration*.

The conditional expectation operator can then be defined in several steps.

5.4.1 Conditional Probability

First, the probability density functions need to be discussed further. If X is a random variable with density function $f(x)$, and if x_0 is one possible value of this random variable, then for *small* dx, we have

$$P\left(|x - x_0| \leq \frac{dx}{2}\right) \approx f(x_0)\, dx \tag{5.10}$$

This is the probability that the x will fall in a small neighborhood of x_0. The neighborhood is characterized by the "distance" dx.

These quantities are shown in Figure 5.3. Note that although $f(x)$ is a nonlinear curve in this figure, for small dx it can be approximated reasonably well by a straight line. Then, the rectangle in Figure 5.3 would be close to the probability that x will fall within a small neighborhood of x_0 represented by the quantity dx. If these probabilities are based on some information set I_t, then $f(x)$ is called a *conditional density*. The dependence on I_t is formally denoted by $f(x|I_t)$. If the $f(x)$ is not based on any particular information, the I_t term is dropped and the density is written as $f(x)$.

Consider a simple example. The odds of a stock market crash will be an example of unconditional probability. The odds of a crash given that one has entered a severe recession can be represented by a conditional probability. In this particular case the information is the knowledge that a severe recession has begun. The use of such information may certainly lead to a revision of the (unconditional) probability of a crash.

5.4.1.1 *Conditional Expectation Operator*

The second step in defining a conditional expectation is the "averaging" operator. In fact, every forecast is an average of possible future values. The values that the random variable can assume in the future are weighted by the probabilities associated with these values, and an average is obtained.

Hence, the operation of conditional expectation involves calculating a weighted sum. Since the possible outcomes are likely to be not only infinite, but also uncountably many, this sum is represented by an integral.

[5] An operator is a function that maps function into functions. That is it takes as input a function and produces as output another function.

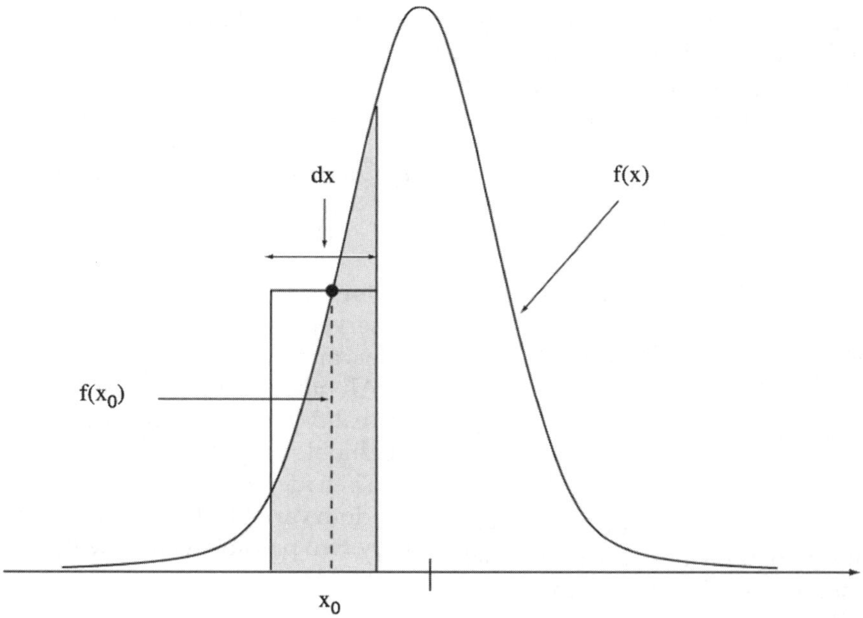

FIGURE 5.3 Approximating probabilities using rectangles.

The conditional expectation (forecast) of some random variable S_t, given the information available at time u, is given by

$$\mathbb{E}[S_t | I_u] = \int_{-\infty}^{\infty} S_t f(S_t | I_u) \, dS_t, \quad u < t \tag{5.11}$$

In this expression, the right-hand side should be read as follows: for a given t, the sum of all possible values that S_t might assume are weighted by the corresponding probabilities $[f(S_t | I_u) \, dS_t]$ and summed. The averaging is done by using probabilities conditional on I_u. This way, any information that one has gets incorporated in the forecast.

5.4.2 Properties of Conditional Expectations

First note a convenient notation. Often, the expectation conditional on an information set, I_t, is written compactly as

$$\mathbb{E}[\cdot | I_u] = \mathbb{E}_t[\cdot] \tag{5.12}$$

The t subscript in $\mathbb{E}_t$ indicates that in the averaging operation one uses all information available up to time t.

The conditional expectation operator $\mathbb{E}_t$ has the following properties.

1. The conditional expectation of the sum of two random variables is the sum of conditional expectations:

$$\mathbb{E}_u[S_t + F(t)] = \mathbb{E}_u[S_t] + \mathbb{E}_u[F(t)] \tag{5.13}$$

   According to this, one can form separate forecasts of random variables and then add these to get a forecast of their *total*.
2. Suppose the most recent information set is I_t, but one is interested in forecasting the expectation $\mathbb{E}_{t+T}[S_{t+T+u}], T > 0, u > 0$. That is, one would like to say something about the forecast of a possible forecast. Since the information set

I_{t+T} is unavailable at time t, the conditional expectation $\mathbb{E}_{t+T}[S_{t+T+u}]$ is unknown. In other words, $\mathbb{E}_{t+T}[S_{t+T+u}]$ is itself a random variable. A property of conditional expectations is that the expectation of this future expectation equals the present forecast of S_{t+T+u}:

$$\mathbb{E}_t\left[\mathbb{E}_{t+T}\left[S_{t+T+u}\right]\right] = \mathbb{E}_t\left[S_{t+T+u}\right] \tag{5.14}$$

According to this, recursive application of conditional expectation operators always equals the conditional expectation with respect to the smaller information set:

$$\mathbb{E}\left[\mathbb{E}\left[\cdot \mid I_t\right] \mid I_0\right] = \mathbb{E}\left[\cdot \mid I_0\right] \tag{5.15}$$

where I_0 is contained in I_t.

Finally, if the conditioning information set I_t is empty, then one obtains the "unconditional" expectation operator $\mathbb{E}$. This means that $\mathbb{E}$ has properties similar to the conditional expectation operator.

5.5 SOME IMPORTANT MODELS

This section discusses some important models for random variables. These models are useful not only in theory, but also in practical applications of asset pricing. In this section we also extend the notion of a random variable to a random process.

5.5.1 Binomial Distribution in Financial Markets

Consider a trader who follows the price of an exchange-traded derivative asset $F(t)$ in real time, using a service such as Reuters, Telerate, or Bloomberg.

The price $F(t)$ changes continuously over time, but the trader is assumed to have limited scope of attention and checks the market price every Δ. We assume that Δ is a small time interval. More importantly, we assume that at any time t there are two possibilities:

1. There is either an *uptick*, and prices increase according to

$$\Delta F(t) = F(t+\Delta) - F(t) = +a\sqrt{\Delta} \tag{5.16}$$

2. Or, there is a *downtick* and prices decrease by

$$\Delta F(t) = F(t+\Delta) - F(t) = -a\sqrt{\Delta} \tag{5.17}$$

where $\Delta F(t)$ represents the change in the observed price during the "small" time interval Δ.

All other outcomes that may very well occur in reality are assumed for *the time being* to have negligible probability.

Then for fixed t, Δ, the ΔFt becomes a binomial random variable. In particular, $\Delta F(t)$ can assume only two possible values with the probabilities

$$P\left(\Delta F(t) = +a\sqrt{\Delta}\right) = p \tag{5.18}$$

$$P\left(\Delta F(t) = -a\sqrt{\Delta}\right) = 1 - p \tag{5.19}$$

The time index t starts from t_0 and increases by multiples of Δ:

$$t = t_0, t_0 + \Delta, \ldots, t_0 + n\Delta \tag{5.20}$$

At each time point a new $F(t)$ is observed. Each increment $\Delta F(t)$ will equal either $+a\sqrt{\Delta}$ or $-a\sqrt{\Delta}$. If the $\Delta F(t)$ are *independent* of each other, the sequence of increments $\Delta F(t)$ will be called a *binomial stochastic process*, or simply a *binomial process*.[6]

Note that these assumptions are somewhat artificial for actual markets. On a given trading day, even in markets with very high turnover, there are many time periods where $\Delta F(t)$ does not change. Or, in some special circumstances, it may change by more than an up- or downtick. However, such complications will be dealt with later. For the time being, we consider the simpler case of binomial processes.

[6]Remember that a stochastic process is a sequence of random variables indexed by time.

5.5.2 Limiting Properties

An important element of the discussion involving the binomial process $\Delta F(t)$ is that the two possible values assumed by each $\Delta F(t)$ depend on the parameter Δ. This dependence permits a discussion of the *limiting behavior* of the binomial process. We can ask a number of questions that will eventually relate to pricing derivative products.

One important question is the following: What does a typical path followed by the $\Delta F(t)$ look like? Clearly, such a trajectory will be made of a sequence of $+a\sqrt{\Delta}$ and $-a\sqrt{\Delta}$'s. If the probability of each of these outcomes is exactly equal to 1/2, then a realization of $\{\Delta F(t), t = t_0, t_0 + \Delta, \ldots\}$ will, as Δ gets smaller, converge to an extremely erratic trajectory that fluctuates between $+a\sqrt{\Delta}$ and $-a\sqrt{\Delta}$.

In fact, as Δ gets smaller, two things happen. First, the observation points come nearer, and second, $\left|a\sqrt{\Delta}\right|$ gets smaller.

The $\Delta F(t)$ is the increment in the price process. What kind of a path is followed by $F(t)$ itself? First, note that if $F(t)$ represents the price of a derivative product at time t, then it will equal the sum of all up- and downticks since t_0. As $\Delta \to 0, F(t)$ will be given by

$$F(t) = F(t_0) + \int_{t_0}^{t} dF(s) \tag{5.21}$$

That is, beginning from an *initial price* $F(t_0)$, we obtain the price at time t by simply adding all subsequent *infinitesimal* changes. Clearly, in continuous time, there are an uncountable number of such infinitesimal changes—hence the use of integral notation. Also, at the limit, the notation for "small" incremental changes $\Delta F(t)$ is replaced by $dF(t)$, which represents infinitesimal changes.

Finally, consider the following question. At the limit, the infinitesimal changes $dF(t)$ are still very erratic. Would the trajectories of $F(t)$ be of *bounded variation*?[7] The question is important, because otherwise the Riemann–Stieltjes way of constructing integrals cannot be exploited and a new definition of integral would be needed.

Another important point is the following: the integral in (5.21) is taken with respect to a random process, and not with respect to a deterministic variable as is the case in standard calculus. Clearly, this integral is itself a random variable. Can such an integral be successfully defined? Can we use the Riemann–Stieltjes procedure of approximations by appropriate rectangles to construct an integral with respect to a random process? These questions lead to the Ito integral and will be answered in Chapter 9.

5.5.3 Moments

One question that we answer here concerns the moments of a binomial process.

Let t be fixed. Then the expected value and the variance of $\Delta F(t)$ are defined as follows:

$$\mathbb{E}[\Delta F(t)] = p\left(a\sqrt{\Delta}\right) + (1-p)\left(-a\sqrt{\Delta}\right) \tag{5.22}$$

$$\mathbb{V}[\Delta F(t)] = p\left(a\sqrt{\Delta}\right)^2 + (1-p)\left(-a\sqrt{\Delta}\right)^2 - \mathbb{E}[\Delta F(t)]^2 \tag{5.23}$$

If we have a 50–50 chance of an uptick at any time t, then

$$p = \frac{1}{2} \tag{5.24}$$

and the expected value will equal 0 while the variance is given by $a^2\Delta$.

It is important to realize that the variance of the binomial process is proportional to Δ. As Δ approaches zero, a variance that is proportional to Δ will go toward zero with the same speed. This means that if we think of Δ as a small but nonnegligible quantity, then the variance will also be nonnegligible.

[7] See Chapter 3.

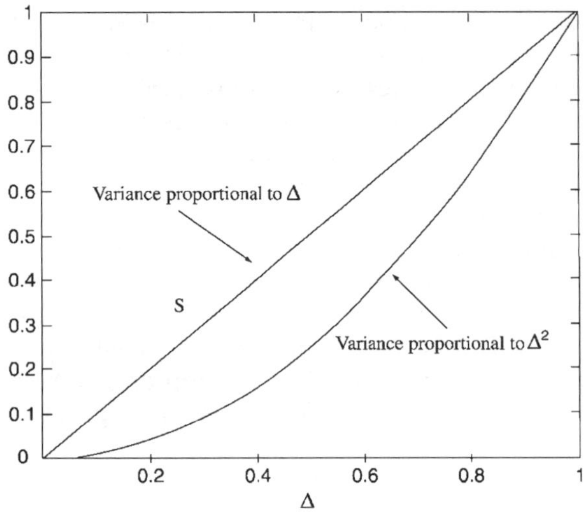

FIGURE 5.4 Difference between a variance proportional to Δ and one proportional to Δ^2.

In contrast, if $\Delta F(t)$ had instead fluctuated between, say, $+a\Delta$ and $-a\Delta$ the variance would be proportional to Δ^2. For "small" Δ, the value of Δ^2 would be much smaller. When $\Delta \to 0$, the variance would go to zero significantly faster. Under such conditions, it can be maintained without any contradiction that the variance of $\Delta F(t)$ is negligible, while Δ itself is not.

Heuristically speaking, a random variable with a variance proportional to Δ^2 will be approximately constant in infinitesimal time intervals.

Figure 5.4 illustrates the difference between a variance proportional to Δ(the 45° line) and one proportional to Δ^2. The latter becomes negligible for small Δ.

This last point is also relevant for higher-order moments of the binomial process. Again assume that $p = 0.5$ for simplicity. Then the expected value is zero and the third moment will be given by

$$\mathbb{E}\left[\Delta F(t)\right]^3 = p\left(+a\sqrt{\Delta}\right)^3 + (1-p)\left(-a\sqrt{\Delta}\right)^3 \tag{5.25}$$

With $p = 0.5$, the third moment equals zero.

The fourth-order moment is obtained as

$$\mathbb{E}\left[\Delta F(t)\right]^4 = \left(+a\sqrt{\Delta}\right)^4 = a^4\Delta^2 \tag{5.26}$$

As $\Delta \to 0$, the fourth-order moment will become negligible. It is proportional to a power of Δ that goes to zero faster than the time interval itself.

These observations imply that for small intervals Δ, higher-order moments of a binomial random variable that assumes values proportional to can be ignored.

5.5.4 The Normal Distribution

Now consider the following experiment with the random variable $F(t)$ discussed in the previous section. We ask the computer to calculate *many* realizations of $F(t)$. Then, beginning from the same initial point $F(0)$, we plot these trajectories.

Beginning from $t_0 = 0$, in the *immediate* future, $F(t)$, has only two possible values:

$$F(0+\Delta) = \begin{cases} F(0) + a\sqrt{\Delta} & \text{with probability } p \\ F(0) - a\sqrt{\Delta} & \text{with probability } 1-p \end{cases} \tag{5.27}$$

Hence, $F(t)$ itself is binomial at $t = 0 + \Delta$.

But if we let some more time pass, and then look at $F(t)$ at, say, $t = 2\Delta, F(t)$ will assume one of three possible values. More precisely, we have the following possibilities:

$$F(2\Delta) = \begin{cases} F(0) + a\sqrt{\Delta} + a\sqrt{\Delta} & \text{with probability } p^2 \\ F(0) - a\sqrt{\Delta} + a\sqrt{\Delta} & \text{with probability } p(1-p) \\ F(0) - a\sqrt{\Delta} - a\sqrt{\Delta} & \text{with probability } (1-p)^2 \end{cases} \tag{5.28}$$

That is, $F(2\Delta)$ may equal $F(0) + 2a\sqrt{\Delta}, F(0) - 2a\sqrt{\Delta}$, or $F(0)$. Of these, the last outcome is most likely if there is a 50–50 chance of an uptick.

Now consider possible values of $F(t)$ once some more time elapses. Several more combinations of upticks and downticks become possible.

For example, by the time $t = 5\Delta$, one possible but extreme outcome may be

$$F(5\Delta) = F(0) + a\sqrt{\Delta} + a\sqrt{\Delta} + a\sqrt{\Delta} + a\sqrt{\Delta} + a\sqrt{\Delta} \tag{5.29}$$

$$F(0) + 5a\sqrt{\Delta} \tag{5.30}$$

Another extreme may be to get five downticks in a row:

$$F(5\Delta) = F(0) - a\sqrt{\Delta} - a\sqrt{\Delta} - a\sqrt{\Delta} - a\sqrt{\Delta} - a\sqrt{\Delta} \tag{5.31}$$

More likely are combinations of upticks and downticks. For example,

$$F(5\Delta) = F(0) - a\sqrt{\Delta} + a\sqrt{\Delta} - a\sqrt{\Delta} + a\sqrt{\Delta} + a\sqrt{\Delta} \tag{5.32}$$

or

$$F(5\Delta) = F(0) - a\sqrt{\Delta} + a\sqrt{\Delta} + a\sqrt{\Delta} - a\sqrt{\Delta} + a\sqrt{\Delta} \tag{5.33}$$

are two different sequences of price changes, each resulting in the same price at time $t = 5\Delta$.

There are several other possibilities. In fact, we can consider the general case and try to find the total number of possible values $F(n\Delta)$ can take. Obviously, as $n \to \infty, F(n\Delta)$ may take any of a possibly infinite number of values. A similar conclusion can be reached if $\Delta \to 0$ and $n \to \infty$ while the product Δn remains constant. In this case, we are considering a fixed time interval and subdividing it into finer and finer partitions.[8] For the case in which Δ was constant and $n \to \infty$, the time period under consideration increased indefinitely, and we looked at a limiting $F(t)$ projected toward a "distant" future.

One question is: What happens to the distribution of the random variable $F(n\Delta)$ as $n \to \infty$ and Δ remains fixed? A somewhat different question is: What happens to the *distribution* of $F(n\Delta)$ as $\Delta \to 0$ while $n\Delta$ is fixed?[9]

[8] In fact, this latter type of convergence is of interest to us. These types of experiments fall in the domain of weak convergence and give us an approximate distribution for a whole sequence of random variables observed during an interval.

[9] Here, too, $n \to \infty$.

Now, remember that at the origin $F(t)$ was binomial, but a little farther away the number of possible outcomes grew and it became multinomial. The probability distribution also changed accordingly. How does the form of the distribution change as $n \to \infty$? What would it look like at the limit?

Questions such as these fall in the domain of "convergence of random variables." There are two different ways one can investigate this issue. The first approach is that of the central limit theorem. The second is called weak convergence.

According to the central limit theorem, the distribution of $F(n\Delta)$ approaches the normal distribution as $n\Delta \to \infty$.

Assume that $p = 0.5$ and that

$$F(0) = 0 \tag{5.34}$$

Then, for fixed Δ and "large" n, the distribution of $F(n\Delta)$ can be approximated by a normal distribution with mean 0 and variance $a^2 n\Delta$. The approximating density function will be given by

$$g(F(n\Delta) = x) = \frac{1}{\sqrt{2\pi a^2 n\Delta}} e^{-\frac{1}{2a^2 n\Delta}(x)^2} \tag{5.35}$$

The corresponding *distribution* function does not have a *closed-form* formula. It can only be represented as an integral.

The convergence in distribution is illustrated in Figure 5.5. It is important for practical asset pricing to realize the meaning of this convergence in distribution. We observe a sequence of random variables indexed by n.[10] As n increases, the distribution function of the nth random variable starts to resemble the normal distribution.[11]

It is the notion of weak convergence that describes the way distributions of whole sequences of random variables converge.

[10] That is, we have a stochastic process.

[11] Again, we emphasize that we are dealing with the distribution of $F(n\Delta)$ and not with the whole sequence $\{F(0), F(\Delta), F(2\Delta), \ldots, F(n\Delta), \ldots\}$.

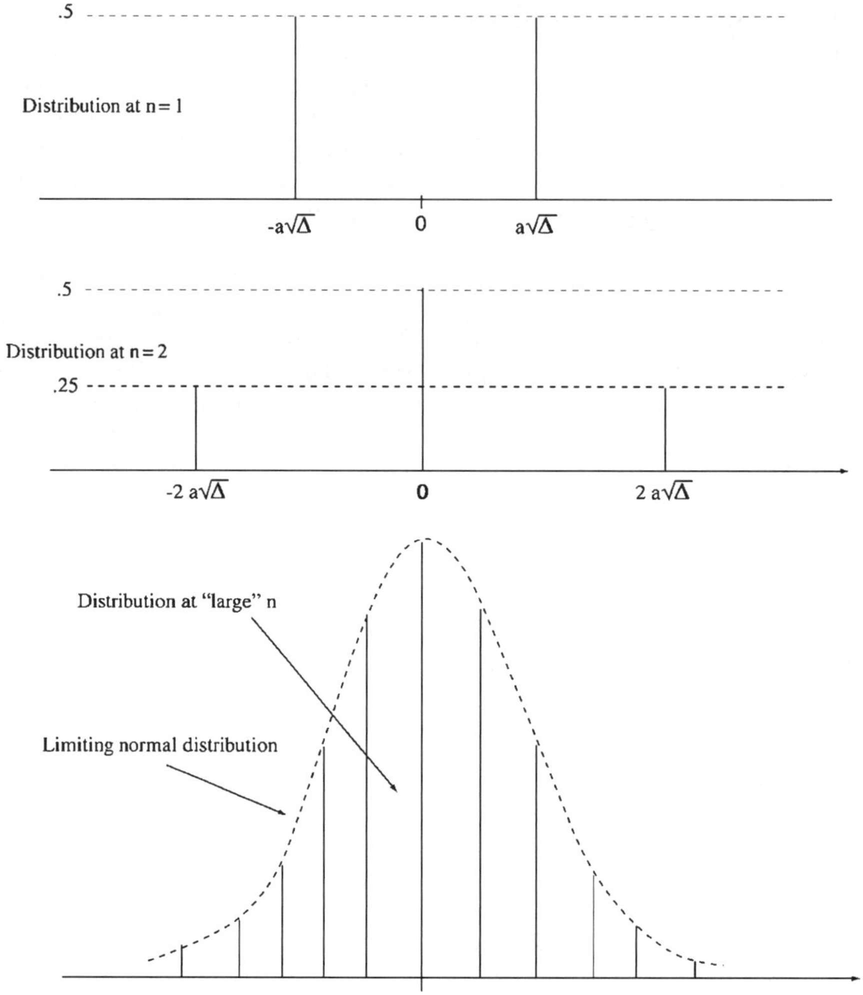

FIGURE 5.5 Illustration of the convergence in distribution.

5.5.5 The Poisson Distribution

In dealing with continuous-time stochastic processes, we need *two* building blocks. One is the continuous-time equivalent of the normal distribution known as Brownian motion or, equivalently, as the Wiener process. As the discussion in the previous section indicates, trajectories of these processes are likely to be continuous.

This implies that the Gaussian model is useful when new information arriving during infinitesimal periods is itself infinitesimal. As illustrated for the binomial approximation, with $\Delta \to 0$, the values assumed by $\Delta F(t)$ become smaller and the variance of new information given by

$$\mathbb{V}(\Delta F(t)) = a^2 \Delta \tag{5.36}$$

goes to zero.

That is, in infinitesimal intervals, the $F(t)$ cannot "jump." Changes are incremental, and at the limit they converge to zero.

Continuous-time versions of the normal distribution are very useful in asset pricing. Under some conditions, however, they may not be sufficient to approximate trajectories of asset prices observed in some financial markets. We may need a model for prices that show "jumps" as well. There were many examples of such "jumps" during the October 1987 crash of stock markets around the world.

How can we represent such phenomena?

The Poisson distribution is the second building block. A Poisson-distributed random process consists of jumps at unpredictable *occurrence times* $t_i, i = 1, 2, \ldots$. The jump times are assumed to be independent of one another, and each jump is assumed to be of the same size.[12] Further, during a small time interval Δ, the probability of observing *more than* one jump is negligible. The total number of jumps observed up to time t is called a *Poisson counting process* and is denoted by N_t.

For a Poisson process, the probability of a jump during a small interval Δ will be approximated by

$$P(\Delta N_t = 1) \approx \lambda \Delta \tag{5.37}$$

where λ is a positive constant called the intensity.

Note the contrast with normal distribution. For a normally distributed variable, the probability of obtaining a value exactly equal to zero is nil. Yet with the Poisson distribution, if Δ is "small," this probability is approximately

$$P(\Delta N_t = 0) \approx 1 - \lambda \Delta \tag{5.38}$$

Hence, during a small interval, there is a "high" probability that no jump will occur. Thus, the trajectory of a Poisson process will consist of a continuous path broken by occasional jumps.

The probability that during a finite interval Δ there will be n jumps is given by

$$P(\Delta N_t = n) = \frac{e^{-\lambda\Delta}(\lambda\Delta)^n}{n!} \tag{5.39}$$

which is the corresponding distribution.

[12] Both of these assumptions can be altered. However, to keep the Poisson characteristic, the jump times need to be independent.

5.6 EXPONENTIAL DISTRIBUTION

The exponential distribution with parameter θ has the following probability distribution function

$$f(x) = \frac{1}{\theta}\exp(-\frac{x}{\theta}), \quad x \geq 0 \tag{5.40}$$

By integration we can see that its cumulative distribution function is given by

$$F(x) = 1 - \exp(-x/\theta), \quad x \geq 0 \tag{5.41}$$

This is the distribution of the times between jumps of a Poisson process with rate $\frac{1}{\theta}$. It is easy to show that the parameter of exponential distribution is also its mean.

Figure 5.6 shows two exponential distributions with different means. It is easy to observe that the exponential distribution with the higher mean has fatter tail.

Another interesting quality about the exponential distribution is that it has what is known as the forgetfulness property. In terms of a telephone switchboard, this property states that the probability of a call in a given time interval is not affected by the fact that no calls may have taken place in the preceding interval(s).

5.7 GAMMA DISTRIBUTION

Gamma distribution is a two parameter distribution. There are three different parameterizations of the gamma distribution:

- a shape parameter α and a scale parameter β;
- a shape parameter α and a rate parameter θ that is an inverse of scale parameter $\theta = \frac{1}{\beta}$;
- a shape parameter α and a mean parameter μ that is $\alpha\beta$.

Both parameters are positive real numbers. The gamma distribution is frequently used to model waiting times. It arises in processes for which the waiting times between Poisson distributed events are relevant.

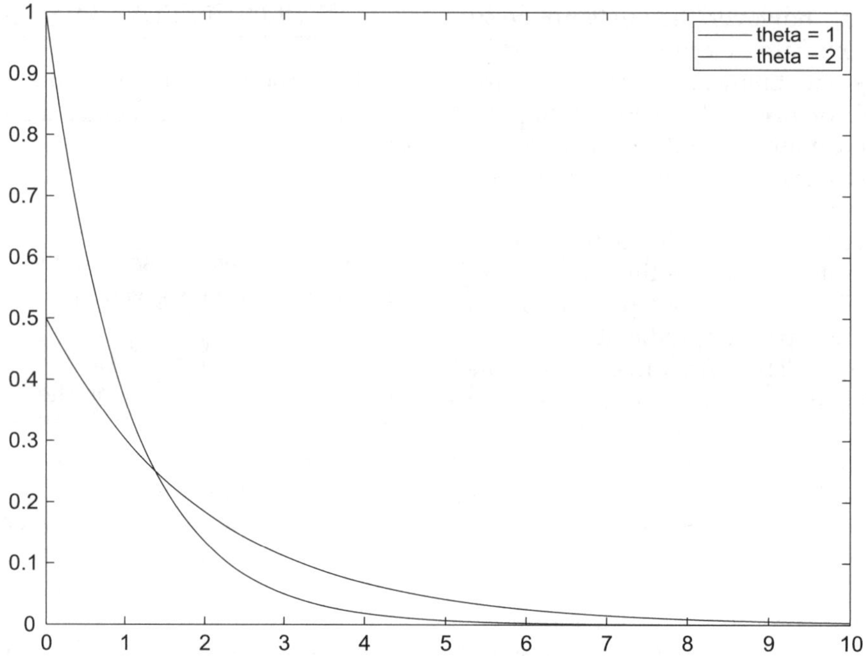

FIGURE 5.6 Exponential distribution with two different means.

A gamma random variable has the following probability distribution function:

$$f(x) = \frac{1}{\Gamma(\alpha)\beta^{\alpha}} x^{\alpha-1} e^{-\frac{x}{\beta}} \tag{5.42}$$

where α is the shape parameter and β is the scale parameter.

It is not surprising that its distribution is similar to the result of the exponential distribution, if α is an integer then $f(x)$ represents the sum of α independent exponential random variables, each of which has a mean of β, which is equivalent to a rate parameter $\theta = \frac{1}{\beta}$. The gamma distribution is also used to model errors in multilevel Poisson regression models, because the combination of the Poisson distribution and a gamma distribution is a negative binomial distribution. The chi-squared distribution $\chi^2(d)$ is a special case of the gamma distribution, gamma $\left(\frac{1}{2}d, 2\right)$.

In case of having large scale parameter in gamma distribution, its distribution converges to a normal distribution. The advantage is that in the case of gamma distribution the density is over positive real numbers.

In Figure 5.7, we plot a gamma distribution with scale parameter 100 and shape parameter 1 versus a normal distribution with mean 100 and standard deviation 10. Observe similarity in their shapes. Characteristic function of a gamma random variable is given by

$$\phi(u) = \left(\frac{1}{1 - iu\beta}\right)^{\alpha} \tag{5.43}$$

5.8 MARKOV PROCESSES AND THEIR RELEVANCE

The discussion thus far has dealt basically with random variables. Yet, this concept is too simple to be useful in finance, although it does constitute a building block for more complex models. In finance, what we really need is a

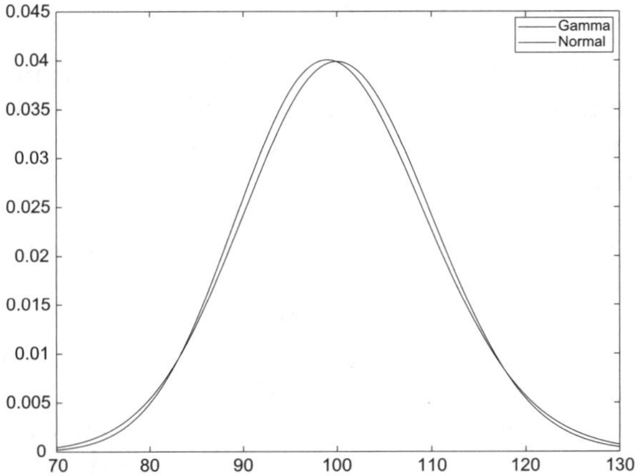

FIGURE 5.7 Normal distribution versus gamma distribution.

model of a *sequence* of random variables, and often those that are observed over continuous time.

Sequences of random variables $\{X_t\}$ indexed by an index t, where t is either discrete, $t = 0, 1, \ldots$, or continuous, $t \in [0, \infty)$, are called stochastic processes. A stochastic process is assumed to have a well-defined joint distribution function,

$$F(x_1, \ldots, x_t) = P(X_1 \le x_1, \ldots, X_t \le x_t)$$

as $t \to \infty$. In case the index t is continuous, one is dealing with uncountably many random variables and clearly the joint distribution function of such a process should be carefully "constructed", as will be illustrated for Wiener process.

In this section, we discuss in detail a class of stochastic processes that plays an important role in derivative asset pricing; namely, the Markov processes. Our discussion, which will be in discrete time, will try to motivate some important aspects of stochastic processes and will also clarify some notions that will be used later in dealing with continuous-time models for interest rate derivatives.[13]

Definition 13. A discrete time process, $\{X_1, \ldots, X_t, \ldots\}$, with joint probability distribution function, $F(x_1, \ldots, x_t)$, is said to be a Markov process if the implied conditional probabilities satisfy

$$P(X_{t+s} \le x_{t+s} | x_t, \ldots, x_1) = P(X_{t+s} \le x_{t+s} | x_t) \tag{5.44}$$

where $0 < s$ and $P(\cdot | I_t)$ is the probability conditional on the information set I_t.

The assumption of Markovness has more than just theoretical relevance in asset pricing. In heuristic terms, and in discrete time $t = 1, 2, \ldots$, a Markov process, $\{X_t\}$, is a sequence of random variables such that knowledge of its past is totally irrelevant for any statement concerning the $X_{t+s}, 0 < s$, given the last observed value, x_t. In other words, any probability statement about some future $X_{t+s}, 0 < s$, will depend only on the latest observation x_t and on nothing observed earlier.[14]

5.8.1 The Relevance

How do these notions help a market practitioner?

Suppose the X_t represents a variable such as instantaneous spot rate r_t. Then, assuming that r_t is Markov means that the (expected) future behavior of r_{t+s} depends only on the latest observation and that a condition such as (5.40) will be valid. We can then proceed as follows.

We split changes in interest rates into expected and unexpected components:

$$r_{t+\Delta} - r_t = \mathbb{E}[(r_{t+\Delta} - r_t) | I_t] + \sigma(I_t, t) \Delta W_t \tag{5.45}$$

[13]It is quite important that the process one is modeling in finance is a Markov process. The Feynman–Kac theorem that we will see in later chapters will be valid only for such processes. Yet, it can be shown that short-term interest rate processes are, in general, not Markov. This imposes limitations on the numerical methods that can be applied for short-rate processes.

[14]We are not talking about the dependence of means or variances of X_{t+s} only. The more distant past should not influence statements concerning the whole probabilistic behavior of a Markov process.

where the ΔW_t is some unpredictable random variable with variance Δ. Then, the $\sigma(I_t, t)\sqrt{\Delta}$ will be the standard deviation of interest rate increments. The first term on the right-hand side will represent expected change in interest rate movements, and the second term will represent the part that is unpredictable given I_t.

It turns out that if r_t is a Markov process, and if I_t contains only the current and past values of r_t, then the conditional mean and variance will be functions of r_t only and we can write:

$$\mathbb{E}\left[(r_{t+\Delta} - r_t)|\, I_t\right] = \mu(r_t, t)\,\Delta \tag{5.46}$$

and

$$\sigma(I_t, t) = \sigma(r_t, t) \tag{5.47}$$

These steps will be discussed in more detail when we develop the notion of stochastic differential equations in Chapter 11. There, letting $\Delta \to 0$, we obtain a standard stochastic differential equation for r_t and write it as

$$dr_t = \mu(r_t, t)\, dt + \sigma(r_t, t)\, dW_t \tag{5.48}$$

With such a model, one can then proceed to parameterize the $\mu(r_t, t)$ and $\sigma(r_t, t)$ and hence obtain a model that captures the dynamics of interest rates.

But, if interest rates were not Markov, these steps cannot be followed, since the conditional mean and variance of the spot rate could potentially depend on observations other than the immediate past.[15]

Hence, the assumption of Markovness appears to be quite relevant in pricing derivatives, at least in case of interest rate derivatives.

[15] Also, if interest rates are not Markov, a very important correspondence between a class of partial differential equations (PDE) and a class of conditional expectations cannot be established. Monte Carlo methods cease to become equivalent to the PDEs commonly used in the field of interest rate derivatives.

5.8.2 The Vector Case

There is another relevant issue concerning multivariate Markov processes. We prefer to discuss it again in discrete time, $t, t+\Delta, \ldots,$ and using interest rates as our motivating variables.

Below we will show that, although two processes can be *jointly* Markov, when we model *one* of these processes in a univariate setting, it will, in general, cease to be a Markov process.

The relevance of this can best be discussed in fixed-income. There (and elsewhere) a central concept is the *yield curve*. The so-called classical approach attempts to model yield curve using a single interest rate process, such as the r_t discussed above. On the other hand, the more recent Heath-Jarrow-Morton (HJM) approach, consistent with Black–Scholes philosophy, models it using k separate forward rates, which are assumed to be Markov jointly. But as we will see below, the univariate dynamics of one element of a k-dimensional Markov process will, in general, not be Markov. Hence, Markovness can be maintained in HJM methodology, but may fail in a short-rate-based approach.

Suppose we have a bivariate process, $[r_t, R_t]$, where the r_t represents the "short" rate and the R_t is the "long" rate. Suppose also that jointly they are Markov:

$$\begin{bmatrix} r_{t+\Delta} \\ R_{t+\Delta} \end{bmatrix} = \begin{bmatrix} \alpha_1 r_t + \beta_1 R_t \\ \alpha_2 r_t + \beta_2 R_t \end{bmatrix} + \begin{bmatrix} \sigma_1 W^1_{t+\Delta} \\ \sigma_2 W^2_{t+\Delta} \end{bmatrix} \tag{5.49}$$

where $W^1_{t+\Delta}, W^2_{t+\Delta}$ are two error terms independent of each other, and of the past $W^1_s, W^2_s, s \le t$. The $\{\alpha_i, \beta_i, \sigma_i\}$ are constant coefficients. According to system (5.46), current short and long rates depend only on the latest observations of r_t and R_t.[16]

Clearly, this is a special case. But, it is sufficient to make the point. We derive a univariate model for r_t implied by the Markovian system in (5.46).

[16] Here the W^i_t do not represent Wiener processes. They are any independent, identically distributed random variables with no dependence on a past.

This derivation is of interest itself because the recursive method utilized here is a standard tool in solving difference equations in other contexts as well.

In order to obtain a univariate model, consider the equation implied by the first row:

$$r_{t+\Delta} = \alpha_1 r_t + \beta_1 R_t + \sigma_1 W^1_{t+\Delta} \tag{5.50}$$

Substitute for the R_t term implied by the second row of the system

$$R_t = \alpha_2 r_{t-\Delta} + \beta_2 R_{t-\Delta} + \sigma_2 W^2_t \tag{5.51}$$

to get

$$r_{t+\Delta} = \alpha_1 r_t + \beta_1 \left[\alpha_2 r_{t-\Delta} + \beta_2 R_{t-\Delta} + \sigma_2 W^2_t\right] + \sigma_1 W^1_{t+\Delta} \tag{5.52}$$

Rearranging:

$$r_{t+\Delta} = \alpha_1 r_t + \beta_1 \alpha_2 r_{t-\Delta} + \beta_1 \beta_2 R_{t-\Delta} + \left[\beta_1 \sigma_2 W^2_t + \sigma_1 W^1_{t+\Delta}\right] \tag{5.53}$$

Now, there is another $R_{t-\Delta}$ on the right-hand side, but this can also be substituted out by using the second row written for time $t - \Delta$:

$$R_{t-\Delta} = \alpha_2 r_{t-2\Delta} + \beta_2 R_{t-2\Delta} + \sigma_2 W^2_{t-\Delta} \tag{5.54}$$

Proceeding this way, and assuming that the coefficients of $R_{t-k\Delta}$ become negligible as k increases, we will obtain an equation for r_t that can be written as:

$$r_{t+\Delta} - r_t = a_0 r_t + a_1 r_{t-\Delta} + a_2 r_{t-2\Delta} + \cdots + \left[b_0 W^1_{t+\Delta} + b_1 W^2_t + \cdots b_2 W^2_{t-\Delta}\right] \tag{5.55}$$

Obviously, such an r_t process cannot be Markov. For one, a forecast of the $r_{t+\Delta} - r_t$ would depend on $r_s, s < t$ in addition to the last observed r_t. Hence, a *univariate* dynamic that assumes Markovian behavior for the short rate, r_t, will not represent interest rate dynamics correctly, although the joint behavior or the short and long rates *is* Markov by assumption.

Thus, even though in a bivariate world the r_t was Markov, when modeled by itself, it does not satisfy the assumption of Markovness.

Obviously, the reverse is also true. Any non-Markov process in a univariate world can be converted into a Markov process by increasing the dimensionality of the problem. This suggests that one can assume that forward rates are Markov, yet at the same time assuming Markovness for spot rates could, in general, be inaccurate. This point will play an important role in modeling interest-sensitive securities. Within the context of yield curve dynamics, this point suggests working with k-dimensional Markov processes rather than non-Markovian univariate models.

5.9 CONVERGENCE OF RANDOM VARIABLES

The notion of *convergence* has several uses in asset pricing. Some of these are theoretical, others practical. The binomial example of the previous section introduced the notion of convergence as a way of approximating a complicated random variable with a simpler model. As $\Delta \to 0$, the approximation improved. In this section, we provide a more systematic treatment of these issues. Again, the discussion here should be considered a brief and heuristic introduction.

5.9.1 Types of Convergence and their Uses

In pricing financial securities, a minimum of three different convergence criteria are used.

The first is "**mean square convergence**". This is a criterion utilized to define the Ito integral. The latter is utilized in characterizing stochastic differential equations (SDEs). As a result, mean square convergence plays a fundamental role in numerical calculations involving SDEs.

Definition 14. Let $X_0, X_1, \ldots, X_n, \ldots$ be a sequence of random variables. Then X_n is said to converge to X in mean square if

$$\lim_{n\to\infty} \mathbb{E}\,[X_n - X]^2 = 0 \qquad (5.56)$$

According to this definition, the random approximation error ε_n defined by

$$\varepsilon_n = X_n - X \qquad (5.57)$$

will have a smaller and smaller variance as n goes to infinity.

Note that for finite n, the variance of ε_n may be small, but not necessarily zero. This has an important implication. In doing numerical calculations, one may have to take such approximation errors into account explicitly. One way of doing this is to use the standard deviation of ε_n as an estimate.

5.9.1.1 *Relevance of Mean Square Convergence*

Mean square (m.s.) convergence is important because the Ito integral is defined as the mean square limit of a certain sum. If one uses other definitions of convergence, this limit may not exist.

We would like to discuss this important point further. Consider a more "natural" extension of the notion of limit used in standard calculus.

Definition 15. A random variable X_n converges to X almost surely (a.s.) if, for arbitrary $\delta > 0$,

$$P\left(\left|\lim_{n\to\infty} X_n - X\right| > \delta\right) = 0 \qquad (5.58)$$

This definition is a natural extension of the limiting operation used in standard calculus. It says that as n goes to infinity, the difference between the two random variables becomes negligibly small. In the case of mean square convergence, it was the variance that converged to zero. Now, it is the difference between X_n and X. In the limit, the two random variables are almost the same.

5.9.1.2 *Example*

Let S_t be an asset price observed at equidistant time points:

$$t_0 < t_0 + \Delta < t_0 + 2\Delta < \cdots < t_0 + n\Delta = T \quad (5.59)$$

Define the random variable X_n, indexed by n:

$$X_n = \sum_{i=0}^{n} S_{t_0+i\Delta}\left[S_{t_0+(i+1)\Delta} - S_{t_0+i\Delta}\right] \qquad (5.60)$$

Here represents the increment in the asset price at time $t_0 + i\Delta$. The observations begin at time t_0 and are recorded every Δ minutes.

Note that X_n is similar to a Riemann–Stieltjes sum. It is as if an interval $[t_0, T]$ is partitioned into n subintervals and the X_n is defined as an approximation to

$$\int_{t_0}^{T} S_t dS_t \qquad (5.61)$$

But there is a fundamental difference. The sum X_n now involves random processes. Hence, in taking a limit of (5.57), a new type of convergence criterion should be used. The standard definition of limit from calculus is not applicable.

Which (random) convergence criterion should be used?

It turns out that if S_t is a Wiener process, then X_n will not converge almost surely,[17] but a mean square limit will exist. Hence, the type of approximation one uses will make a difference. This important point is taken up during the discussion of the Ito integral in later chapters.

5.9.2 Weak Convergence

The notion of m.s. convergence is used to find approximations to values assumed by random variables. As some parameter n goes to infinity, values assumed by some random variable X_n can be approximated by values of some limiting random variable X.

[17]The same result applies if, in addition, S_t displays occasional jumps.

In the case of *weak convergence* (the third kind of convergence), what is being approximated is not the value of a random variable X_n, but the probability associated with a sequence $X_0, \ldots, X_n$. Weak convergence is used in approximating the distribution function of families of random variables.

Definition 16. Let X_n be a random variable indexed by n with probability distribution P_n. We say that X_n converges to X weakly and

$$\lim_{n\to\infty} P_n = P \tag{5.62}$$

where P is the probability distribution of X if

$$\mathbb{E}^{P_n}\left[f(X_n)\right] \to \mathbb{E}^{P}\left[f(X)\right] \tag{5.63}$$

where $f(\cdot)$ is any bounded, continuous, real-valued function; $\mathbb{E}^{P_n}\left[f(X_n)\right]$ is the expectation of a function of X_n under probability distribution P_n; $\mathbb{E}^{P}\left[f(X)\right]$ is the expectation of a function of X under probability distribution P.

According to this definition, a random variable X_n converges to X weakly if functions of the two random variables have *expectations* that are close enough. Thus, X_n and X do not necessarily assume values that are very close, yet they are governed by arbitrarily close probabilities as $n \to \infty$.

5.9.2.1 Relevance of Weak Convergence

We are often interested in values assumed by a random variable as some parameter n goes to infinity. For example, to define an Ito integral, a random variable with a simple structure is first constructed. This random variable will depend on some parameter n. In the second step, one shows that as $n \to \infty$, this simple variable converges to the Ito integral in the m.s. sense.

Hence, in defining an Ito integral, *values* assumed by a random variable are of fundamental interest, and mean square convergence needs to be used.

At other times, such specific values may not be relevant. Instead, one may be concerned only with expectations—i.e., some sort of average—of random processes.

For example, $F(S_T, T)$ may denote the random price of a derivative product at expiration time T. The derivative is written on the underlying asset S_T. We know that if there are no-arbitrage opportunities, then there exists a "risk-neutral" probability $\mathbb{Q}$ such that, under some simplifying assumptions, the value of the derivative at time t is given by

$$F(t) = e^{-r(T-t)}\mathbb{E}_t^{\mathbb{Q}}\left[F(S_T, T)\right] \tag{5.64}$$

Thus, instead of being concerned with the exact future value of S_T, we need to calculate the expectation of some function $F(\cdot)$ of S_T. Using the concept of weak convergence, an approximation of S_T can be utilized. This may be desirable if it is more convenient to work with than the actual random variable S_T. For example, S_T may be a continuous-time random process, whereas may be a random sequence defined over small intervals that depend on some parameter n. If the work is done on computers, it will be easier to work with than S_T. This idea was utilized earlier in obtaining a binomial approximation to a continuous normally distributed process.

5.9.2.2 An Example

Consider a time interval $[0, 1]$ and let $t \in [0, 1]$ represent a particular time.[18] Suppose we are given n observations $\varepsilon_i, i = 1, 2, \ldots, n$ drawn independently from the uniform distribution $U(0, 1)$.[19]

Next define the random variables $X_i(t), i = 1, \ldots, n$ by

[18] We may, for example, let the expiration time of some derivative contract be 1, while 0 represents the *present*.

[19] This means that

$$P(\varepsilon_i \leq t) = t \tag{5.65}$$

for any $0 \leq t \leq 1$.

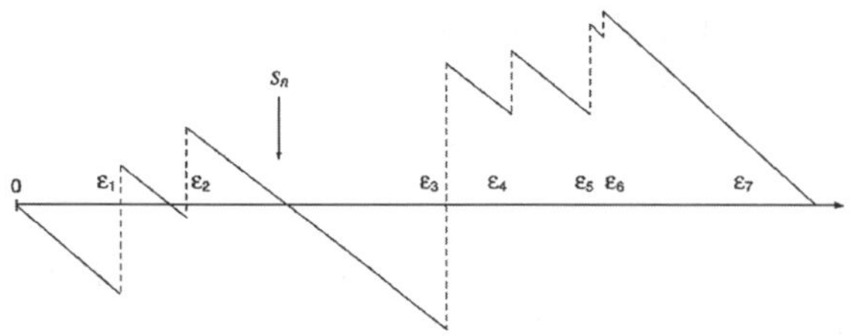

FIGURE 5.8 Construction of a Gaussian process in discrete times.

$$X_i(t) = \begin{cases} 1 & \text{if} \varepsilon_i < t \\ 0 & \text{otherwise} \end{cases} \tag{5.66}$$

According to this, $X_i(t)$ is either 0 or 1, depending on the t and on the value assumed by ε_i.

Using $X_i(t), i = 1, \ldots, n$, we define the random variable $S_n(t)$:

$$S_n(t) = \frac{1}{\sqrt{n}} \sum_{i=1}^{n} (X_i(t) - t) \tag{5.67}$$

Figure 5.8 displays this construction for $n = 7$. Note that $S_n(t)$ is a piecewise continuous function with jumps at ε_i (see Figure 5.7).

As $n \to \infty$, the jump points become more frequent and the "oscillations" of $S_n(t)$ more pronounced. The sizes of the jumps, however, will diminish. At the limit $n \to \infty$, $S_n(t)$ will be very close to a normally distributed random variable for each t. Interestingly, the process will be continuous at the limit, the initial and the endpoints being identically equal to zero.[20]

Clearly, what happens here is that as $n \to \infty$, the $S_n(t)$ starts to behave more and more like a normally distributed process. For large n, we may find a limiting Gaussian process more convenient to work with than $S_n(t)$.

It should also be emphasized that in this example, as n increases, the number of points at which $S_n(t)$ changes will increase. In applications where we go from small discrete intervals toward continuous-time analysis, this would often be the case.

[20]Such a process is called a Brownian bridge.

5.10 CONCLUSIONS

This chapter briefly reviewed some basic concepts of probability theory.

We spent a minimal amount of time on the standard definitions of probability. However, we made a number of important points.

First, we characterized normally distributed random variables and Poisson processes as two basic building blocks.

Second, we discussed an important binomial process. This example was used to introduce the important notion of convergence of stochastic processes. The binomial example discussed here also happens to have practical implications, since it is very similar to the *binomial tree models* routinely used in pricing financial assets.

5.11 REFERENCES

In the remainder of this book, we do not require any further results on probability than what is reviewed here. However, a financial market participant or a finance student will always benefit from a good understanding of the theory of stochastic processes. An excellent introduction is Ross (1993). Liptser and Shiryayev (1977) is an excellent advanced introduction. Cinlar (1978) is another source for the intermediate level. The book by Brzezniak and Zastawniak (1999) is a good source for introductory stochastic processes. See also the new book by Ross (1999).

5.12 EXERCISES

1. You are given two discrete random variables X, Y that assume the possible values 0, 1 according to the following joint distribution:

	$\mathbb{P}(Y=1)$	$\mathbb{P}(Y=0)$
$\mathbb{P}(Y=1)$	0.2	0.4
$\mathbb{P}(Y=0)$	0.15	0.25

(a) What are the marginal distributions of X and Y?
(b) Are X and Y independent?
(c) Calculate $\mathbb{E}[X]$ and $\mathbb{E}[Y]$.
(d) Calculate the conditional distribution $\mathbb{P}[X|Y=1]$.
(e) Obtain the conditional expectation $\mathbb{E}[X|Y=1]$ and the conditional variance $\mathbb{V}[X|Y=1]$.

2. We let the random variable X_n be a binomial process,

$$X_n = \sum_{i=1}^{n} B_i$$

where each B_i is independent and is distributed according to

$$B_i = \begin{cases} 1 \text{ with probability } p \\ 0 \text{ with probability } 1-p \end{cases}$$

(a) Calculate the probabilities $\mathbb{P}\left[X_4 > k\right]$ for $k = 0, 1, 2, 4$ and plot the distribution function.
(b) Calculate the expected value and the variance of X_n for $n = 3$.

3. We say that Z is exponentially distributed with parameter $\lambda > 0$ if the distribution function of Z is given by:

$$\mathbb{P}(Z < z) = 1 - e^{-\lambda x}$$

(a) Determine and plot the density function of Z.
(b) Calculate $\mathbb{E}[Z]$.
(c) Obtain the variance of Z.
(d) Suppose Z_1 and Z_2 are both distributed as exponential and are independent. Calculate the distribution of their sum:

$$S = Z_1 + Z_2$$

(e) Calculate the mean and the variance of S.

4. A random variable Z has Poisson distribution if

$$p(k) = \mathbb{P}(Z < k) = \frac{\lambda^k e^{-\lambda}}{k!}$$

for $k = 0, 1, 2, \ldots$

(a) Use the expansion

$$e^{\lambda} = 1 + \lambda + \frac{\lambda^2}{2} + \frac{\lambda^3}{3!} + \cdots$$

to show that

$$\sum_{k=0}^{\infty} p(k) = 1$$

(b) Calculate the mean $\mathbb{E}[Z]$ and the variance $\mathbb{V}(X)$.

5. Suppose a random variable is defined as $Z = Z_1 + Z_2$, where Z_1 and Z_2 are two independent random variables following Poisson distribution as

$$p_i(Z_i = k) = P(Z_1 = k) = \frac{\lambda_i^k e^{-\lambda_i}}{k!} \tag{5.68}$$

for $k = 0, 1, 2 \ldots$ and $i = 1, 2$. Calculate the distribution of Z.

6. Generate a random walk using a binomial process with $p = 0.5$. Assume that at each step the random walk takes on values 1 (with probability p), and -1 (with probability $1-p$). Perform 1000 random walk steps and calculate expected value and variance of a step of the random walk. Calculate the sum of the random walk over all 1000 steps. Repeat 500 times and compute the mean and variance of the random walk.

CHAPTER

6

Martingales and Martingale Representations

OUTLINE

Financial Derivatives, Third Edition.
http://dx.doi.org/10.1016/B978-0-12-384682-2.00006-2

6.1 INTRODUCTION

Martingales are one of the central tools in the modern theory of finance. In this chapter we introduce the basics of martingale theory. However, this theory is vast, and we only emphasize those aspects that are directly relevant to pricing financial derivatives.

We begin with a comment on notation. In this chapter, we use the notation ΔW_t or ΔS_t to represent "small" changes in W_t or S_t. Occasionally, we may also use their incremental versions dW_t, dS_t, which represent stochastic changes during infinitesimal intervals. For the time being, the reader can interpret these differentials as "infinitesimal" stochastic changes observed over a continuous-time axis. These concepts will be formally defined in Chapter 9.

To denote a small interval, we use the symbols h or Δ. An infinitesimal interval, on the other hand, is denoted by dt. In later chapters, we show that these notations are not equivalent. An operation such as

$$\mathbb{E}\left[S_{t+\Delta} - S_t\right] = 0$$

where Δ is a "small" interval, is well defined. Yet, writing

$$\mathbb{E}\left[dS_t\right] = 0$$

is informal, since dS_t is only a symbolic expression, as we will see in the definition of the Ito integral.

6.2 DEFINITIONS

Martingale theory classifies observed time series according to the way they "trend." A stochastic process behaves like a martingale if its trajectories display no discernible trends or periodicities. A process that, on the average, increases is called a *submartingale*. The term *supermartingale* represents processes that, on the average, decline. This section gives formal definitions of these concepts. First, some notation.

6.2.1 Notation

Suppose we observe a family of random variables indexed by time index t. We assume that time is continuous and deal with continuous-time stochastic processes. Let the observed process be denoted by $\{S_t, t \in [0, \infty]\}$. Let $\{I_t, t \in [0, \infty]\}$ represent a family of information sets that become continuously available to the decision maker as time passes.[1] With $s < t < T$, this family of information sets will satisfy

$$I_s \subseteq I_t \subseteq I_T \tag{6.1}$$

The set $\{I_t, t \in [0, \infty]\}$ is called a *filtration*.

In discussing martingale theory (and throughout the rest of this book), we occasionally need to consider values assumed by some stochastic process at particular points in time. This is often accomplished by selecting a sequence $\{t_i\}$ such that

$$0 = t_0 < t_1 < \cdots < t_{k-1} < t_k = T \tag{6.2}$$

represent various time periods over a continuous-time interval $[0, T]$. Note the way the initial value and the endpoint of the interval are handled in this notation. The symbol t_0 is assigned to the initial point, whereas t_k is the "new" symbol for T. In this notation, as $k \to \infty$, and $\left(t_i - t_{i-1}\right) \to 0$, the interval $[0, T]$ would be partitioned into finer and finer pieces.

Now consider the random price process S_t during the finite interval $[0, T]$. At some particular time t_i, the value of the price process will be S_{t_i}. If the value of S_t is included in the information set I_t at each $t \geq 0$, then it is said that $\{S_t, t \in [0, T]\}$ is adapted to $\{I_t, t \in [0, T]\}$. That is, the value S_t will be known, given the information set I_t.

We can now define continuous-time martingales.

[1] Depending on the problem at hand, the I_t will represent different types of information. The most natural use of I_t will be to represent the information one can obtain from the realized prices in financial markets up to time t.

6.2.2 Continuous-Time Martingales

Using different information sets, one can conceivably generate different "forecasts" of a process $\{S_t\}$. These forecasts are expressed using conditional expectations. In particular,

$$\mathbb{E}_t [S_T] = \mathbb{E} [S_T | I_t] \tag{6.3}$$

is the formal way of denoting the forecast of a future value, S_T of S_t, using the information available as of time t. $\mathbb{E}_u [S_t], u < t$, would denote the forecast of the same variable using a smaller information set as of or earlier than time u.

The defining property of a martingale relates to these conditional expectations.

Definition 17. We say that a process $\{S_t, t \in [0, \infty]\}$ is a martingale with respect to the family of information sets I_t and with respect to the probability P, if, for all $t > 0$,

1. St is known, given I_t. (S_t is I_t-adapted.)
2. Unconditional "forecasts" are finite:

$$\mathbb{E} |S_t| < \infty \tag{6.4}$$

3. And if

$$\mathbb{E}_t [S_T] = S_t \quad \text{for all } t < T \tag{6.5}$$

with probability 1. That is, the best forecast of unobserved future values is the last observation on S_t.

Here, all expectations $\mathbb{E}[\cdot], \mathbb{E}_t[\cdot]$ are assumed to be taken with respect to the probability P.

According to this definition, martingales are random variables whose future variations are completely unpredictable given the current information set. For example, suppose S_t is a martingale and consider the forecast of the *change* in S_t over an interval of length $u > 0$:

$$\mathbb{E}_t [S_{t+u} - S_t] = \mathbb{E}_t [S_{t+u}] - \mathbb{E}_t [S_t] \tag{6.6}$$

But $\mathbb{E}_t [S_t]$ is a forecast of a random variable whose value is already "revealed" (since $S(t)$ is by definition I_t-adapted). Hence, it equals S_t. If S_t is a martingale, $\mathbb{E}_t [S_{t+u}]$ would also equal S_t. This gives

$$\mathbb{E}_t [S_{t+u} - S_t] = 0 \tag{6.7}$$

i.e., the best forecast of the *change* in S_t over an arbitrary interval $u > 0$ is zero. In other words, the directions of the future movements in martingales are impossible to forecast. This is the fundamental characteristic of processes that behave as martingales. If the trajectories of a process display clearly recognizable long- or short-run "trends," then the process is not a martingale.[2]

Before closing this section, we reemphasize a *very* important property of the definition of martingales. A martingale is always defined *with respect to* some information set, *and* with respect to some probability measure. If we change the information content and/or the probabilities associated with the process, the process under consideration may cease to be a martingale.

The opposite is also true. Given a process X_t that does not behave like a martingale, we may be able to modify the relevant probability measure P and convert X_t into a martingale.

6.3 THE USE OF MARTINGALES IN ASSET PRICING

According to the definition above, a process S_t is a martingale if its future movements are completely unpredictable given a family of information sets. Now, we know that stock prices or bond prices are *not* completely unpredictable. The price of a discount bond is expected to *increase* over time. In general, the same is true for stock prices. They are expected to increase on the average. Hence, if B_t represents the price of a discount bond maturing at time $T, t < T$,

$$B_t < \mathbb{E}[B_u], \quad t < u < T$$

[2]A sample path of a martingale may still contain patterns that look like short-lived trends. However, these up or down trends are completely random and do not have any systematic character.

Clearly, the price of a discount bond does not move like a martingale.

Similarly, in general, a risky stock S_t will have a positive expected return and will not be a martingale. For a small interval Δ, we can write

$$\mathbb{E}[S_{t+\Delta} - S_t] \approx \mu\Delta \tag{6.8}$$

where μ is a positive rate of expected return.[3]

A similar statement can be made about futures or options. For example, options have "time value," and, as time passes, the price of European-style options will decline ceteris paribus. Such a process is a *supermartingale*.[4]

If asset prices are more likely to be sub- or supermartingales, then why such an interest in martingales?

It turns out that although most financial assets are not martingales, one can *convert* them into martingales. For example, one can find a probability distribution $\mathbb{Q}$ such that bond or stock prices discounted by the risk-free rate become martingales. If this is done, equalities such as

$$\mathbb{E}_t^{\mathbb{Q}}\left[e^{-ru}B_{t+u}\right] = B_u, \quad 0 < u < T - t \tag{6.9}$$

for bonds, or

$$\mathbb{E}_t^{\mathbb{Q}}\left[e^{-ru}S_{t+u}\right] = S_t, \quad 0 < u \tag{6.10}$$

for stock prices, can be very useful in pricing derivative securities.

One important question that we study in later chapters is how to obtain this conversion. There are, in fact, two ways of converting submartingales into martingales.

The first method should be obvious. We can subtract an *expected trend* from $e^{-rt}S_t$ or $e^{-rt}B_t$. This would make the *deviations* around the trend completely unpredictable. Hence, the "transformed" variables would be martingales.

This methodology is equivalent to using the so-called representation results for martingales. In fact, Doob–Meyer decomposition implies that, under some general conditions, an arbitrary continuous-time process can be decomposed into a martingale and an increasing (or decreasing) process. Elimination of the latter leaves the martingale to work with. Doob–Meyer decomposition is handled in this chapter.

The second method is more complex and, surprisingly, more useful. Instead of transforming the submartingale directly, we can transform its *probability distribution*. That is, if one had

$$\mathbb{E}_t^{\mathbb{P}}\left[e^{-ru}S_{t+u}\right] > S_t \quad 0 < u \tag{6.11}$$

where $\mathbb{E}_t^{\mathbb{P}}[\cdot]$ is the conditional expectation calculated using a probability distribution $\mathbb{P}$, we may try to find an "equivalent" probability $\mathbb{Q}$, such that the new expectations satisfy

$$\mathbb{E}_t^{\mathbb{Q}}\left[e^{-ru}S_{t+u}\right] = S_t \quad 0 < u \tag{6.12}$$

and the $e^{-rt}S_t$ becomes a martingale.

Probability distributions that convert equations such as (6.12) into equalities such as (6.13) are called *equivalent martingale measures*. They will be treated in Chapter 14.

If this second methodology is selected to convert arbitrary processes into martingales, then the transformation is done using the *Girsanov theorem*. In financial asset pricing, this method is more promising than the Doob–Meyer decompositions.

6.4 RELEVANCE OF MARTINGALES IN STOCHASTIC MODELING

In the absence of arbitrage possibilities, market equilibrium suggests that we can find a synthetic probability distribution $\mathbb{Q}$ such that all properly discounted asset prices S_t behave as

[3] The approximation here is in the sense of dropping higher-order terms involving Δ in a Taylor series expansion of $\mathbb{E}_t\left[S_{t+\Delta} - S_t\right]$,

$$\mathbb{E}_t\left[S_{t+\Delta} - S_t\right] = \mu\Delta + o(\Delta)$$

where $o(\Delta)$ represents all higher-order terms of the corresponding Taylor series expansion.

[4] Deep in the money, American puts may have negative time value.

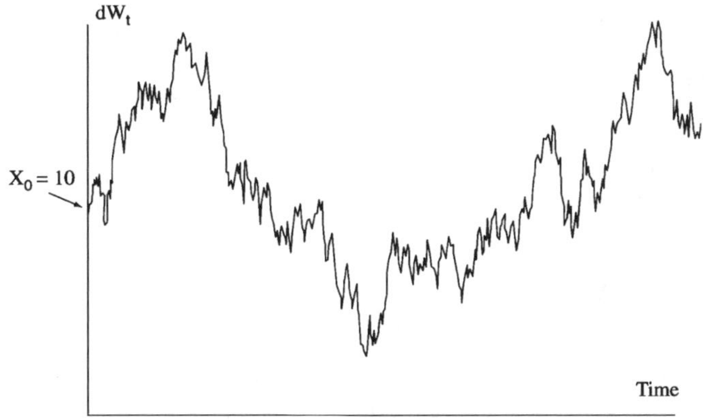

FIGURE 6.1 An example of a continuous martingale.

martingales:

$$\mathbb{E}_t^{\mathbb{P}}\left[e^{-ru}S_{t+u}\middle| I_t\right] = S_t \quad u > 0 \tag{6.13}$$

Because of this, martingales have a fundamental role to play in practical asset pricing.

But this is not the only reason why martingales are useful tools. Martingale theory is very rich and provides a fertile environment for discussing stochastic variables in continuous time. In this section, we discuss these useful technical aspects of martingale theory.

Let X_t represent an asset price that has the martingale property with respect to the filtration $\{I_t\}$ and with respect to the probability $\mathbb{Q}$,

$$\mathbb{E}_t^{\mathbb{Q}}\left[X_{t+\Delta}| I_t\right] = X_t \tag{6.14}$$

where $\Delta > 0$ represents a small time interval. What type of trajectories would such an X_t have in continuous time?

To answer this question, first define the *martingale difference* ΔX_t,

$$\Delta X_t = X_{t+\Delta} - X_t \tag{6.15}$$

and then note that since X_t is a martingale,

$$\mathbb{E}_t^{\mathbb{Q}}\left[\Delta X_t| I_t\right] = 0 \tag{6.16}$$

As mentioned earlier, this equality implies that increments of a martingale should be totally unpredictable, no matter how small the time interval Δ is. But, since we are working with continuous time, we can indeed consider *very* small Δ's. Martingales should then display very irregular trajectories. In fact, X_t should not display any trends discernible by inspection, even during infinitesimally small time intervals Δ. If it did, it would become predictable.

Such irregular trajectories can occur in two different ways. They can be *continuous*, or they can display *jumps*. The former leads to *continuous martingales*, whereas the latter are called *right-continuous martingales*.

Figure 6.1 displays an example of a continuous martingale. Note that the trajectories are continuous, in the sense that for $\Delta \to 0$,

$$P(\Delta X_t > \varepsilon) \to 0, \quad \text{for all } \varepsilon > 0 \tag{6.17}$$

Figure 6.2 displays an example of a right-continuous martingale. Here, the trajectory is interrupted with occasional jumps.[5] What makes the trajectory *right*-continuous is the way jumps are modeled. At jump times t_0, t_1, t_2, the martingale is continuous rightwards (but not leftwards).

[5]Note that the process still does not have a trend.

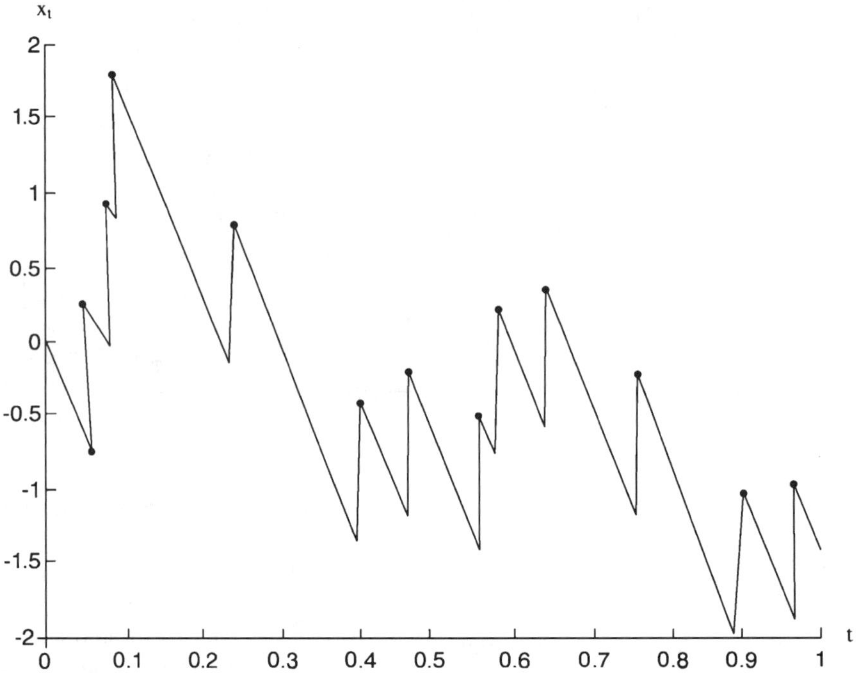

FIGURE 6.2 An example of a right continuous martingale.

This irregular behavior and the possibility of incorporating jumps in the trajectories is certainly desirable as a theoretical tool for representing asset prices, especially given the arbitrage theorem.

But martingales have significance beyond this. In fact, suppose one is dealing with a continuous martingale X_t that also has a finite second moment

$$\mathbb{E}\left[X_t^2\right] < \infty \tag{6.18}$$

for all $t > 0$.

Such a process has finite variance and is called a *continuous square integrable martingale*. It is significant that one can represent all such martingales by running the Brownian motion at a modified time clock. [See Karatzas and Shreve (1991)]. In other words, the class of continuous square integrable martingales is very close to the Brownian motion. This suggests that the unpredictability of the changes and the absence of jumps are two properties of Brownian motion in continuous time.

Note what this essentially means. If the continuous square integrable martingale is appropriate for modeling an asset price, one may as well assume normality for small increments of the price process.

6.4.1 An Example

We will construct a martingale using two independent Poisson processes observed during "small intervals" Δ.

Suppose financial markets are influenced by "good" and "bad" news. We ignore the content of the news, but retain the information on whether it is "good" or "bad."

The N_t^G and N_t^B denote the total number of instances of "good" and "bad" news, respectively, until time t. We assume further that the way news arrives in financial markets is totally unrelated to past data, and that the "good" news and the "bad" news are independent.

Finally, during a small interval Δ, at most *one* instance of good news or one instance of bad news can occur, and the probability of this occurrence is the same for both types of news. Thus, the probabilities of incremental changes $\Delta N_t^G, \Delta N_t^B$ during Δ are assumed to be given approximately by

$$P\left(\Delta N_t^G = 1\right) = P\left(\Delta N_t^B = 1\right) \approx \lambda\Delta \qquad (6.19)$$

Then the variable M_t, defined by

$$M_t = N_t^G - N_t^B \qquad (6.20)$$

will be a martingale.

To see this, note that the increments of M_t over small intervals Δ will be given by

$$\Delta M_t = \Delta N_t^G - \Delta N_t^B \qquad (6.21)$$

Apply the conditional expectation operator:

$$\mathbb{E}\left[\Delta M_t\right] = \mathbb{E}\left[\Delta N_t^G\right] - \mathbb{E}\left[\Delta N_t^B\right] \qquad (6.22)$$

But, approximately,

$$\mathbb{E}\left[\Delta N_t^G\right] \approx 1 \times \lambda\Delta + 0 \times (1 - \lambda\Delta) \qquad (6.23)$$

$$\approx \lambda\Delta \qquad (6.24)$$

and similarly for $\mathbb{E}\left[\Delta N_t^B\right]$. This means that

$$\mathbb{E}\left[\Delta M_t\right] \approx \lambda\Delta - \lambda\Delta = 0 \qquad (6.25)$$

Hence, increments in M_t are unpredictable given the family I_t. It can be shown that M_t satisfies other (technical) requirements of martingales. For example, at time t, we know the "good" or "bad" news that has already happened. Hence, M_t is I_t-adapted.

Thus, as long as the probability of "good" and "bad" news during Δ is given by the same expression $\lambda\Delta$ for both N_t^G and N_t^B, the process M_t will be a martingale with respect to I_t and these probabilities.

However, if we assume that "good" news can occur with a slightly greater probability than "bad" news,

$$P\left(\Delta N_t^G = 1\right) \approx \lambda^G\Delta > P\left(\Delta N_t^G = 1\right) \approx \lambda^B\Delta \qquad (6.26)$$

then M_t will cease to be a martingale with respect to It, since

$$\mathbb{E}\left[\Delta N_t^G\right] \approx \lambda^G\Delta - \lambda^B\Delta > 0 \qquad (6.27)$$

(In fact, M_t will be a submartingale.) Hence, changing the underlying probabilities or the information set may alter martingale characteristics of a process.

6.5 PROPERTIES OF MARTINGALE TRAJECTORIES

The properties of the trajectories of continuous square integrable martingales can be made more precise.

Assume that $\{X_t\}$ represents a trajectory of a continuous square integrable martingale. Pick a time interval $[0, T]$ and consider the times $\{t_i\}$:

$$t_0 = 0 < t_1 < t_2 < \cdots < t_{n-1} < t_n = T \qquad (6.28)$$

We define the *variation* of the trajectory as

$$V^1 = \sum_{i=1}^{n} \left|X_{t_i} - X_{t_{i-1}}\right| \qquad (6.29)$$

Heuristically, V^1 can be interpreted as the length of the trajectory followed by X_t during the interval $[0, T]$.

The quadratic variation is given by

$$V^2 = \sum_{i=1}^{n} \left|X_{t_i} - X_{t_{i-1}}\right|^2 \qquad (6.30)$$

One can similarly define *higher-order* variations. For example, the fourth-order variation is defined as

$$V^4 = \sum_{i=1}^{n} |X_{t_i} - X_{t_{i-1}}|^4 \tag{6.31}$$

Obviously, the V^1 or V^2 are different measures of how much X_t varies over time. The V^1 represents the sum of absolute changes in X_t observed during the subintervals $t_i - t_{i-1}$. The V^2 represents the sums of squared changes.

When X_t is a continuous martingale, the V^1, V^2, V^3, V^4 happen to have some very important properties.

We recall some relevant points. Remember that we want X_t to be continuous and to have a nonzero variance. As mentioned earlier, this means two things. First, as the partitioning of the interval $[0, T]$ gets finer and finer, "consecutive" X_t's get nearer and nearer, for any $\varepsilon > 0$

$$P\left(|X_{t_i} - X_{t_{i-1}}| > \varepsilon\right) \to 0 \tag{6.32}$$

if $t_i \to t_{i-1}$, for all i. Second, as the partitions get finer and finer, we still want

$$P\left(\sum_{i=1}^{n} |X_{t_i} - X_{t_{i-1}}|^2 > 0\right) = 1 \tag{6.33}$$

This is true because X_t is after all a random process with nonzero variance.

Now consider some properties of V^1 and V^2.

First, note that even though X_t is a continuous martingale, and X_{t_i} approaches $X_{t_{i-1}}$ as the subinterval $[t_i - t_{i-1}]$ becomes smaller and smaller, this does not mean that V^1 also approaches zero. The reader may find this surprising. After all, V^1 is made of the sum of such incremental changes:

$$V^1 = \sum_{i=1}^{n} |X_{t_i} - X_{t_{i-1}}| \tag{6.34}$$

As X_{t_i} approaches $X_{t_{i-1}}$, would not V^1 go toward zero as well?

Surprisingly, the opposite is true. As $[0, T]$ is partitioned into finer and finer subintervals, changes in X_t get smaller. But, at the same time, the *number* of terms in the sum defining V^1 increases. It turns out that in the case of a continuous-time martingale, the second effect dominates and the V^1 goes toward infinity. The trajectories of continuous martingales have *infinite* variation, except for the case when the martingale is a constant.

This can be shown heuristically as follows. We have

$$\sum_{i=1}^{n} |X_{t_i} - X_{t_{i-1}}|^2 < \max_i |X_{t_i} - X_{t_{i-1}}| \times \sum_{i=1}^{n} |X_{t_i} - X_{t_{i-1}}| \tag{6.35}$$

because the right-hand side is obtained by factoring out the "largest" $X_{t_i} - X_{t_{i-1}}$.[6] This means that

$$V^2 < \max_i |X_{t_i} - X_{t_{i-1}}| \, V^1 \tag{6.36}$$

As $t_i \to t_{i-1}$ for all i, the continuity of the martingale implies that "consecutive" X_{t_i}'s will get very near each other. At the limit,

$$\max_i |X_{t_i} - X_{t_{i-1}}| \to 0 \tag{6.37}$$

This, according to Eq. (6.37), means that unless V^1 gets very large, V^2 will go toward zero in some probabilistic sense. But this is not allowed because X_t is a stochastic process with a nonzero variance, and consequently $V^2 > 0$ even for very fine partitions of $[0, T]$. This implies that we must have $V^1 \to \infty$.

Now consider the same property for higher-order variations. For example, consider V^4 and apply the same "trick" as in (6.37):

$$V^4 < \max_i |X_{t_i} - X_{t_{i-1}}|^2 \, V^2 \tag{6.38}$$

[6]The notation $\max_i \left|X_{t_i} - X_{tli-1}\right|$ should be read as choosing the largest observed increment out of all incremental changes in X_{t_i}.

As long as V^2 converges to a well-defined random variable,[7] the right-hand side of (6.39) will go to zero. The reason is the same as above. The X_t is a continuous martingale and its increments get smaller as the partition of the interval $[0, T]$ becomes finer. Hence, as $t_i \to t_{i-1}$ for all i:

$$\max_i \left| X_{t_i} - X_{t_{i-1}} \right|^2 \to 0 \qquad (6.39)$$

This means that V_4 will tend to zero. The same argument can be applied to all variations of order greater than two.

For formal proofs of such arguments, the reader can consult Karatzas and Shreve (1991). Here we summarize the three properties of the trajectories:

- The variation V^1 will converge to infinity in some probabilistic sense and the continuous martingale will behave very irregularly.
- The quadratic variation V^2 will converge to some well-defined random variable. This means that regardless of how irregular the trajectories are, the martingale is square integrable and the sums of *squares* of the increments over small subperiods converge. This is possible because the square of a small number is even smaller. Hence, though the sum of increments is "too large" in some probabilistic sense, the sum of *squared* increments is not.
- All higher-order variations will vanish in some probabilistic sense. A heuristic way of interpreting this is to say that higher-order variations do not contain much information beyond those in V^1 and V^2.

These properties have important implications. First, we see that V^1 is not a very useful quantity to use in the calculus of continuous square integrable martingales, whereas the V^2 can be used in a meaningful way. Second, higher-order variations can be ignored if one is certain that the underlying process is a *continuous* martingale.

[7] And does not converge to infinity.

These themes will reappear when we deal with the differentiation and integration operations in stochastic environments. A reader who remembers the definition of the Riemann–Stieltjes integral can already see that the same methodology cannot be used for integrals taken with respect to continuous square integrable martingales. This is the case since the Riemann–Stieltjes integral uses the equivalent of V^1 in deterministic calculus and considers finer and finer partitions of the interval under consideration. In stochastic environments such limits do not converge.

Instead, stochastic calculus is forced to use V^2. We will discuss this in detail later.

Define $||d|| = \max_i |t_i - t_{i-1}|$. Then

$$[X, X]_T = \lim_{||d|| \to 0} V^2 \qquad (6.40)$$

If $X(t)$ is a jump process, its quadratic variation could be random, which implies it is path dependent, unlike Brownian motion, which we can use to show that $[W, W](T) = T$. For quadratic covariation of two processes X_1 and X_2 on $[0, T]$, we first define

$$QC(X) = \sum_{i=0}^{m-1} (X_1(t_{i+1}) - X_1(t_i))(X_2(t_{i+1}) - X_2(t_i)) \qquad (6.41)$$

Then

$$[X_1, X_2]_T = \lim_{||d|| \to 0} QC(X) \qquad (6.42)$$

6.6 EXAMPLES OF MARTINGALES

In this section, we consider some examples of continuous-time martingales.

6.6.1 Example 1: Brownian Motion

Suppose X_t represents a continuous process whose increments are normally distributed. Such a process is called a (generalized) Brownian motion. We observe a value of X_t for each t. At every instant, the infinitesimal change in X_t is denoted by dX_t. Incremental changes in X_t are assumed to be independent across time.

Under these conditions, if Δ is a small interval, the increments ΔX_t during Δ will have a normal distribution with mean $\mu\Delta$ and variance $\sigma^2\Delta$.[8]

$$\Delta X_t \sim \mathcal{N}\left(\mu\Delta, \sigma^2\Delta\right) \tag{6.44}$$

The fact that increments are uncorrelated can be expressed as

$$\mathbb{E}\left[(\Delta X_u - \mu\Delta)(\Delta X_t - \mu\Delta)\right] = 0, \quad u \neq t \tag{6.45}$$

Leaving aside formal aspects of defining such a process X_t, here we ask a simple question: is X_t a martingale?

The process X_t is the "accumulation" of infinitesimal increments over time, that is,

$$X_{t+T} = X_0 + \int_0^{t+T} dX_u \tag{6.46}$$

Assuming that the integral is well defined, we can calculate the relevant expectations.[9]

Consider the expectation taken with respect to the probability distribution given in (6.41), and given the information on X_t observed up to time t:

$$\mathbb{E}_t\left[X_{t+T}\right] = \mathbb{E}_t\left[X_t + \int_t^{t+T} dX_u\right] \tag{6.47}$$

But at time t, future values of ΔX_{t+T} are predictable because all changes during small intervals Δ have expectation equal to $\mu\Delta$. This means

$$\mathbb{E}_t\left[\int_t^{t+T} dX_u\right] = \mu T \tag{6.48}$$

[8]It is not clear why the variance of ΔX_t should be proportional to Δ. For example, is it possible that

$$\mathbb{V}(\Delta X_t) = \sigma^2\Delta^2 \tag{6.43}$$

This question is more complicated to answer than it seems. It will be at the core of the next chapter.

[9]We have not yet defined integrals of random incremental changes.

So,

$$\mathbb{E}_t\left[\int_t^{t+T} dX_u\right] = X_t + \mu T \tag{6.49}$$

Clearly, $\{X_t\}$ is not a martingale with respect to the distribution in Eq. (6.41) and with respect to the information on current and past X_t.

But, this last result gives a clue on how to generate a martingale with $\{X_t\}$. Consider the new process:

$$Z_t = X_t - \mu t \tag{6.50}$$

It is easy to show that Z_t is a martingale:

$$\mathbb{E}_t\left[Z_{t+T}\right] = \mathbb{E}_t\left[X_{t+T} - \mu(t+T)\right] \tag{6.51}$$

$$= \mathbb{E}_t\left[X_t + \left(X_{t+T} - X_t\right)\right] - \mu(t+T) \tag{6.52}$$

which means

$$\mathbb{E}_t\left[Z_{t+T}\right] = X_t + \mathbb{E}\left[\left(X_{t+T} - X_t\right)\right] - \mu(t+T) \tag{6.53}$$

But the expectation on the right-hand side is equal to μT, as shown in Eq. (6.46). This means

$$\mathbb{E}_t\left[Z_{t+T}\right] = X_t - \mu(t+T) \tag{6.54}$$

$$= Z_t \tag{6.55}$$

That is, Z_t is a martingale.

Hence, we were able to transform X_t into a martingale by subtracting a deterministic function. Also, note that this deterministic function was *increasing* over time. This result holds in more general settings as well.

6.6.2 Example 2: A Squared Process

Now consider a process S_t with uncorrelated increments during small intervals Δ:

$$\Delta S_t \sim \mathcal{N}\left(0, \sigma^2\Delta\right) \tag{6.56}$$

where the initial point is given by

$$S_0 = 0 \tag{6.57}$$

Define a new, random variable:

$$Z_t = S_t^2 \tag{6.58}$$

According to this, Z_t is a nonnegative random variable equaling the square of S_t. Is Z_t a martingale?

The answer is no because the squares of the increments of Z_t are predictable. Using a "small" interval Δ, consider the expectation of the increment in Z_t:

$$\begin{aligned}\mathbb{E}_t\left[S_{t+\Delta}^2 - S_t^2\right] &= \mathbb{E}_t\left[\mathbb{E}\left[S_t - (S_t - S_{t+\Delta})^2 - S_t^2\right]\right] \\ &= \mathbb{E}_t\left[S_{t+\Delta} - S_t\right]^2\end{aligned}$$

The last equality follows because increments in S_t are uncorrelated with current and past S_t. As a result, the cross product terms drop. But this means that

$$\mathbb{E}\left[\Delta Z_t\right] = \sigma^2 \Delta \tag{6.59}$$

which proves that increments in Z_t are predictable. Z_t cannot be a martingale.

But, using the same approach as in Example 1, we can "transform" the Z_t with a mean change and obtain a martingale. In fact, the following equality is easy to prove:

$$\mathbb{E}_t\left[Z_{t+T} - \sigma^2 (T + t)\right] = Z_t - \sigma^2 t \tag{6.60}$$

By subtracting $\sigma^2 t$ from Z_t, we obtain a martingale.

This example again illustrates the same principle. If somehow a stochastic process is not a martingale, then by subtracting a proper "mean,"[10] it can be transformed into one.

This brings us to the point made earlier. In financial markets one cannot expect the observed market value of a risky security to equal its expected value discounted by the risk-free rate. There has to be a risk premium. Hence, any risky asset price, when discounted by the risk-free rate, will not be a martingale. But the previous discussion suggests that such securities prices can perhaps be transformed into one. Such a transformation would be very convenient for pricing financial assets.

6.6.3 Example 3: An Exponential Process

The third example is more complicated and will only be partially dealt with here.

Again assume that X_t is as defined in Example 1 and consider the transformation

$$S_t = e^{\left(\alpha X_t - \frac{\alpha^2}{2} t\right)} \tag{6.61}$$

where α is any real number. Suppose the mean of X_t is zero. Does this transformation result in a martingale?

The answer is yes. We shall prove it in later chapters.[11] However, notice something odd. The X_t is itself a martingale. Why is it that one still has to subtract the function of time $g(t)$,

$$g(t) = \frac{\alpha^2}{2} t \tag{6.62}$$

in order to make sure that S_t is a martingale? Were not the increments of X_t impossible to forecast anyway?

The answers to these questions have to do with the way one takes derivatives in stochastic environments. This is treated in later chapters.

6.6.4 Example 4: Right-Continuous Martingales

We consider again the Poisson counting process N_t discussed in this chapter. Clearly, N_t will increase over time, since it is a counting process and the number of jumps will grow as time passes. Hence, N_t cannot be a martingale. It has a clear upward trend.

Yet, the *compensated Poisson* process denoted by N_t^*,

$$N_t^* = N_t - \lambda t \tag{6.63}$$

[10] That is, by subtracting from it a function of time, say $g(t)$.

[11] Once we learn about Ito's Lemma.

will be a martingale. Clearly, the N_t^* also has increments that are unpredictable. It is a right-continuous martingale. Its variance is finite, and it is square integrable.

6.7 THE SIMPLEST MARTINGALE

There is a simple martingale that one can generate that is used frequently in pricing complicated interest rate derivatives. We work with discrete-time intervals.

Consider a random variable Y_T with probability distribution P. Y_T will be revealed to us at some future date T. Suppose we keep getting new information denoted by I_t concerning Y_T as time passes, $t, t+1, \ldots T-1, T$, such that:

$$I_t \subseteq I_{t+1} \subseteq \cdots \subseteq I_{T-1} \subseteq I_T \tag{6.64}$$

Next, consider successive "forecasts," denoted by M_t, of the same Y_T made at different times,

$$M_t = \mathbb{E}^P\left[Y_T | I_t\right] \tag{6.65}$$

with respect to some probability P.

It turns out that the sequence of forecasts, $\{M_t\}$, is a martingale. That is, for $0 < s$:

$$\mathbb{E}^P\left[M_{t+s} | I_t\right] = M_t \tag{6.66}$$

This result comes from the recursive property of conditional expectations, which we will see several times in later chapters. For any random variable Z, we can write:

$$\mathbb{E}^P[\mathbb{E}^P[Z|I_{t+s}]|I_t] = \mathbb{E}^P[Z|I_t], \quad s > 0 \tag{6.67}$$

which says, basically, that the best forecast of a future forecast is what we forecast now. Applying this to $Z = [M_{t+s}]$, we have

$$\mathbb{E}^P[M_{t+s}|I_t] = \mathbb{E}^P[\mathbb{E}^P[Y_T|I_{t+s}]|I_t] \tag{6.68}$$

which is trivially true.

But M_{t+s} is itself a forecast. Using (6.64) on the right-hand side of (6.65),

$$\mathbb{E}^P[\mathbb{E}^P[Y_T|I_{t+s}]|I_t] = \mathbb{E}^P[Y_T|I_t] = M_t \tag{6.69}$$

Thus, M_t is a martingale.

6.7.1 An Application

There are many financial applications of the logic used in the previous section. We deal with one common case.

Most derivatives have random payoffs at finite expiration dates T. Many do not make any interim payouts until expiration either. Suppose this is the case and let the expiration payoff be dependent on some underlying asset price S_T and denoted by

$$G_T = f(S_T) \tag{6.70}$$

Next, consider the investment of \$1 that grows at the constant, continuously compounded rate r_s until time T:

$$B_T = e^{\int_t^T r_s ds} \tag{6.71}$$

This is a sum to be received at time T and may be random, if r_s is stochastic. Here B_T is assumed to be known.

Finally, consider the ratio G_T/B_T, which is a relative price. In this ratio, we have a random variable that will be revealed at a fixed time T. As we get more information on the underlying asset, S_t, successive conditional expectations of this ratio can be calculated until the G_T/B_T is known exactly at time T. Let the successive conditional expectations of this ratio, calculated using different information sets, be denoted by M_t,

$$M_t = \mathbb{E}^P\left[\left.\frac{G_T}{B_T}\right| I_t\right] \tag{6.72}$$

where I_t denotes, as usual, the information set available at time t, and P is an appropriate probability.

According to the previous result, these successive conditional expectations should form a martingale:

$$M_t = \mathbb{E}^P \left[M_{t+s} | I_t \right], \quad s > 0 \tag{6.73}$$

6.7.2 A Remark

Suppose r_t is stochastic and G_T is the value at time T of a default-free pure discount bond. If T is the maturity date, then

$$G_T = 100 \tag{6.74}$$

the par value of the bond.

Then, the M_t is the conditional expectation of the discounted payoff at maturity under the probability P. It is also a martingale *with respect to P*, according to the discussion in the previous section.

The intersecting question is whether we can take M_t as the arbitrage-free price of the discount bond at time t? In other words, letting the T-maturity default-free discount bond price be denoted by $B(t,T)$, and assuming that $B(t,T)$ is arbitrage-free, we can say that

$$B(t,T) = M_t \tag{6.75}$$

In the second half of this book we will see that, if the expectation is calculated under a probability $\mathbb{P}$, and if this probability is the *real-world probability*, then M_t will *not*, in general, equal the fair price $B(t,T)$.

But, if the probability used in calculating M_t is selected judiciously as an arbitrage-free "equivalent" probability $\mathbb{Q}$, then

$$B(t,T) = M_t \tag{6.76}$$

$$= \mathbb{E}^{\mathbb{Q}} \left[\frac{100}{B_T} \middle| I_t \right] \tag{6.77}$$

that is, the M_t will correctly price the zero-coupon bond.

The mechanics of how this could be selected will be discussed in later chapters. But, already the idea that martingales are critical tools in dynamic asset pricing should become clear. It should also be clear that we can define several M_t using different probabilities, and they will all be martingales (with respect to their particular probabilities). Yet, only one of these martingales will equal the arbitrary-free price of $B(t,T)$.

6.8 MARTINGALE REPRESENTATIONS

The previous examples showed that it is possible to transform a wide variety of continuous-time processes into martingales by subtracting appropriate means.

In this section, we formalize these special cases and discuss the so-called Doob–Meyer decomposition.

First, a fundamental example will be introduced. The example is important for (at least) three reasons.

The first reason is practical. By working with a partition of a continuous-time interval, we illustrate a practical method used to price securities in financial markets.

Second, it is easier to understand the complexities of the Ito integral if one begins with such a framework.

And finally, the example provides a concrete discussion of a probability space and how one can assign probabilities to various trajectories associated with asset prices.

6.8.1 An Example

Suppose a trader observes at times t_i,

$$t_0 < t_1 < t_2 < \cdots < t_{k-1} < t_k = T \tag{6.78}$$

the price of a financial asset S_t.

If the intervals between the times t_{i-1} and t_i are very small, and if the market is "liquid," the price of the asset is likely to exhibit at most one uptick or one downtick during a typical $t_i - t_{i-1}$.

We formalize this by saying that at each instant t_i, there are only two possibilities for change:

$$\Delta S_{t_i} = \begin{cases} 1 & \text{with probability } p \\ -1 & \text{with probability } (1-p) \end{cases} \tag{6.79}$$

It is assumed that these changes are independent of each other. Further, if $p = 1/2$, then the expected value of ΔS_{t_i} will equal zero. Otherwise the mean of price changes is nonzero.

Given these conditions, we first show how to construct the *underlying probability space.*

We observe ΔS_t at k distinct time points.[12] We begin with the notion of probability. The $\{p, (1-p)\}$ refers to the probability of a change in S_{t_i} and is only a (marginal) probability distribution. What is of interest is the probability of a *sequence* of price changes. In other words, we would like to discuss probabilities associated with various "trajectories."[13] Doing this requires constructing a probability space.

Given that a typical object of interest is a *sample path,* or trajectory, of price changes, we first need to construct a *set* made of all possible paths. This space is called a *sample space.* Its elements are made of sequences of +1's and −1's. For example, a typical sample path can be

$$\{\Delta S_{t_1} = -1, \ldots, \Delta S_{t_k} = +1\} \tag{6.80}$$

Since k is finite, given an initial point S_{t_0} we can easily determine the trajectory followed by the asset price by adding incremental changes. This way we can construct the set of all possible trajectories, i.e., the *sample space.*

Next we define a *probability* associated with these trajectories. When the price changes are independent (and when k is finite), doing this is easy. The probability of a certain sequence is found by simply multiplying the probabilities of each price change.

[12]Note the important assumption that k is finite.

[13]For example, the trader may be interested in the length of the current uptrend or downtrend in asset prices.

For example, the particular sequence ΔS^* that begins with +1 at time t_0 and alternates until time t_k,

$$\Delta S^* = \{\Delta S_{t_1} = +1, \Delta S_{t_1} = -1, \ldots, \Delta S_{t_k} = -1\} \tag{6.81}$$

will have the probability (assuming k is even)

$$P(\Delta S^*) = p^{k/2}(1-p)^{k/2} \tag{6.82}$$

The probability of a trajectory that continuously declines during the first $k/2$ periods, then continuously increases until time t_k, will also be the same.

Since k is finite, there are a finite number of possible trajectories in the sample space, and we can *assign* a probability to every one of these trajectories.

It is worth repeating what enables us to do this. The finiteness of k plays a role here, since with a finite number of possible trajectories this assignment of probabilities can be made one by one. Pricing derivative products in financial markets often makes the assumption that k is finite and exploits this property of generating probabilities.

Another assumption that simplifies this task is the *independence* of successive price changes. This way the probability of the whole trajectory can be obtained by simply multiplying the probabilities associated with each incremental change.

Up to this point, we have dealt with the sequence of *changes* in the asset price. Derivative securities are, in general, written on the price itself. For example, in the case of an option written on the S&P500, our interest lies with the *level* of the index, not the *change.*

One can easily obtain the level of the asset price from subsequent changes, given the opening price S_{t_0}:

$$S_{t_k} = S_{t_0} + \sum_{i=1}^{k} (S_{t_i} - S_{t_{i-1}}) \tag{6.83}$$

Note that since a typical S_{t_k} is made of the *sum* of ΔS_{t_i}'s, probabilities such as (6.79) can be used to obtain the probability distribution of the S_{t_k} as well. In doing this we would simply add the probabilities of different trajectories that lead to the same S_{t_k}.[14]

To be more precise, the highest possible value for S_{t_k} is $S_{t_0} + k$. This value will result if all incremental changes $\Delta S_{t_i}, i = 1, \ldots, k$ are made of $+1$'s. The probability of this outcome is

$$P\left(S_{t_k} = S_0 + k\right) = p^k \tag{6.84}$$

Similarly, the lowest possible value is $S_0 - k$. The probability of this is given by

$$P\left(S_{t_k} = S_0 - k\right) = (1-p)^k \tag{6.85}$$

In these extreme cases, there is only *one* trajectory that gives $S_{t_k} = S_0 + k$ or $S_{t_k} = S_0 - k$.

In general, the price would be somewhere between these two extremes. Of the k incremental changes observed, m would be made of $+1$'s and $k - m$ made of -1's, with $m \leq k$. The S_{t_k} will assume the value

$$S_{t_k} = S_0 + m - (k - m) \tag{6.86}$$

Note that there are several possible trajectories that eventually result in the same value for S_{t_k}. Adding the probabilities associated with all these combinations, we obtain

$$P\left(S_{t_k} = S_0 + m - (k-m)\right) = C_k^{m-k} p^m (1-p)^{k-m} \tag{6.87}$$

where

$$C_k^{m-k} = \frac{k!}{m!\,(k-m)!}$$

This probability is given by the *binomial distribution*. As $k \to \infty$, this distribution converges to normal distribution.[15]

6.8.1.1 Is S_{t_k} a Martingale?

Is the $\{S_{t_k}\}$ defined in Eq. (6.80) a martingale, with respect to the information set consisting of the increments in "past" price changes ?

[14]Addition of probabilities is permitted if the underlying events are mutually exclusive. In this particular case, different trajectories satisfy this condition by definition.

[15]This is an example of weak convergence.

Consider the expectations under the probabilities given in (6.84)

$$\begin{aligned}\mathbb{E}\left[S_{t_k} \mid S_{t_0}, \Delta S_{t_1}, \ldots, \Delta S_{t_{k-1}}\right] \\ = S_{t_{k-1}} + \left[(+1)p + (-1)(1-p)\right]\end{aligned} \tag{6.88}$$

where the second term on the right-hand side is the expectation of ΔS_{t_k}, the unknown increment given the information at time I_{t_k}. Clearly, if $p = 1/2$, this term is zero, and we have

$$\mathbb{E}\left[S_{t_k} \mid S_{t_0}, \Delta S_{t_1}, \ldots, \Delta S_{t_{k-1}}\right] = S_{t_{k-1}} \tag{6.89}$$

which means that $\{S_{t_k}\}$ will be a martingale with respect to the information set generated by past price changes *and* with respect to this particular probability distribution.

If $p \neq 1/2$, the $\{S_{t_k}\}$ will cease to be a martingale with respect to $\{I_{t_k}\}$. However, the centered process Z_{t_k}, defined by

$$Z_{t_k} = \left[S_{t_0} + (1-2p)\right] + \sum_{i=1}^{k}\left[\Delta S_{t_i} + (1-2p)\right] \tag{6.90}$$

or

$$Z_{t_k} = S_{t_k} + (1-2p)(k+1) \tag{6.91}$$

will again be a martingale with respect to I_{t_k}.[16]

6.8.2 Doob–Meyer Decomposition

Consider the case where the probability of an uptick at any time t_i is somewhat greater than the probability of a downtick for a particular asset, so that we expect a general upward trend in observed trajectories:

$$1 > p > 1/2 \tag{6.92}$$

Then, as shown earlier,

$$\mathbb{E}^P\left[S_{t_k} \mid S_{t_0}, S_{t_1}, \ldots, S_{t_{k-1}}\right] = S_{t_{k-1}} - (1-2p) \tag{6.93}$$

[16]It can be checked that the expectation of $\{Z_{t_k}\}$, conditional on past $\{Z_{t_k}\}$, will equal $\{Z_{t_{k-1}}\}$.

which means,

$$\mathbb{E}^P\left[S_{t_k}\middle|\, S_{t_0}, S_{t_1}, \ldots, S_{t_{k-1}}\right] > S_{t_{k-1}} \tag{6.94}$$

since $2p > 1$, according to (6.89). This implies that $\{S_{t_k}\}$ is a *submartingale*.

Now, as shown earlier, we can write

$$S_{t_k} = -(1-2p)(k+1) + Z_{t_k} \tag{6.95}$$

where Z_{t_k} is a martingale. Hence, we decomposed a submartingale into two components. The first term on the right-hand side is an increasing deterministic variable. The second term is a martingale that has a value of $S_{t_0} + 1 - 2p$ at time t_0. The expression in (6.92) is a simple case of Doob–Meyer decomposition.[17]

6.8.2.1 *The General Case*

The decomposition of an upward-trending submartingale into a deterministic trend and a martingale component was done for a process observed at a finite number of points during a continuous interval. Can a similar decomposition be accomplished when we work with *continuously* observed processes?

The Doob–Meyer theorem provides the answer to this question. We state the theorem without proof.

Let $\{I_t\}$ be the family of information sets discussed above.

Theorem 3. *If $X_t, 0 \le t \le \infty$ is a right-continuous submartingale with respect to the family $\{I_t\}$, and if $\mathbb{E}[X_t] < \infty$ for all t, then X_t admits the decomposition*

$$X_t = M_t + A_t \tag{6.96}$$

where M_t is a right-continuous martingale with respect to probability P, and A_t is an increasing process measurable with respect to I_t.

This theorem shows that even if continuously observed asset prices contain occasional jumps and trend upwards at the same time, then we can convert them into martingales by subtracting a process observed as of time t.

If the original continuous-time process does not display any jumps, but is continuous, then the resulting martingale will also be continuous.

6.8.2.2 *The Use of Doob Decomposition*

The fact that we can take a process that is not a martingale and convert it into one may be quite useful in pricing financial assets. In this section we consider a simple example.

We assume again that time $t \in [0, T]$ is continuous. The value of a call option C_t written on the underlying asset S_t will be given by the function

$$C_T = \max[S_T - K, 0] \tag{6.97}$$

at expiration date T.

According to this, if the underlying asset price is above the strike price K, the option will be worth as much as this spread. If the underlying asset price is below K, the option has zero value.

At an earlier time $t, t < T$, the exact value of C_T is unknown. But we can calculate a forecast of it using the information I_t available at time t,

$$\mathbb{E}^P[C_T|\, I_t] = \mathbb{E}^P[\max[S_T - K, 0]|\, I_t] \tag{6.98}$$

where the expectation is taken with respect to the distribution function that governs the price movements.

Given this forecast, one may be tempted to ask if the fair market value C_t will equal a properly discounted value of $\mathbb{E}^P[\max[S_T - K, 0]|\, I_t]$.

For example, suppose we use the (constant) risk-free interest rate r to discount $\mathbb{E}^P[\max[S_T - K, 0]|\, I_t]$, to write

$$C_t = e^{-r(T-t)}\mathbb{E}^P[\max[S_T - K, 0]|\, I_t] \tag{6.99}$$

Would this equation give the fair market value C_t of the call option?

The answer depends on whether or not $e^{-rt}C_t$ is a martingale with respect to the pair I_t, P. If it is, we have

$$\mathbb{E}^P\left[e^{-rT}C_t\middle|\, I_t\right] = e^{-rt}C_t \tag{6.100}$$

[17] This term is often used for martingales in continuous time. Here we are working with a discrete partition of a continuous-time interval.

or, after multiplying both sides of the equation by e^{-rt},

$$\mathbb{E}^P\left[e^{-r(T-t)}C_T\middle|I_t\right] = C_t \quad (6.101)$$

Then $e^{-rt}C_t$ will be a martingale.

But can we expect $e^{-rt}S_t$ to be a martingale under the true probability P?

As discussed in Chapter 2, under the assumption that investors are risk-averse, for a typical risky security we have

$$\mathbb{E}^P\left[e^{-r(T-t)}\middle|I_t\right] > S_t \quad (6.102)$$

That is,

$$e^{-rt}S_t \quad (6.103)$$

will be a submartingale.

But, according to Doob–Meyer decomposition, we can decompose the

$$e^{-rt}S_t \quad (6.104)$$

to obtain

$$e^{-rt}S_t = A_t + Z_t \quad (6.105)$$

where A_t is an increasing I_t measurable random variable, and Z_t is a martingale with respect to the information I_t.

If the function A_t can be obtained explicitly, we can use the decomposition in (6.102), along with (6.99), to obtain the fair market value of a call option at time t.

However, this method of asset pricing is rarely pursued in practice. It is more convenient and significantly easier to convert asset prices into martingales, not by subtracting their drift, but instead by changing the underlying probability distribution P.

6.9 THE FIRST STOCHASTIC INTEGRAL

We can use the results thus far to define a new martingale.

Let $H_{t_{i-1}}$ be any random variable adapted to $I_{t_{i-1}}$.[18] Let Z_t be a martingale with respect to I_t and to some probability measure P. Then the process defined by

$$M_{t_k} = M_{t_0} + \sum_{i=1}^{k} H_{t_{i-1}}\left[Z_{t_i} - Z_{t_{i-1}}\right] \quad (6.106)$$

will also be a martingale with respect to I_t.

The idea behind this representation is not difficult to describe. Z_t is a martingale and has unpredictable increments. The fact that $H_{t_{i-1}}$ is $I_{t_{i-1}}$-adapted means $H_{t_{i-1}}$ are "constant", given $I_{t_{i-1}}$. Then, increments in Z_{t_i} will be uncorrelated with $H_{t_{i-1}}$ as well. Using these observations, we can calculate

$$\mathbb{E}_{t_0}\left[M_{t_k}\right] = M_{t_0} + \mathbb{E}_{t_0} \times \left[\sum_{i=1}^{k} \mathbb{E}_{t_{i-1}}\left[H_{t_{i-1}}\left(Z_{t_i} - Z_{t_{i-1}}\right)\right]\right] \quad (6.107)$$

But increments in Z_{t_i} are unpredictable as of time t_{i-1}.[19] Also, I_t is adapted. This means we can move the operator "inside" to get

$$H_{t_{i-1}}\left(Z_{t_i} - Z_{t_{i-1}}\right) = 0$$

This implies

$$\mathbb{E}_{t_0}\left[M_{t_k}\right] = M_{t_k} \quad (6.108)$$

M_t thus has the martingale property.

It turns out that M_t defined this way is the first example of a stochastic integral. The question is whether we can obtain a similar result when $\sup_i\left[t_i - t_{i-1}\right]$ goes to zero. Using some analogy, can we obtain an expression such as

$$M_t = M_0 + \int_0^t H_u dZ_u \quad (6.109)$$

where dZ_u represents an infinitesimal stochastic increment with zero mean given the information at time t?

[18] We remind the reader that this means, given the information in $I_{t_{i-1}}$, that the value of $H_{t_{i-1}}$ will be known exactly.

[19] Remember that $\mathbb{E}_{t_0}[\mathbb{E}_{t_{i-1}}[\cdot]] = \mathbb{E}_{t_0}[\cdot]$.

The question that we will investigate in the next few chapters is whether such an integral can be defined meaningfully. For example, can the Riemann–Stieltjes approximation scheme be used to define the stochastic integral in (6.106)?

6.9.1 Application to Finance: Trading Gains

Stochastic integrals have interesting applications in financial theory. One of these applications is discussed in this section.

We consider a decision maker who invests in both a riskless and a risky security at *trading times* t_i:

$$0 = t_0 < t_1 < \cdots < t_n = T$$

Let $\alpha_{t_{i-1}}$ and $\beta_{t_{i-1}}$ be the *number* of shares of riskless and risky securities held by the investor right before time t_i trading begins. Clearly, these random variables will be $I_{t_{i-1}}$-adapted.[20] α_{t_0} and β_{t_0} are the nonrandom initial holdings. Let B_{t_i} and S_{t_i} denote the prices of the riskless and risky securities at time t_i.

Suppose we now consider trading strategies that are *self-financing*. These are strategies where time t_i investments are financed solely from the proceeds of time t_{i-1} holdings. That is, they satisfy

$$\alpha_{t_{i-1}} B_{t_i} + \beta_{t_{i-1}} S_{t_i} = \alpha_{t_i} B_{t_i} + \beta_{t_i} S_{t_i} \tag{6.110}$$

where $i = 1, 2, \ldots, n$.

According to this strategy, the investor can sell his holdings at time t_i for an amount equal to the left-hand side of the equation, and with all of these proceeds purchase $\alpha_{t_i}, \beta_{t_i}$ units of riskless and risky securities. In this sense his investment today is completely financed by his investment in the previous period.

[20]At time t_i, the investor knows his holdings of riskless and risky securities.

We can now substitute recursively for the left-hand side using Eq. (6.107) for $t_{i-1}, t_{i-2}, \ldots$, and using the definitions

$$B_{t_i} = B_{t_{i-1}} + \left[B_{t_i} - B_{t_{i-1}}\right]$$
$$S_{t_i} = S_{t_{i-1}} + \left[S_{t_i} - S_{t_{i-1}}\right]$$

We obtain

$$\alpha_{t_0} B_{t_0} + \beta_{t_0} S_{t_0} + \sum_{j=1}^{i-1} [\alpha_{tj} [B_{t_{j+1}} - B_{t_j}] + \beta_{t_j} [S_{t_{j+1}} - S_{t_j}]] = \alpha_{t_i} B_{t_i} + \beta_{t_i} S_{t_i} \tag{6.111}$$

where the right-hand side is the wealth of the decision maker *after* time *ti* trading.

A close look at the expression (6.108) indicates that the left-hand side has exactly the same setup as the *stochastic integral* discussed in the previous section. Indeed, the α_{t_j} and β_{t_j} are $I_{t_{j-1}}$-adapted, and they are multiplied by increments in securities prices.

Hence, stochastic integrals are natural models for formulating intertemporal budget constraints of investors.

6.10 MARTINGALE METHODS AND PRICING

Doob–Meyer decomposition is a Martingale Representation Theorem. These types of results at the outset seem fairly innocuous. Given any *submartingale* C_t, they say that we can decompose it into two components. One is a "known" trend, given the information at time t; the other is a martingale with respect to the same information set and the probability $\mathbb{P}$. This statement is equivalent, under some technical conditions, to the representation:

$$C_T = C_t + \int_t^T D_s ds + \int_t^T g\left(C_s\right) dM_s \tag{6.112}$$

where the D_s is known, given the information set I_s, the $g\left(\cdot\right)$ is a nonanticipative function of C_s, and

M_s is a martingale, given the information sets $\{I_s\}$ and the probability $\mathbb{P}$.[21]

In this section, we show that this theorem is an abstract version of some very important market practices and that it suggests a general methodology for martingale methods in financial modeling.

First, some motivation for what is described below.

Suppose we would like to price a derivative security whose price is denoted by C_t. At expiration, its payoff is C_T. We have seen in Chapter 2 that a properly normalized C_t can be combined with a martingale measure $\mathbb{Q}$ to yield the pricing equation:

$$\frac{C_t}{B_t} = \mathbb{E}_t^{\mathbb{Q}}\left[\frac{C_T}{B_T}\right] \tag{6.113}$$

It turns out that this equation can be obtained from (6.109). Note that in Eq. (6.110), it is as if we are applying the conditional expectation operator $\mathbb{E}_t^{\mathbb{Q}}[\cdot]$ to both sides of Eq. (6.109) after *normalizing* the C_t by B_t, and then letting

$$\mathbb{E}_t^{\mathbb{Q}}\left[\int_t^T \widetilde{D}_s ds\right] = 0 \tag{6.114}$$

$$\mathbb{E}_t^{\mathbb{Q}}\left[\int_t^T g\left(\frac{C_s}{B_s}\right) ds\right] = 0 \tag{6.115}$$

where the $\widetilde{D}$ is the trend of the *normalized* C_t, i.e., of the ratio C_t/B_t.

This suggests a way of obtaining the pricing Eq. (6.110). Given a derivative security C_t, if we can write a martingale representation for it, we can then try to find a normalization that can satisfy the conditions in (6.111) and (6.112) under the risk-neutral measure. We can use this procedure as a general way of pricing derivative securities.

In the next section we do exactly that. First, we show how a martingale representation can be obtained for a derivative security's price C_t. Then, we look at the implications of this representation and explain the notion of a self-financing portfolio.

[21] As we will see later, the nonanticipative nature of the function $g(\cdot)$ implies that $g(C_s)$ and dM_s are uncorrelated.

6.11 A PRICING METHODOLOGY

We proceed in discrete time by letting $h > 0$ represent a small, finite interval and we subdivide the period $[t, T]$ into n such intervals as in the previous section. The C_t and S_t represent the current price of a derivative security and the price of the underlying asset, respectively. The C_t is the unknown of the problem below. The T is the expiration date. At expiration, the derivative will have a market value equal to its payoff,

$$C_T = G(S_T) \tag{6.116}$$

where the function $G(\cdot)$ is known and the S_T is the (unknown) price of the underlying asset at time T.

The discrete equivalent of the martingale representation in (6.109) is then given by the following equation:

$$C_T = C_t + \sum_{i=1}^{n} D_{t_i}\Delta + \sum_{i=1}^{n} g\left(C_{t_i}\right)\Delta M_{t_i} \tag{6.117}$$

where ΔM_{t_i} means

$$\Delta M_{t_i} = M_{t_{i+1}} - M_{t_i}$$

and n is such that

$$t_0 = t < \cdots < t_n = T \tag{6.118}$$

How could this representation be of any use in determining the arbitrage-free price of the derivative security C_t?

6.11.1 A Hedge

The first step in such an endeavor is to construct a synthetic "hedge" for the security C_t.

We do this by using the standard approach utilized in Chapter 2. Let B_t be the risk-free borrowing and lending at the short-rate r, assumed to be constant. Let the S_{t_i} be the price of the underlying security observed at time t_i. Thus, the pair is known at time t_i.

Now, suppose we select the $\alpha_{t_i}, \beta_{t_i}$ as in the previous section, to form a replicating portfolio:

$$C_{t_i} = \alpha_{t_i} B_{t_i} + \beta_{t_i} S_{t_i} \tag{6.119}$$

where the $\alpha_{t_i}, \beta_{t_i}$ are the "weights" of the replicating portfolio that ensure that its value matches the C_{t_i}. Note that we know the terms on the right-hand side, given the information at time t_i. Hence, the $\{\alpha_{t_i}, \beta_{t_i}\}$ are *nonanticipative*. We can now apply the martingale representation theorem using this "hedge," i.e., the replicating portfolio.

6.11.2 Time Dynamics

We now consider changes in C_{t_i} during the period $[t, T]$. We can write trivially:

$$C_T = C_t + \sum_{i=0}^{n} \Delta C_{t_i} \tag{6.120}$$

because $\Delta C_{t_i} = C_{t_{i+1}} - C_{t_i}$. Or, using the replicating portfolio:

$$C_T = C_t + \sum_{i=1}^{n} \Delta\left[\alpha_{t_i} B_{t_i} + \beta_{t_i} S_{t_i}\right] \tag{6.121}$$

$$= C_t + \sum_{i=1}^{n} \Delta\left[\alpha_{t_i} B_{t_i}\right] + \sum_{i=1}^{n} \Delta\left[\beta_{t_i} S_{t_i}\right] \tag{6.122}$$

where the Δ represents the operation of taking first differences.

Now, recall that the "change" in a product, $u.v$, can be calculated using the "product rule":

$$d\,(u.v) = du.v + u.dv \tag{6.123}$$

Applying this to the second and third terms on the right-hand side of (6.119)[22]

$$\sum_{i=0}^{n} \Delta\left[\alpha_{t_i} B_{t_i}\right] = \sum_{i=0}^{n} \left(\Delta\alpha_{t_i}\right) B_{t_{i+1}} + \sum_{i=1}^{n} \alpha_{t_i}\left(\Delta B_{t_i}\right) \tag{6.124}$$

and

$$\sum_{i=0}^{n} \Delta\left[\alpha_{t_i} S_{t_i}\right] = \sum_{i=0}^{n} \left(\Delta\alpha_{t_i}\right) B_{t_{i+1}} + \sum_{i=1}^{n} \alpha_{t_i}\left(\Delta B_{t_i}\right) \tag{6.125}$$

where we used the notation,

$$\Delta\left[\alpha_{t_i} B_{t_i}\right] = \left[\alpha_{t_{i+1}} B_{t_{i+1}}\right] - \left[\alpha_{t_i} B_{t_i}\right]$$
$$\Delta\alpha_{t_i} = \alpha_{t_{i+1}} - \alpha_{t_i}$$
$$\Delta\beta_{t_i} = \beta_{t_{i+1}} - \beta_{t_i}$$

and

$$\Delta B_{t_i} = B_{t_{i+1}} - B_{t_i}$$
$$\Delta S_{t_i} = S_{t_{i+1}} - S_{t_i}$$

Thus (6.119) can be rewritten as:

$$C_T = C_t + \sum_{i=0}^{n} \left(\Delta\alpha_{t_i}\right) B_{t_{i+1}} + \sum_{i=1}^{n} \alpha_{t_i}\left(\Delta B_{t_i}\right) + \sum_{i=0}^{n} \left(\Delta\alpha_{t_i}\right) S_{t_{i+1}} + \sum_{i=1}^{n} \alpha_{t_i}\left(\Delta S_{t_i}\right) \tag{6.126}$$

Regrouping,

$$C_T = C_t + \sum_{i=0}^{n} \left[\left(\Delta\alpha_{t_i}\right) B_{t_{i+1}} + \left(\Delta\alpha_{t_i}\right) S_{t_{i+1}}\right] + \sum_{i=0}^{n} \left[\alpha_{t_i}\left(\Delta S_{t_i}\right) + \alpha_{t_i}\left(\Delta S_{t_i}\right)\right] \tag{6.127}$$

[22] Another way of obtaining the equations below is by simple algebra. Given

$$\Delta[\alpha_{t_i} B_{t_i}] = \alpha_{t_{i+1}} B_{t_{i+1}} - \alpha_{t_i} B_{t_i}$$

note that we can add and subtract $\alpha_{t_i} B_{t_i}$ on the right-hand side, factor out similar terms, and obtain:

$$\alpha_{t_{i+1}} B_{t_{i+1}} - \alpha_{t_i} B_{t_i} = (\alpha_{t_{i+1}} - \alpha_{t_i}) B_{t_{i+1}} + \alpha_{t_i}(B_{t_{i+1}} - B_{t_i}) = \Delta\alpha_{t_i} B_{t_{i+1}} + \alpha_{t_i} \Delta B_{t_i}$$

Now consider the terms on the right-hand side of this expression. The C_t is the unknown of the problem. We are, in fact, looking for a method to determine an arbitrage-free value for this term that satisfies the pricing Eq. (6.110). The two other terms in the brackets need to be discussed in detail.

Consider the first bracketed term. Given the information set at time t_{i+1}, every element of this bracket will be known. The $B_{t_{i+1}}, S_{t_{i+1}}$ are prices observed in the markets, and the $\Delta\alpha_{t_i}, \Delta\beta_{t_i}$ is the *rebalancing* of the replicating portfolio as described by the financial analyst. Hence, the first bracketed term has some similarities to the D_t term in the martingale representation (6.109).

The second bracketed term will be unknown, given the information set I_{t_i}, because it involves the price changes that occur *after* t_i, and hence may contain new information not contained in I_{t_i}. However, although unknown, these price changes are, in general, *predictable*. Thus we cannot expect the second term to play the role of dM_t in the martingale representation theorem. The second bracketed term will, in general, have a nonzero drift and will fail to be a martingale.

Accordingly, at this point we cannot expect to apply an expectation operator $\mathbb{E}_t^{\mathbb{P}}[\cdot]$, where $\mathbb{P}$ is real-life probability, to Eq. (6.124) and hope to end up with something like

$$C_t = \mathbb{E}_t^{\mathbb{P}}[C_T]$$

The bracketed terms in (6.124) will not, in general, vanish under such an operation. But, at this point, there are two tools available to us.

First, we can divide the $\{C_t, B_t, S_t\}$ in (6.124) by another arbitrage-free price, and write the martingale representation not for the actual prices, but instead for *normalized* prices. Such a *normalization*, if done judiciously, may ensure that any drift in the C_t process is "compensated" by the drift of this normalizing variable. This may indeed be quite convenient, given that we may want to discount the future payoff, C_T, anyway.

Second, when we say that the second bracketed term is, in general, predictable, and hence not a martingale, we say this with respect to the real-world probability. We can invoke the *Girsanov theorem* and switch probability distributions. In other words, we could work with risk-neutral probabilities.[23]

We now show how these steps can be applied to Eq. (6.124).

6.11.3 Normalization and Risk-Neutral Probability

In order to implement the steps discussed above, we first "normalize" every asset by an appropriately chosen price. In this case, a convenient normalization is to divide by the corresponding value of B_t and define

$$\widetilde{C}_t = \frac{C_t}{B_t}, \quad \widetilde{S}_t = \frac{S_t}{B_t}, \quad \widetilde{B}_t = \frac{B_t}{B_t} = 1 \qquad (6.128)$$

Notice immediately that the $\widetilde{B}_t$ is a constant and does not grow over time. We will have

$$\Delta\widetilde{B}_t = 0, \quad \text{for all } t_i \qquad (6.129)$$

The normalization by B_t has clearly eliminated the trend in this variable. But there is more.

Consider next the expected change in normalized during an infinitesimal interval dt. We can write in continuous time,

$$dB_t = rB_t dt \qquad (6.130)$$

because the yield to instantaneous investment, B_t, is the risk-free rate r. We now use this in:

$$d\widetilde{S}_t = d\frac{S_t}{B_t} = \frac{dS_t}{B_t} - \widetilde{S}_t\frac{dB_t}{B_t} \qquad (6.131)$$

$$= \frac{dS_t}{S_t}\widetilde{S}_t - \widetilde{S}_t r dt \qquad (6.132)$$

where we substituted r for dB_t/B_t.[24] Remember from Chapter 2 that under the no-arbitrage condition, and with money market normalization,

[23] Girsanov theorem will be discussed in detail in Chapters 12 and 13. The discussion here provides a motivation.

[24] Because B_t is deterministic and S_t enters linearly, there is no Ito correction term here.

the expected return from S_t will be the risk-free return r:

$$\mathbb{E}_t^{\mathbb{Q}}\left[d\widetilde{S}_t\right] = \mathbb{E}_t^{\mathbb{Q}}\left[\frac{dS_t}{S_t}\widetilde{S}_t\right] - \widetilde{S}_t r dt \quad (6.133)$$

$$= r\widetilde{S}_t dt - \widetilde{S}_t r dt = 0 \quad (6.134)$$

where the $\mathbb{Q}$ is the risk-neutral probability, obtained from state-prices as discussed in Chapter 2. Hence normalized S_t also has zero mean under $\mathbb{Q}$.

We can now use the discrete-time equivalent of this logic to eliminate the unwanted bracketed terms in (6.124). We start by writing

$$\widetilde{C}_T = \widetilde{C}_t + \sum_{i=0}^{n}\left[\left(\Delta\alpha_{t_i}\right)\widetilde{B}_{t_{i+1}} + \left(\Delta\alpha_{t_i}\right)\widetilde{S}_{t_i}\right] + \sum_{i=0}^{n}\left[\alpha_{t_i}\left(\Delta\widetilde{S}_{t_i}\right)\right] \quad (6.135)$$

with the new restriction that under the risk-neutral probability $\mathbb{Q}$,

$$\mathbb{E}_t^{\mathbb{Q}}\left[\Delta\widetilde{S}_t\right] = 0 \quad (6.136)$$

Thus, applying the operator $\mathbb{E}_t^{\mathbb{Q}}[\cdot]$ to Eq. (6.132) gives:

$$\mathbb{E}_t^{\mathbb{Q}}\left[\widetilde{C}_T\right] = \widetilde{C}_t + \mathbb{E}_t^{\mathbb{Q}}\left\{\sum_{i=0}^{n}\left[\left(\Delta\alpha_{t_i}\right)\widetilde{B}_{t_{i+1}} + \left(\Delta\alpha_{t_i}\right)\widetilde{S}_{t_i}\right]\right\} + \mathbb{E}_t^{\mathbb{Q}}\left[\sum_{i=0}^{n}\left[\alpha_{t_i}\left(\Delta\widetilde{S}_{t_i}\right)\right]\right] \quad (6.137)$$

$$= \widetilde{C}_t + \mathbb{E}_t^{\mathbb{Q}}\left\{\sum_{i=0}^{n}\left[\left(\Delta\alpha_{t_i}\right)\widetilde{B}_{t_{i+1}} + \left(\Delta\alpha_{t_i}\right)\widetilde{S}_{t_i}\right]\right\} + 0 \quad (6.138)$$

Clearly, if we can eliminate the bracketed term, we will get the desired result

$$C_t = B_t\mathbb{E}_t^{\mathbb{Q}}\left[\frac{C_T}{B_T}\right] \quad (6.139)$$

the arbitrage-free value of the unknown C_t.

So, how do we eliminate this last bracketed term in Eq. (6.136)? We do this by choosing the $\{\alpha_{t_i}, \beta_{t_i}\}$, so that

$$\mathbb{E}_t^{\mathbb{Q}}\left\{\sum_{i=0}^{n}\left[\left(\Delta\alpha_{t_i}\right)\widetilde{B}_{t_{i+1}} + \left(\Delta\alpha_{t_i}\right)\widetilde{S}_{t_i}\right]\right\} \quad (6.140)$$

that is, by making sure that the replicating portfolio is self-financing. In fact, the last equality will be obtained if we had

$$\alpha_{t_{i-1}}B_{t_i} + \beta_{t_{i-1}}S_{t_i} = \alpha_{t_i}B_{t_i} + \beta_{t_i}S_{t_i} \quad (6.141)$$

for all i. That is, the time t_{i+1} value of the portfolio chosen at time t_i is exactly sufficient to readjust the weights of the portfolio. Note that this last equation is written for the nonnormalized prices. This can be done because whatever the normalization we used, it will cancel out from both sides.

6.11.4 A Summary

We can now summarize the calculations from the point of view of asset pricing.

First, the tools. The calculations in the previous section depend basically on three important tools. The first is the martingale representation theorem. This says that, given a process, we can decompose it into a known trend and a martingale. This result, although technical in appearance, is in fact quite intuitive. Given any time series, one can in principle separate it into a trend and deviations around this trend. Market participants who work with real-world data and who estimate such trend components routinely are, in fact, using a crude form of martingale representation theorem.

The second tool that we used was the normalization. Martingale representation theorem is applied to the normalized price, instead of the observed price. This conveniently eliminates some unwanted terms in the martingale representation theorem.

The third tool was the measure change. By calculating expectations using the risk-neutral probability, we made sure that the remaining unwanted terms in the martingale representation vanished. In fact, utilization of the risk-neutral measure had the effect of changing the *expected trend* of the S_t process, and the normalization made sure that this new trend was eliminated by the growth in B_t. As a result of all this, the normalized C_t ended up having no trend at all and became a martingale. This gives the pricing Eq. (6.124), if one uses self-financing replicating portfolios.

6.12 CONCLUSIONS

This chapter dealt with martingale tools. Martingales were introduced as processes with no recognizable time trends. We discussed several examples that will be useful in later chapters.

This chapter also introduced ways of obtaining martingales from processes that have positive (or negative) time trends.

We close this chapter with a discussion that illustrates why theoretical concepts introduced here are relevant to a practitioner.

Let S_t be the price of an asset observed by a trader at time t. During infinitesimal periods, the trader receives new unpredictable information on S_t. These are denoted by

$$dS_t = \sigma_t dW_t$$

where σ_t is volatility and dW_t is an increment of Brownian motion. Note that volatility has a time subscript, and consequently changes over time. Also note that dS_t has no predictable drift component.

Over a longer period, such unpredictable information will accumulate. After an interval T, the asset price becomes

$$S_{t+T} = S_t + \int_t^{t+T} \sigma_u dW_u$$

This equation has the same form as (6.106). If every incremental news is unpredictable, then the sum of incremental news should also be unpredictable (as of time t). But this means that S_t should be a martingale, and we must have

$$\mathbb{E}_t\left[\int_t^{t+T} \sigma_u dW_u\right] = 0$$

This is an important property of stochastic integrals. But it is also a restriction imposed on financial market participants by the way information flows in markets. Martingale methods are central in discussing such equalities. They are also essential for practitioners.

6.13 REFERENCES

A reader willing to learn more about martingale arithmetic should consult the introductory book by Williams (1991). The book is very readable and provides details on the mechanics of all major martingale results using simple models. Revuz and Yor (1994) is an excellent advanced text on martingales. The survey by Shiryayev (1984) is an intermediate-level treatment that contains most of the recent results. For trading gains and stochastic integrals, the reader may consult Cox and Huang (1989). Dellacherie and Meyer (1980) is a comprehensive source on martingales. Musiela and Rutkowski (1997) is an excellent and comprehensive source on martingale methods in asset pricing.

6.14 EXERCISES

1. Let Y be a random variable with

$$\mathbb{E}[Y] < \infty$$

(a) Show that the M_t defined by

$$M_t = \mathbb{E}[Y|I_t]$$

is a martingale.

(b) Does this mean that every conditional expectation is a martingale, given the increasing sequence of information sets $\{I_0 \subseteq \cdots \subseteq I_t \subseteq I_{t+1} \subseteq \cdots\}$.

2. Consider the random variable:

$$X_n = \sum_{i=1}^{n} B_i$$

where each B_i is obtained as a result of the toss of a fair coin:

$$B_i = \begin{cases} +1 & \text{Head} \\ -1 & \text{Tail} \end{cases}$$

We let $n = 4$ and consider X_4.

(a) Calculate the $\mathbb{E}[X_4 | I_1], \mathbb{E}[X_4 | I_2], \mathbb{E}[X_4 | I_4]$.

(b) Let

$$Z_i = \mathbb{E}[X_4 | I_i]$$

Is $Z_i, i = 1, \ldots, 4$ a martingale?

(c) Now define:

$$V_i = B_i + \sqrt{i}$$

and

$$\widetilde{X}_n = \sum_{i=1}^{n} V_i$$

Is V_i a martingale?

(d) Can you convert V_i into a martingale by an appropriate transformation?

(e) Can you convert V_i into a martingale by changing the probabilities associated with a coin toss?

3. Let W_t be a Wiener process and t denote the time. Are the following stochastic processes martingales?

(a) $X_t = 2W_t + t$

(b) $X_t = W_t^2$

(c) $X_t = W_t^2 - 2\int_0^t sW_s ds$

4. You are given the representation:

$$M_T(X_t) = M_0(X_0) + \int_0^T g(t, X_t)\, dW_t$$

where the equality holds, given the sequence of information sets $\{I_t\}$. The underlying process X_t is known to follow the SDE:

$$dX_t = \mu dt + \sigma dW_t$$

Determine the $g(\cdot)$ in the above representation for the case where $M(\cdot)$ is given by:

(a) $M_T(X_T) = W_T$.

(b) $M_T(X_T) = W_T^2$.

(c) $M_T(X_T) = e^{W_T}$.

5. Given the representation:

$$M_T(X_t) = M_0(X_0) + \int_0^T g(t, X_t)\, dW_t$$

can you determine the $g(\cdot)$, if the $M_T(X_T)$ is the payoff of a plain vanilla European call option at expiration? That is, if $M_T(X_T)$ is given by:

$$M_T(X_T) = \max[X_T - K, 0]$$

where $0 < K < \infty$ is the strike price. Where is the difficulty?

6. Let X be a process with $\mathbb{E}(X_t) = 0$ for all $t \geq 0$ and independent increments (that is, $X_t - X_s$ is independent from $\mathcal{I}_s$ for any s, t, with $0 \leq s < t < \infty$).

(a) Show that X is a martingale.

(b) Suppose X has independent increments but that $t \to \mathbb{E}(X_t)$ is not constant. Is X again a martingale? Justify your answer.

7. Consider the process $X_t = W_t^3$. Compute the expectation numerically by simulating values of W_t in order to check if the process is a martingale.

CHAPTER

7

Differentiation in Stochastic Environments

OUTLINE

7.1 INTRODUCTION

Differentiation in deterministic environments was reviewed in Chapter 3. The derivative of a function $f(x)$ with respect to x gave us information about the rate at which $f(\cdot)$ would respond to a small change in x, denoted by dx. This response was calculated as

$$df = f_x dx \tag{7.1}$$

where f_x is the derivative of $f(x)$ with respect to x.

We need similar concepts in stochastic environments as well. For example, given the variations in the price of an underlying asset S_t, how would the price of, say, a call option written on S_t react? In deterministic environments one would use "standard" rules of differentiation to investigate such questions. But, in pricing financial assets, we deal with *stochastic* variables, and the notion of risk plays a central role. Can similar formulas be used when the underlying variables are continuous-time stochastic processes?

The notion of differentiation is closely linked to models of *ordinary differential equations* (ODE), where the effect of a change in a variable on another set of variables can be modeled explicitly. In fact, (vector) differential equations are formal ways of modeling the dynamics of deterministic

Financial Derivatives, Third Edition.
http://dx.doi.org/10.1016/B978-0-12-384682-2.00007-4

processes, and the existence of the derivative is necessary for doing this.

Can differential equations be used in modeling the dynamics of asset prices as well? The first difficulty in doing this results from the randomness of asset prices. The way heat is transferred in a metal rod may be approximated reasonably well by a *deterministic* model. But, in the case of pricing *derivative* assets, the randomness of the underlying instrument is essential. After all, it is the desire to eliminate or take risk that leads to the existence of derivative assets. In deterministic environments, where everything can be fully predicted, there will be no risk. Consequently, there will be no need for financial derivative products. But if randomness is essential, how would one define differentiation in a stochastic environment?

Can one simply attach random error terms to ordinary differential equations and use them in pricing financial derivatives? Or are there new difficulties in defining stochastic differential equations (SDE) as well?

This chapter treats differentiation in stochastic environments using the stochastic differential equations as the underlying model. We first construct the SDE from scratch, and then show the difficulties of importing the differentiation formulas directly from deterministic calculus.

More precisely, we first show under what conditions the behavior of a continuous-time process, S_t, can be approximated using the dynamics described by the *stochastic differential equation*

$$dS_t = a(S_t, t)\, dt + b(S_t, t)\, dW_t \tag{7.2}$$

where dW_t is an innovation term representing unpredictable events that occur during the infinitesimal interval dt. The $a(S_t, t)$ and the $b(S_t, t)$ are the *drift* and the *diffusion* coefficients, respectively. They are I_t-adapted.

Second, we study the properties of the innovation term dW_t, which drives the system and is the source of the underlying randomness. We show that W_t is a very irregular process and that its derivative does not exist in the sense of deterministic calculus. Hence, increments such as dS_t or dW_t have to be justified by some other means.

Constructing the SDE from scratch has a side benefit. This is one way we can get familiar with methods of continuous-time stochastic calculus. It may provide a bridge between discrete-time and continuous-time calculations, and several misconceptions may be eliminated this way.

7.2 MOTIVATION

This section gives a heuristic comparison of differentiation in deterministic and stochastic environments.

Let S_t be the price of a security, and let $F(S_t, t)$ denote the price of a derivative instrument written on S_t. A stockbroker will be interested in knowing dS_t, the next instant's incremental change in the security price. On the other hand, a derivatives desk needs dF_t, the incremental change in the price of the derivative instrument written on S_t. How can one calculate the dF_t, departing from some estimate of dS_t?

What is of interest here is not how the underlying instrument changes, but instead, how the financial derivative *responds* to change in the price of the underlying asset. In other words, a "chain rule" needs to be utilized. If the rules of standard calculus are applicable, a market participant can use the formula

$$dF_t = \frac{\partial F}{\partial S} dS_t \tag{7.3}$$

or, in the partial derivative notation,

$$dF_t = F_s dS_t \tag{7.4}$$

But are the rules of deterministic calculus really applicable? Can this chain rule be used in stochastic environments as well?

Below we show that the rules of differentiation are indeed different in stochastic environments. We proceed with the discussion by utilizing a function $f(x)$ of x.

As discussed in Chapter 3, standard differentiation is the limiting operation defined as

$$\lim_{h \to 0} \frac{f(x+h) - f(x)}{h} = f_x \tag{7.5}$$

where the limit satisfies

$$f_x < \infty \tag{7.6}$$

Here, $f(x+h)-f(x)$ represents the change in the function as x changes by h. Hence, if x represents time, then the derivative is the rate at which $f(x)$ is changing during an infinitesimal interval.[1] In this case, time is a deterministic variable and one can use "standard" calculus.

But what if the x in $f(x)$ is a random variable moving along a continuous time axis? Can one define the derivative in a similar fashion and use standard rules?

The answer to this question is, in general, no. We begin with a heuristic discussion of this important issue.

Suppose $f(x)$ is a function of a random process x.[2] Now suppose we want to expand $f(x)$ around a known value of x, say x_0.[3] A Taylor series expansion will yield

$$\begin{aligned} f(x) = f(x_0) + f_x(x_0)(x - x_0) \\ + \frac{1}{2} f_{xx}(x_0)(x - x_0)^2 \\ + \frac{1}{3!} f_{xxx}(x_0)(x - x_0)^3 + R(x, x_0) \end{aligned} \tag{7.7}$$

where $R(x, x_0)$ represents all the remaining terms of the Taylor series expansion. Note that this remainder is made of three types of terms: partial derivatives of $f(x)$ of order higher than 3, factorials of order higher than 3, and powers of $(x - x_0)$ higher than 3.

[1]By dividing $f(x+h) - f(x)$ by h, we obtain a ratio. This ratio tells us how much $f(x)$ changes per h. Hence, the derivative is a rate of change.

[2]For the sake of notational simplicity, we omit the time subscript on x.

[3]The interested reader is referred back to Chapter 3 for a review of Taylor series expansions.

Now switch to a Taylor series approximation and consider the terms on the right-hand side other than $R(x, x_0)$.

The $f(x)$ can be rewritten as $f(x_0 + \Delta x)$, if we let

$$\Delta x = x - x_0 \tag{7.8}$$

Then the Taylor series approximation will have the form[4]

$$\begin{aligned} f(x_0 + \Delta x) - f(x_0) \approx f_x(\Delta x) + \frac{1}{2} f_{xx}(\Delta x)^2 \\ + \frac{1}{3}(\Delta x)^3 \end{aligned} \tag{7.9}$$

On the right-hand side of this representation, Δx represents a "small" change in the random variable x. Note that although this change is considered to be small, we do not want it to be so small that it becomes negligible. After all, our purpose is to evaluate the effect of a change in x on the $f(x)$, and this cannot be done by considering negligible changes in x. Hence, in a potential approximation of the right-hand side, we would like to *keep* the term $\frac{1}{2} f_{xx}(\Delta x)^2$.

Consider the second term. If the variable x were deterministic, one could have said that the term $(\Delta x)^2$ is small. This could have been justified by keeping the size of Δx nonnegligible, yet small enough that its square $(\Delta x)^2$ is negligible. In fact, if Δx was small, the square of it would be even smaller and at some point would become negligible. However, in the present case, x is a random variable. So, changes in x will also be random. Suppose these changes have zero mean. Then a random variable is random, because it has a positive variance:

$$\mathbb{E}[\Delta x]^2 > 0 \tag{7.10}$$

But read literally, this equality means that, "on the average," the size of $(\Delta x)^2$ is nonzero. In other words, as soon as x becomes a random variable, treating $(\Delta x)^2$ as if it were zero will be equivalent

[4]In the following, for notational simplicity, we omit the arguments of $f_x(x_0)$, $f_{xx}(x_0)$, $f_{xxx}(x_0)$.

to equating its variance to zero. This amounts to approximating the random variable x by a non-random quantity and will defeat our purpose. After all, we are trying to find the effect of a random change in x on $f(x)$.

Hence, as long as x is random, the right-hand side of the Taylor series approximation must keep the second-order term.

On the other hand, note that while keeping the first- and second-order terms in Δx on the right-hand side as required, one can still make a reasonable argument to drop the term that contains the third- and higher-order powers of Δx. This would not cause any inconsistency if higher-order moments are negligible.[5]

As a result, one candidate for a Taylor-style approximation can be written as

$$f(x_0 + \Delta x) - f(x_0) \approx f_x(\Delta x) + \frac{1}{2} f_{xx} \mathbb{E}\left[(\Delta x)^2\right] \tag{7.11}$$

where the $(\Delta x)^2$ is replaced with its expectation. This is equivalent to replacing the term with its "average" value as a method of approximation. In the second part of this book, we introduce tools that take exactly this direction.

A second possibility is to use, instead of $\mathbb{E}\left[(\Delta x)^2\right]$, some appropriate limit of the random variable $(\Delta x)^2$ as the time interval under consideration goes to zero. Such approximations were discussed in Chapter 4. It turns out that under some conditions, these two procedures would result in the same expression. In fact, if h represents the time period during which the change Δx is observed, and if h is "small," under some conditions $\sigma^2 h$ may be close enough to $(\Delta x)^2$ in the mean square sense.

[5]Some readers may remember the discussion involving variations of continuous-time martingales in Chapter 6. There, we showed that for continuous square integrable martingales, the first variation was infinite and the quadratic variation converged to a meaningful random variable, while the higher-order variations all vanished. Hence, if the x is a continuous square integrable martingale, the higher-order terms in Δx can be set equal to zero in some approximate sense.

Thus, we have two possible approximating equations, depending on whether x is random or not.

- If x is random, we can write

$$f(x_0 + \Delta x) - f(x_0) \approx f_x(\Delta x) + \frac{1}{2} f_{xx} \mathbb{E}\left[(\Delta x)^2\right] \tag{7.12}$$

or

$$f(x_0 + \Delta x) - f(x_0) \approx f_x(\Delta x) + \frac{1}{2} f_{xx} \mathbb{E}\left[x^*\right] \tag{7.13}$$

where x^* is the mean square limit of $(\Delta x)^2$.

- Once x becomes deterministic, we can assume that $(\Delta x)^2$ is negligible for small Δx and use

$$f(x_0 + \Delta x) - f(x_0) \approx f_x(\Delta x) \tag{7.14}$$

One result of all this is the way differentiation is handled in deterministic and stochastic environments.

For example, in the case of Eq. (7.13), we can try to divide both sides by Δx and obtain the approximation

$$\frac{f(x_0 + \Delta x) - f(x_0)}{\Delta x} \approx f_x \tag{7.15}$$

But with stochastic Δx, it is not clear whether we can ignore the third term, let $\Delta x \to 0$ in (7.8),

$$\lim_{\Delta x \to 0} \frac{f(x + \Delta x) - f(x)}{\Delta x} \approx f_x + \frac{1}{2} f_{xx} \frac{(\Delta x)^2}{\Delta x}$$

and define a derivative. This is discussed next.

7.3 A FRAMEWORK FOR DISCUSSING DIFFERENTIATION

The concept of differentiation deals with incremental changes in infinitesimal intervals. In applications to financial markets, changes in asset prices over incremental *time* periods are of interest. In addition, these changes are assumed

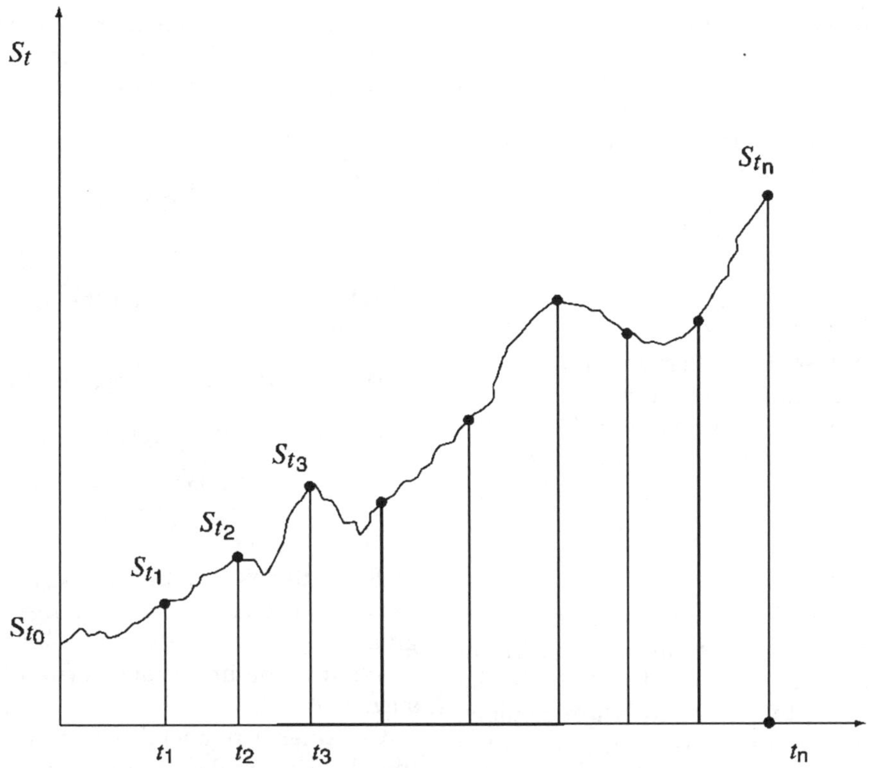

FIGURE 7.1 Discrete time construction of a path based on quantities observed on finite intervals.

to be random. Thus, in stochastic calculus, the concept of derivative has to use some type of probabilistic convergence.[6]

The natural framework to utilize for discussing differentiation is the stochastic differential equation (SDE):

$$d(S_t, t) = a(S_t, t)\,dt + b(S_t, t)\,dW_t \tag{7.16}$$

In order to understand the way differentiation can proceed in stochastic environments, the SDE will be "constructed" from scratch. The construction will proceed from discrete time to continuous time.

We will consider a time interval $t \in [0, T]$.

Consider Figure 7.1. The x axis, $[0, T]$, is partitioned into n intervals of equal length h. In terms of the notation used in previous chapters, we consider intervals given by the partitions

$$0 = t_0 < t_1 < \cdots < t_n = T \tag{7.17}$$

One major difference in this chapter is that we have, for all k,

$$t_k - t_{k-1} = h \tag{7.18}$$

which means that

$$t_k = kh \tag{7.19}$$

[6]Remember that in probabilistic convergence we are interested in finding a random variable X^*, to which a sequence or family of random variables X_n converges. For "large" n, the limiting random variable X^* can then be used as an approximation for X_n, since often the limiting variable would be easier to handle than X_n itself.

Thus, we have the relation

$$n = \frac{T}{h} \tag{7.20}$$

We define the following quantities observed during these finite intervals:

$$S_k = S(kh) \tag{7.21}$$

and

$$\Delta S_k = S(kh) - S((k-1)h) \tag{7.22}$$

The latter represents the change in the security price $S(t)$ during a finite interval h.

Now pick a particular interval k. As long as the corresponding expectations exist, we can always define a random variable ΔW_k in the following fashion:

$$\Delta W_k = [S_k - S_{k-1}] - \mathbb{E}_{k-1}[S_k - S_{k-1}] \tag{7.23}$$

Here, the symbol $\mathbb{E}_{k-1}[\cdot]$ represents the expectation conditional on information available at the end of interval $k-1$. The ΔW_k is the part in $[S_k - S_{k-1}]$ that is totally unpredictable, given the information available at the end of the $(k-1)$th interval. The first term on the right-hand side represents actual change in the asset price $S(t)$ during the kth interval. The second term is the change that a market participant would have predicted, given the information set I_{k-1}.[7] We call unpredictable components of new information "innovations."

Note the following properties of the innovation terms.

- ΔW_k is unknown at the end of the interval $(k-1)$. It is observed at the end of interval k. In the terminology of measure theory, ΔW_k is said to be measurable with respect to I_k. That is, given the set I_k, one can tell the exact value of ΔW_k.
- Values of ΔWk are unpredictable, given the information set of time $k-1$:

$$\mathbb{E}[\Delta W_k] = 0, \quad \text{for all } k \tag{7.24}$$

[7]If the information set is completely uninformative about the future movements in $S(t)$, then this prediction will be zero. Under these conditions, $[S_k - S_{k-1}]$ will itself be the unpredictable component.

- ΔW_k represents changes in a martingale process and is called a martingale difference. The accumulated error process W_k will be given by

$$W_k = \Delta W_1 + \cdots + \Delta W_k \tag{7.25}$$

$$= \sum_{i=1}^{k} \Delta W_i \tag{7.26}$$

where we assume that the initial point W_0 is zero.

We can show that W_k is a martingale:

$$\mathbb{E}_{k-1}[W_k] = \mathbb{E}_{k-1}[\Delta W_1 + \cdots + \Delta W_k] \tag{7.27}$$

$$= \Delta W_1 + \cdots \Delta W_{k-1} + \mathbb{E}_{k-1}[\Delta W_k] = W_{k-1} \tag{7.28}$$

The latter is true because $\mathbb{E}_{k-1}[\Delta W_k]$ equals zero and the $\Delta W_i, i = 1, \ldots, k-1$ are known, given I_{k-1}.

What is the importance of random variables such as ΔW_k?

Consider a financial market participant. For this decision maker, the important information contained in asset prices is indeed ΔW_k. These unpredictable "news" occur continuously and can be observed "live" on all major networks such as Reuters or Bloomberg. Hence, "live" movements in asset prices will be dominated by ΔW_k. This implies that to discuss differentiation in stochastic environments, one needs to study the properties of ΔW_k. In particular, we intend to show that under some fairly acceptable assumptions, ΔW_k^2 and its infinitesimal equivalent dW_t^2 cannot be considered as "negligible" in Taylor-style approximations.

7.4 THE "SIZE" OF INCREMENTAL ERRORS

The innovation term ΔW_k represents an unpredictable change. $(\Delta W_k)^2$ is its square. In deterministic environments, the concept of differentiation deals with terms such as ΔW_k, and squared changes are considered as negligible.

Indeed, in deterministic calculus, terms such as $(\Delta W_k)^2$ do not show up during the differentiation process.[8] On the other hand, in stochastic calculus, one in general has to take into account the variation in the second-order terms. This section deals with a formal approximation of these terms.

There are two ways of doing this. One is the method used in courses on stochastic processes. The second is the one discussed in Merton (1990). We use Merton's approach because it permits a better understanding of the economics behind the assumptions that will be made along the way. Merton's approach is to study the characteristics of the information flow in financial markets and to try to model this information flow in some precise way.

We first need to define some notation.

Let the (unconditional) variance of ΔW_k be denoted by V_k:

$$\mathbb{V}^k = \mathbb{E}_0\left[\Delta W_k^2\right] \tag{7.29}$$

The variance of cumulative errors is defined as:

$$\mathbb{V} = \mathbb{E}_0\left[\sum_{k=1}^{n} \Delta W_k\right]^2 = \sum_{k=1}^{n} \mathbb{V}^k \tag{7.30}$$

where the property that ΔW_k are uncorrelated across k is used and the expectations of cross product terms are set equal to zero.

We now introduce some assumptions, following Merton (1990).

Assumption 1.

$$\mathbb{V} > A_1 > 0 \tag{7.31}$$

where A_1 is independent of n.

This assumption imposes a lower bound on the volatility of security prices. It says that when the period $[0, T]$ is divided into finer and finer subintervals,[9]

$$n \to \infty \tag{7.32}$$

and the variance of cumulative errors, $\mathbb{V}$, will be positive. That is, more and more frequent observations of securities prices will not eliminate *all* the "risk." Clearly, most financial market participants will accept such an assumption. Uncertainty of asset prices never vanishes, even when one observes the markets during finer and finer time intervals.

Assumption 2.

$$\mathbb{V} < A_2 < \infty \tag{7.33}$$

where A_2 is independent of n.

This assumption imposes an upper bound on the variance of cumulative errors and makes the *volatility* bounded from above. As the time axis is chopped into smaller and smaller intervals, more frequent trading is allowed. Such trading does not bring unbounded instability to the system. A large majority of market participants will agree with this assumption as well. After all, allowing for more frequent trading and having access to online screens does not lead to infinite volatility.

For the third assumption, define

$$\mathbb{V}_{\max} = \max_k\left[\mathbb{V}^k, k = 1, \ldots, n\right] \tag{7.34}$$

That is, $\mathbb{V}_{\max}$ is the variance of the asset price during the most volatile subinterval. We now have

Assumption 3.

$$\frac{\mathbb{V}^k}{\mathbb{V}_{\max}} > A_3, \quad 0 < A_3 < 1 \tag{7.35}$$

with A_3 independent of n.

According to this assumption, uncertainty of financial markets is not *concentrated* in some special periods. Whenever markets are open, there exists at *least* some volatility. This assumption rules out lottery-like uncertainty in financial markets.

[8] They are confined to higher-order derivatives.

[9] Remember that the subintervals have the same length h.

Now we are ready to discuss a very important property of $(\Delta W_k)^2$.

The following proposition is at the center of stochastic calculus.

Proposition 1. *Under assumptions 1, 2, and 3, the variance of ΔW_k is proportional to h,*

$$\mathbb{E}[\Delta W_k]^2 = \sigma_k^2 h \tag{7.36}$$

where σ_k is a finite constant that does not depend on h. It may depend on the information at time $k-1$.

According to this proposition, asset prices become less volatile as h gets smaller.

Since this is a central result, we provide a sketch of the proof.

Proof. Use assumption 3:

$$\mathbb{V}^k > A_3 \mathbb{V}_{\max} \tag{7.37}$$

Sum both sides over all intervals:

$$\sum_{k=1}^{n} \mathbb{V}^k > n A_3 \mathbb{V}_{\max} \tag{7.38}$$

Assumption 2 says that the left-hand side of this is bounded from above:

$$A_2 > \sum_{k=1}^{n} \mathbb{V}^k > n A_3 \mathbb{V}_{\max} \tag{7.39}$$

Now divide both sides by nA_3:

$$\frac{1}{n}\frac{A_2}{A_3} > \mathbb{V}_{\max} \tag{7.40}$$

Note that $n = \frac{T}{h}$. Then,

$$\frac{1}{n}\frac{A_2}{A_3} > \mathbb{V}_{\max} > \mathbb{V}^k \tag{7.41}$$

$$\frac{h}{T}\frac{A_2}{A_3} > \mathbb{V}^k \tag{7.42}$$

This gives an upper bound on V_k that depends only on h. We now obtain a lower bound that also depends only on h. We know that

$$\sum_{k=1}^{n} \mathbb{V}^k > A_1 \tag{7.43}$$

is true. Then,

$$n\mathbb{V}_{\max} > \sum_{k=1}^{n} \mathbb{V}^k > A_1 \tag{7.44}$$

Divide (7.44) by n:

$$\mathbb{V}_{\max} > \frac{A_1}{n} \tag{7.45}$$

Then,

$$\mathbb{V}_{\max} > \frac{A_1}{T} h \tag{7.46}$$

Use assumption 3:

$$\mathbb{V}^k > A_3 \mathbb{V}_{\max} > \frac{A_3 A_1}{T} h \tag{7.47}$$

This means that

$$\mathbb{V}^k > \frac{A_3 A_1}{T} h \tag{7.48}$$

Therefore,

$$\frac{h}{T}\frac{A_2}{A_3} > \mathbb{V}^k > \frac{A_3 A_1}{T} h \tag{7.49}$$

Clearly the variance term $\mathbb{V}^k$ has upper and lower bounds that are proportional to h, regardless of what n is. This means that we should be able to find a constant σ_k depending on k, such that $\mathbb{V}^k$ is proportional to h, and ignoring the (smaller) higher-order terms in h, write:

$$\mathbb{V}^k = \mathbb{E}[\Delta W_k]^2 = \sigma_k^2 h \tag{7.50}$$

□

7.5 ONE IMPLICATION

This proposition has several implications. An immediate one is the following. First, remember that if the corresponding expectations exist, one can always write

$$S_k - S_{k-1} = \mathbb{E}_{k-1}\left[S_k - S_{k-1}\right] + \sigma_k \Delta W_k \tag{7.51}$$

where ΔW_k now has variance h.[10] After dividing both sides by h:

$$\frac{S_k - S_{k-1}}{h} = \frac{\mathbb{E}_{k-1}\left[S_k - S_{k-1}\right]}{h} + \frac{\sigma_k \Delta W_k}{h} \quad (7.52)$$

But, according to the proposition,

$$\mathbb{E}\left[\Delta W_k\right]^2 = h \quad (7.53)$$

Suppose we use this to justify the approximation:

$$\Delta W_k^2 \approx h \quad (7.54)$$

(In Chapter 9 we show that this approximation is valid in the sense of mean square convergence.)

In Chapter 3, when we defined the standard notion of derivative, we let h go to zero. Suppose we do the same here and *pretend* we can take the "limit" of the random variable:

$$\lim_{h \to 0} \frac{W_{(k-1)h+h} - W_{(k-1)h}}{h} \quad (7.55)$$

Then, this could be interpreted as a time derivative of W_t. The approximation in (7.54) indicates that this derivative may not be well defined:

$$\lim_{h \to 0} \frac{\left|W_{(k-1)h+h} - W_{(k-1)h}\right|}{h} \to \infty$$

Figure 7.2 shows this graphically. We plot the function $f(h)$:

$$\frac{h^{1/2}}{h}$$

Clearly, as h gets smaller $f(h)$ goes to infinity. A well-defined limit does not exist.

Of course, the argument presented here is heuristic. The limiting operation was applied to random variables rather than deterministic functions, and it is not clear how one can formalize this. But the argument is still quite instructive because it shows that the fundamental characteristic of unpredictable "news" in infinitesimal intervals, namely, that

$$\mathbb{E}\left[\sigma_k \Delta W_k\right]^2 = \sigma_k^2 h$$

may lead to insurmountable difficulties in defining a stochastic equivalent of the time derivative.

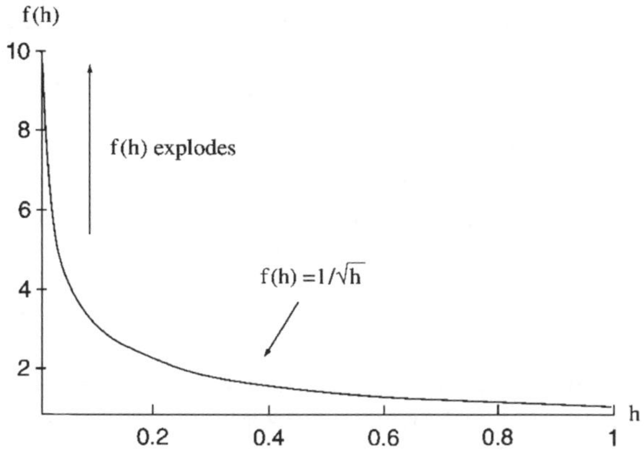

FIGURE 7.2

7.6 PUTTING THE RESULTS TOGETHER

Up to this point we have accomplished two things. First, we saw that one can take any stochastic process S_t and write its variation during some finite interval h as

$$S_k - S_{k-1} = \mathbb{E}_{k-1}\left[S_k - S_{k-1}\right] + \sigma_k \Delta W_k \quad (7.56)$$

where the term ΔW_k is unpredictable, given the information at the beginning of the time interval.[11]

Second, we showed that if h is "small," the unpredictable innovation term has a variance

[10] In this equation, the parameter σ_k is explicitly made into a coefficient of the ΔW_k term. This is a trivial transformation, because the term $\sigma_k \Delta W_k$ will now have a variance equal to $\sigma^2 h$.

[11] Assuming that the corresponding expectations exist.

that is proportional to the length of the time interval, h:

$$\mathbb{V}[\Delta W_k] = h \tag{7.57}$$

In order to obtain a stochastic difference equation defined over finite intervals, we need a third and final step. We need to approximate the first term on the right-hand side of (7.56),

$$\mathbb{E}_{k-1}\left(S_k - S_{k-1}\right) \tag{7.58}$$

This term is a conditional expectation, or a forecast of a change in asset prices. The magnitude of this change depends on the latest information set and on the length of the time interval one is considering. Hence, $\mathbb{E}_{k-1}\left(S_k - S_{k-1}\right)$ can be written as

$$\mathbb{E}_{k-1}\left(S_k - S_{k-1}\right) = A\left(I_{k-1}, h\right) \tag{7.59}$$

where $A(\cdot)$ represents some function. Viewed this way, it is clear that if $A(\cdot)$ is a smooth function of h, it will have a Taylor series expansion around $h = 0$,

$$A\left(I_{k-1}, h\right) = A\left(I_{k-1}, 0\right) + a\left(I_{k-1}\right) h + R\left(I_{k-1}, h\right) \tag{7.60}$$

Here, $a\left(I_{k-1}\right)$ is the first derivative of $A\left(I_{k-1}, h\right)$ with respect to h evaluated at $h = 0$. The $A(Ik - 1, h)$ is the remainder of the Taylor series expansion.[12]

Now, if $h = 0$, time will not pass and the predicted change in asset prices will be zero. In other words,

$$A\left(I_{k-1}, 0\right) = 0 \tag{7.61}$$

Also, the convention in the literature dealing with ordinary stochastic differential equations is that any deterministic terms having powers of h greater than one are small enough to be ignored.[13]

[12]Given I_{k-1}, we are dealing with nonrandom quantities, and the derivatives in the Taylor series expansion can be taken in a standard fashion.

[13]Since h^2 is a deterministic function, this is consistent with the standard calculus, which ignores all second-order terms in differentiation.

Thus, as in standard calculus, we can let

$$R\left(I_{k-1}, 0\right) \approx 0 \tag{7.62}$$

and obtain the first-order Taylor series approximation:

$$\mathbb{E}_{k-1}\left(S_k - S_{k-1}\right) \approx a\left(I_{k-1}, kh\right) h \tag{7.63}$$

Utilizing these results together, we can rewrite (7.56) as a stochastic difference equation:[14]

$$S_{kh} - S_{(k-1)h} \approx a\left(I_{k-1}, kh\right) h + \sigma_k \left[W_{kh} - W_{(k-1)h}\right] \tag{7.64}$$

In later chapters, we let $h \to 0$ and obtain the infinitesimal version of (7.56), which is the stochastic differential equation (SDE):

$$dS_t = a\left(I_t, t\right) dt + \sigma_t dW_t \tag{7.65}$$

This stochastic differential equation is said to have a *drift* $a(I_t, t)$ and a *diffusion* σ_t component.

7.6.1 Stochastic Differentials

At several points in this chapter we had to discuss limits of random increments. The need to obtain formal definitions for incremental changes such as dS_t, dW_t is evident.

How can these terms be made more explicit?

It turns out that to do this we need to define the fundamental concept of the Ito integral. Only with the Ito integral can we formalize the notion of *stochastic differentials* such as dS_t, dW_t, and hence give a solid interpretation of the tools of stochastic differential equations. This, however, has to wait until Chapter 9.

7.7 CONCLUSION

Differentiation in standard calculus cannot be extended in a straightforward fashion to

[14]Here, we are reintroducing the h in the notation for S_k and W_k. This shows the dependence of these terms on h explicitly.

stochastic derivatives, because in infinitesimal intervals the variance of random processes does not equal zero. Further, when the flow of new information obeys some fairly mild assumptions, continuous-time random processes become very erratic and time derivatives may not exist. In small intervals, ΔW_k dominates h. As the latter becomes smaller, the ratio of ΔWk to h is likely to get larger in absolute value. A well-defined limit cannot be found.

On the other hand, the difficulty of defining the differentials notwithstanding, we needed few assumptions to construct a SDE. In this sense, a stochastic differential equation is a fairly general representation that can be written down for a large class of stochastic processes. It is basically constructed by decomposing the change in a stochastic process into both a predictable part and an unpredictable part, and then making some assumptions about the smoothness of the predictable part.

7.8 REFERENCES

The proof that, under the three assumptions, unpredictable errors will have a variance proportional to h, is from Merton (1990). The chapter in Merton (1990) on the mathematics of continuous-time finance could at this point be useful to the reader.

7.9 EXERCISES

1. We consider the random process S_t, which plays a fundamental role in Black–Scholes analysis:

$$S_t = S_0 e^{\mu t + \sigma W_t}$$

where W_t is a Wiener process with $W_0 = 0, \mu$ is a "trend" factor, and

$$(W_t - W_s) \sim \mathcal{N}(0, (t-s))$$

which says that the increments in W_t have zero mean and a variance equal to $t - s$. Thus, at t the variance is equal to the time that elapsed since W_s is observed. We also know that these Wiener increments are independent over time. According to this, S_t can be regarded as a random variable with log-normal distribution. We would like to work with the possible trajectories followed by this process. Let $\mu = 0.01, \sigma = 0.15$, and $t = 1$. Subdivide the interval $[0, 1]$ into four subintervals and select four numbers randomly from:

$$x \sim \mathcal{N}(0, 0.25)$$

(a) Construct the W_t and S_t over the $[0, 1]$ using these random numbers. Plot the W_t and S_t. (You will obtain piecewise linear trajectories that will approximate the true trajectories.)

(b) Repeat the same exercise with a subdivision of $[0, 1]$ into eight intervals.

(c) What is the distribution of

$$\log\left(\frac{S_t}{S_{t-\Delta}}\right)$$

for "small" $0 < \Delta$.

(d) Let $\Delta = 0.25$. What does the term

$$\frac{\log S_t - \log S_{t-0.25}}{0.25}$$

represent? In what units is it measured? How does this random variable change as time passes?

(e) Now let $\Delta = 0.000001$. How does the random variable change as time passes?

$$\frac{\log S_t - \log S_{t-\Delta}}{\Delta}$$

(f) If $\Delta \to 0$, what happens to the trajectories of the "random variable"

$$\frac{\log S_t - \log S_{t-\Delta}}{\Delta}$$

(g) Do you think the term in the previous question is a well-defined random variable?

2. Show that $\{Y(t), t \geq 0\}$ is a martingale when

$$Y(t) = \exp\left\{cW(t) - \frac{c^2 t}{2}\right\} \tag{7.66}$$

where c is an arbitrary constant. What is $\mathbb{E}(Y(t))$?

3. Consider the equation $X_t = X_0 + \mu dt + \sigma W_t$, with $Y_t = e^{X_t}$. Assuming $\mu = 0.1, \sigma = 0.15$ and a time period $t = 0.5$, generate paths for X_t and Y_t by generating random normal variables W_t. Compute the expectation of X_t and Y_t. Plot histograms of X_t and Y_t. What distribution does it look like they follow?

CHAPTER

8

The Wiener Process, Lévy Processes, and Rare Events in Financial Markets

OUTLINE

8.1 INTRODUCTION

At every instant of an *ordinary* trading day, there are three states of the world: prices may go up by one tick, decrease by one tick, or show no change. In fact, the price of a liquid instrument rarely changes by more than a minimum tick. Hence, pricing financial assets in continuous time may proceed quite realistically with just three states of the world, as long as one ignores "rare" events. Unfortunately, most markets for financial assets and derivative products may from time to time exhibit "extreme" behavior. These periods are exactly

Financial Derivatives, Third Edition.
http://dx.doi.org/10.1016/B978-0-12-384682-2.00008-6

when we have the greatest need for accurate pricing.

What makes an event "extreme" or "rare"? Is turbulence in financial markets the same as "rare events"? In this chapter we intend to clarify the probabilistic structure of rare events and contrast them with the behavior of Wiener processes. In particular, we discuss the types of events that a Wiener process is capable of characterizing. This discussion naturally leads to the characterization of rare events.

We show that "rare events" have something to do with the discontinuity of observed price processes. This is not the same as turbulence. Increased variance or volatility can be accounted for by continuous-time stochastic processes.

What distinguishes rare events is the way their size and their probability of occurrence changes (or does not change) with the observation interval. In particular, as the interval of observation, h, gets smaller, the *size* of normal events also gets smaller. This is, after all, what makes them "ordinary." In one month, several large price changes may be observed. In a week, fewer are encountered. Observing a number of large price jumps during a period of a few minutes is even less likely. Often, the events that occur during an "ordinary" minute are not worth much attention. This is the main characteristic of "normal" events. They become unimportant as $h \to 0$.

On the other hand, because they are ordinary, even in a very small time interval h, their probability of occurrence is not zero. During small time intervals, there is always a nonzero probability that some "nonnoticeable" news will arrive.

A *rare event* is different. By definition, it is supposed to occur infrequently. In continuous time, this means that as $h \to 0$, its probability of occurrence goes to zero. Yet, its size may not shrink. A market crash such as the one in 1987 is "rare." On a given day, during a very short period, there is negligible probability that one will observe such a crash. But when it occurs, its size may not be very different whether one looks at an interval of 10 min or an interval of a full trading day.

The previous chapter established one important result. Under some very mild assumptions, the surprise component of asset prices, $\sigma_t \Delta W_t$, had a variance

$$\mathbb{E}\,[\sigma_t \Delta W_t]^2 = \sigma_t^2 h \tag{8.1}$$

during a small interval.

In heuristic terms, this means that unpredictable changes in the asset price will have the expected size $\sigma_t \sqrt{h}$.[1]

But remember how a "standard deviation" is obtained: one multiplies possible sizes with the corresponding probabilities. It is the product of *two* terms, the probability multiplied by the "size" of the event. A variance proportional to h can be obtained either by probabilities that depend on h while the size is independent, or by probabilities that are independent of h while the size is dependent.[2]

The first case corresponds to rare events, and the second to normal events.

8.1.1 Relevance of the Discussion

This chapter is focused on the distinction between rare and normal events. The reader may be easily convinced that, from a technical point, such a distinction is important—especially if the existence of rare events implies discontinuous paths for asset prices. But are there *practical applications* of such discontinuities? Would pricing financial assets proceed differently if rare events existed?

The answers to these questions are in general affirmative. One has to use *different* formulas if asset prices exhibit jump discontinuities. This will indeed affect the pricing of financial assets.

[1] "The expected size" refers only to the absolute value of the change. Because surprises are, by definition, unpredictable, one knows nothing about the sign of these changes.

[2] Or by a combination of the two.

As an example, consider recent issues in risk management. One issue is capital requirements. How much capital should a financial institution put aside to cover losses due to adverse movements in the market?

The answer depends on how much "value" is at risk. There are several ways of calculating such *value-at-risk measures*, but they all try to measure changes in a portfolio's value when some underlying asset price moves in some *extreme* fashion.

During such an exercise it is very important to know if there exist rare events that cause prices to jump discontinuously. If such jumps are not likely, value-at-risk calculations can proceed using the normal distribution. Price changes can be modeled as outcomes of normally distributed random processes, and, under appropriate conditions, the value-at-risk will also be normally distributed. It would then be straightforward to attach a probability to the amount one can lose under some extreme price movement.

On the other hand, if sporadic jumps are a systematic part of asset price changes, then value-at-risk calculations become more complicated. Attaching a probability to the amount one is likely to lose in extreme circumstances requires modeling the "rare event" process as well.

8.2 TWO GENERIC MODELS

There are two basic building blocks in modeling continuous-time asset prices. One is the Wiener process, or Brownian motion. This is a continuous stochastic process and can be used if markets are dominated by "ordinary" events while "extremes" occur only infrequently, according to the probabilities in the tail areas of a normal distribution. The second is the Poisson process, which can be used for modeling systematic jumps caused by rare events. The Poisson process is discontinuous.

By combining these two building blocks appropriately, one can generate a model that is suitable for a particular application.

Before discussing rare and normal events, this section reviews these two building blocks.

8.2.1 The Wiener Process

In continuous time, "normal" events can be modeled using the Wiener process, or Brownian motion. A Wiener process is appropriate if the underlying random variable, say W_t, can only change continuously. With a Wiener process, during a small time interval h, one in general observes "small" changes in W_t, and this is consistent with the events being ordinary.

There are several ways one can discuss a Wiener process.

One approach was introduced earlier. Consider a random variable ΔW_{t_i} that takes one of the two possible values $\sqrt{h}$ or $-\sqrt{h}$ at instances

$$0 = t_0 < t_1 < \cdots < t_n = T \tag{8.2}$$

where for all i,

$$t_i - t_{i-1} = h \tag{8.3}$$

Suppose ΔW_{t_i} is independent of ΔW_{t_j} for $i \neq j$. Then the sum

$$W_{t_n} = \sum_{i=1}^{n} \Delta W_{t_i} \tag{8.4}$$

will converge weakly to a Wiener process as n goes to infinity. Heuristically, this means that the Wiener process will be a good approximating model for the sum on the right-hand side.[3]

In this definition, a Wiener process is obtained as the limit, in some probabilistic sense, of a sum of independent, identically distributed random variables. The important point to note is that possible outcomes for these increments are *functions*

[3] As n goes to infinity, the expression on the right-hand side will be a sum of a very large number of random variables that are independent of one another and that are all of infinitesimal size. Under some conditions, the distribution of the sum will be approximately normal. This is typical of central limit theorems, or, in continuous time, of weak convergence.

of h, the length of subintervals. As $h \to 0$, changes in W_t become smaller. With this approach, we see that the Wiener process will have a Gaussian (normal) distribution.

One can also approach the Wiener process as a continuous square integrable martingale. In fact, suppose W_t is a process that is continuous, has finite variance[4], and has increments that are unpredictable, given the family of information sets $\{I_t\}$[5]. Then, according to a famous theorem by Lévy, these properties are sufficient to guarantee that the increments in W_t are normally distributed with mean zero and variance $\sigma^2 dt$.

The formal definition of Wiener processes approached as martingales is as follows.

Definition 18. A Wiener process W_t, relative to a family of information sets $\{I_t\}$, is a stochastic process such that:

1. W_t is a square integrable martingale with $W_0 = 0$ and

$$\mathbb{E}\left[(W_t - W_s)^2\right] = t - s, \quad s \le t \tag{8.5}$$

2. The trajectories of W_t are continuous over t.

This definition indicates the following properties of a Wiener process:

- W_t has uncorrelated increments because it is a martingale, and because every martingale has unpredictable increments.
- W_t has zero mean because it starts at zero, and the mean of every increment equals zero.
- W_t has variance t.
- Finally, the process is continuous; that is, in infinitesimal intervals, the movements of W_t are infinitesimal.

Note that, in this definition, nothing is said about increments being normally distributed. When the martingale approach is used, the normality follows from the assumptions stated in the definition.[6]

The Wiener process is the natural model for an asset price that has unpredictable increments but nevertheless moves over time continuously. Before we discuss this point, however, we need to clarify a possible confusion.

8.2.1.1 Wiener Process or Brownian Motion?

The reader may have noticed the use of the term *Brownian motion* to describe processes such as W_t. Do the terms Brownian motion and Wiener process refer to the same concept, or are there any differences?

The definition of Wiener process given earlier used the fact that W_t was a square integrable martingale. But nothing was said about the distribution of W_t.

We now give the definition of Brownian motion.

Definition 19. A random process $B_t, t \in [0, T]$, is a (standard) Brownian motion if:

- The process begins at zero, $B_0 = 0$.
- B_t has stationary, independent increments.
- B_t is continuous in t.
- The increments $B_t - B_s$ have a normal distribution with mean zero and variance $|t - s|$:

$$(B_t - B_s) \sim \mathcal{N}(0, |t - s|) \tag{8.6}$$

This definition is, in many ways, similar to that of the Wiener process. There is, however, a crucial difference. W_t was assumed to be a martingale, while no such statement is made about B_t. Instead, it is posited that B_t has a normal distribution.

These appear to be very important differences. In fact, the reader may think that W_t is much more general than the Brownian motion, since no assumption is made about its distribution.

This first impression is not correct. The well-known Lévy theorem states that there are no differences between the two processes.

[4] That is, it is square integrable.

[5] This also means that the increments are uncorrelated over time.

[6] This is the famous Lévy theorem.

Theorem 4. *Any Wiener process W_t relative to a family I_t is a Brownian motion process.*

This theorem is very explicit. We can use the terms Wiener process and Brownian motion interchangeably. Hence, no distinction will be made between these two concepts in the remaining chapters.

8.2.2 The Poisson Process

Now consider a quite different type of random environment. Suppose N_t represents the total number of extreme shocks that occur in a financial market until time t. Suppose these major events occur in an unpredictable fashion.

The increments in N_t can have only one of two possible values. Either they will equal zero, meaning that no new major event has occurred, or they will equal one, implying that some major event has occurred. Given that major events are "rare," increments in N_t that have size 1 should also occur rarely.

We use the symbol dN_t to represent incremental changes in N_t during an infinitesimal time period of length dt. Consider the following characterization of the incremental changes in N_t[7]:

$$dN_t = \begin{cases} 1 \text{ with probability } \lambda\, dt \\ 0 \text{ with probability } 1 - \lambda\, dt \end{cases} \tag{8.7}$$

Note that here we have increments in N_t that can assume two possible values during an infinitesimal interval dt. The critical difference from the case of Brownian motion is that, this time, the size of Poisson outcomes does *not* depend on dt. Instead, the probabilities associated with the outcomes are functions of dt. As the observation period goes toward zero, the increments of Brownian motion become smaller[8], while the movements in N_t remain of the same size.

The reader would recognize N_t as the Poisson counting process. Assuming that the rate of occurrence of these events during dt is λ, the process defined as

$$M_t = N_t - \lambda t \tag{8.8}$$

will be a discontinuous square integrable martingale.[9] It is interesting to note that

$$\mathbb{E}\,[M_t] = 0 \tag{8.9}$$

and

$$\mathbb{E}\,[M_t]^2 = \lambda t \tag{8.10}$$

Thus, although the trajectories of M_t are discontinuous, the first and second moments of M_t and W_t have the same characterization. In particular, over small time intervals of length h, both processes have increments with variance proportional to h.[10]

We emphasize the following points.

The trajectories followed by the two processes are very different. One is continuous, the other is of the pure "jump" type.

Second, the probability that M_t will show a jump during a very small interval goes to zero. Heuristically, this means that the trajectories of M_t are *less* irregular than the trajectories of W_t, because the Poisson counting process is constant

[7] At this point, the use of dN_t and dt instead of ΔN_t and h should be considered symbolic. In later chapters, it is hoped that the meaning of the notation dN_t and dt will become clearer.

[8] At a speed proportional to $\sqrt{h}$.

[9] M_t is called a compensated Poisson process. The λt is referred to as the compensatory term. It "compensates" for the positive trend in N_t and converts it into a "trendless" process M_t.

[10] A heuristic way of calculating the variance of dM_t is as follows:

$$\mathbb{E}\left[dM_t\right]^2 = 1^2 \lambda\, dt + 0^2 \left[1 - \lambda\, dt\right] \tag{8.11}$$

which gives

$$\mathbb{E}\left[dM_t\right]^2 = \lambda\, dt \tag{8.12}$$

This is heuristic because we do not know whether we can treat increments such as dM_t as "objects" similar to standard random variables. To make the discussion precise, one must begin with a finite subdivision of the time interval, and then present some type of limiting argument.

"most of the time." Although M_t displays discrete jumps, it will not have unbounded variation. W_t, on the other hand, displays infinitesimal changes, but these changes are uncountably many. As a result, the variation becomes unbounded. Hence, it may be more difficult to define integrals such as

$$\int_{t_0}^{T} f(W_t)\, dW_t \tag{8.13}$$

than integrals with respect to M_t:

$$\int_{t_0}^{T} f(M_t)\, dM_t \tag{8.14}$$

Indeed, it is true that, in general, the Riemann–Stieltjes definition may be applied to this latter integral.

8.2.3 Examples

Figure 8.1 displays a Poisson process generated by a computer. First, a $\lambda = 13.4$ was selected. Next, $h = 0.001$ was fixed. The computer was asked to generate a trajectory for the Poisson counting process $Nt, t \in [0, 1]$. This trajectory is also displayed. We note the following characteristics of the Poisson paths:

- The trajectory has a positive slope. (Hence, N_t is not a martingale.)
- Changes occur in equal jumps of size 1.
- The trajectory is constant between these jumps.
- In this particular example, there are 14 jumps, which is very close to the mean.

Figure 8.2 displays a mixture of the Poisson and Wiener processes. First, a trajectory was drawn from the Poisson process. Next, the computer was asked to generate a trajectory from a standard Wiener process with variance $h = 0.001$. The two trajectories were added to each other.

We see the following characteristics of this sample path:

- The path shows occasional jumps, due to the Poisson component.
- Between jumps, the process is not constant; it fluctuates randomly. This is due to the Wiener component.
- The noise introduced by the Wiener process is much smaller than the jumps due to the Poisson process. This may change if we select a Wiener process with higher variance. Then, it could be very difficult to distinguish between Poisson jumps and noise caused by the Wiener component.

8.2.4 Lévy Processes

The class of the Lévy process consists of all stochastic processes with stationary, independent increments. The Lévy-Khintchine theorem (Sato, 2001) provides a characterization of Lévy processes in terms of the characterization of the underlying process. It states that there exists a measure ν such that for all $u \in \mathbb{R}$ and t nonnegative, the characteristic function of a Lévy process can be written as

$$\mathbb{E}(e^{iuX_t}) = \exp(t\phi(u)) \tag{8.15}$$

where

$$\phi(u) = i\gamma u - \frac{1}{2}\sigma^2 u^2 + \int_{-\infty}^{+\infty} (e^{iuy} - 1 - iuy\mathbb{1}_{\{|y|\leq 1\}})d\nu(y) \tag{8.16}$$

Here γ and σ are real numbers, ν is a measure on $\mathbb{R}$ such that $\nu(0) = 0$, and

$$\int_{-\infty}^{+\infty} \min(1, x^2) d\nu(x) \tag{8.17}$$

is bounded. Assume a Lévy process, $\{X_t\}_{t\geq 0}$, of the following form:

$$X_t = (r - q + \omega)t + Z_t \tag{8.18}$$

This process has a drift term controlled by ω and a pure jump component $\{Z_t\}_{t\geq 0}$. In the case of variance gamma process, the Lévy measure associated to the pure jump component can be written

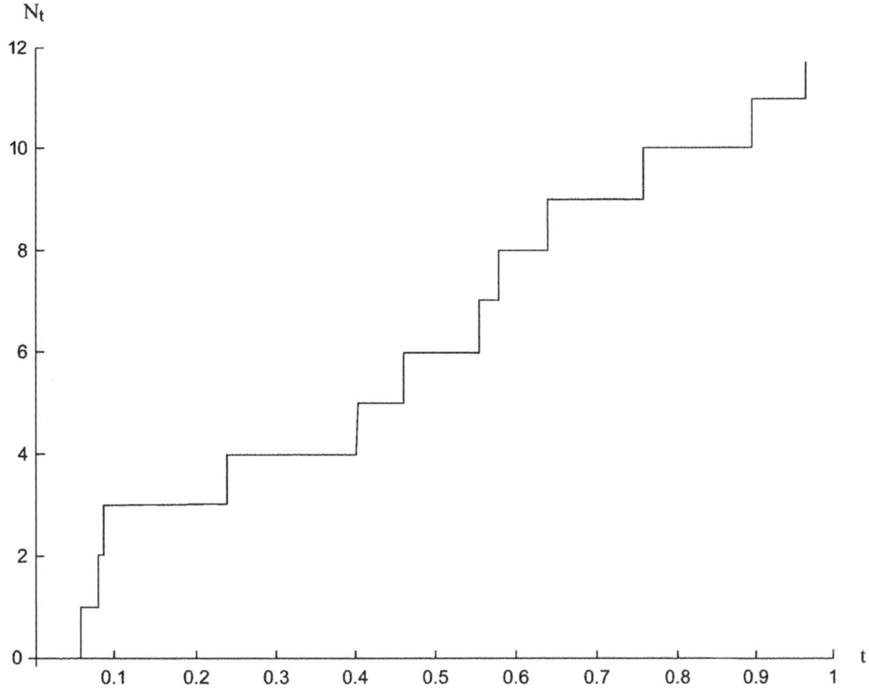

FIGURE 8.1 A Poisson process sample path.

as $d\nu(y) = k(y)dy$, where $k(y)$ is defined as

$$k(y) = \frac{e^{-\lambda_p y}}{\nu y} \mathbb{1}_{y>0} + \frac{e^{-\lambda_n |y|}}{\nu |y|} \mathbb{1}_{y<0} \tag{8.19}$$

where

$$\lambda_p = \left(\frac{\theta^2}{\sigma^4} + \frac{2}{\sigma^2 \nu} \right)^{\frac{1}{2}} - \frac{\theta}{\sigma^2}$$

$$\lambda_n = \left(\frac{\theta^2}{\sigma^4} + \frac{2}{\sigma^2 \nu} \right)^{\frac{1}{2}} + \frac{\theta}{\sigma^2}$$

8.2.5 Back to Rare Events

Compared to events that occur in a routine fashion, a rare event is by definition something that has a "large" size. This classification seems obvious, but at a closer look is not very easy to justify. Consider the Wiener process. A stochastic differential equation that is driven by a Wiener process amounts to assuming that in small intervals of length h unexpected price changes occur with a variance of $\sigma^2 h$, where the σ may depend on the available information as well. Further, the distribution of these unexpected price changes is normal.

A normal distribution has tails that extend to infinity. With *small* but nonzero h, there is a positive probability that a very large, unexpected price change will occur. Hence, with a nonzero h, the Wiener process seems to be perfectly capable of introducing "big" events in the stochastic differential equations. Why would we then need another discussion of "rare" or big events?

The problem with characterizing rare events using a Wiener process is the following. As h goes to zero, the tails of the normal distribution carry less and less weight. At the limit, $h = 0$, these tails have completely vanished. In fact, the whole

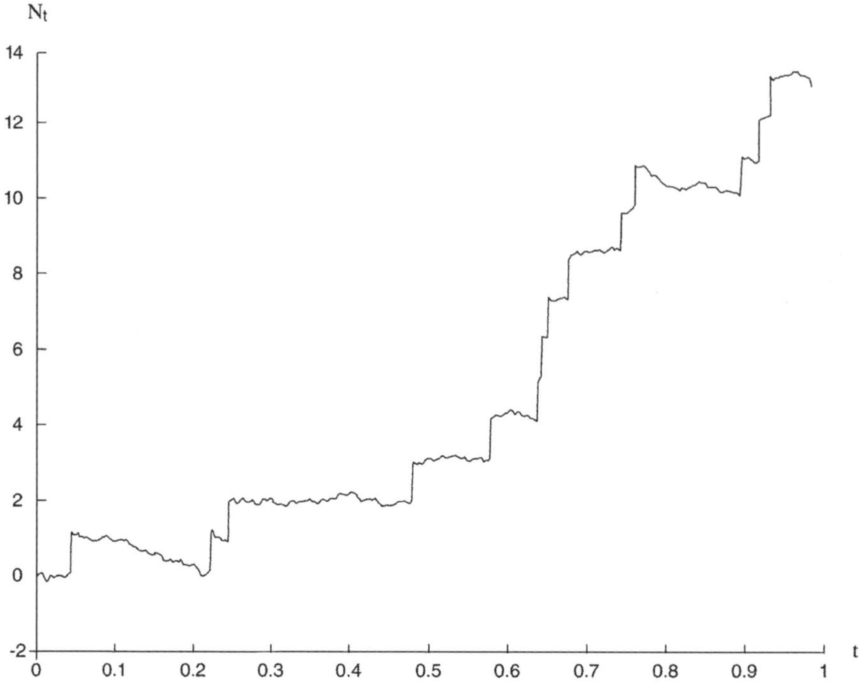

FIGURE 8.2 A sample path generated by a mixture of the Poisson and Wiener processes.

distribution has concentrated on zero. This is to be expected because the Wiener process is continuous with probability one. As $h \to 0$, the *size* of price changes represented by the Wiener process *has* to become smaller and smaller. In this sense, the Wiener process is not suitable for representing situations where, in an extremely short interval, prices can move in some extreme fashion.

What we need is a disturbance term that is capable of generating large events in extremely small intervals. In other words, we need a process that may exhibit jumps. Such a process will have outcomes that do not depend on h, and as h gets small, the size of the outcomes will not shrink.

Thus, "rare" events correspond to occasional jumps in the sample paths of the process.

Several markets in derivatives exhibit jumps in prices. This is more often the case in commodities, where a single news item is more likely to carry important information for the underlying commodity. Reports on crops, for example, are likely to cause jumps in futures on the same commodity. In the case of financial derivatives, this is less likely. The weight of a single news item in determining the price or interest rate or currency derivatives is significantly smaller, although present.

In the following sections, we characterize normal and rare events, and learn ways of modeling price series that are likely to exhibit occasional jumps.

8.3 SDE IN DISCRETE INTERVALS, AGAIN

A deeper analysis of normal vs rare events is best done by considering a stochastic differential equation in finite intervals.[11]

[11] Remember from Chapter 7 that in order to obtain the SDE in discrete intervals, we used several approximations. For small but noninfinitesimal h, such equations hold in an approximate sense only.

Consider again the SDE that was introduced for discrete intervals of equal size h in Chapter 7:

$$S_k - S_{k-1} = a\left(S_{k-1}, k\right) h + \sigma\left(S_{k-1}, k\right) \Delta W_k, \quad k = 1, 2, \ldots n \tag{8.20}$$

where the $a\left(S_{k-1}, k\right) h$ is the drift component which determines how, on the average, the increment $S_k - S_{k-1}$ is expected to behave during the next interval. ΔW_k is the innovation term, determining the "surprise" component of asset prices. It was shown that, under some assumptions, the variance of the innovation term is proportional to h, the length of the interval. The term $\sigma\left(S_{k-1}, k\right)^2$ is the factor of proportionality.

In order to study "normal" and "rare" events in more detail, we make a further simplifying assumption.[12]

Assumption 4. ΔW_k can assume only a finite number of possible values. The possible outcomes of ΔW_k and their corresponding probabilities are[13]

$$\sigma_k \Delta W_k = \begin{cases} \omega_1 & \text{with probability } p_1 \\ \omega_2 & \text{with probability } p_2 \\ \ldots & \\ \omega_m & \text{with probability } p_m \end{cases} \tag{8.21}$$

Although it is not clear *which* event will occur, the set of possible events is known by all agents. A typical ω_i represents a possible outcome of the innovation term $\sigma_k \Delta W_k$, while p_i denotes the associated probability. The parameter m is the total number of possible outcomes. It is an integer.[14]

There are two types of ω_i. The first three represent "normal" outcomes. For example, ω_1 may represent an uptick, ω_2 may be a downtick, and the ω_3 may represent "no change" in asset prices. In real time, these are certainly routine developments in financial markets.

The remaining possibilities, $\omega_4, \omega_5, \ldots,$ are reserved for various types of special events that may occur rarely. For example, if the underlying security is a derivative written on grain futures, ω_4 may be the effect of a major drought, the ω_5 may be the effect of an unusually positive crop forecast, and so on. Clearly, if such possibilities refer to extreme price changes, and if they are rare, then they must lead to price changes greater than one tick. Otherwise, price changes are caused by normal events $\omega_1, \omega_2, \omega_3$.

This setup will be used in the next section to determine the probabilistic structure of rare events.

8.4 CHARACTERIZING RARE AND NORMAL EVENTS

Under assumptions 1−3 of the previous chapter, an important result was proven. It was shown that the variance of $\sigma_k \Delta W_k$,

$$\mathbb{E}\left[\sigma_k \Delta W_k\right]^2 = \sigma_k^2 h \tag{8.22}$$

was proportional to the observation interval h, where σ_k was a known parameter, given the information set I_{k-1}.

This result can be exploited further if we use assumption 4. In fact, a very explicit characterization of rare and normal events can be given this way, although the reader may find the notation a

[12]Here also we follow Merton (1990).

[13]There are two reasons that we introduce this assumption. First, the distinction between rare and normal events will be much easier to introduce if the possibilities are *finite*. Second, actual asset pricing in financial markets often proceeds with either binomial or trinomial *trees*. In the case of binomial trees, the market participant assumes that, at any instant, there are only two possible moves for the price. With trinomial trees, possible moves are raised to three. Hence, in practical situations, the total number of possible states is selected as finite anyway.

[14]Both ω_i and p_i can very well be made to depend on the information set I_k. However, this would add a k subscript to these variables and make the notation more cumbersome. To avoid this, we make ω_i and p_i independent of k.

bit unpleasant. However, this is a small price to pay if a useful characterization of rare and normal events is eventually obtained.

According to assumption 4, ΔW_k can assume only a finite number of values. In terms of ω_i and the corresponding probabilities, p_i, we can explicitly write the variance as

$$\mathbb{V}\left[\sigma_k \Delta W_k\right] = \sum_{i=1}^{m} p_i \omega_i^2$$

Using the important proposition of the previous chapter, this means

$$\sum_{i=1}^{m} p_i \omega_i^2 = \sigma_k^2 h \tag{8.23}$$

where the parameter m is the number of possible states. The left-hand side of Eq. (8.17) is simply the weighted average of squared deviations from the mean, which in this case is zero. The "weights" are probabilities associated with possible outcomes.[15]

Now, the left-hand side of (8.17) is a sum of m finite, nonnegative numbers. If the sum of such numbers is proportional to h, and if each element is positive (or zero), then *each* term in the sum should also be proportional to h or should equal zero. In other words, *each* $p_i \omega_i^2$ will be given by

$$p_i \omega_i^2 = c_i h \tag{8.24}$$

where $0 < c_i$ is some factor of proportionality.[16]

Equation (8.18) says that all terms such as are linear functions of h. Then, one can visualize the p_i and the ω_i as two functions of h, whose product is proportional to h. That is,

$$p_i = p_i(h) \tag{8.25}$$

[15]We show the potential dependence of ω_i, p_i on the information that becomes available as time passes by adding the k subscript to σ_k.

[16]In general, c_i will depend on k as well. To keep notation simple, we eliminate the k subscript.

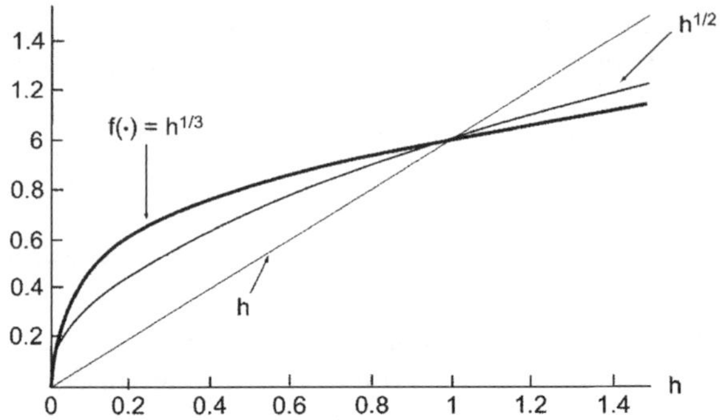

FIGURE 8.3 Plot of h^{r_i} for different choices for r_i.

and

$$\omega_i = \omega_i(h) \tag{8.26}$$

such that

$$p_i(h)\omega_i(h)^2 = c_i h \tag{8.27}$$

We follow Merton (1990) and assume specific exponential forms for these functions $p_i(h)$ and $\omega_i(h)$:

$$\omega_i(h) = \overline{\omega}_i h^{r_i} \tag{8.28}$$

and

$$p_i(h) = \overline{p}_i h^{q_i} \tag{8.29}$$

where r_i and q_i are nonnegative constants. $\overline{\omega}_i$ and $\overline{p}_i$ are constants that may depend on i or k, but are independent of h, the size of the observation interval.

Figure 8.3 displays some choices for h^{r_i}. Three examples are shown: the case when $r_i = 1$ (not allowed in this particular discussion), the case when $r_i = 0.5$, and the case when $r_i = 1/3$. In particular, we see that for small $h, h^{r_i} > h$.

According to Eqs. (8.22) and (8.23), both the size and the probability of the event may depend on the interval length, h. As h gets larger, then the (absolute) magnitude of the observed price change and its probability will get larger, except when r_i or q_i are zero.

To characterize rare and normal events we use the parameters r_i and q_i. Both of these parameters are nonnegative. r_i governs how fast the size of the event goes to zero as the observation interval gets smaller. q_i governs how fast the probability goes to zero as the observation interval decreases. It is, of course, possible that r_i or q_i vanish, although they cannot do so at the same time.[17]

We now show explicitly how restrictions on the parameters r_i, q_i can distinguish between rare and normal events.

The variance of ΔW_k in (8.18) is made of terms such as

$$p_i\omega_i^2 = \overline{\omega}_i^2\overline{p}_i h^{2r_i} h^{q_i} \tag{8.30}$$

But we know that each $p_i\omega_i^2$ is proportional to h as well:

$$p_i\omega_i^2 = c_i h \tag{8.31}$$

Hence,

$$\overline{\omega}_i^2\overline{p}_i h^{2r_i+q_i} = c_i h \tag{8.32}$$

But this implies that

$$q_i + 2r_i = 1 \tag{8.33}$$

and

$$c_i = \overline{\omega}_i^2\overline{p}_i \tag{8.34}$$

Thus, the parameters q_i, r_i must satisfy the restrictions

$$0 \leq r_i \leq \frac{1}{2} \tag{8.35}$$

and

$$0 \leq q_i \leq 1 \tag{8.36}$$

We find that there are, in fact, only two cases of interest—namely,

$$r_i = \frac{1}{2}, \quad q_i = 0 \tag{8.37}$$

and

$$r_i = 0, \quad q_i = 1 \tag{8.38}$$

The first case leads to events that we call "normal." The second is the case of "rare" events. We discuss these in turn.

[17]Remember that the product of ω_i and p_i must be proportional to h. If both r_i and q_i equal zero, these products will not depend on h, and this is not allowed.

8.4.1 Normal Events

The condition for "normal" events is

$$\frac{1}{2} \geq r_i \geq 0 \tag{8.39}$$

To interpret this, consider what happens when we select $r_i = 1/2$.

First, we know that the q_i must equal zero.[18] As a result, the functions that govern the size and the probability of the outcome ω_i become, respectively,

$$\omega_i = \overline{\omega}_i h^{1/2} = \overline{\omega}_i\sqrt{h} \tag{8.41}$$

According to this, the sizes of events having $r_i = 0.5$ will get smaller as the interval length h gets smaller. On the other hand, their probability does not depend on h. These outcomes are "small" but have a constant probability of occurrence as observation intervals get smaller. They are "ordinary."

Now suppose all possible outcomes for ΔW_k are of this type and have $r_i = 0.5$. Then the sample paths of the resulting W_t process will have a number of interesting properties.

8.4.1.1 Continuous Paths

If there are no rare events, then all ω_i will have $r_i = 0.5$, and their size

$$\omega_i = \overline{\omega}_i\sqrt{h} \tag{8.42}$$

will shrink as h gets smaller. At the same time, as h goes to zero, the values of ω_i approach each other. This means that the process W_k will, in the limit, be continuous. The steps taken by ΔW_k will approach zero:

$$\lim_{h\to 0}\omega_i = \lim_{h\to 0}\overline{\omega}_i\sqrt{h} \tag{8.43}$$

This will be true for every "normal" event, $\overline{\omega}_i$. In the limit, the trajectories of W_t will be such that

[18]Remember that

$$q_i + 2r_i = 1 \tag{8.40}$$

and that q_i cannot be negative.

one could *plot* the data without lifting one's hand. Each incremental value will have infinitesimal size.

On the other hand, since $q_i = 0$ for "normal" events, the probabilities of these ω_i will *not* tend to zero as $h \to 0$. In fact, the probability of these events will be independent of h:

$$p_i = \overline{p}_i \tag{8.44}$$

It is in this sense that normal events can generate continuous-time paths.

8.4.1.2 Smoothness of Sample Paths

The sample paths of an innovation term that has outcomes with $r_i = 1/2$ are continuous. But they are not smooth.

First, remember what smoothness means within the context of a deterministic function. Heuristically, a function will be "smooth" if it does not change abruptly. In other words, suppose we select a point x_0, where the function $f(x)$ is evaluated. $f(x)$ will be smooth at x_0, if for small h, the ratio

$$\frac{f\left(x_0+h\right)-f\left(x_0\right)}{h} \tag{8.45}$$

stays finite as h get smaller and smaller. That is, the function is smooth if it has a derivative at that point.

Is the same definition of smoothness valid for nondeterministic functions such as W_t as well?

In the particular case discussed here, there are a finite number m of possible values that ΔW_k can assume. The sizes of these events are all proportional to $h^{1/2}$. In other words, as time passes, the new events that affect prices will cause changes of the order $\sqrt{h}$.

At any time t, the unexpected *rate* of change of prices can be written as

$$\frac{f\left(x_0+h\right)-f\left(x_0\right)}{h} \tag{8.46}$$

for some i. Taking limits,

$$\lim_{h\to 0} \frac{W_{t+h}-W_t}{h} = \lim_{h\to 0} \frac{\omega_i}{h} \tag{8.47}$$

or, after substituting for ω_i,

$$= \lim_{h\to 0} \overline{\omega}_i \frac{h^{\frac{1}{2}}}{h} \tag{8.48}$$

$$= \overline{\omega}_i \lim_{h\to 0} \frac{1}{h^{\frac{1}{2}}} \to \infty \tag{8.49}$$

This means that as the interval h gets smaller, the W_t starts to change at an *infinite* rate. Asset prices will behave continuously but erratically. (Here we assumed, without any loss of generality, that $\overline{\omega}_i$ was positive.).

This concludes the discussion of trajectories that are generated by events of normal size. We now consider paths generated by rare events.

8.4.2 Rare Events

Assume that for some event ω_i, the parameter r_i equals zero. Then the corresponding q_i equals 1, and the probability of this particular outcome will by definition be given by

$$p_i = \overline{p}_i h \tag{8.50}$$

The events ω_i that have a $r_i = 0, q_i = 1$ are "rare" events, since, according to this equation, their probability vanishes as $h \to 0$.

On the other hand, the size of the events will be given by

$$\omega_i = \overline{\omega}_i \tag{8.51}$$

that is, they will *not* depend on the length of the interval h.

We make the following observations concerning rare events.

8.4.2.1 Sample Paths

Sample paths of an innovation term that contains rare events will be discontinuous. In fact, the sizes of those ω_i with $q_i = 1$ do not depend on h. As h goes to zero, ΔW_k will from time to time assume values that do not get any smaller. The size of unexpected price changes will be independent of h. When such rare outcomes occur, W_t will have a jump.

On the other hand, if $q_i = 1$, the probability of these jumps *will* depend on h, and as the latter gets smaller, the probability of observing a jump will also go down. Hence, although the trajectory contains jumps, these jumps are not common.

Clearly, if the random variable ΔW_k contains jumps, its sample paths will not be continuous. One would need a model other than the Wiener process to capture the behavior of such random shocks.

8.4.2.2 Further Comments

What can be said of the remaining values for r_i and q_i? In other words, consider the ranges

$$0 < r_i < \frac{1}{2} \tag{8.52}$$

and

$$0 < q_i < \frac{1}{2} \tag{8.53}$$

What types of sample paths would the W_t possess if the possible outcomes have r_i and q_i within these ranges?

It turns out that for all r_i, q_i within these ranges, the sample paths will be continuous but nonsmooth, just as in the case of a Wiener process.

This is easy to see. As long as $0 < r_i < 0.5$ is satisfied, the size of ω_i will be a function of h. As $h \to 0, \omega_i$ will go to zero. In terms of *size*, they are not rare events.

Note that for such outcomes the corresponding probabilities also go to zero. Thus, these outcomes are not observed frequently. But given that their size will get smaller, they are not qualified as rare events.

8.5 A MODEL FOR RARE EVENTS

What type of models can one use to represent asset prices if there are rare events?

Consider what is needed. Our approach tries to represent asset prices by an equation that decomposes observed changes into two components: one that is predictable, given the information at that time, and another that is unpredictable. In small intervals of length h, we write

$$S_k - S_{k-1} = a\left(S_{k-1}, k\right) h + \sigma\left(S_{k-1}, k\right) \Delta W_t \tag{8.54}$$

As h gets smaller, we obtain the continuous-time version valid for infinitesimal intervals:

$$dS_t = a\left(S_t, t\right) dt + \sigma\left(S_t, t\right) dW_t \tag{8.55}$$

In later chapters we study the SDEs more precisely and show what the differentials such as dS_t or dW_t really mean.

There is no need to adopt a different representation in order to take into account rare events. These also occur unexpectedly, and their variance is also proportional to h, the time interval. In fact, the only difference from the case of a Wiener process occurs in the continuity of sample paths. Hence, the same SDE representation can be used with a simple modification. What is needed is a *new model* for the random, unpredictable errors dW_t.

In the case of rare events, the defining factors are that the size of the event is not infinitesimal even when h is, while its probability does become negligible with $h \to 0$. Accordingly, the new innovation term should be able to represent (random) jumps that occur rarely in asset prices. Further, the model should be flexible enough to capture any potential variation of the probability of occurrence of such jumps.

One can be more specific. First, split the error term in two. It is clear from the previous discussion that changes in asset prices will be a mixture of normal events that occur in a continuous fashion, and of jumps that occur sporadically. We denote the first component by ΔW_k. The second component is denoted by the symbol ΔN_k. To make this more precise, assume that the event is a jump in asset prices of size 1. At any instant $k-1$, one has

$$N_k - N_{k-1} = \begin{cases} 1 \text{ with probability } \lambda h \\ 0 \text{ with probability } (1 - \lambda h) \end{cases} \tag{8.56}$$

where λ does not depend on the information set available at time $k-1$. We let

$$\Delta N_k = N_k - N_{k-1} \tag{8.57}$$

Such ΔN_k represent jumps of size 1 that occur with a constant rate λ.[19]

It is clear that N_k can be modeled using a Poisson counting process. A Poisson process has the following properties:

1. During a small interval h, at most one event can occur with probability very close to 1.
2. The information up to time t does not help to predict the occurrence (or the nonoccurrence) of the event in the next instant h.[20]
3. The events occur at a constant rate λ.

In fact, the Poisson process is the only process that satisfies all these conditions simultaneously. It seems to be a good candidate for modeling jump discontinuities. We may, however, need two modifications.

First, the rate of occurrence of jumps in a certain asset price may change over time. The Poisson process has a *constant* rate of occurrence and cannot accommodate such behavior. Some adjustment is needed.

Second, the increments in N_t have nonzero mean. The SDE approach deals with innovation terms with zero mean only. Another modification is needed to eliminate the mean of dN_t.

Consider the modified variable

$$J_t = (N_t - \lambda_t) \tag{8.58}$$

The increments ΔJ_k will have zero mean and will be unpredictable. Further, if we multiply the J_t by a (time-dependent) constant, say $\sigma_2(S_{k-1}, k)$, the size of the jumps will be time-dependent. Hence, $\sigma_2(S_{k-1}, k)\,\Delta J_k$ is an appropriate candidate to represent unexpected jumps in asset prices.

This means that if the market for a financial instrument is affected by sporadic rare events, the stochastic differential equations can be written as

$$\begin{aligned} S_k - S_{k-1} &= a(S_{k-1}, k)\,h + \sigma_1(S_{k-1}, k)\,\Delta W_t \\ &\quad + \sigma_2(S_{k-1}, k)\,\Delta J_t \end{aligned} \tag{8.59}$$

As h gets small, this becomes

$$dS_t = a(S_t, t)\,dt + \sigma_1(S_t, t)\,dW_t + \sigma_2(S_t, t)\,dJ_t \tag{8.60}$$

This stochastic differential equation will be able to handle "normal" and "rare" events simultaneously.

Finally, note that the jump component dJ_t and the Wiener component dW_t have to be statistically independent at every instant t. As h gets smaller, the size of "normal" events has to get smaller, while the size of "rare" events remains the same. Under these conditions the two types of events cannot be "related" to each other. Their instantaneous correlation must be zero.

8.6 MOMENTS THAT MATTER

The distinction between "normal" and "rare" events is important for one other reason.

Practical work with observed data proceeds either directly or indirectly by using appropriate "moments" of the underlying processes. In Chapter 5, we defined the term "moment" as representing various expectations of the underlying process. For example, the simple expected value $\mathbb{E}[X_t]$ is the first moment. The variance

$$\mathbb{V}[X_t] = \mathbb{E}[X_t - \mathbb{E}[X_t]]^2 \tag{8.61}$$

is the second (centered) moment. Higher-order (centered) moments are obtained by

$$\mathbb{E}[X_t - \mathbb{E}[X_t]]^k \tag{8.62}$$

where $k > 2$.

[19]The rate of occurrence of the jump during an interval h can be calculated by dividing the corresponding probability λh by h.

[20]As $h \to 0$, this probability will become 1.

As mentioned earlier, moments give information about the process under consideration. For example, variance is a measure of how volatile the prices are. The third moment is a measure of the skewness of the distribution of price changes. The fourth moment is a measure of heavy tails.

In this section, we show that when dealing with changes over infinitesimal intervals, in the case of normal events only the first *two* moments matter. Higher-order moments are of marginal significance. However, for rare events, all moments need to be taken into consideration.

Consider again the case where the unpredictable surprise components are made of m possible events denoted by ω_i.

The first two moments of such an unpredictable error term will be given by[21]

$$\mathbb{E}[\sigma_1 \Delta W_k + \sigma_2 \Delta J_k] = [p_1\omega_1 + \cdots + p_m\omega_m] = 0 \tag{8.63}$$

$$\mathbb{V}[\sigma_1 \Delta W_k + \sigma_2 \Delta J_k] = [p_1\omega_1^2 + \cdots + p_m\omega_m^2] \tag{8.64}$$

where the independence of ΔW_k and ΔJ_k is implicitly used.

Now consider the magnitude of these moments when all events are of the "normal" type, having a size proportional to $h^{1/2}$. That is, consider the case when all $q_i = 0$.

The first moment is a weighted sum of m such values. Unless it is zero, it will be proportional to $h^{1/2}$:

$$\mathbb{E}[\sigma_1 \Delta W_k] = h^{1/2}\left[p_1\overline{\omega}_1 + \cdots + p_m\overline{\omega}_m\right] \tag{8.65}$$

As we divide this by h, we obtain the average rate of unexpected changes in prices. Clearly, for small h the $\sqrt{h}$ is larger than h, and the expression

$$\frac{\mathbb{E}[\Delta W_k]}{h} \tag{8.66}$$

gets larger as h gets smaller. We conclude that when the first moment is not equal to zero, it is "large" and cannot be ignored, even in small intervals h.

[21]In the remaining part of this section, $\sigma_i(S_t, t), i = 1, 2$ will be abbreviated as σ_i.

The same is true for the second moment. The variance of an unpredictable change in prices contains terms such as ω_i^2. When the ω_i are of normal type, their size is proportional to $h^{1/2}$. Hence, the variance will be proportional to h:

$$\mathbb{V}[\sigma_1 \Delta W_k] = h\left[\sum_{i=1}^{m} p_i\overline{\omega}_i^2\right] \tag{8.67}$$

As we divide this by h, we obtain the average rate of variance. Clearly, the h will cancel out and the rate of variance remains constant as h gets smaller.

This means that the variance does not become negligible as $h \to 0$. In the case of "normal" events, the variance provides significant information about the underlying randomness even during an infinitesimal interval h.

Now consider what happens with higher-order moments,

$$\mathbb{E}[\sigma_1 \Delta W_k]^n = \left[p_1\omega_1^n + \cdots + p_m\omega_m^n\right] \tag{8.68}$$

with $n > 2$.

Here, when the events under consideration are of the normal type, raising the ω_i to a power of n will result in terms such as

$$\omega_i^n = \overline{\omega}_1^n \left(h^{1/2}\right)^n \tag{8.69}$$

But when $n > 2$, for small h we have

$$h^{n/2} < h \tag{8.70}$$

Consequently, as we divide higher-order moments by h, we obtain the corresponding rate:

$$\frac{\mathbb{E}[\sigma_1 \Delta W_k]^n}{h} = h^{(n-2)/2}\sum_{i=1}^{m}\overline{\omega}_i^n \tag{8.71}$$

This rate will depend on h positively. As h gets smaller, $h^{(n-2)/2}$ will converge to zero.[22]

Consequently, for small h, higher-order moments of unpredictable price changes will not carry any useful information, if the underlying events are all of the "normal" type. A probabilistic model that depends only on *two* parameters, one representing the first moment and the second representing the variance, will be *sufficient* to capture all the relevant information in price data for small "h". The Wiener process is then a natural choice if there are no rare events.

If there are, the situation is different.

Suppose all events are rare. By definition, rare events assume values ω_i that do not depend on h. For the second moment, we obtain

$$\mathbb{E}\left[\sigma_2 \Delta J_k\right]^2 = h\left[\sum_{i=1}^{m} \omega_i^2 \overline{p}_i\right] \tag{8.72}$$

where the ω_i *do not depend on h*. As we divide the right-hand side of the last equation by h, it will become independent of h. Hence, variance cannot be considered negligible. Here, there is no difference from Wiener processes.

However, the higher-order moments will be given by

$$\mathbb{E}\left[\sigma_2 \Delta J_k\right]^n = h\left[\sum_{i=1}^{m} \omega_i^n \overline{p}_i\right] \tag{8.73}$$

This is the case because, with rare events, the probabilities are proportional to h, and the latter can be factored out. With $n > 2$, higher-order moments are also of order h. As we divide higher-order moments of ΔJ_k by h, they will *not* get any smaller as $h \to 0$. Unlike Wiener processes, higher-order moments of ΔJ_t cannot be ignored over infinitesimal time intervals. This means that if prices are affected by rare events, higher-order moments may provide useful information to market participants.

[22]When n is greater than 2, the exponent of h will be positive.

This discussion illustrates when it is appropriate to limit the innovation terms of SDEs to Wiener processes. If one has enough conviction that the events at the roots of the volatility in financial markets are of the "normal" type, then a distribution function that depends only on the first two moments will be a reasonable approximation. The assumption of normality of dW_t will be acceptable in the sense of making little difference for the end results, because in small intervals the data will depend on the first two moments anyway. However, if rare events are a systematic part of the data, the use of a Wiener process may not be appropriate.

8.7 CONCLUSIONS

In the next two chapters, we formalize the notion of stochastic differential equations. This chapter and the previous one laid out the groundwork for SDEs. We showed that the dynamics of an asset price can always be captured by a stochastic differential equation,

$$dS_t = a\left(S_t, t\right) dt + \left[\sigma_1\left(S_t, t\right) dW_t + \sigma_2\left(S_t, t\right) dJ_t\right] \tag{8.74}$$

where the first term on the right-hand side is the expected change in S_t , and the second term in brackets is the surprise component, unpredictable given the information at time t. The stochastic differentials were not defined formally, so the discussion proceeded using "small" increments, ΔS_k and ΔW_k.

The unpredictable components of SDEs are made of two parts: dW_t captures events of insignificant size that happen regularly; dJ_t captures "large" events that occur rarely.

In small intervals, the random variable W_t is described fully by the first- and second-order moments. Higher-order moments do not provide any additional information. Hence, assuming normality and letting W_t be the Wiener process provides a good approximation for such events.

Rare events cannot be captured by the normal distribution. If they are likely to affect the financial market under consideration, the unexpected components should be complemented by the dJ_t process. The Poisson process would represent the properties of such a term reasonably well.

Given that the market participant can pick the parameters $\sigma_1(S_t, t)$ and $\sigma_2(S_t, t)$ at will, the combination of the Wiener and Poisson processes can represent all types of disturbances that may affect financial markets.

8.8 RARE AND NORMAL EVENTS IN PRACTICE

In this section, we treat how the distinction between *normal* and *rare* events will exhibit itself in practical modeling of asset price dynamics. In particular, is this distinction only a theoretical curiosity, or can it be made more concrete by explicitly taking into account the abovementioned discussion?

The answer to the last question is yes. This is best seen within the class of binomial pricing models dealt with in Chapter 2. We discuss two binomial models, one being driven by a random term representing "normal" events, the other that incorporates "rare" events.

First, we need to review the standard binomial model for a financial asset price. We work with an underlying stock price S_t, although a process such as instantaneous spot-rate r_t could also be considered.

8.8.1 The Binomial Model

We are interested in discretizing the behavior of a continuous-time process S_t, over time interval $[0, T]$, $T < \infty$. We also want this discretization to be "systematic" and "simple." As usual, we divide the time interval of length T into n subintervals of equal length Δ such that:

$$t_0 = 0 < t_1 < \cdots < t_n = T \tag{8.75}$$

with

$$n\Delta = T \tag{8.76}$$

This gives the discrete time points $\{t_i\}$.

We next model the values of S_t at these specific time points, t_i. For the sake of notational simplicity, we denote these by S_i:

$$S_i = S_{t_i}, \quad i = 0, 1, \ldots, n \tag{8.77}$$

The binomial model implies that once it reaches a certain *state* or node, at every discrete point i, the immediate movement in S_i will be limited to only two *up* and *down* states, which depend on two parameters denoted by u_i and d_i.[23]

The way these two parameters are chosen depends on the types of movements S_t is believed to exhibit. We will discuss two cases.

In the first case, the sizes of u_i and d_i will be made to depend on the Δ, whereas the probabilities associated with them will be independent of Δ. In the second case, the reverse will be true. The u_i and d_i will be independent of Δ, while the probabilities of up and down states will depend on it. Clearly, the first will correspond to the case of "normal" events and will eventually be captured by variables driven by the Wiener process. The second will correspond to "rare" events and will lead to a Poisson-type behavior.

8.8.2 Normal Events

Suppose the S_i has an instantaneous percentage trend represented by the parameter μ, and an instantaneous precentage volatility of σ. For both cases considered below, we assume that S_i

[23] These states are labeled as "up" and "down," but in practice, both of the movements may be up, down, or one of them may stay the same. This choice of the terms should be regarded only as a symbolic way of naming the two states. Also, the parameters u_i and d_i may also depend on the S_i observed at that node or even at earlier nodes. Here we adopt the simpler case of state-independent up and down movements.

evolves according to the following:

$$S_{i+1} = \begin{cases} u_i S_i \text{ with probability } p_i \\ d_i S_i \text{ with probability } 1 - p_i \end{cases} \tag{8.78}$$

For the case where S_i is influenced only by "normal" events, the growth coefficients u_i and d_i can be chosen as:[24]

$$u_i = e^{\sigma\sqrt{\Delta}} \text{ for all } i \tag{8.79}$$

$$d_i = -e^{\sigma\sqrt{\Delta}} \text{ for all } i \tag{8.80}$$

and the probability p_i can be chosen as:

$$p_i = \frac{1}{2}\left[1 + \frac{\mu}{\sigma}\sqrt{\Delta}\right] \tag{8.81}$$

First, some comments. The parameters u_i, d_i, and p_i are chosen so that they are the same at every *node i*. This is the case because on the right-hand side of Eqs (8.74–8.76) there is no dependence on $S_i, i = 1, \ldots, i$. According to this, the dynamics of S_i are discretized in a fashion that is homogeneous across time. Clearly, this need not be so, and more complex u_i, d_i, or p_i can be selected as long as the dependence on Δ is kept as modeled in (8.74–8.76). Thus, in this particular case we can even remove the *i* subscript from u_i, d_i.

Second, and more important for our purposes, note what happens to parameters u_i, d_i, and p_i as Δ goes to zero.

From the definitions of these parameters we see that as $\Delta \to 0$ the u_i, d_i go toward zero. Hence, with a parameterization such as in Eq. (8.73), the movements in S_t become negligible over infinitesimal intervals. Yet, the probability of these moments go to $1/2$, a constant:

$$\lim_{\Delta \to 0} \frac{1}{2}\left[1 + \frac{\mu}{\sigma}\sqrt{\Delta}\right] = \frac{1}{2} \tag{8.82}$$

Clearly, this way of parameterizing a binomial model is consistent with the notion that the events that drive the S_i over various nodes of the tree are "normal." These events occur frequently, even in small intervals, but their size is small.

[24]This is not the only choice that will characterize "normal" events.

8.8.3 Rare Events

Now we keep the same characterization of the binomial setup, except change the way u_i, d_i, and p_i are modeled. In particular, we change the dependence on the time interval Δ.

Thus, in place of Eqs. (8.74–8.76) we assume that the parameters of the model are now given by:

$$u_i = \widehat{u} \text{ for all } i \tag{8.83}$$

$$d_i = e^{\alpha d} \text{ for all } i \tag{8.84}$$

and the probability, p_i, of an "up" movement is chosen as:

$$p_i = \lambda\Delta \tag{8.85}$$

where $0 < \lambda$ and $0 < \alpha$ are two parameters to be calibrated according to the "size" and probability of jumps that one is expecting in S_i. The $\widehat{u} \neq 1$ is also a positive constant. It represents the behavior of S_i when there is a jump. d_i is the case of no jump.

Consider the implications of this type of binomial behavior. As Δ, the time interval, is made smaller and smaller, the probability p_i of the "up" state will approach zero, whereas the probability of the "down" state will approach one. This means that S_i becomes less likely to exhibit "up" changes, $\widehat{u}$, as we consider smaller and smaller time intervals. As $\Delta \to 0$, the S_i will follow a stable path during a finite interval. Yet, even with very small Δ, there is a small probability that a "rare" event will occur because, according to (8.80):

$$P\left(S_{i+1} = e^{\alpha\Delta} S_i\right) = 1 - \lambda\Delta \tag{8.86}$$

which, depending on Δ, is perhaps very close to one.

This is the case because in small intervals:

$$d_i = e^{\alpha\Delta} \tag{8.87}$$

$$\approx 1 \tag{8.88}$$

with Δ close to zero.

Clearly, this way of modeling the binomial parameters is more in line with the rare event characterization discussed earlier in this chapter.

8.8.4 The Behavior of Accumulated Changes

The discussion above dealt with possible ways of modeling the probability and the size of a discretized two-state process S_i as a function of the discretization interval Δ. We were mainly interested in what happened to *one-step* movements in S_i as Δ is made smaller and smaller.

There is another interesting question that we can ask: Leaving aside the one-step changes, how do the accumulated movements in S_i behave as "time" passes?

In other words, instead of looking at the probability of one-step changes in S_i as i increases, we might be interested in looking at the behavior of

$$\frac{S_{i+n\Delta}}{S_i} \tag{8.89}$$

for some integer, $n > 1$, which, in a sense, represents the accumulated changes in S_i after n successive periods of length Δ has passed.

First, some comments about why we need to investigate the behavior of such a random variable.

Clearly, the modeling of S_i as a two-state process may be a reasonable approximation for the immediate future, especially if the Δ is small, but may still leave the market practitioner in the dark if the trading or investment horizon is in a *more distant future* that occurs after n steps of length Δ. For example, the interest of the market professional may be in the value of $S_T, t < T$, at expiration, rather than the immediate S_t, and the modeling of immediate one-step probabilities may not say much about this.

Hence, a market professional may be interested in the probabilistic behavior of the expiration point value S_T, as well as in its immediate behavior. And the probabilistic behavior of the accumulated changes may be quite different than the p_i that governs the immediate changes in S_i. This is the case because in n periods, the S_i may assume many values different from just u_iS_i or d_iS_i.

Thus, we consider the probabilistic behavior of the ratio:

$$\frac{S_{i+n\Delta}}{S_i} \tag{8.90}$$

which depends on the way the main parameters of the binomial-tree are modeled. The discussion will proceed in terms of an integer-valued random variable Z, which represents the number of "up" movements observed between points i and $i+n$. According to this, if beginning at point i, S_i experiences only "up" movements, then $Z = n$. If only half of the movements are up, then $Z = n/2$, and so on.

We investigate the probabilistic behavior of the logarithm of S_{i+n}/Si, instead of ratio itself, because this will linearize the random effects of u_i, d_i in terms of Z.[25]

Before we proceed further, we eliminate the i subscript from u_i, d_i, p_i, given that at least in this section they are assumed to be constant.

We can now write:

$$\log \frac{S_{i+n}}{S_i} = Z \log u + (n - Z) \log d \tag{8.91}$$

$$= Z \log \frac{u}{d} + n \log d \tag{8.92}$$

As discussed in the previous paragraph, this last equation is now a linear function of the random variable Z.

With a linear equation we can calculate the mean and variance of the random variable $\log(S_{i+1}/S_i)$ easily:

$$\mathbb{E}\left[\log \frac{S_{i+n}}{S_i}\right] = \log \frac{u}{d}\mathbb{E}[Z] + n \log d \tag{8.93}$$

[25]The u_i, d_i are multiplicative parameters. Taking the log converts a product into a sum, which is easier to analyze in asymptotic theory. Central limit theorems are formulated, in general, in terms of sums.

$$\mathbb{V}\left[\log \frac{S_{i+n}}{S_i}\right] = \left[\log \frac{u}{d}\right]^2 \mathbb{V}[Z] \tag{8.94}$$

But we know that the $\mathbb{E}[Z]$ is simply np and the $\mathbb{V}[Z]$ is $np(1-p)$.[26]

Replacing these:

$$\mathbb{E}\left[\log \frac{S_{i+n}}{S_i}\right] = \log \frac{u}{d} np + n \log d \tag{8.96}$$

$$\mathbb{V}\left[\log \frac{S_{i+n}}{S_i}\right] = \left[\log \frac{u}{d}\right]^2 np\,(1-p) \tag{8.97}$$

Here remember that

$$n = \frac{T}{\Delta} \tag{8.98}$$

Replacing this and the values of u, d, p in both (8.91) and (8.92) we can get the asymptotic equivalents of the mean and the variance. In other words, with u, d, p, given in Eqs. (8.74–8.76), the first order approximation gives:

$$\log \frac{u}{d} np + n \log d \approx \mu T \tag{8.99}$$

$$\left[\log \frac{u}{d}\right]^2 np\,(1-p) \approx \sigma^2 T \tag{8.100}$$

This is equivalent to a process that takes steps of expected size $\mu\Delta$ over $[0, T]$, and whose volatility is equal to $\sigma\sqrt{\Delta}$ at each step. Hence, the mean and variance of the rate of change of S_i modeled this way will be proportional to Δ. Such stochastic processes are called geometric processes.

One can also get the approximate (asymptotic) probability distribution of $\log\left(S_{i+n}/S_i\right)$. First, note that the $\log\left(S_{i+n}/S_i\right)$ is in fact logarithmic change in the underlying process:

$$\log\left(\frac{S_{i+n}}{S_i}\right) = \log S_{i+n} - \log S_i \tag{8.101}$$

It can be shown that if we adopt the parametrization in (8.74) and (8.75) that corresponds to normal events, then the distribution of $\left[\log S_{i+n} - \log S_i\right]$ is approximated, as $\Delta \to 0$ by

$$\left[\log S_{i+n} - \log S_i\right] \sim \mathcal{N}\left(\mu\,(n\Delta), \sigma^2\Delta\right) \tag{8.102}$$

That is, $\left[\log S_{i+n} - \log S_i\right]$ is approximately normally distributed.

If, on the other hand, the parameterization in (8.78) and (8.79) that corresponds to rare events is adopted, then the distribution of $\left[\log S_{i+n} - \log S_i\right]$ will, as $\Delta \to 0$, approximately be given by:

$$\left[\log S_{i+n} - \log S_i\right] \sim \text{Poisson} \tag{8.103}$$

These are two examples of Central Limit Theorem, where the sum of a large number of random variables starts having a recognizable distribution.

What causes this divergence between the applications of central limit theorems?

It turns out that, in order for a properly scaled sum of independent random variables to converge to a normal distribution, each element of the sum must be asymptotically negligible. The condition for asymptotic negligibility is exactly the one that distinguishes normal events from rare events. Thus, with the choice of parameters for u_i, d_i, p_i for rare events, the events are likely to be asymptotically nonnegligible, and convergence will be toward a Poisson distribution.

8.9 REFERENCES

The discussion characterizing rare events is covered in Merton (1990). The assumption

[26] The expected value is easy to calculate. If we have n independent trials, each with a probability p of "up," then the total number of expected "up" movements will be np. The $\mathbb{V}[Z]$ is slightly more complicated. The variance of Z for a single trial is:

$$p\,(1-p)^2 + (1-p)\,(0-p)^2 = p\,(1-p) \tag{8.95}$$

For n trials, the variance of the n independent movements will be n times $p(1-p)$, or $np(1-p)$

that innovation terms have a finite number of possible values simplified the discussion significantly. A reader interested in the formal arguments justifying the statements made in this chapter can consider Bremaud (1979). Bremaud adopts a martingale approach to discuss the dynamics of point processes, which can be labeled as generalizations of Poisson processes.

8.10 EXERCISES

1. Show that as $n \to \infty$:
   (a) $1\left(1-\frac{1}{n}\right)\ldots\left(1-\frac{k-1}{n}\right) \to 1$
   (b) $\left(1-\frac{\lambda}{n}\right)^n \to e^{-\lambda}$
   (c) $\left(1-\frac{\lambda}{n}\right)^k \to 1$
2. Let the random variable X_n have a binomial distribution:

$$X_n = \sum_{i=1}^{n} B_i$$

where each B_i is independent and is distributed according to

$$B_i = \begin{cases} 1 \text{ with probability } p \\ 0 \text{ with probability } 1-p \end{cases}$$

We can look at X_n as the cumulated sum of a series of events that occur over time. The events are the individual B_i. Note that there are two parameters of interest here. Namely, the p and the n. The first governs the probability of each "event" B_i, whereas the second governs the number of events. The question is, what happens to the distribution of X_n as the number of events go to infinity? There are two interesting cases, and the questions below relate to these.

   (a) Suppose now, $n \to \infty$, while $p \to 0$ such that $\lambda = np$ remains constant. That is, the probability of getting a $B_i = 1$ goes to zero as n increases. But, the expected "frequency" of getting a one remains the same. This clearly imposes a certain speed of convergence on the probability. What is the probability $P\left(X_n = k\right)$? Write the implied formula as a function of p, n, and k.
   (b) Substitute $\lambda = np$ to write $P\left(X_n = k\right)$ as a function of the three terms shown in Question 1.
   (c) Let $n \to \infty$ and obtain the Poisson distribution:

$$P\left(X_n = k\right) = \frac{\lambda^k e^{-\lambda}}{k!}$$

   (d) Remember that during this limiting process, the $p \to 0$ at a certain speed. How do you interpret this limiting probability? Where do rare events fit in?
3. Let T_α denote the first time the Wiener process hits α. When $\alpha > 0$, compute $P(T_\alpha \leq t)$.
4. Generate $n = 1000$ binomial random variables with $p = 0.001$. Consider the process B_i which is equal to 1 with probability p and 0 with probability $1-p$. Run 10,000 trials and plot a histogram of the terminal value of X_i.

CHAPTER

9

Integration in Stochastic Environments

OUTLINE

9.1 INTRODUCTION

One source of practical interest in differentiation and integration operations is the need to obtain *differential equations*. Differential equations are used to describe the dynamics of physical phenomena. A simple linear differential equation will be of the form

$$\frac{dX_t}{dt} = AX_t + By_t, \quad t \geq 0 \tag{9.1}$$

where dX_t/dt is the derivative of X_t with respect to t, and where y_t is an exogenous variable. A and B are parameters.[1]

Ordinary differential equations are necessary tools for practical modeling. For example, an engineer may think that there is some variable y_t that, together with the past values of X_t,

[1] If $B = 0$, the equation is said to be homogeneous. When y_t is independent of t, the system becomes autonomous. Otherwise, it is nonautonomous.

Financial Derivatives, Third Edition.
http://dx.doi.org/10.1016/B978-0-12-384682-2.00009-8

determines future changes in X_t. This relationship is approximated by the differential equation, which can be utilized in various applications.[2]

The following agenda is used to obtain the ordinary differential equation. First, a notion of derivative is defined. It is shown that for most functions of interest denoted by X_t, this derivative exists. Once existence is established, the agenda proceeds with approximating dX_t/dt using Taylor series expansions. After taking into consideration any restrictions imposed by the theory under consideration, one gets the differential equation.

At the end of the agenda, the *fundamental theorem of calculus* is proved to show that there is a close correspondence between the notions of integral and derivative. In fact, integral denotes a sum of increments, while derivative denotes a rate of change. It seems natural to expect that if one adds changes dX_t in a variable X_t, with initial value $X_0 = 0$, one would obtain the latest value of the variable:

$$\int_0^t dX_u = X_t \tag{9.2}$$

This suggests that for every differential equation, we can devise a corresponding *integral equation*.

In stochastic calculus, application of the same agenda is *not* possible. If unpredictable "news" arrives continuously, and if equations representing the dynamics of the phenomena under consideration are a function of such noise, a meaningful notion of derivative cannot be defined.

Yet, under some conditions, an *integral* can be obtained successfully. This permits replacing *ordinary* differential equations by *stochastic* differential equations

$$dX_t = a_t dt + \sigma_t dW_t, \quad t \in [0, \infty) \tag{9.3}$$

where future movements are expressed in terms of differentials dX_t, dt, and dW_t instead of derivatives such as dX_t/dt. These differentials are defined using a new concept of integral. For example, as h gets smaller, the increments

$$X_{t+h} - X_t = \int_t^{t+h} dX_u \tag{9.4}$$

can be used to give meaning to dX_t. In fact, at various earlier points, we made use of differentials such as dS_t or dW_t but never really discussed them in any precise fashion. The definition of the Ito integral will permit doing so.

Now, consider the SDE, which represents dynamic behavior of some asset price S_t:

$$dS_t = a(S_t, t)\, dt + \sigma(S_t, t)\, dW_t, \quad t \in [0, \infty) \tag{9.5}$$

After we take integrals on both sides, this equation implies that

$$\int_0^t dS_u = \int_0^t a(S_u, u)\, du + \int_0^t \sigma(S_u, u)\, dW_u \tag{9.6}$$

where the last term on the right-hand side is an integral with respect to increments in the Wiener process W_t.

The interpretation of the integrals on the right-hand side of (9.6) is not immediate. As discussed in Chapters 5–7, increments in W_t are too erratic during small intervals h. The rate of change of the W_t was, on the average, equal to $h^{-1/2}$, and this became larger as h became smaller.[3] If these increments are too erratic, would not their sum be infinite?

This chapter intends to show how this seemingly difficult problem can be solved.

9.1.1 The Ito Integral and SDEs

Obtaining a formal definition of the Ito integral will make the notion of a stochastic differential equation more precise. Once the integral

$$\int_0^t \sigma(S_u, u)\, dW_u \tag{9.7}$$

[2]For example, the engineer may have in mind some desired future path for X_t. Then the issue is to find the proper $\{y_t\}$ which will ensure that X_t follows this path.

[3]By the average rate of change we mean the standard deviation of $W_{t+h} - W_t$ divided by h. In Chapter 6 it was shown that under fairly general assumptions, the standard deviations of unpredictable shocks were proportional to $h^{1/2}$.

is defined in some precise way, then one could integrate both sides of the SDE in (9.5):

$$S_{t+h} - S_t = \int_t^{t+h} a(S_u, u)\, du + \int_t^{t+h} \sigma(S_u, u)\, dW_u \tag{9.8}$$

where h is some finite time interval.

From here, one can obtain the *finite difference approximation* that we used several times in Chapters 7 and 8. Indeed, if h is small, $a(S_u, u)$ and $\sigma(S_u, u)$ may not change very much during $u \in [t, t+h]$, especially if they are smooth functions of S_u and u. Then, we could rewrite this equation as:

$$S_{t+h} - S_t \approx a(S_t, t) \int_t^{t+h} du + \sigma(S_t, t) \int_t^{t+h} dW_u \tag{9.9}$$

Taking the integrals in a straightforward way, we would obtain the finite difference approximation:

$$S_{t+h} - S_t \approx a(S_t, t)\, h + \sigma(S_t, t) \left[W_{t+h} - W_t\right] \tag{9.10}$$

Rewriting,

$$\Delta S_t \approx a(S_t, t)\, h + \sigma(S_t, t)\, \Delta W_t \tag{9.11}$$

This is the SDE representation in finite intervals that we often used in previous chapters. The representation is an *approximation* for at least two reasons. First, the $\mathbb{E}_t\left[S_{t+h} - S_t\right]$ was set equal to a first-order Taylor series approximation with respect to h:

$$\mathbb{E}_t\left[S_{t+h} - S_t\right] = a(S_t, t)\, h$$

Second, the $a(S_u, u), \sigma(S_u, u), u \in [t, t+h]$ were approximated by their value at $u = t$. Both of these approximations require some smoothness conditions on $a(S_u, u)$ and $\sigma(S_u, u)$. All these imply that when we write

$$dS_t = a(S_t, t)\, dt + \sigma(S_t, t)\, dW_t \tag{9.12}$$

we, in fact, mean that in the *integral equation*,

$$\int_t^{t+h} dS_u = \int_t^{t+h} a(S_u, u)\, du + \int_t^{t+h} \sigma(S_u, u)\, dW_u \tag{9.13}$$

the second integral on the right-hand side is defined in the Ito sense and that as $h \to 0$,

$$\int_t^{t+h} \sigma(S_u, u)\, dW_u \approx \sigma(S_t, t)\, dW_t \tag{9.14}$$

That is, the diffusion terms of the SDEs are in fact Ito integrals approximated during infinitesimal time intervals.

For these approximations to make sense, an integral with respect to W_t should first be defined formally. Second, we must impose conditions on the way $a(S_t, t)$ and $\sigma(S_t, t)$ move over time. In particular, we cannot allow these I_t-measurable parameters to be too erratic.

9.1.2 The Practical Relevance of the Ito Integral

In practice, the Ito integral is used less frequently than stochastic differential equations. Practitioners almost never use the Ito integral *directly* to calculate derivative asset prices. As will be discussed later, arbitrage-free prices are calculated either by using partial differential equation methods or by using martingale transformations. In neither of these cases is there a need to calculate any Ito integrals directly.

It may thus be difficult at this point to see the practical relevance of this concept from the point of view of, say, a trader. It may appear that defining the Ito integral is essentially a theoretical exercise, with no practical implications. A practitioner may be willing to accept that the Ito integral exists and prefer to proceed directly into using SDEs.

The reader is cautioned against this. Understanding the definition of the Ito integral is important for (at least) two reasons. First, as mentioned earlier, a stochastic differential equation can be defined only in terms of the Ito integral. To understand the real meaning behind the SDEs, one has to have some understanding of the Ito integral. Otherwise, errors can be made in applying SDEs to practical problems.

This brings us to the second reason why the Ito integral is relevant. Given that SDEs are defined for infinitesimal intervals, their use in finite intervals may require some *approximations*. In fact, the approximation in (9.14) may not be valid if h is not "small." Then a new approximation will have to be defined using the Ito integral.

This point is important from the point of view of pricing financial derivatives, since in practice one always does calculations using finite intervals. For example, "one day" is clearly not an infinitesimal interval, and the utilization of SDEs for such periods may require approximations. The precise form of these approximations will be obtained by taking into consideration the definition of Ito integral.

To summarize, the ability to go from a stochastic difference equation defined over the finite intervals,

$$\Delta S_k = a_k h + \sigma_k \Delta W_k, \quad k = 1, 2, \ldots n \qquad (9.15)$$

to stochastic differential equations,

$$dS_t = a(S_u, u)\, dt + \sigma(S_u, u)\, dW_u, \quad t \in [0, \infty) \qquad (9.16)$$

and vice versa, is the ability to interpret dW_t by defining $\int_t^{t+h} \sigma(S_u, u)\, dW_u$ in a meaningful manner. This can only be done by constructing a stochastic integral.

9.2 THE ITO INTEGRAL

The Ito integral is one way of defining sums of uncountable and unpredictable random increments over time. Such an integral cannot be obtained by utilizing the method used in the Riemann–Stieltjes integral. It is useful to see why this is so.

As seen earlier, increments in a Wiener process, dW_t, represent random variables that are unpredictable, even in the immediate future. The value of the Wiener process at time t, written as W_t, is then a sum of an uncountable number of independent increments:

$$W_t = \int_0^t dW_u \qquad (9.17)$$

(Remember that at time zero, the Wiener process has a value of zero. Hence, $W_0 = 0$.) This is the simplest stochastic integral one can write down.

A more relevant stochastic integral is obtained by integrating the innovation term in the SDE:

$$\int_0^t \sigma(S_u, u)\, dW_u \qquad (9.18)$$

The integrals in (9.17) and (9.18) are summations of very erratic random variables, since two shocks that are $\varepsilon > 0$ apart from each other, dW_t and $dW_{t+\varepsilon}$, are still uncorrelated. The question that arises is whether the sum of such erratic terms can be meaningfully defined. After all, the sum of so many (uncountable) erratic elements can very well be unbounded.

Consider again the way standard calculus defines the integral.

9.2.1 The Riemann–Stieltjes Integral

Suppose we have a nonrandom function $F(x_t)$, where x_t is a deterministic variable of time $F(\cdot)$ is continuous and differentiable, with the derivative

$$\frac{dF(x_t)}{dx_t} = f(x_t) \qquad (9.19)$$

In this particular case where the derivative $f(\cdot)$ exists, the Riemann–Stieltjes integral can be written in two ways:

$$\int_0^T f(x_t)\, dx_t = \int_0^T dF(x_t) \qquad (9.20)$$

The integral on the left-hand side is taken with respect to x_t, where t varies from 0 to T. Then, the value of $f(\cdot)$ at each x_t is multiplied by the infinitesimal increment dx_t. These (uncountably many) values are used to obtain the integral. This notation is in general preserved for the Riemann integral.

In the notation on the right-hand side, the integral is taken with respect to $F(\cdot)$. Increments in $F(\cdot)$ are used to obtain the integral. We can complicate the latter notation further. For example, we may be interested in calculating the integral

$$\int_0^T g(x_t)\, dF(x_t) \tag{9.21}$$

Here, we have an integral of a function $g(x_t)$ taken with respect to $F(\cdot)$.

Similar notation occurs when we deal with expectations of random variables. For example, $F(\cdot)$ may represent the distribution function of a random variable x_t, and we may want to calculate the expected value of some $g(x_t)$ for *fixed* t[4]:

$$\mathbb{E}\left[g(x_t)\right] = \int_{-\infty}^{\infty} g(x_t)\, dF(x_t) \tag{9.22}$$

Heuristically, in this integral, x_t is varied from minus to plus infinity, and the corresponding values of $g(\cdot)$ are averaged using the increments in $dF(\cdot)$. $dF(\cdot)$ in this case represents the probability associated with those values.

Note the important difference between the integrals in (9.21) and (9.22). In the first case, it is the t that moves from 0 to T. The value of x_t for a particular t is left unspecified. It could very well be a random variable. This would make the integral itself a random variable.

The integral in (9.22) is quite different. The t is constant, and it is x_t that goes from minus to plus infinity. The integral is not a random variable.

For the case when there are no random variables in the picture, the Riemann–Stieltjes integral was defined as a limit of some infinite sum. The integral would exist as long as this limit was well defined. To highlight differences with Ito integral, we review Riemann–Stieltjes methodology once again.

Suppose we would like to calculate

$$\int_0^T g(x_t)\, dF(x_t)$$

The formal calculation using the Riemann–Stieltjes methodology is based on the familiar construction where the interval $[0, T]$ is *partitioned* into n smaller intervals using the times

$$t_0 = 0 < t_1 < \cdots < t_n \tag{9.23}$$

Then the finite Riemann sum V_n is defined:

$$V_n = \sum_{i=0}^{n-1} g\left(x_{t_{i+1}}\right)\left[F\left(x_{t_{i+1}}\right) - F\left(x_{t_i}\right)\right] \tag{9.24}$$

The right-hand side of this equation is a sum of elements such as

$$g\left(x_{t_{i+1}}\right)\left[F\left(x_{t_{i+1}}\right) - F\left(x_{t_i}\right)\right] \tag{9.25}$$

which is the product of $g\left(x_{t_i}\right)$ and $\left[F\left(x_{t_{i+1}}\right) - F\left(x_{t_i}\right)\right]$. The first term represents $g(\cdot)$ evaluated at a point x_{t_i}. The second term resembles the increments $dF(x_t)$. Each element $g\left(x_{t_{i+1}}\right)\left[F\left(x_{t_{i+1}}\right) - F\left(x_{t_i}\right)\right]$ can be visualized as a rectangle with base $\left[F\left(x_{t_{i+1}}\right) - F\left(x_{t_i}\right)\right]$ and height $g\left(x_{t_{i+1}}\right)$.

V_n is the sum of all such rectangles. If consecutive $t_i, i = 0, \ldots, n$ are not very distant from each other—that is, if we have a *fine* partition of $[0, T]$—this approximation may work reasonably well. In other words, if the function $g(\cdot)$ is integrable, then the limit

$$\lim_{\sup_i |t_{i+1}-t_i| \to 0} \sum_{i=0}^{n-1} g\left(x_{t_{i+1}}\right)\left[F\left(x_{t_{i+1}}\right) - F\left(x_{t_i}\right)\right] = \int_0^T g(x_t)\, dF(x_t) \tag{9.26}$$

will exist and will be called the Riemann–Stieltjes integral. The reader should read this equality as a definition. The integral is defined as the limit of the sums on the right-hand side.[5] The sums V_n are called Riemann sums.[6]

[4]When the function $g(\cdot)$ is the square or the cube of x_t, this integral will simply be the second or third moment.

[5]That is, if this limit converges.

[6]There are many different ways rectangles can approximate the area under a curve. One can pick the base of the rectangle the same way, but change the height of the rectangle to either $g(x_{t_i})$ or to $g\left(\frac{x_{t_{i+1}}+x_{t_i}}{2}\right)$.

9.2.2 Stochastic Integration and Riemann Sums

Hence, the value of the Riemann–Stieltjes integral can be approximated using rectangles with a "small" base and varying heights. Can we adopt similar reasoning in the case of stochastic integration?

We can ask this question more precisely by considering the SDE written over finite intervals of *equal* length h[7]:

$$S_k - S_{k-1} = a\left(S_{k-1}, k\right) h + \sigma\left(S_{k-1}, k\right) \Delta W_k, \quad k = 1, 2, \ldots, n \tag{9.27}$$

Suppose we sum the increments ΔS_k on the left-hand side of (9.27):

$$\sum_{k=1}^{n-1} \left[S_k - S_{k-1}\right] = \sum_{k=1}^{n-1} \left[a\left(S_{k-1}, k\right) h\right] + \sum_{k=1}^{n-1} \sigma\left(S_{k-1}, k\right) \left[\Delta W_k\right] \tag{9.28}$$

Can we use a methodology similar to the Riemann–Stieltjes approach and define an integral with respect to the random variable S_t as (some type of) a limit

$$\int_0^T dS_u = \lim_{n \to \infty} \left\{ \sum_{k=1}^{n-1} \left[a\left(S_{k-1}, k\right) h\right] + \sum_{k=1}^{n-1} \sigma\left(S_{k-1}, k\right) \left[\Delta W_k\right] \right\} \tag{9.29}$$

where, as usual, it is assumed that $T = nh$?

[7]By considering intervals of equal length, the partition of $[0, T]$ can be made finer with $n \to \infty$. Otherwise, the condition $\sup_i \left|t_i - t_{i-1}\right| \to 0$ has to be used.

The first term on the right-hand side of (9.29) does not contain any random terms once information in time k becomes available. More importantly, the integral is taken with respect to increments in time h. By definition, time is a smooth function and has "finite variation." This means that the same procedure used for the Riemann–Stieltjes case can be applied to define an integral such as[8]

$$\int_0^T a\left(S_u, u\right) du = \lim_{n \to \infty} \sum_{k=1}^{n-1} \left[a\left(S_{k-1}, k\right) h\right] \tag{9.31}$$

However, the second term on the right-hand side of (9.28) contains random variables even after I_{k-1} is revealed. In fact, as of time $k-1$, the term

$$W_k - W_{k-1} \tag{9.32}$$

is a random variable, and the sum

$$\sum_{k=1}^{n-1} \sigma\left(S_{k-1}, k\right) \left[W_k - W_{k-1}\right] \tag{9.33}$$

is an integral with respect to a *random variable*.

We can ask several questions:

- Which notion of limit should be used? The question is relevant because the sum in (9.33) is random and, in the limit, should converge to a random variable. The deterministic notion of limit utilized by Riemann–Stieltjes methodology cannot be used here.
- Under what conditions would such a limit converge (i.e. do the sums in (9.33) really have a meaningful limit)?
- What are the properties of the limiting random variable?

[8]The sum on the right-hand side can be written in more detailed form as

$$\lim_{n \to \infty} \sum_{k=1}^{n-1} \left[a\left(S_{(k-1)h}, kh\right)\right] \left[(k)h - (k-1)h\right] \tag{9.30}$$

with $kh = t_k$.

We limit our attention to a particular integral determined by the error terms in the SDEs. It turns out that, under some conditions, it is possible to define a stochastic integral as the limit *in mean square* of the random sum:

$$\sum_{k=1}^{n-1} \sigma\left(S_{k-1}, k\right)\left[W_k - W_{k-1}\right] \tag{9.34}$$

This integral would be a random variable.

The use of mean square convergence implies that the difference between the sum

$$\sum_{k=1}^{n-1} \sigma\left(S_{k-1}, k\right)\left[W_k - W_{k-1}\right] \tag{9.35}$$

and the random variable called the Ito integral,

$$\int_0^T \sigma\left(S_u, u\right) dW_u \tag{9.36}$$

has a variance that goes to zero as n increases toward infinity. Formally:

$$\lim_{n\to\infty} \mathbb{E}\left[\sum_{k=1}^{n-1} \sigma\left(S_{k-1}, k\right)\left[W_k - W_{k-1}\right] - \int_0^T \sigma\left(S_u, u\right) dW_u\right]^2 = 0 \tag{9.37}$$

9.2.3 Definition: The Ito Integral

We can now provide a definition of the Ito integral within the context of stochastic differential equations.

Definition 21. Consider the finite interval approximation of the stochastic differential equation

$$S_k - S_{k-1} = a\left(S_{k-1}, k\right) h + \sigma\left(S_{k-1}, k\right) \left[W_k - W_{k-1}\right], \quad k = 1, 2, \ldots, n \tag{9.38}$$

where $\left[W_k - W_{k-1}\right]$ is a standard Wiener process with zero mean and variance h. Suppose that the $\sigma\left(S_t, t\right)$ are nonanticipative, in the sense that they are independent of the future, and that the random variables $\sigma\left(S_t, t\right)$ are "non-explosive:"

$$\mathbb{E}\left[\int_0^T \sigma\left(S_t, t\right)^2 dt\right] < \infty \tag{9.39}$$

Then the Ito integral

$$\int_0^T \sigma\left(S_t, t\right) dW_t \tag{9.40}$$

is the mean square limit,

$$\sum_{k=1}^{n-1} \sigma\left(S_{k-1}, k\right)\left[W_k - W_{k-1}\right] \to \int_0^T \sigma\left(S_u, u\right) dW_u \tag{9.41}$$

as $n \to \infty (h \to 0)$.[9]

According to this definition, as the number of intervals goes to infinity and the length of each interval becomes infinitesimal, the finite sum will approach the random variable represented by the Ito integral. Clearly, the definition makes sense only if such a limiting random variable exists. The assumption that $\sigma\left(S_{k-1}, k\right)$ is nonanticipating turns out to be a fundamental condition for the existence of such a limit.[10]

To summarize, we see three major differences between deterministic and stochastic integrations. First, the notion of limit used in stochastic integration is different. Second, the Ito integral is defined for nonanticipative functions only. And third, while integrals in standard calculus are defined using the actual "paths" followed by functions, stochastic integrals are defined within *stochastic equivalence*. It is essentially these differences that make some rules of stochastic calculus different from standard calculus.

The following example illustrates the utilization of mean square convergence in defining the Ito integral. In a second example, we show why the Ito integral cannot be defined "pathwise."

[9]Remember that $[0, T]$ is partitioned into n equal intervals, with $T = nh$.

[10]One technical point is whether the limiting random variable, that is, the Ito integral, depends on the choice of how one partitions the $[0, T]$. It can be shown that the choice of partition does not influence the value of the Ito integral.

9.2.4 An Expository Example

The Ito integral is a limit. It is the mean square limit of a certain finite sum. Thus, in order for the Ito integral to exist, some appropriate sums must converge.

Given proper conditions, one can show that Ito sums converge and that the corresponding Ito integral exists. Yet it is, in general, not possible to *explicitly* calculate the mean square limit. This can be done only in some special cases. In this section, we consider an example where the mean square limit can be evaluated explicitly.[11]

Suppose one has to evaluate the integral

$$\int_0^T x_t dx_t \tag{9.42}$$

where it is known that $x_0 = 0$.

If x_t was a deterministic variable, one could calculate this integral using the finite sums defined in (9.24). To do this, one would first partition the interval $[0, T]$ into n smaller subintervals all of size h using

$$t_0 = 0 < t_1 < \cdots < t_n \tag{9.43}$$

where, as usual, $T = nh$ and for any $i, t_{i+1} - t_i = h$.[12] Second, one would define the sums

$$V_n = \sum_{i=0}^{n-1} x_{t_{i+1}} \left[x_{t_{i+1}} - x_{t_i}\right] \tag{9.44}$$

and let n go to infinity. The result is well known. The Riemann–Stieltjes integral of (9.42) with $x_0 = 0$ will be given by

$$\int_0^T x_t dx_t = \frac{1}{2} x_T^2 \tag{9.45}$$

This situation can easily be seen in Figure 9.1, where we consider an arbitrary function of time x_t and use a single rectangle to obtain the area under the curve.[13]

If x_t is a Wiener process, the same approach cannot be used. First of all, the V_n must be modified to

$$V_n = \sum_{i=0}^{n-1} x_{t_i} \left[x_{t_{i+1}} - x_{t_i}\right] \tag{9.46}$$

In other words, the first x_t has to be evaluated at time t_i instead of at t_{i+1}, because otherwise these terms will fail to be *nonanticipating*. The x_{t_i} will be *unknown* as of time t_i, and will be correlated with the increments $\left[x_{t_{i+1}} - x_{t_i}\right]$. In the case of the Riemann–Stieltjes integral, one could use either type of sum and still get the same answer in the end. In the case of stochastic integration, results will change depending on whether one used $x_{t_{i+1}}$ or x_{t_i}. As will be seen later, it is a fundamental condition of the Ito integral that the integrands be nonanticipating.

Second, V_n is now a random variable and simple limits cannot be taken. In taking the limit of V_n, one has to use a probabilistic approach. As mentioned earlier, the Ito integral uses the mean square limit.

Thus, we have to determine a limiting random variable V such that

$$\lim_{n\to\infty} \mathbb{E}\left[V_n - V\right]^2 \to 0 \tag{9.47}$$

Or, equivalently,

$$\lim_{n\to\infty} \mathbb{E}\left[\sum_{i=0}^{n-1} x_{t_i} \Delta x_{t_i} - V\right]^2 \to 0 \tag{9.48}$$

where for simplicity we let

$$\Delta x_{t_i} = \left[x_{t_{i+1}} - x_{t_i}\right] \tag{9.49}$$

Below we calculate this limit explicitly.

[11] This is in contrast to a proof where it is shown that the limit "exists."

[12] Equal-sized subintervals are a convenience. The same result can be shown with unequal $t_{i+1} - t_i$ as well.

[13] A single rectangle works because the function being integrated, $f(x_t)$, is just the 45-degree line, $f(x_t) = x_t$.

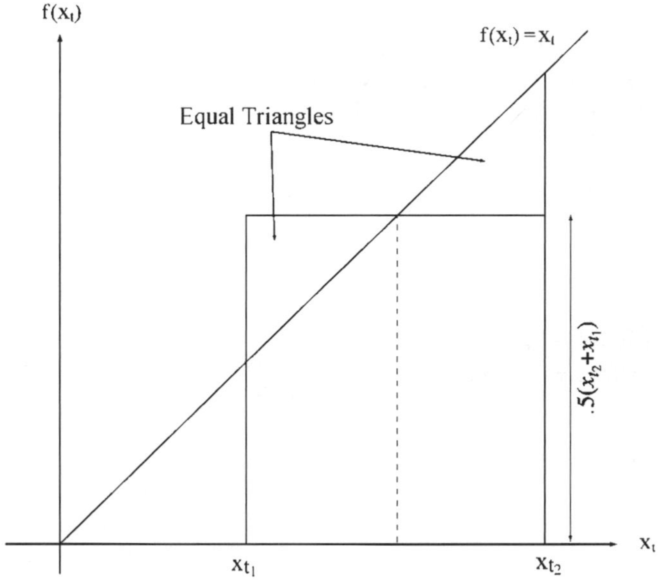

FIGURE 9.1 Use a single rectangle to obtain the area under the curve in case of $f(x_t) = x_t$.

9.2.4.1 *Explicit Calculation of Mean Square Limit*

We intend to calculate the limiting random variable V step by step to clarify the meaning of the Ito integral as a mean square limit of a random sum. The first step is to manipulate the terms inside V_n.

We begin by noting that for any a and b we have

$$(a+b)^2 = a^2 + b^2 + 2ab \tag{9.50}$$

or

$$ab = \frac{1}{2}\left[(a+b)^2 - a^2 - b^2\right] \tag{9.51}$$

From (9.44), and letting $a = x_{t_i}$ and $b = \Delta x_{t_i}$ gives

$$V_n = \frac{1}{2}\sum_{i=0}^{n-1}\left[\left(x_{t_i} + \Delta x_{t_{i+1}}\right)^2 - x_{t_i}^2 - \Delta x_{t_{i+1}}^2\right] \tag{9.52}$$

But

$$x_{t_i} + \Delta x_{t_{i+1}} = x_{t_{i+1}} \tag{9.53}$$

which gives:

$$V_n = \frac{1}{2}\left[\sum_{i=0}^{n-1} x_{t_{i+1}}^2 - \sum_{i=0}^{n-1} x_{t_i}^2 - \sum_{i=0}^{n-1}\Delta x_{t_{i+1}}^2\right] \tag{9.54}$$

Now the first and second summations in (9.54) are the same except for the very first and last elements. Canceling similar terms, and noting that $x_0 = 0$ by definition,[14]

$$V_n = \frac{1}{2}\left[x_T^2 - \sum_{i=0}^{n-1}\Delta x_{t_{i+1}}^2\right] \tag{9.55}$$

Note that x_T is independent of n, and consequently the mean square limit of V_n will be determined by the mean square limit of the term $\sum_{i=0}^{n-1}\Delta x_{t_{i+1}}^2$.

In other words, we now have to find the Z in

$$\lim_{n\to\infty}\mathbb{E}\left[\sum_{i=0}^{n-1}\Delta x_{t_{i+1}}^2 - Z\right]^2 = 0 \tag{9.56}$$

[14]By construction $t_n = T$.

In this expression, there are two "squares" on the left-hand side. One is due to the random variable itself, and the other to the type of limit we are using. Hence, the limit will involve fourth powers of $\Delta x_{t_{i+1}}$.

First, we calculate the expectation:

$$\mathbb{E}\left[\sum_{i=0}^{n-1} \Delta x_{t_{i+1}}^2\right] \tag{9.57}$$

This will be a good candidate for Z. Taking expectations in a straightforward way,

$$\mathbb{E}\left[\sum_{i=0}^{n-1} \Delta x_{t_{i+1}}^2\right] = \sum_{i=0}^{n-1} \mathbb{E}\left[\Delta x_{t_{i+1}}^2\right] = \sum_{i=0}^{n-1} \left[t_{i+1} - t_i\right] \tag{9.58}$$

which simplifies to

$$\sum_{i=0}^{n-1} \left[t_{i+1} - t_i\right] = T \tag{9.59}$$

Now using this as a candidate for Z, we can evaluate the expectation

$$\begin{aligned}
&\mathbb{E}\left[\sum_{i=0}^{n-1} \Delta x_{t_{i+1}}^2 - T\right]^2 \\
&= \mathbb{E}\left[\sum_{i=0}^{n-1} \Delta x_{t_{i+1}}^4 + 2\sum_{i=0}^{n-1}\sum_{j<i}^{n-1}\left[\Delta x_{t_{i+1}}^2\right]\right. \\
&\quad \left.\times \left[\Delta x_{t_{j+1}}^2\right] + T^2 - 2T\sum_{i=0}^{n-1}\left[\Delta x_{t_{i+1}}^2\right]\right]
\end{aligned} \tag{9.60}$$

We consider the components of the right-hand side of (9.60) individually. Realizing that Wiener process increments are independent,

$$\mathbb{E}\left[\Delta x_{t_{i+1}}^2 \Delta x_{t_{j+1}}^2\right] = \left(t_{i+1} - t_i\right)\left(t_{j+1} - t_j\right) \tag{9.61}$$

and

$$\mathbb{E}\left[\Delta x_{t_{i+1}}^4\right] = 3\left(t_{i+1} - t_i\right)^2 \tag{9.62}$$

we obtain

$$\begin{aligned}
\mathbb{E}\left[\sum_{i=0}^{n-1} \Delta x_{t_{i+1}}^2 - T\right]^2 &= \sum_{i=0}^{n-1} 3\left(t_{i+1} - t_i\right)^2 \\
&\quad + 2\sum_{i=0}^{n-1}\sum_{j<i}^{n-1}\left(t_{i+1} - t_i\right)\left(t_{j+1} - t_j\right) \\
&\quad + T^2 - 2T\sum_{i=0}^{n-1}\left(t_{i+1} - t_i\right)
\end{aligned} \tag{9.63}$$

Now we use the fact that $t_{i+1} - t_i = h$, for all i, since all intervals are the same size. We have the following:

$$\sum_{i=0}^{n-1} 3\left(t_{i+1} - t_i\right)^2 = 3nh^2 \tag{9.64}$$

$$2\sum_{i=0}^{n-1}\sum_{j<i}^{n-1}\left(t_{i+1} - t_i\right)\left(t_{j+1} - t_j\right) = n\left(n-1\right)h^2 \tag{9.65}$$

and

$$T^2 - 2T\sum_{i=0}^{n-1}\left(t_{i+1} - t_i\right) = -T^2 = -n^2h^2 \tag{9.66}$$

Put all these together,

$$\mathbb{E}\left[\sum_{i=0}^{n-1} \Delta x_{t_{i+1}}^2 - T\right]^2 = 3nh^2 + n(n-1)h^2 - n^2h^2 \tag{9.67}$$

which means that

$$\mathbb{E}\left[\sum_{i=0}^{n-1} \Delta x_{t_{i+1}}^2 - T\right]^2 = 2nh^2 = 2Th \tag{9.68}$$

This implies that as $n \to \infty$, the size of the intervals will go to zero, and

$$\lim_{h\to 0} \mathbb{E}\left[\sum_{i=0}^{n-1} \Delta x_{t_{i+1}}^2 - T\right]^2 = \lim_{h\to 0} 2hT = 0 \quad (9.69)$$

Thus, the mean square limit of $\sum_{i=0}^{n-1} \Delta x_{t_{i+1}}^2$ is T.

Going back to V_n,

$$V_n = \frac{1}{2}\left[x_T^2 - \sum_{i=0}^{n-1} \Delta x_{t_{i+1}}^2\right] \quad (9.70)$$

we find the mean square limit of V_n by using the mean square limit of just obtained:

$$\lim_{n\to\infty} \mathbb{E}\,[V_n]^2 = \frac{1}{2}\left[x_T^2 - T\right] \quad (9.71)$$

The term on the right-hand side is the Ito integral,

$$\int_0^T x_t\, dx_t \quad (9.72)$$

We see that the Ito integral results in a different expression from that in standard calculus. The Ito integral is given by

$$\int_0^T x_t\, dx_t = \frac{1}{2}\left[x_T^2 - T\right] \quad (9.73)$$

In the case of the Riemann integral, there was no additional term T.

This is one example where the Ito integral can be calculated explicitly using mean square limits. We find that the Ito integral is the limiting random variable $\frac{1}{2}\left[x_T^2 - T\right]$.

9.2.4.2 An Important Remark

In the previous section it was shown that

$$\lim_{h\to 0} \mathbb{E}\left[\sum_{i=0}^{n-1} \Delta x_{t_{i+1}}^2 - T\right]^2 = 0 \quad (9.74)$$

It is interesting to convert this into integral notation.

Assume that x_t is a Wiener process and consider the integral

$$\int_0^T (dx_t)^2 \quad (9.75)$$

which can be interpreted as the sum of squared increments in x_t.

If this integral exists in the Ito sense, then by definition,

$$\lim_{h\to 0} \mathbb{E}\left[\sum_{i=0}^{n-1} \Delta x_{t_{i+1}}^2 - \int_0^T (dx_t)^2\right]^2 = 0 \quad (9.76)$$

But we know that

$$\int_0^T dt = T \quad (9.77)$$

Putting the equalities (9.74), (9.76), and (9.77) together, we obtain a result that may seem a bit "unusual" to one who is used to working with standard calculus:

$$\int_0^T (dx_t)^2 = \int_0^T dt \quad (9.78)$$

where the equality holds in the mean square sense. It is in this sense that if W_t represents a Wiener process, for infinitesimal dt, one can write:

$$dW_t^2 = dt \quad (9.79)$$

In fact, in all practical calculations dealing with stochastic calculus, it is a common practice to replace the terms involving dW_t^2 by dt. The preceding discussion traces the logic behind this procedure. The equality should be interpreted in the sense of mean square convergence.

9.3 PROPERTIES OF THE ITO INTEGRAL

Consider the stochastic differential equation

$$dS_t = a\,(S_t, t)\, dt + \sigma\,(S_t, t)\, dW_t \quad (9.80)$$

Integrating this equation over an interval $[0, T]$, we obtain

$$\int_0^T dS_t = \int_0^T a(S_t, t)\, dt + \int_0^T \sigma(S_t, t)\, dW_t \tag{9.81}$$

where the second integral on the right-hand side is defined in the Ito sense. What can we say about the properties of this integral?

9.3.1 The Ito Integral is a Martingale

It turns out that the Ito integral is a martingale. This property is useful in modeling the innovation terms of asset prices in financial theory and for practical calculations of asset prices.

Models that describe the dynamic behavior of asset prices contain innovation terms that represent unpredictable news. As a result, an integral of the form [15]

$$\int_t^{t+\Delta} \sigma_u\, dW_u \tag{9.82}$$

is a sum of unpredictable disturbances that affect asset prices during an interval of length Δ. Now, if each increment is unpredictable given the information set at time t, the sum of these increments should also be unpredictable. This makes the integral shown in (9.82) a *martingale difference*:

$$\mathbb{E}\left[\int_t^{t+\Delta} \sigma_u\, dW_u\right] = 0 \tag{9.83}$$

Then, the integral

$$\int_0^t \sigma_u\, dW_u \tag{9.84}$$

becomes a martingale:

$$\mathbb{E}_s\left[\int_t^{t+\Delta} \sigma_u\, dW_u\right] = \int_0^s \sigma_u\, dW_u \tag{9.85}$$

[15]We are simplifying the notation by letting $\sigma(S_u, u) = \sigma_u$.

Hence, the existence of unpredictable innovation terms in equations describing the dynamics of asset prices coincides well with the martingale property of the Ito integral. The condition that ensures this martingale property is the one that requires σ_t be nonanticipative given the information set I_t.

We consider two cases of interest.

9.3.1.1 Case 1

Assume that the volatility parameter $\sigma(S_t, t)$ is a constant independent of the level of asset price S_t, and of time t:

$$\sigma(S_t, t) = \sigma \tag{9.86}$$

Then the Ito integral will be identical to the Riemann integral and will be given by

$$\int_t^{t+\Delta} \sigma\, dW_t = \sigma\left[W_{t+\Delta} - W_t\right] \tag{9.87}$$

Consider a forecast of the integral

$$\mathbb{E}_s\left[\int_t^{t+\Delta} \sigma\, dW_u \,\middle|\, \int_0^t \sigma\, dW_u\right] = \int_0^t \sigma\, dW_u$$
$$= \sigma\left(W_t - W_0\right) \tag{9.88}$$

where $\Delta > 0$. This is the case because increments in the Wiener process have zero mean and are uncorrelated:

$$\mathbb{E}_s\left[\sigma\left(W_{t+\Delta} - W_0\right) | \left(W_{t+\Delta} - W_0\right)\right] \tag{9.89}$$
$$= \mathbb{E}_s\left[\sigma\left(W_{t+\Delta} - W_t\right) + \sigma\left(W_t - W_0\right) | \right.$$
$$\left. \times \left(W_{t+\Delta} - W_0\right)\right] \tag{9.90}$$
$$= \sigma\left(W_t - W_0\right) \tag{9.91}$$

We see again that the Ito integral has the martingale property.[16]

Thus, when σ is constant, the Riemann and Ito integrals will coincide and both will be martingales.

9.3.1.2 Case 2

On the other hand, if σ depends on S_t, which in turn depends on W_t, the Ito integral diverges

[16]Remember that $W_0 = 0$.

from the Riemann integral and remains a martingale, whereas the Riemann integral ceases to be one.

For example, if the price of the underlying asset has a geometric distribution with the diffusion term

$$\sigma(S_t, t) = \sigma S_t \tag{9.92}$$

then the Ito integral will be different from the Riemann integral, and using Riemann sums to approximate the Ito integral may lead to self-contradiction.

This is illustrated by the following example.

9.3.1.3 An Example

Suppose asset prices follow the SDE

$$dS_t = a(S_t, t)\,dt + \sigma(S_t, t)\,dW_t, \quad 0 \leq t \tag{9.93}$$

where the drift and diffusion parameters are given as

$$a(S_t, t) = \mu S_t \tag{9.94}$$

and

$$\sigma(S_t, t) = \sigma S_t \tag{9.95}$$

That is, both parameters are proportional to the last observed asset price S_t.

Consider again a small interval of length Δ and integrate this SDE:

$$\int_t^{t+\Delta} dS_u = \int_t^{t+\Delta} \mu S_u du + \int_t^{t+\Delta} \sigma S_u dW_u \tag{9.96}$$

Note that the term $\sigma S(W_t)$ depends on W_t indirectly, through S_t.[17]

Now consider what happens when we try to approximate the second integral on the right-hand side using Riemann sums.

One approximation used by Riemann sums uses the values of the Wiener process observed at "midpoints" of subintervals. This amounts to calculating first the terms

$$\sigma S\left(\frac{W_{t+\Delta} + W_t}{2}\right) \tag{9.97}$$

and then multiplying these by the "base" of the rectangle, $W_{t+\Delta} - W_t$.

Riemann sums would then involve terms such as

$$\sigma S\left(\frac{W_{t+\Delta} + W_t}{2}\right)[W_{t+\Delta} - W_t] \tag{9.98}$$

Clearly, the expectation of such terms is not zero, since the argument of $S(\cdot)$ and the base of the rectangle contains terms that are correlated.

We consider the simple case where the SDE is given by

$$dS_t = \sigma W_t dW_t \tag{9.99}$$

The innovation terms in this equation will be of the form

$$\int_t^{t+\Delta} \sigma W_u dW_u \tag{9.100}$$

To approximate such an integral with a Riemann sum, a rectangle with base $W_{t+\Delta} - W_t$ and height $\sigma\left(\frac{W_{t+\Delta}+W_t}{2}\right)$ may be used[18]:

$$\int_t^{t+\Delta} \sigma W_u dW_u \approx \left[\sigma \frac{W_{t+\Delta} + W_t}{2}\right](W_{t+\Delta} - W_t) \tag{9.101}$$

But, applying the conditional expectation operator $\mathbb{E}_t[\cdot]$ to the right-hand side,

$$\mathbb{E}\left[\left(\frac{W_{t+\Delta} + W_t}{2}\right)(W_{t+\Delta} - W_t)\Big| W_t\right]$$

$$= \mathbb{E}\left[\frac{1}{2}\left(W_{t+\Delta}^2 + W_t^2\right)\Big| W_t\right] \tag{9.102}$$

$$= \frac{1}{2}\Delta \tag{9.103}$$

and $\Delta \neq 0$. This means that the approximating sum has a conditional expectation that is *not* equal to zero. It is *predictable*. Clearly, this contradicts the claim that the integral on the left-hand side represents an innovation term.

[17] Here we abuse the notation in writing $S(W_t) = S_t$. But it simplifies the exposition.

[18] For simplicity, we use *one* rectangle. In fact, much finer partitions of the interval $[t, t + \Delta]$ can be used.

If such correlations are not zero, evaluating the Ito integral using Riemann sums will imply innovation disturbance terms with nonzero expectations:

$$\mathbb{E}\left[\int_t^{t+\Delta} \sigma_s dW_s\right] \neq 0, \quad 0 < \Delta \tag{9.104}$$

In order to preserve the *nonanticipating* property of $\sigma(S_t, t)$, approximation of the Ito integral must use rectangles such as

$$\sigma(S_t, t)(W_{t+\Delta} - W_t) \tag{9.105}$$

where the terms $\sigma(S_t, t)$ will, by definition, be uncorrelated with the increments ΔW_t.

The preceding discussion shows that the Riemann integral is not consistent with assumptions made in asset pricing models, except in the very special case when

$$\sigma(S_t, t) = \sigma(t) \tag{9.106}$$

There is an additional comment that relates to the same point. If the functions being integrated are not nonanticipating, then there will be no guarantee that the partial sums used to construct the Ito integral will converge in mean square to a meaningful random variable. Hence, there is an even more fundamental problem than losing the martingale property: the integral may not exist.

The next section discusses this point briefly.

9.3.2 Pathwise Integrals

In stochastic calculus, one occasionally encounters the statement that stochastic integrals cannot be defined *pathwise*. What does this mean?

Consider the binomial process $S_{t_{i+1}} - S_{t_i}, i = 1, 2, \ldots, n$, measured over discrete intervals of length Δ during a period $[0, T]$:

$$S_{t_{i+1}} - S_{t_i} = \begin{cases} \sqrt{\Delta} & \text{with probability } p \\ -\sqrt{\Delta} & \text{with probability } 1 - p \end{cases} \tag{9.107}$$

where, as usual, $T = n\Delta$.

A typical *path* of this process will be a sequence of $\sqrt{\Delta}$ and $-\sqrt{\Delta}$ following each other. For example, a typical realization may look like

$$\left\{\sqrt{\Delta}, \sqrt{\Delta}, -\sqrt{\Delta}, \sqrt{\Delta}, \ldots\right\} \tag{9.108}$$

Suppose a financial analyst has to approximate an integral of the form

$$\int_0^T f(S_t)\, dS_t$$

using a finite sum such as:

$$V_n = \sum_{i=0}^{n-1} f\left(S_{t_{i+1}}\right)\left[S_{t_{i+1}} - S_{t_i}\right] \tag{9.109}$$

Suppose V_n is calculated using a particular path for S_t. For example, consider the path where plus and minus alternate:

$$\left\{\sqrt{\Delta}, -\sqrt{\Delta}, \sqrt{\Delta}, -\sqrt{\Delta}, \ldots, \sqrt{\Delta}\right\} \tag{9.110}$$

Replacing the $S_{t_{i+1}} - S_{t_i}$ in V_n with these observed values, we get

$$V_n = \left[f\left(\sqrt{\Delta}\right)\left(\sqrt{\Delta}\right) + f(0)\left(-\sqrt{\Delta}\right) + f\left(\sqrt{\Delta}\right)\left(\sqrt{\Delta}\right) + \cdots\right] \tag{9.111}$$

The value of V_n depends on a particular trajectory of S_t. If V_n converges, it can be called a pathwise integral.

It turns out that there is no guarantee that such pathwise integrals converge in stochastic environments. We consider a simple example.

Let the functions $f(\cdot)$ in V_n be given by

$$f\left(S_{t_{i+1}}\right) = sign\left(S_{t_{i+1}} - S_{t_i}\right) \tag{9.112}$$

In other words, $f(\cdot)$ assumes the value of plus or minus one, depending on the sign of $S_{t_{i+1}} - S_{t_i}$.

This means that all elements in V_n are positive, so

$$V_n = \sum_{i=0}^{n-1} \sqrt{\Delta} = n\sqrt{\Delta} \tag{9.113}$$

Using $T = n\Delta$,

$$V_n = \frac{T}{\sqrt{\Delta}} \tag{9.114}$$

Clearly, as $\Delta \to 0$, V_n will go to infinity.

If such paths have a positive probability of occurrence, then the *pathwise* sum V_n cannot converge in any probabilistic sense.

This example is important for two reasons.

First, we see the meaning of a pathwise integral. In calculating the integral pathwise, we did not use the *probabilities* associated with $\Delta S_{t_{i+1}}$. The integral was calculated using the actual realization of the process. The Ito integral, on the other hand, is calculated using mean square convergence, and the integral is determined within stochastic equivalence.

Second, we see the importance of using nonanticipative functions as $f(\cdot)$. In fact, because $f(\cdot)$ was able to "see the future," it anticipated the sign of $S_{t_{i+1}} - S_{t_i}$. That made all the elements in the summation sign positive and led to an exploding V_n as n increased.

9.3.3 Itô Isometry

For any $X_t(\omega)$ which is square integrable, the following holds

$$\mathbb{E}\left(\left(\int_0^t X_u(\omega)dW_u(\omega)\right)^2\right) = \mathbb{E}\left(\int_0^t X_u(\omega)^2 du\right) \tag{9.115}$$

This is called Itô Isometry.

9.3.3.1 An Example on Itô Isometry

The following stochastic differential equation is called the Ornstein–Uhlenbeck equation

$$dX_t = \mu X_t dt + \sigma dB_t \tag{9.116}$$

where μ and σ are real constants. We can use Itô's lemma to solve it. In order to do that we use the integrating factor $e^{-\mu t}$ and then apply Itô's lemma to $e^{-\mu t}X_t$ to compute $d(e^{-\mu t}X_t)$ to get

$$\begin{aligned} d(e^{-\mu t}X_t) &= (-\mu e^{-\mu t}X_t \\ &\quad + e^{-\mu t}\mu X_t)dt + e^{-\mu t}\sigma dB_t && (9.117) \\ &= e^{-\mu t}\sigma dB_t && (9.118) \end{aligned}$$

Then integrating it to get

$$e^{-\mu t}X_t = X_0 + \int_0^t e^{-\mu s}\sigma dB_s \tag{9.119}$$

or equivalently

$$X_t = e^{\mu t}X_0 + \int_0^t e^{\mu(t-s)}\sigma dB_s \tag{9.120}$$

Having solved for X_t, we can now find expressions for its mean and variance that is $\mathbb{E}(X_t)$ and $Var(X_t)$. Since $\int_0^t e^{\mu(t-s)}\sigma dB_s$ is a stochastic integral, then $\int_0^t e^{\mu(t-s)}\sigma dB_s$ is a martingale. Hence,

$$\mathbb{E}(X_t) = e^{\mu t}X_0 + 0 = e^{\mu t}X_0 \tag{9.121}$$

and

$$\begin{aligned} \mathbb{E}(X_t^2) &= \mathbb{E}\left(e^{2\mu t}X_0^2 + 2e^{\mu t}X_0\int_0^t e^{\mu(t-s)}\sigma dB_s \right. \\ &\quad \left. + \left(\int_0^t e^{\mu(t-s)}\sigma dB_s\right)^2\right) && (9.122) \\ &= e^{2\mu t}X_0^2 + 2e^{\mu t}X_0\mathbb{E}\left(\int_0^t e^{\mu(t-s)}\sigma dB_s\right) \\ &\quad + \mathbb{E}\left(\left(\int_0^t e^{\mu(t-s)}\sigma dB_s\right)^2\right) && (9.123) \end{aligned}$$

As in the expectation, the second term is a stochastic integral, and therefore a martingale and using Itô isometry on the last term we get

$$\begin{aligned} \mathbb{E}(X_t^2) &= e^{2\mu t}X_0^2 + 0 + \mathbb{E}\left(\int_0^t e^{2\mu(t-s)}\sigma^2 dt\right) && (9.124) \\ &= e^{2\mu t}X_0^2 + \int_0^t e^{2\mu(t-s)}\sigma^2 dt && (9.125) \\ &= e^{2\mu t}X_0^2 + \frac{\sigma^2}{2\mu}(e^{2\mu t} - 1) && (9.126) \end{aligned}$$

Therefore

$$var(X_t) = \mathbb{E}\left(X_t^2\right) - (\mathbb{E}(X_t))^2 \quad (9.127)$$

$$= \frac{\sigma^2}{2\mu}(e^{2\mu t} - 1) \quad (9.128)$$

9.4 OTHER PROPERTIES OF THE ITÔ INTEGRAL

The Itô integral has some other properties.

9.4.1 Existence

One can ask the question: when does the Itô integral of a general random function $f(S_t, t)$,

$$\int_0^t f(S_u, u)\, dS_u \quad (9.129)$$

where $\{S_t\}$ is given by (9.6), exist?

It turns out that if the function $f(\cdot)$ is continuous, and if it is nonanticipating, this integral exists. In other words, the finite sums

$$\sum_{i=0}^{n-1} f(S_{t_i}, t_i)\left[S_{t_{i+1}} - S_{t_i}\right] \quad (9.130)$$

converge in mean square to "some" random variable that we call the Ito integral.[19]

9.4.2 Correlation Properties

It should not be forgotten that the Ito integral is a random variable. (More precisely, it is a random process.) Therefore, it will have various moments.

The martingale property gives the first moment of the integral of a nonanticipating $f(\cdot)$ with respect to a Wiener process

$$\mathbb{E}\left[\int_0^T f(W_t, t)\, dW_t\right] = 0 \quad (9.131)$$

[19]Although it may exist, determining such a limit explicitly is not guaranteed.

where W_t is a Wiener process. The second moments are given by the variance and covariances

$$\mathbb{E}\left[\int_0^T f(W_t, t)\, dW_t \int_0^T g(W_t, t)\, dW_t\right]$$
$$\int_0^T \mathbb{E}\left[f(W_t, t)\, g(W_t, t)\right] dW_t \quad (9.132)$$

and

$$\mathbb{E}\left[\int_0^T f(W_t, t)\, dW_t\right]^2 = \mathbb{E}\left[\int_0^T f(W_t, t)^2\, dW_t\right] \quad (9.133)$$

Note the recurring use of the equivalence $dW_t^2 = dt$ discussed earlier.

9.4.3 Addition

The Ito integral also has some properties similar to those of the Riemann–Stieltjes integral.

In particular, the integral of the sum of two (random) functions of S_t in (9.6) is equal to the sum of their integrals:

$$\int_0^T \left[f(S_t, t) + g(S_t, t)\right] dS_t = \int_0^T f(S_t, t)\, dS_t + \int_0^T g(S_t, t)\, dS_t \quad (9.134)$$

9.5 INTEGRALS WITH RESPECT TO JUMP PROCESSES

What complicated the definition of a stochastic integral was the extreme irregularity both of continuous-time martingales and of the Wiener process. This made a pathwise definition of the integral impossible.

Do we have the same problem if we have a stochastic integral with respect to some jump process? Could one use the Riemann–Stieltjes integral when dealing with, say, Poisson processes?

Surprisingly, the answer to this question is affirmative under some conditions.

Suppose a process M_t is a martingale that exhibits finite jumps only and has no Wiener component. Trajectories of such an M_t will exhibit occasional jumps, but otherwise will be very smooth. Then, one could define a V_n,

$$V_n = \sum_{i=0}^{n-1} f\left(M_{t_i}\right)\left[M_{t_{i+1}} - M_{t_i}\right] \tag{9.135}$$

pathwise.

This V_n will converge, and the variation of the process M_t will be finite with probability one. Under these conditions, we say that V_n converges pathwise.

9.6 CONCLUSION

This chapter dealt with the definition of the Ito integral.

From the point of view of a practitioner, there are two important points to keep in mind. First, the error terms in stochastic differential equations are defined in the sense of the Ito integral. Numerical calculations must obey the conditions set by this definition. Second, the stochastic differential equations routinely used in asset pricing are also defined in the sense of the Ito integral.

Above all, we saw that the Ito integral is the mean square limit of some random sums. These random sums are carefully put together so that the resulting integral is a martingale.

We also discussed several examples and showed that the rules of integration are in general very different in stochastic environments, when compared with the deterministic case. This was the result of using mean square convergence.

Fortunately, in evaluating Ito integrals, the direct route of obtaining the mean square limit will rarely be used. Instead, Ito integrals can be evaluated in a more straightforward fashion using a result called Ito's Lemma. This will be discussed in the next chapter, where we will also discuss further examples of evaluating the Ito integral.

9.7 REFERENCES

There are several excellent sources on the derivation of the Ito integral. Karatzas and Shreve (1991) and Revuz and Yor (1994) were already mentioned. Two additional sources that the reader may find a bit easier to read are Oksendal (2000) and Protter (1990). The former source could be a very good manual for quantitatively oriented practitioners and for beginning graduate students. It is well written and easy to understand. Technicalities are avoided as much as possible.

9.8 EXERCISES

1. Let W_t be a Wiener process defined over $[0, T]$ and consider the integral:

$$\int_0^t W_t^2 dW_s$$

Use the subdivision of $[0, t]$:

$$t_0, t_1, \ldots, t_{n-1}, t_n$$

in the following:

(a) Write the approximation of the above integral as three different Riemann sums.
(b) Write the integral in discrete time using an Ito sum.
(c) Calculate the expectation of the three Riemann sums.
(d) Calculate the expectation of the Ito sum.

2. Show that given

$$t_0, t_1, \ldots, t_{n-1}, t_n$$

and

$$W_{t_0}, W_{t_1}, \ldots, W_{t_{n-1}}, W_{t_n}$$

we can always write:

$$\sum_{j=1}^{n}\left[t_j W_{t_j} - t_{j-1} W_{t_{j-1}}\right] = \sum_{j=1}^{n}\left[t_j\left(W_{t_j} - W_{t_{j-1}}\right)\right] + \sum_{j=1}^{n}\left[\left(t_j - t_{j-1}\right) W_{t_j}\right] \quad (9.136)$$

How is this different from the standard formula for the differentiation of products:

$$d(uv) = (du)v + u(dv)$$

3. Now use this information to show that:

$$\int_0^t s dW_s = tW_t - \int W_s ds$$

4. In the above equation there are two integrals. Which integral is defined in the sense of Itô only?
5. Can we say that this is a change of variables?
6. Can we say that this is an application of integration by parts?
7. Let Y be a random variable defined by

$$Y(\omega) = \int_0^1 W_s(\omega) ds \quad (9.137)$$

where W is a standard Wiener process. Compute $\mathbb{E}(Y)$ and $\mathbb{E}(Y^2)$.

8. Write a program that numerically calculates the value of the following stochastic integral via simulation:

$$X_t = \int_0^t W_t^2 dW_s \quad (9.138)$$

9. Consider the stochastic process, X_t which satisfies

$$X_t = \int_0^t \gamma_s dB_s \quad (9.139)$$

where γ_s is stochastic and B_s is a standard Brownian motion. Find an expression for $Cov(X_s, X_t)$ for $s < t$ (Itô isometry application).

10. Show that

$$\int_0^t B_s dB_s = \frac{1}{2}B_t^2 - \frac{1}{2}t.$$

Observe that the term $-\frac{1}{2}t$ shows that the stochastic integral does not yield the same result as deterministic calculus and behaves differently.

CHAPTER

10

Itô's Lemma

OUTLINE

10.1 INTRODUCTION

As discussed earlier, in stochastic environments a formal notion of derivative does not exist. Shocks to asset prices are assumed to be unpredictable, and in continuous time they become "too erratic." The resulting asset prices may be continuous, but they are not smooth. Stochastic differentials need to be used in place of derivatives.

Ito's rule provides an analytical formula that simplifies handling stochastic differentials and leads to explicit computations. It is the main topic of this chapter.

We begin by discussing various types of derivatives.

Financial Derivatives, Third Edition.
http://dx.doi.org/10.1016/B978-0-12-384682-2.00010-4

10.2 TYPES OF DERIVATIVES

Suppose we have a function $F(S_t, t)$ depending on two variables S_t and t, where S_t itself varies with time t. Further, assume that S_t is a random process.

In standard calculus, where all variables are deterministic, there are three sorts of derivatives that one can talk about.

The first are the *partial derivatives* of $F(S_t, t)$, denoted by

$$F_s = \frac{\partial F(S_t, t)}{\partial S_t}, \quad F_t \frac{\partial F(S_t, t)}{\partial t} \tag{10.1}$$

The second is the *total derivative* dealing with differentials:

$$dF_t = F_s dS_t + F_t dt \tag{10.2}$$

In (10.2), dF_t is used as a shorthand notation for $dF(S_t, t)$. This should not be confused with F_t, the partial of $F(\cdot)$ with respect to t.

The third is the chain rule:

$$\frac{dF(S_t, t)}{dt} = F_s \frac{dS_t}{dt} + F_t \tag{10.3}$$

A financial market participant may be interested in these derivatives for various reasons.

The partial derivative has no direct real-life counterpart, but gives "multipliers"that can be used in evaluating responses of asset prices to observed changes in risk factors. For example, F_s measures the response of $F(S_t, t)$ to a small change in S_t only. As such, F_s is a hypothetical concept, since the only way a continuous random variable S_t can change is if some time passes. Hence, in reality, t has to change as well. Partial derivatives abstract from such questions. Because they are simple multipliers, there is no difference between the way stochastic and deterministic environments define partial derivatives.

A classical example of the use of partial derivatives occurs in *delta hedging*. Suppose a market participant knows the functional form of $F(S_t, t)$. Then, this mathematical formula can be differentiated only with respect to S_t, in order to find the partial derivative F_s. This F_s is a measure of how much the derivative asset price will change *per* unit change in S_t. In this sense, one does not have any of the difficulties encountered in defining a time derivative for Wiener processes. What is under investigation is not how $F(S_t, t)$ moves over time, but how $F(\cdot)$ responds to a "small"hypothetical change in S_t, with time fixed.

The total derivative is a more "realistic"notion. It is assumed that both time t and the underlying security price S_t change, and then the total response of $F(S_t, t)$ is calculated. The result is the (stochastic) differential dF_t. This is clearly a very useful quantity to the market participant. It represents the observed change in the price of the derivative asset during an interval dt.

The chain rule is quite similar to the total derivative. In classical calculus, the chain rule expresses the rate of change of a variable as a chain effect of some initial variation. In stochastic calculus, we know that operations such as $dF_t/dt, dS_t/dt$ cannot be defined for continuous-time square integrable martingales, or Brownian motion. But a stochastic equivalent of the chain rule can be formulated in terms of absolute changes such as dF_t, dS_t, dt, and the Ito integral can be used to justify these terms. Thus, in stochastic calculus, the term "chain rule"will refer to the way *stochastic differentials* relate to one another. In other words, a stochastic version of total differentiation is developed.

10.2.1 Example

We discuss a simple example before going into Ito's formula. The example will help clarify the mechanics of taking various derivatives. Let $F(r_t, t)$ be the price of a T-bill that matures at time T, and let r_t be a fixed, continuously compounding risk free rate. Then

$$F(r_t, t) = e^{-r_t(T-t)}100 \tag{10.4}$$

Let us calculate the partial derivatives F_r, F_t:

$$F_r = \frac{\partial F}{\partial r_t} = -(T-t)\left[e^{-r_t(T-t)}100\right] \tag{10.5}$$

and

$$F_t = \frac{\partial F}{\partial t} = r_t \left[e^{-r_t(T-t)} 100 \right] \tag{10.6}$$

Note that these partials will be the same regardless of whether r_t is deterministic or random. By taking these partial derivatives, we are simply calculating the rate of change of $F(\cdot)$, with respect to small hypothetical changes in r_t or in t.

On the other hand, the total derivative relates to the actual occurrence of random events. In standard calculus, with *nonrandom* r_t, the total derivative of this particular $F(\cdot)$ will be given by

$$dF(r_t, t) = -(T-t)\left[e^{-r_t(T-t)}100\right]dr_t + r_t\left[e^{-r_t(T-t)}100\right]dt \tag{10.7}$$

This example suggests that when r_t is random, we may be able to define the counterpart of total derivative, using the Ito integral, which gives a meaning to stochastic differentials such as dr_t. This intuition is correct, and the result is Ito's formula. However, with stochastic r_t, not only does the interpretation of dr_t change,[1] but the formula will also be different.

10.3 ITO'S LEMMA

The stochastic version of the chain rule is known as Ito's Lemma. Let S_t be a continuous-time process which depends on the Wiener process W_t. Suppose we are given a function of S_t, denoted by $F(S_t, t)$, and suppose we would like to calculate the change in $F(\cdot)$ when dt amount of time passes. Clearly, passing time would influence the $F(S_t, t)$ in two different ways. First, there is a *direct* influence through the t variable in $F(S_t, t)$. Second, as time passes, one obtains new information about W_t and observes a new increment, dS_t. This will also make $F(\cdot)$ change. The sum of these two effects is represented by the stochastic differential $dF(S_t, t)$ and is given by the stochastic equivalent of the chain rule.

[1]Recall that such quantities are defined in terms of mean square convergence and within stochastic equivalence.

Let the random process S_t be observed in continuous time. We again partition the time interval $[0, T]$ into n equal pieces, each with length h, and use the finite difference approximation. However, we write this as an equality

$$\Delta S_k = a_k h + \sigma_k \Delta W_k, \quad k = 1, 2, \ldots, n \tag{10.8}$$

using the mean square equivalence between the left- and right-hand side as $h \to 0$. This notation will be preserved throughout this chapter. Also note that we shortened the notation for $a\left(S_{k-1}, k\right)$ to a_k and for $\sigma\left(S_{k-1}, k\right)$ to σ_k.

We calculate Ito's formula in this setting, using the Taylor series. Recall the Taylor series expansion of a smooth (i.e., infinitely differentiable) function $f(x)$ around some arbitrary point x_0,

$$f(x) = f(x_0) + f_x(x_0)(x - x_0) + \frac{1}{2} f_{xx}(x_0)(x - x_0)^2 + R \tag{10.9}$$

where R denotes the remainder.

We apply this formula to $F(S_t, t)$. At the outset, $F(\cdot)$ has to be a smooth function of S_t.[2] But there are two additional complications. First, the Taylor series formula in (10.9) is valid for a $f(x)$ which is a function of a single variable x, while $F(S_t, t)$ depends on two variables, S_t and t. Second, the formula in (10.9) is valid for deterministic variables, while S_t is a random process. Before using Taylor series, these complications must be addressed.

The extension of a univariate Taylor series formula to two variables is straightforward. One adds the partials with respect to the second variable. With two variables, cross partials should be included as well.

The applicability of the Taylor series formula to a random environment is a deeper issue. First, it should be remembered that some of the

[2]Incidentally, some readers may wonder if this "smoothness"does not contradict the extreme irregularity of S_t. $F(\cdot)$ can be a smooth function of St and still be a very irregular stochastic process. Irregularity here is in the sense of how $F(\cdot)$ changes over time. It is not a statement about how St relates to $F(\cdot)$.

terms in Taylor series are partial derivatives. With respect to these, one does not have any difficulty with differentiation in stochastic environments. Second, we have differentials such as dS_t. Here, we do need an adjustment, which is in terms of the interpretation of the equality and not in the Taylor series expansion itself. The formula for Taylor series expansion will remain the same, but the meaning of the equality sign would change. The equality would have to be interpreted in the context of mean square convergence.

We apply the Taylor series formula to $F(S_k, k), k = 1, 2, \ldots,$ where the S_k is assumed to obey

$$\Delta S_k = a_k h + \sigma_k \Delta W_k \tag{10.10}$$

First, fix k. Given the information set I_{k-1}, S_{k-1} is a known number. Next, apply Taylor's formula to expand $F(S_k, k)$ around S_{k-1} and $k-1$:

$$\begin{aligned} F\left(S_k, k\right) = {} & F\left(S_{k-1}, k-1\right) + F_s\left[S_k - S_{k-1}\right] \\ & + F_t[h] + \frac{1}{2} F_{ss}\left[S_k - S_{k-1}\right]^2 \\ & + \frac{1}{2} F_{tt}[h] + F_{st}\left[h\left(S_k - S_{k-1}\right)\right] + R \end{aligned} \tag{10.11}$$

where the partials $F_s, F_{ss}, F_t, F_{tt}, F_{st}$ are all evaluated at $S_{k-1}, k-1$. R represents the remaining terms of the Taylor series expansion. Here we are keeping the F_t, F_{st}, F_{tt} notation for convenience, although these partials are with respect to k.

Transpose $F(S_{k-1}, k-1)$ and relabel the increments in (10.11) as follows:

$$F\left(S_k, k\right) - F\left(S_{k-1}, k-1\right) = \Delta F(k) \tag{10.12}$$

$$S_k - S_{k-1} = \Delta S_k \tag{10.13}$$

Notice that Eq. (10.11) already uses the increment for the time variable:

$$kh - \left(k-1\right)h = h \tag{10.14}$$

Now substitute these into (10.11):

$$\begin{aligned} \Delta F(k) = {} & F_s \Delta S_k + F_t[h] + \frac{1}{2} F_{ss}\left[\Delta S_k\right]^2 \\ & + \frac{1}{2} F_{tt}[h] + F_{st}\left[h \Delta S_k\right] + R \end{aligned} \tag{10.15}$$

But we know that the dynamics of St are governed by Eq. (10.10), and that we have

$$\Delta S_k = a_k h + \sigma_k \Delta W_k \tag{10.16}$$

We can substitute the right-hand side of this for ΔS_k in the Taylor series expansion of (10.11):

$$\begin{aligned} \Delta F(k) = {} & F_s\left[a_k h + \sigma_k \Delta W_k\right] \\ & + F_t[h] + \frac{1}{2} F_{ss}\left[a_k h + \sigma_k \Delta W_k\right]^2 \\ & + \frac{1}{2} F_{tt}[h] + F_{st}\left[h\left(a_k h + \sigma_k \Delta W_k\right)\right] + R \end{aligned} \tag{10.17}$$

What does this equation mean? On the left-hand side, $\Delta F(k)$ indicates the total change in $F(S_k, k)$ due to changing k and S_k. Hence, if $F(S_k, k)$ is the price of a derivative security, on the left-hand side we have the change in the derivative asset's price during a short interval. This change is explained by the terms on the right-hand side.

The *first-order* effects are the effects of time, represented by $F_t[h]$, and the effects of change in the underlying asset's price, $F_{st}\left[h\left(a_k h + \sigma_k \Delta W_k\right)\right]$. In the latter we again see that changes in security prices have predictable and unpredictable components. Second-order effects are those changes that are represented, for the time being, by squared terms and by cross products. Higher-order terms are grouped in the remainder R.

In order to obtain a chain rule in stochastic environments, the terms on the right-hand side will be classified as negligible and nonnegligible. It will then be shown that in "small" time intervals, negligible terms can be dropped from the right-hand side and a chain rule formula obtained. In addition, as $h \to 0$, a limiting argument can be used and a precise formula obtained in the mean square sense. This formula is known as Ito's Lemma.

The first step of this derivation is to separate the terms on the right-hand side. This requires an explicit criterion for deciding which terms are negligible. Afterward, one can consider the size of the terms on the right-hand side of (10.11) individually and decide which ones are to be dropped.

10.3.1 The Notion of "Size" in Stochastic Calculus

This section discusses the convention used in determining which variables can be classified as "negligible" in stochastic calculus.

In standard calculus, the Taylor series expansion of some function $f(S)$ around S_0 gives

$$f(S) - f(S_0) = \Delta f = f_s(S_0)\,\Delta S + \frac{1}{2} f_{ss}(S_0)\,\Delta S^2 + \frac{1}{3} f_{sss}(S_0)\,\Delta S^3 \tag{10.18}$$

where R is the remainder. But the formula for total derivatives is just

$$df = f_s dS \tag{10.19}$$

This is equivalent to assuming that while in the Taylor series expansion (10.18), ΔS is small and nonnegligible, the terms involving $(\Delta S)^2, (\Delta S)^3, \ldots$ are smaller and can be ignored as $\Delta S \to 0$. Consequently, in the limit, the term $f_s dS$ is preserved, while all other terms are dropped. The result is the (total) differentiation formula (10.19).

To see why such a convention makes sense, note that as ΔS gets smaller, terms such as $(\Delta S)^2, (\Delta S)^3, \ldots$ get small *faster*. This is shown in Figure 10.1, where the functions

$$g_1(\Delta S) = \Delta S \tag{10.20}$$

and

$$g_2(\Delta S) = [\Delta S]^2 \tag{10.21}$$

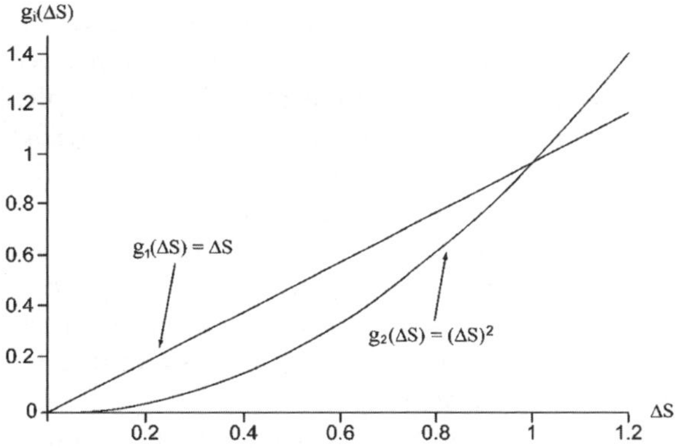

FIGURE 10.1 Illustration of convergence of ΔS versus $(\Delta S)^2$.

are graphed. Note that the function $g_2(\Delta S)$ approaches zero much faster than the function $g_1(\Delta S)$ as ΔS gets smaller and smaller.

Thus, in standard calculus, all terms involving powers of dS higher than one are assumed to be negligible and are dropped from total derivatives. The question is whether we can do the same in stochastic calculus.

The answer to this important question is no. In stochastic settings, the time variable t is still deterministic. So, with respect to the *time* variable, the same criterion of smallness as in deterministic calculus can be applied. Any terms involving powers of dt higher than one may be considered negligible.

On the other hand, the same rationale cannot be used for a stochastic differential such as dS_t^2. Chapter 9 already showed that, in the mean square sense, we have

$$dW_t^2 = dt \tag{10.22}$$

Hence, terms involving dS_t^2 are likely to have sizes of order dt, which was considered as nonnegligible. If terms involving dt are preserved in Taylor approximations, the same must apply to squares of stochastic differentials.

We further emphasize this important point. If ΔS_t is a random increment with mean zero, then $\mathbb{E}[\Delta S_t]^2$ will be the variance of this increment. Since ΔSt is random, its variance will be positive. But variance is the "size" of a typical $(\Delta St)^2$. Hence, on average, assuming that $(\Delta St)^2$ is negligible will be equivalent to assuming that its variance is approximately zero—that S_t is, approximately, not random. This is a contradiction, and it defeats the purpose of using SDEs in markets for derivative products. After all, the objective is to price risk, and risk is generated by unexpected news.

Hence, in contrast to deterministic environments, terms such as $(\Delta S_k)^2$ cannot be ignored in stochastic differentiation.

Given that the terms of size h are of first order, and that these are by convention not small, the following rule will be used to distinguish negligible terms from nonnegligible ones.

CONVENTION: Given a function $g(\Delta W_k, h)$ dependent on the increments of the Wiener process W_t, and on the time increment, consider the ratio

$$\frac{g(\Delta W_k, h)}{h} \tag{10.23}$$

If this ratio vanishes (in the m.s. sense) as $h \to 0$, then we consider $g(\Delta W_k, h)$ as negligible in small intervals. Otherwise, $g(\Delta W_k, h)$ is nonnegligible.

This convention amounts to comparing various terms with h. In particular, if the mean square limit of the function $g(\Delta W_k, h)$ is proportional to h^r with $r > 1$, it will go toward zero faster than h (i.e., the square of a small number is smaller than the number itself). On the other hand, if $r < 1$, then the mean square limit of $g(\Delta W_k, h)$ will be proportional to a larger power of h than h itself.[3]

The following discussion uses this convention in deciding which terms of a stochastic Taylor series expansion can be considered small.

[3]Here, it should not be forgotten that the function $g(\Delta W_k, h)$ depends on powers of ΔW_k, and that these also determine whether the ratio in (10.23) becomes negligible as h gets smaller. Such is the case when we deal with cross-product terms of Taylor series expansions.

10.3.2 First-Order Terms

Now consider Eq. (10.11) again:

$$\begin{aligned}\Delta F(k) = {} & F_s\left[a_k h + \sigma_k \Delta W_k\right] + F_t[h] \\ & + \frac{1}{2} F_{ss}\left[a_k h + \sigma_k \Delta W_k\right]^2 \\ & + \frac{1}{2} F_{tt}[h]^2 + F_{st}[h]\left[a_k h + \sigma_k \Delta W_k\right]\end{aligned} \tag{10.24}$$

Here, the terms that contain h or ΔS_k are clearly first-order increments that are not negligible. As $F_s\left[a_k h + \sigma_k \Delta W_k\right]$ or $F_t h$ are divided by h, and h is made smaller and smaller, these terms do not vanish. For example, the ratios

$$\lim_{h\to 0} \frac{F_s a_k h}{h} = F_s a_k \tag{10.25}$$

and

$$\lim_{h\to 0} \frac{F_t h}{h} = F_t \tag{10.26}$$

are clearly independent of h, and do not vanish as h gets smaller.

On the other hand, we already know that the ratio

$$\frac{F_s \Delta W_k}{h} \tag{10.27}$$

gets larger (in a probabilistic sense) as h becomes smaller, since the term ΔW_k is of the order $h^{1/2}$.

All first-order terms in (10.24) are thus nonnegligible.

10.3.3 Second-Order Terms

Now divide the second-order terms on the right-hand side of (10.24) by h, and consider the ratio

$$\frac{F_{tt} h^2}{2h} \tag{10.28}$$

This term remains proportional to h, since in the numerator we have an increment that depends on h^2, a power of h higher than one, and the increment is *not* random. Hence, this term is negligible:

$$\lim_{h \to 0} F_{tt} h = 0 \qquad (10.29)$$

Next, consider the second-order term that depends on $[\Delta S_k]^2$,

$$\frac{1}{2} F_{ss} [\Delta S_k]^2$$

Substituting for ΔS_k, expanding the square, and dividing by h,

$$\frac{1}{2} F_{ss} \left[\frac{a_k h^2}{h} + \frac{(\sigma_k \Delta W_k)^2}{h} + \frac{2 a_k \sigma_k h \Delta W_k}{h} \right]^2 \qquad (10.30)$$

In this equation, the first term is "small." The numerator contains a power of h greater than one, and the term is not random. The third term is also "small." It involves a cross product (see next section). The second term, on the other hand, contains the random variable $(\Delta W_k)^2$. This is the square of a random variable with mean zero that is unpredictable given the past. Its variance was shown to be

$$\mathbb{V} [\sigma_k \Delta W_k] = \sigma_k^2 h \qquad (10.31)$$

It was also shown that in the mean square sense discussed earlier,

$$dW_t^2 = dt \qquad (10.32)$$

Thus, ΔW_k^2 is a term that cannot be considered negligible, since by definition we are dealing with stochastic S_k, and the nonzero variance of ΔS_k implies:

$$\sigma_k > 0 \qquad (10.33)$$

Consequently, using the criterion of negligibility, we write for small h:

$$\frac{1}{2} F_{ss} \left[\frac{a_k h^2}{h} + \frac{(\sigma_k \Delta W_k)^2}{h} + \frac{2 a_k \sigma_k h \Delta W_k}{h} \right]^2 \approx \frac{1}{2} F_{ss} \sigma_k^2 \qquad (10.34)$$

Again, this approximation should be interpreted in the m.s. sense. That is, in small intervals, the difference between the two sides of equality (10.34) has a variance that will tend to zero as $h \to 0$.

Before one can write the Ito formula, the remaining terms of the Taylor series expansion in (10.24) must also be discussed.

10.3.4 Terms Involving Cross Products

The terms in (10.24) involving cross products are also negligible in small intervals, under the assumption that the unpredictable components do not contain any "jumps." The argument rests on the continuity of the sample paths for S_t.

Consider the following cross-product term in (10.24) and divide it by h:

$$\frac{F_{st} [h] [a_k h + \sigma_k \Delta W_k]}{h} = F_{st} [a_k h + \sigma_k \Delta W_k] \qquad (10.35)$$

The right-hand side of (10.35) depends on ΔW_k. As $h \to 0$, ΔW_k goes to zero. In particular, ΔW_k becomes negligible, because as $h \to 0$ its variance goes to zero. That is, W_k does not change at the limit $h = 0$. This is another way of saying that the Wiener process has continuous sample paths.

As long as the processes under consideration are continuous and do not display any jumps, terms involving cross products of ΔW_k and h would be negligible, according to the convention adopted earlier.

10.3.5 Terms in the Remainder

All the terms in the remainder R contain powers of h and of ΔW_k greater than two. According to the convention adopted earlier, if the unpredictable shocks are of "normal" type—i.e., there are no "rare events"—powers of ΔW_k greater than two will be negligible. In fact, it was shown in Chapter 8 that continuous martingales and Wiener processes have higher-order moments that are negligible as $h \to 0$.

10.4 THE ITO FORMULA

We can now summarize the discussion involving the terms in (10.24). As $h \rightarrow 0$ and we drop all negligible terms, we obtain the following result:

Theorem 5 (Ito's Lemma). *Let $F(S_t, t)$ be a twice-differentiable function of t and of the random process S_t:*

$$dS_t = a_t dt + \sigma_t dW_t$$

with well-behaved drift and diffusion parameters, a_t, σ_t.[4] *Then we have*

$$dF_t = \frac{\partial F}{\partial S_t} dS_t + \frac{\partial F}{\partial t} dt + \frac{1}{2} \sigma_t \frac{\partial^2 F}{\partial S_t^2} dt \tag{10.36}$$

or, after substituting for dS_t using the relevant SDE,

$$dF_t = \left[\frac{\partial F}{\partial S_t} + \frac{\partial F}{\partial t} dt + \frac{1}{2} \sigma_t \frac{\partial^2 F}{\partial S_t^2} \right] dt + \frac{\partial F}{\partial S_t} \sigma_t dW_t \tag{10.37}$$

where the equality holds in the mean square sense.

In situations that call for the Ito formula, one will in general be given an SDE that drives the process S_t:

$$dS_t = a(S_t, t) dt + \sigma_t (S_t, t) dW_t \tag{10.38}$$

Thus, the Ito formula can be seen as a vehicle that takes the SDE for S_t and determines the SDE that corresponds to $F(S_t, t)$. In fact, Eq. (10.37) is a stochastic differential equation for $F(S_t, t)$.

Ito's formula is clearly a very useful tool to have in dealing with financial derivatives. The latter are contracts written on underlying assets. Using the Ito formula, we can determine the SDE for financial derivatives once we are given the SDE for the underlying asset. For a market participant who wants to price a derivative asset but is willing to take the behavior of the underlying asset's price as exogenous, Ito's formula is a necessary tool.

[4]With this we mean that the drift and diffusion parameters are not too irregular. Square integrability would satisfy this condition. For notational simplicity, we write $a(S_t, t)$ as a_t and $\sigma(S_t, t)$ as σ_t.

10.5 USES OF ITO'S LEMMA

The first use of Ito's Lemma was just mentioned. The formula provides a tool for obtaining stochastic differentials for functions of random processes.

For example, we may want to know what happens to the price of an option if the underlying asset's price changes. Letting $F(S_t, t)$ be the option price, and S_t the underlying asset's price, we can write

$$dF(S_t, t) = F_s dS_t + F_t dt + \frac{1}{2} F_{ss} \sigma_t^2 dt \tag{10.39}$$

If one has an exact formula for $F(S_t, t)$, one can then take the partial derivatives explicitly and replace them in the foregoing formula to get the stochastic differential, $dF(S_t, t)$. Later in this section, we give some examples of this use of Ito's Lemma.

The second use of Ito's Lemma is quite different. Ito's Lemma is useful in evaluating Ito integrals. This may be unexpected, because Ito's formula was introduced as a tool to deal with stochastic differentials. Under normal circumstances, one would not expect such a formula to be of much use in evaluating Ito integrals. Yet stochastic calculus is different. It is not like ordinary calculus, where integral and derivative are separately defined and then related by the fundamental theorem of calculus. As we pointed out earlier, the differential notation of stochastic calculus is a *shorthand* for stochastic integrals. Thus, it is not surprising that Ito's Lemma is useful for evaluating stochastic integrals.

We give some simple examples of these uses of Ito's Lemma. More substantial examples will be seen in later chapters when derivative asset pricing is discussed.

10.5.1 Ito's Formula as a Chain Rule

A discussion of some simple examples may be useful in getting familiar with the terms introduced by Ito's formula.

10.5.1.1 Example 1

Consider a function of the *standard* Wiener process W_t given by

$$F(W_t, t) = W_t^2 \quad (10.40)$$

Remember that W_t has a drift parameter 0 and a diffusion parameter 1.

Applying the Ito formula to this function,

$$dF_t = \frac{1}{2}[2dt] + 2W_t dW_t \quad (10.41)$$

or

$$dF_t = dt + 2W_t dW_t \quad (10.42)$$

Note that Ito's formula results, in this particular case, in an SDE that has

$$a(I_t, t) = 1 \quad (10.43)$$

and

$$\sigma(I_t, t) = 1 \quad (10.44)$$

Hence, the drift is constant and the diffusion depends on the information set I_t.

10.5.1.2 Example 2

Next, we apply Ito's formula to the function

$$F(W_t, t) = 3 + t + e^{W_t} \quad (10.45)$$

We obtain

$$dF_t = dt + e^{W_t} dW_t + \frac{1}{2}e^{W_t} dt \quad (10.46)$$

Grouping,

$$dF_t = \left[\frac{1}{2}e^{W_t} + 1\right] dt + e^{W_t} dW_t \quad (10.47)$$

In this case, we obtain an SDE for $F(S_t, t)$ with I_t-dependent drift and diffusion terms:

$$a(I_t, t) = \left[\frac{1}{2}e^{W_t} + 1\right] \quad (10.48)$$

and

$$\sigma(I_t, t) = e^{W_t} \quad (10.49)$$

10.5.2 Ito's Formula as an Integration Tool

Suppose one needs to evaluate the following Ito integral, which was discussed in Chapter 9:

$$\int_0^t W_s dW_s \quad (10.50)$$

In Chapter 9, this integral was evaluated directly by taking the mean square limit of some approximating sums. That evaluation used straightforward but lengthy calculations. We now exploit Ito's Lemma in evaluating the same integral in a few steps.

Define

$$F(W_t, t) = \frac{1}{2}W_t^2 \quad (10.51)$$

and apply the Ito formula to $F(W_t, t)$:

$$dF_t = 0 + W_t dW_t + \frac{1}{2}dt \quad (10.52)$$

This is an SDE with drift 1/2 and diffusion W_t. Writing the corresponding integral equation,

$$F(W_t, t) = \int_0^t W_s dW_s + \frac{1}{2}\int_0^2 dt \quad (10.53)$$

or, after taking the second integral on the right-hand side, and using the definition of $F(W_t, t)$:

$$\frac{1}{2}W_t^2 = \int_0^t W_s dW_s + \frac{1}{2}t \quad (10.54)$$

Rearranging terms, we obtain the desired result

$$\int_0^t W_s dW_s = \frac{1}{2}W_t^2 - \frac{1}{2}\int_0^2 dt \quad (10.55)$$

which is the same result that was obtained in Chapter 9 using mean square convergence.

It is important to summarize how Ito's formula was exploited to evaluate Ito integrals.

1. We guessed a form for the function $F(W_t, t)$.
2. Ito's Lemma was used to obtain the SDE for $F(W_t, t)$.
3. We applied the integral operator to both sides of this new SDE, and obtained an integral equation.[5] This equation contained integrals that were simpler to evaluate than the original integral.
4. Rearranging the integral equation gave us the desired result.

The technique is indirect but straightforward. The only difficulty is in guessing the exact form of the function $F(W_t, t)$.

This technique of using Ito's Lemma in evaluating integrals will be exploited in the next chapter.

10.5.2.1 *Another Example*

Suppose we need to evaluate

$$\int_0^t s dW_s \tag{10.56}$$

where W_t is again a Wiener process.

We use Ito's Lemma. First, we define a function $F(W_t, t)$:

$$F(W_t, t) = tW_t \tag{10.57}$$

Applying Ito's Lemma to $F(\cdot)$,

$$dF_t = W_t dt + t dW_t + 0 \tag{10.58}$$

Using the definition of dF_t in the corresponding integral equation,

$$\int_0^t d[sW_s] = \int_0^t W_s ds + \int_0^t s dW_s \tag{10.59}$$

[5]In fact, SDE notation is simply a shorthand for integral equations. Hence, this step amounts to writing the SDE in full detail.

Rearranging, we obtain the desired integral:

$$\int_0^t s dW_s = tW_t - \int_0^t W_s ds \tag{10.60}$$

Here the first term on the right-hand side is obtained from

$$\int_0^t d[sW_s] = tW_t - 0 \tag{10.61}$$

Again, the use of Ito's Lemma yields the desired integral in an indirect but straightforward series of operations.

10.6 INTEGRAL FORM OF ITO'S LEMMA

As repeatedly mentioned, stochastic differentials are simply shorthand for Ito integrals over small time intervals. One can thus write the Ito formula in integral form.

Integrating both sides of (10.37), we obtain

$$F(S_t, t) = F(S_0, 0) + \int_0^t \left[F_u + \frac{1}{2}F_{ss}\sigma_u^2\right] du + \int_0^t F_s dS_u \tag{10.62}$$

where use has been made of the equality

$$\int_0^t dF_u = F(S_t, t) - F(S_0, 0) \tag{10.63}$$

We can use the version of the Ito formula shown in (10.62) in order to obtain another characterization. Rearranging (10.62),

$$\int_0^t F_s dS_u = [F(S_t, t) - F(S_0, 0)] - \int_0^t \left[F_u + \frac{1}{2}F_{ss}\sigma_u^2\right] du \tag{10.64}$$

This equality provides an expression where integrals with respect to Wiener processes or other continuous-time stochastic processes are expressed as a function of integrals with respect to time. It should be kept in mind that in (10.62) and (10.64), F_s and F_{ss} depend on u as well.

10.7 ITO'S FORMULA IN MORE COMPLEX SETTINGS

Ito's formula is seen as a way of obtaining the SDE for a function $F(S_t, t)$, given the SDE for the underlying process S_t. Such a tool is very useful when $F(S_t, t)$ is the price of a financial derivative and S_t is the underlying asset. But the Ito formula introduced thus far may end up not being sufficiently general under some plausible circumstances that a practitioner may face in financial markets.

Our discussion has established Ito's formula in a univariate case, and under the assumption that unanticipated news can be characterized using Wiener process increments.

We can visualize two circumstances where this model may not apply. Under some conditions, the function $F(\cdot)$ may depend on more than a single *stochastic* variable S_t. Then a multivariate version of the Ito formula needs to be used. The extension is straightforward, but it is best to discuss it briefly.

The second generalization is more complex. One may argue that financial markets are affected by rare events, and that it is inappropriate to consider error terms made of Wiener processes only. One may want to add jump processes to the SDEs that drive asset prices. The corresponding Ito formula would clearly change. This is the second generalization that we discuss in this section.

10.7.1 Multivariate Case

We now extend the Ito formula to a multivariate framework and give an example. For simplicity, we pick the bivariate case and hope that the reader can readily extend the formula to higher-order systems.

Suppose S_t is a 2×1 vector of stochastic processes obeying the following stochastic differential equation[6]:

$$\begin{bmatrix} dS_1(t) \\ dS_2(t) \end{bmatrix} = \begin{bmatrix} a_1(t) \\ a_2(t) \end{bmatrix} dt + \begin{bmatrix} \sigma_{11}(t) & \sigma_{12}(t) \\ \sigma_{21}(t) & \sigma_{22}(t) \end{bmatrix} \begin{bmatrix} dW_1(t) \\ dW_2(t) \end{bmatrix} \tag{10.65}$$

This means that we have two equations of the following forms:

$$dS_1(t) = a_1(t)dt + \left[\sigma_{11}(t)dW_1(t) + \sigma_{12}(t)dW_2(t)\right] \tag{10.66}$$

and

$$dS_2(t) = a_2(t)dt + \left[\sigma_{21}(t)dW_1(t) + \sigma_{22}(t)dW_2(t)\right] \tag{10.67}$$

where $a_i(t), \sigma_{ij}(t), i = 1, 2, j = 1, 2$, are the drift and diffusion parameters possibly depending on $S_i(t)$, and where $W_1(t)W_2(t)$ are two *independent* Wiener processes.

In this bivariate framework, $S_1(t), S_2(t)$ represent two stochastic processes that are influenced by the same Wiener components. Because the parameters $\sigma_{ij}(t)$ may differ across equations, error terms affecting the two equations may not be identical. Yet, because the $S_1(t)$, $S_2(t)$ have common error components, they will in general be correlated, except for the special case when

$$\sigma_{12}(t) = 0, \quad \sigma_{21}(t) = 0 \tag{10.68}$$

for all t.

Suppose we now have a continuous, twice-differentiable function of $S_1(t)$ and $S_2(t)$ that we denote by $F(S_1(t), S_2(t), t)$. How can we write the stochastic differential dF_t?

The answer is provided by the multivariate form of Ito's Lemma,[7]

$$dF_t = F_t dt + F_{s_1} dS_1 + F_{s_2} dS_2 + \frac{1}{2}\left[F_{s_1 s_1} dS_1^2 + F_{s_2 s_2} dS_2^2 + F_{s_1 s_2} dS_1 dS_2\right] \tag{10.69}$$

[6]There is a slight change in the notation dealing with the time variable t.

[7]In the following equation we write the stochastic differentials without showing their dependence on t.

where the squared differentials $[dS_1]^2, [dS_2]^2$ and the cross-product term $dS_1 dS_2$ need to be equated to their mean square limits.

We already know that dt^2 and cross products such as $dt\,dW_1(t)$ and $dt\,dW_2(t)$ are equal to zero in the mean square sense. This point was discussed in obtaining the univariate Ito's Lemma. The only novelty now is the existence of cross products such as $dW_1(t)dW_2(t)$.[8] Here we have the product of the increments of two independent Wiener processes. Over a finite interval Δ, we expect

$$\mathbb{E}\left[\Delta W_1(t)\Delta W_2(t)\right] = 0 \tag{10.70}$$

Hence, a limiting argument can be constructed so that in the mean square sense:

$$dW_1(t)dW_2(t) = 0 \tag{10.71}$$

This gives the following mean square approximations for $dS_1(t)^2$ and $dS_2(t)^2$:

$$dS_1(t)^2 = \left[\sigma_{11}^2(t) + \sigma_{12}^2(t)\right] dt \tag{10.72}$$

and

$$dS_2(t)^2 = \left[\sigma_{21}^2(t) + \sigma_{22}^2(t)\right] dt \tag{10.73}$$

The cross-product term is given by

$$dS_1(t)^2 dS_2(t)^2 = \left[\sigma_{11}(t)\sigma_{21}(t) + \sigma_{12}(t)\sigma_{22}(t)\right] dt \tag{10.74}$$

These expressions can be substituted into the bivariate Ito formula in (10.69) to eliminate $dS_1(t)^2, dS_2(t)^2$, and $dS_1(t)dS_2(t)$.

10.7.1.1 An Example from Financial Derivatives

Options written on bonds are among the most popular interest rate derivatives. In valuing these derivatives, the yield curve plays a fundamental role. One class of models of interest rate options assumes that the *yield curve* depends on two state variables, r_t representing a short rate and R_t representing a long rate. The price of the interest rate derivative will then be denoted by $F(r_t, R_t, t), t \in [0, T]$.

These interest rates are assumed to follow the following SDEs:

$$dr_t = a_1(t)dt + \left[\sigma_{11}(t)dW_1(t) + \sigma_{12}(t)dW_2(t)\right] \tag{10.75}$$

and

$$dR_t = a_2(t)dt + \left[\sigma_{21}(t)dW_1(t) + \sigma_{22}(t)dW_2(t)\right] \tag{10.76}$$

Thus, the short and the long rates have correlated errors. Over a finite interval of length h, this correlation is given by

$$Corr\,(\Delta r_t, \Delta R_t) = \left[\sigma_{11}(t)\sigma_{21}(t) + \sigma_{12}(t)\sigma_{22}(t)\right] h \tag{10.77}$$

The market participant can select the parameters $\sigma_{ij}(t)$ so that the equations capture the correlation and volatility properties of the observed short and long rates.

In valuing these interest rate options, one may want to know how the option price reacts to small changes in the yield curve, that is, to dr_t and dR_t. In other words, one needs the stochastic differential dF_t. Here the multivariate form of the Ito formula must be used[9]:

$$\begin{aligned} dF_t &= F_t dt + F_r dr + F_R dR \\ &\quad + \frac{1}{2}\left[F_{rr}\left(\sigma_{11}^2 + \sigma_{12}^2\right) + F_{RR}\left(\sigma_{21}^2 + \sigma_{22}^2\right)\right. \\ &\quad + F_{rR}\left(\sigma_{11}\sigma_{21} + \sigma_{12}\sigma_{22}\right)] \, dt \end{aligned} \tag{10.78}$$

The stochastic differential dF_t would measure how the price of an interest rate derivative will change during a small interval dt, and given a small variation in the yield curve, the latter being caused by dr_t and dR_t.

10.7.1.2 Wealth

An investor buys $N_i(t)$ units of the ith asset at a price $P_i(t)$. There are n assets, and both

[8] Terms such as $dS_1(t)dS_2(t)$ will depend on $dW_1(t)dW_2(t)$.

[9] Again, for notational simplicity, we write $\sigma_{ij}(t)$ as σ_{ij}.

the $N_i(t)$ and $P_i(t)$ are continuous-time stochastic processes, potentially a function of the same random shocks.

The total value of the investment is given by the wealth $Y(t)$ at time t:

$$Y(t) = \sum_{i=1}^{n} N_i(t)P_i(t) \tag{10.79}$$

Suppose we would like to calculate the increments in wealth as time passes. We use Ito's Lemma:

$$dY(t) = \sum_{i=1}^{n} N_i(t)dP_i(t) + \sum_{i=1}^{n} dN_i(t)P_i(t) + \sum_{i=1}^{n} dN_i(t)dP_i(t) \tag{10.80}$$

It is clear that if one used the formulas in standard calculus, the last term of the equation would not be present.

10.7.2 Itô's Formula and Jumps

Thus far, the underlying process S_t was always assumed to be a function of random shocks representable by Wiener processes. This assumption may be too restrictive. There may be a jump component to random errors as well. In this section, we provide this extension of the Ito formula.

Suppose we observe a process S_t, which is believed to follow the SDE

$$dS_t = a_t dt + \sigma_t dW_t + dJ_t, \quad t \geq 0 \tag{10.81}$$

where dW_t is a standard Wiener process. The new term dJt represents possible unanticipated jumps. This jump component has zero mean during a finite interval h:

$$\mathbb{E}[\Delta J_t] = 0 \tag{10.82}$$

We need to make this assumption, since this term is part of the unpredictable innovation terms. This assumption is not restrictive, as any predictable part of the jumps may be included in the drift component a_t.

We assume the following structure for the jumps. Between jumps, J_t remains constant. At jump times $\tau_j, j = 1, 2, \ldots$, it varies by some discrete and random amount. We assume that there are k possible types of jumps, with sizes $\{a_i, i = 1, \ldots, k\}$. The jumps occur at a rate λ_t that may depend on the latest observed S_t. Once a jump occurs, the jump type is selected randomly and independently. The probability that a jump of size a_i will occur is given by p_i.[10]

Thus, during a finite but small interval h, the increment ΔJ_t will be given (approximately) by

$$\Delta J_t = \Delta N_t - \left[\lambda_t h \left(\sum_{i=1}^{k} a_i p_i\right)\right] \tag{10.83}$$

where N_t is a process that represents the sum of all jumps up to time t. More precisely, ΔN_t will have a value of a_i if there was a jump during the h, and if the value of the jump was given by a_i. The term $(\sum_{i=1}^{k} a_i p_i)$ is the expected size of a jump, whereas λth represents, loosely speaking, the probability that a jump will occur. These are subtracted from ΔN_t to make ΔJ_t unpredictable.

Under these conditions, the drift coefficient t can be seen as representing the sum of two separate drifts, one belonging to the Wiener continuous component, the other to the pure jumps in S_t,

$$a_t = \alpha_t + \lambda_t \left(\sum_{i=1}^{k} a_i p_i\right) \tag{10.84}$$

where α_t is a drift coefficient of the continuous movements in S_t.

It is worth discussing one aspect of the jump process again. The process has *two* sources of randomness. The occurrence of a jump is a random event. But once the jump occurs, the size of the

[10]In the case of the standard Poisson process, all jumps have size 1. This step is thus redundant.

jump is also random. Moreover, the structure just given assumes that these two sources of randomness are independent of each other.

Under these conditions, the Ito formula is given by

$$dF(S_t,t) = \left[F_t + \lambda_t \sum_{i=1}^{k} (F(S_t + a_i, t) - F(S_t, t))\, p_i + \frac{1}{2} F_{ss}\sigma^2 \right] dt + F_s dS_t + dJ_F \tag{10.85}$$

where dJ_F is given by

$$dJ_F = \left[F(S_t,t) - F\left(S_t^-, t\right)\right] - \lambda_t \left[\sum_{i=1}^{k} (F(S_t + a_i, t) - F(S_t, t))\, p_i \right] dt \tag{10.86}$$

Finally, S_t^- is defined as

$$S_t^- = \lim_{s \to t} S_s, \quad s < t \tag{10.87}$$

That is, it is the value of S at an infinitesimal time before t.

How would one calculate the dJ_F in practice? One would first evaluate the expected change due to possible random jumps, which is the second term on the right-hand side of Eq. (10.86). To do this, one uses both the rate of possible jumps occurring during dt and the expected size of jump in $F(\cdot)$ caused by jumps in S_t. If during that particular time a jump is observed, then the first term on the right-hand side is also included. Otherwise, the term will equal zero.

10.7.3 Itô's Lemma for Semimartingales

A semimartingale is a local martingale plus a finite-variation process (Protter, 1990). In particular, a finite-variation process itself is a semimartingale. It is well known that all diffusion processes are semimartingales. If X is a semimartingale and $f \in C^2$, then $(f(X_t))_{t \geq 0}$ is a semimartingale and

$$f(X_t) - f(X_0) = \int_0^t f'(X_s)dX_s + \frac{1}{2}\int_0^t f''(X_s)dX_s + \sum_{0<s\leq t} (f(X_s) - f(X_{s-}) - f'(X_{s-}))\Delta X_s) \tag{10.88}$$

Let $X = (X_1, \ldots, X_n)$ be an n-tuple of continuous semimartingales, and let $f: \mathbb{R}^n \to \mathbb{R}$ have continuous second-order partial derivatives. Then $f(X)$ is a semimartingale and the following formula holds:

$$f(X_t) - f(X_0) = \sum_{i=1}^{n} \int_0^t \frac{\partial f}{\partial x_i}(X_s)dX_s^i + \frac{1}{2} \sum_{1\leq i,j\leq n} \int_0^t \frac{\partial^2 f}{\partial x_i \partial x_j} d[X^i, X^j]_s \tag{10.89}$$

where $[X^i, X^j]$ is the quadratic covariation of X^i and X^j.

This version of Itô's Lemma can be found in Protter (1990, p. 74). In addition, Protter (1990, p. 68) contains a theorem which implies that if X is a continuous finite-variation process, then, for any semimartingale Y, the quadratic covariation of X and Y, $[X,Y]_t$ is equal to X_0Y_0 for all t. Therefore, $d[W,M]_t = 0$ and $d[M,M]_t = 0$.

10.8 CONCLUSION

Ito's Lemma is the central differentiation tool in stochastic calculus. There are a few basic things to remember. First, the formula helps to determine stochastic differentials for financial derivatives, given movements in the underlying asset. Second, the formula is completely dependent on the definition of the Ito integral. This means that equalities should be interpreted within stochastic equivalence.

Finally, from a practical point of view, the reader should remember that standard formulas used in deterministic calculus give significantly different results than the Ito formula. In particular, if one uses standard formulas, this would amount to assuming that all processes under observation have zero infinitesimal volatility. This is not a pleasant assumption when one is trying to price risk using financial derivatives.

10.9 REFERENCES

The sources recommended for Chapter 9 also apply here. Ito's Lemma and the Ito integral are two topics that are always treated together. One additional source the reader may appreciate is the book by Kushner and Dupuis (2000), which provides several examples of Ito's Lemma with jump processes.

10.10 EXERCISES

1. Differentiate the following functions with respect to the Wiener process W_t and, if applicable, with respect to t.
   - **(a)** $f(W_t) = W_t^2$
   - **(b)** $f(W_t) = \sqrt{W_t}$
   - **(c)** $f(W_t) = e^{\left(W_t^2\right)}$
   - **(d)** $f(W_t, t) = e^{\left(\sigma W_t - \frac{1}{2}\sigma^2 t\right)}$
   - **(e)** $f(W_t) = e^{\sigma W_t}$
   - **(f)** $g = \int_0^t W_s ds$
2. Suppose the $W_{t_i}, i = 1, 2$ are two Wiener processes. Use Ito's Lemma in obtaining appropriate stochastic differential equations for the following transformations.
   - **(a)** $X_t = (W_{t_1})^4$
   - **(b)** $X_t = (W_{t_1} + W_{t_2})^2$
   - **(c)** $X_t = t^2 + e^{W_{t_2}}$
   - **(d)** $X_t = e^{t^2 + W_{t_2}}$
3. Let W_t be a Wiener process. Consider the geometric process S_t again:

$$S_t = S_0 e^{\left(\mu - \frac{1}{2}\sigma^2\right)t + \sigma W_t}$$

   - **(a)** Calculate dS_t.
   - **(b)** What is the "expected rate of change" of S_t?
   - **(c)** If the exponential term in the definition of S_t did not contain the term, what would be the dS_t? What would then be the expected change in S_t?
4. Let W be a Wiener process. Compute $\mathbb{E}(W_t^4)$.
5. Consider the equation $X_t = cos(e^{W_t})$. Write a program that computes the estimated mean and variance of X_t via simulation.

CHAPTER

11

The Dynamics of Derivative Prices

OUTLINE

11.1 INTRODUCTION

The concept of a stochastic differential equation (SDE) was introduced in Chapter 7. In Chapter 9 we used the Ito integral to formalize this concept. The notation

$$dS_t = a\left(S_t, t\right) dt + \sigma\left(S_t, t\right) dW_t, \quad t \in [0, \infty) \tag{11.1}$$

was justified as a symbolic way of writing

$$\int_t^{t+h} dS_u = \int_t^{t+h} a\left(S_u, u\right) du + \int_t^{t+h} \sigma\left(S_u, u\right) dW_u \tag{11.2}$$

when h is infinitesimal.

We repeat some aspects of this derivation. First of all, no concept from financial markets or financial theory was used to obtain (11.1). The basic tools used were the Ito integral and the ability

Financial Derivatives, *Third Edition*.
http://dx.doi.org/10.1016/B978-0-12-384682-2.00011-6

to split some increment in a random price into predictable and unpredictable components.

This brings us to another point. Given that the decomposition in Eq. (11.1) is done using the information set available at time t, then to the extent different players may have access to different sets of information, the SDE in (11.1) may also be different. For example, consider the following extreme case. Suppose a market participant has "inside information" and learns all the random events that influence price changes in advance. Under these (unrealistic) conditions, the diffusion term in (11.1) would be zero. Since the participant knows how dS_t is going to change, he or she can predict this variable perfectly, and $dW_t = 0$ for all t. If we were to write this participant's SDE, we would get

$$dS_t = a^* (S_t, t)\, dt \tag{11.3}$$

whereas for all other market participants,

$$dS_t = a (S_t, t)\, dt + \sigma (S_t, t)\, dt \tag{11.4}$$

In these two equations the drift and the diffusion terms cannot be the same. The error terms that drive the SDEs are different, which makes $a^*(S_t, t)$ different from an $a(S_t, t)$. This example shows that the exact form of the SDE, and hence the definition of the error term dW_t, always depends on the family of information sets $\{I_t, t \in [0, \infty)\}$. If we had access to a different family of information sets I_t^*, we would make different prediction errors, and the probabilistic behavior of the error terms would change. Given a different family of information sets, we may have to denote the errors by dW_t^* instead of dW_t. It may be that dW_t^* has a smaller variance than dW_t.

In stochastic calculus, this property of W_t is formally summarized by saying that the Wiener process W_t is adapted to the family of information sets I_t.

The SDEs are utilized in pricing derivative assets because they give us a formal model of how an underlying asset's price changes over time. But it is also true that the formal derivation of SDEs is compatible with the way dealers behave in financial markets. In fact, on a given trading day, a trader continuously tries to forecast the price of an asset and record the "new events" as time passes. These events always contain some parts that are unpredictable until one observes the dS_t. After that, they become known and become part of the new information set the trader possesses.

This chapter considers some properties of stochastic differential equations.

11.1.1 Conditions on a_t and t

The drift and diffusion parameters of the SDE

$$dS_t = a (S_t, t)\, dt + \sigma (S_t, t)\, dW_t, \quad t \in [0, \infty) \tag{11.5}$$

were allowed to depend on S_t and t. Hence, these parameters are themselves random variables. The point is that, given the information at time t, they are observed by the market participant. Conditional on available information, they "become" constant. This is the consequence of the important assumption that these parameters are I_t-adapted. At several points during the previous chapters, we made assumptions suggesting that these parameters should be well behaved.

It is customary to specify these "regularity conditions" each time an SDE is proposed as a model.

The $a(S_t, t)$ and $\sigma(S_t, t)$ parameters are assumed to satisfy the conditions

$$P\left(\int_0^t |a (S_u, u)|\, du < \infty\right) = 1$$

and

$$P\left(\int_0^t |\sigma (S_u, u)|\, du < \infty\right) = 1$$

These conditions have similar meanings. They require that the drift and diffusion parameters do not vary "too much" over time.

Note that the integrals in these conditions are taken with respect to time. In this sense, they can be defined in the usual context. According to this, the conditions imply that the drift and diffusion parameters are functions of bounded variation with probability one.

In the remainder of this book, we assume that these conditions are always satisfied and never repeat them.

11.2 A GEOMETRIC DESCRIPTION OF PATHS IMPLIED BY SDEs

Consider the stochastic differential equation

$$dS_t = a(S_t, t)\,dt + \sigma(S_t, t)\,dW_t, \quad t \in [0, \infty) \tag{11.6}$$

where the drift and diffusion parameters depend on the level of observed asset price S_t and (possibly) on t.

What type of geometric behavior would such an SDE imply for S_t?

An example is shown in Figure 11.1. We consider small but discrete intervals of length h. We see that, over time, the behavior of S_t can be decomposed into two types of movements. First, there is an *expected* path during the interval. These are indicated by upward- or downward-sloping arrows. Then, at each $t_k = kh$, there is a second movement orthogonal to the predicted changes.[1] These are represented by vertical arrows. Sometimes they are negative; other times they are positive. The actual movement of S_t over time is determined by the sum of these two components and is indicated by the heavy line.

This geometric derivation emphasizes once again that the trajectories of S_t are likely to be very erratic when h becomes infinitesimal.

11.3 SOLUTION OF SDEs

A stochastic differential equation is by definition an *equation*. That is, it contains an unknown. This unknown is the stochastic process S_t. The notion of a solution to an SDE is thus more complicated than it may seem at the outset. What we are searching for is not a number or vector of numbers. It is a random process whose trajectories and the probabilities associated with those trajectories need to be determined exactly.

[1]"Orthogonal" here implies "uncorrelated."

11.3.1 What Does a Solution Mean?

First, consider the finite difference approximation in small, discrete intervals:

$$S_k - S_{k-1} = a(S_{k-1}, k)\,h + \sigma(S_{k-1}, k)\,\Delta W_k, \quad k = 1, 2, \ldots, n \tag{11.7}$$

The solution to this equation is a random process S_t. We are interested in finding a sequence of random variables indexed by k, such that the increments ΔS_k satisfy (11.7). Moreover, we would like to know the moments and the distribution function of a process S_k that satisfies Eq. (11.7). At the outset, it is not clear that, given a particular $a(\cdot)$ and $\sigma(\cdot)$ we could find a sequence of random numbers whose trajectories will satisfy the equality in (11.7) for all k.

More importantly, our purpose is to look for this solution when h, the interval length, goes to zero. If a continuous-time process S_t satisfies the equation

$$\int_0^t dS_u = \int_0^t a(S_u, u)\,du + \int_0^t \sigma(S_u, u)\,dW_u \tag{11.8}$$

for all $t > 0$, then we say that S_t is the solution of

$$dS_t = a(S_t, t)\,dt + \sigma(S_t, t)\,dW_t \tag{11.9}$$

Because the solutions of SDEs are random processes, the nature of these solutions could be quite different when compared with ordinary differential equations. In fact, in stochastic calculus, there can be *two* types of solutions.

11.3.2 Types of Solutions

The first type of solution to an SDE is similar to the case of ordinary differential equations. Given

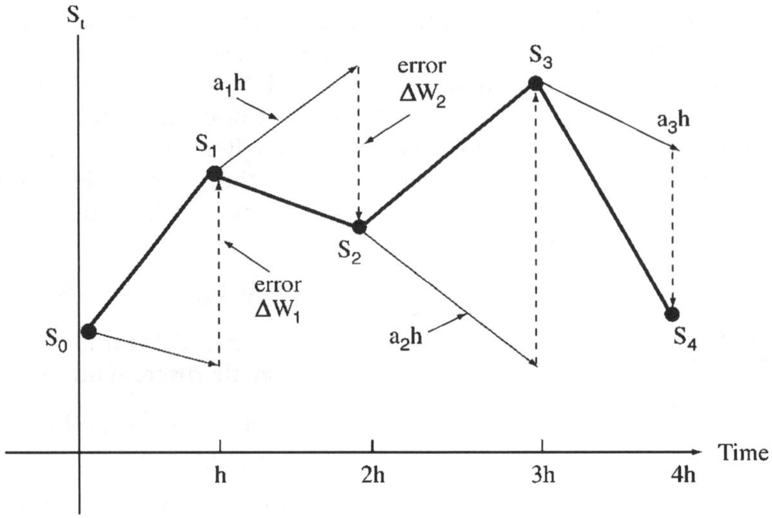

FIGURE 11.1 A sample path of a geometric SDE on discrete intervals of length h.

the drift and diffusion parameters *and* the random innovation term dW_t, we determine a random process S_t, paths of which satisfy the SDE:

$$dS_t = a\,(S_t, t)\,dt + \sigma\,(S_t, t)\,dW_t, \quad t \in [0, \infty) \tag{11.10}$$

Clearly, such a solution S_t will depend on time t, and on the past and contemporaneous values of the random variable W_t, as the underlying integral equation illustrates:

$$\int_0^t dS_u = \int_0^t a\,(S_u, u)\,du + \int_0^t \sigma\,(S_u, u)\,dW_u \tag{11.11}$$

for all $t > 0$. The *solution* determines the exact form of this dependence. When W_t on the right-hand side of (11.8) is given exogenously and S_t is then determined, we obtain the so-called strong solution of the SDE. This is similar to solutions of ordinary differential equations.

The second solution concept is specific to stochastic differential equations. It is called the *weak solution*. In the weak solution, one determines the process $\widetilde{S}_t$

$$\widetilde{S}_t = f\left(t, \widetilde{W}_t\right) \tag{11.12}$$

where $\widetilde{W}_t$ is a Wiener process whose distribution is determined *simultaneously* with $\widetilde{S}_t$. According to this, for the weak solution of SDEs, the "givens" of the problems are *only* the drift and diffusion parameters, $a(\cdot)$ and $\sigma\,(\cdot)$ respectively.

The idea of a weak solution can be explained as follows. Given that solving SDEs involves finding random variables that satisfy Eq. (11.8), one can argue that finding an $\widetilde{S}_t$ and a $\widetilde{W}_t$ such that the pair $\{\widetilde{S}_t, \widetilde{W}_t\}$ satisfies this equation, is also a type of solution to the stochastic differential equation.

In this type of solution, we are given the drift parameter $a(S_t, t)$ and the diffusion parameter $\sigma(S_t, t)$. We then find the processes $\widetilde{S}_t$ and $\widetilde{W}_t$ such that Eq. (11.8) is satisfied. This is in contrast to strong solutions where one does not solve for the Wt, but considers it another given of the problem.

Clearly, there are some potentially confusing points here. First of all, what is the difference between dW_t and $d\widetilde{W}_t$ if both are Wiener processes with zero mean and variance dt? Are these not the same object?

In terms of the form of distribution functions, this is a valid question. The density functions of dWt and $d\widetilde{W}_t$ are given by the same formula.

In this sense, there is no difference between the two random errors. The difference will be in the sequence of information sets that define dW_t and $d\widetilde{W}_t$.[2] Although the underlying densities may be the same, the two random processes could indeed represent very different real-life phenomena if they are measurable with respect to different information sets.

This has to be made more precise because it reemphasizes an important point made earlier—a point that needs to be clarified in order for the reader to understand the structure of continuous-time stochastic models. Consider the following SDE, where the diffusion term contains the exogenously given dW_t:

$$dS_t = a(S_t, t)\, dt + \sigma(S_t, t)\, dW_t \tag{11.13}$$

Heuristically, the error process dW_t symbolizes infinitesimal events that affect prices in a completely unpredictable fashion. The "history" generated by such infinitesimal events is the set of information that we have at time t. This we denoted by I_t.[3]

The strong solution then calculates an S_t that satisfies Eq. (11.13) with dW_t given. That is, in order to obtain the strong solution S_t, we need to know the family I_t. This means that the strong solution S_t will be I_t-adapted.

The weak solution, on the other hand, is not calculated using the process that generates the information set I_t. Instead, it is found along with some process $\widetilde{W}_t$. The process could generate some other information set H_t. The corresponding will not necessarily be I_t-adapted. But it will still be a martingale with respect to histories H_t.[4]

Hence, the weak solution will satisfy

$$dS_t = a(S_t, t)\, dt + \sigma(S_t, t)\, d\widetilde{W}_t \tag{11.14}$$

where the drift and the diffusion components are the same as in (11.8), and where it is adapted to some family of information sets H_t.

11.3.3 Which Solution is to be Preferred?

Note that the strong and the weak solutions have the same drift and diffusion components. Hence, S_t and $\widetilde{S}_t$ will have similar statistical properties. Given some means and variances, we will not be able to distinguish between the two solutions. Yet the two solutions may also be different.[5]

The use of a strong solution implies knowledge of the error process W_t. If this is the case, the financial analyst may work with strong solutions.

Often when the price of a derivative is calculated using a solution to an SDE, one does not know the exact process Wt. One may use only the volatility and (sometimes) the drift component. Hence, in pricing derivative products under such conditions, one works with weak solutions.

11.3.4 A Discussion of Strong Solutions

The stochastic differential equation is, as mentioned earlier, an equation. This means that it contains an unknown that has to be solved. In the case of SDEs, the "unknown" under consideration is a stochastic process. By solving an SDE, we mean determining a process S_t such that the integral equation

$$S_t = S_0 + \int_0^t a(S_u, u)\, du + \int_0^t \sigma(S_u, u)\, dW_u \tag{11.15}$$

is valid for all t. In other words, the evolution of S_t, starting from an initial point S_0, is determined by the two integrals on the right-hand side. The solution process S_t must be such that when these

[2] As we see in Chapter 14, the two Wiener processes may imply different probability measures on dS_t.

[3] As mentioned earlier, mathematicians call such information sets σ-field or σ-algebra.

[4] Because of this martingale property, the Ito integrals that are in the background of SDEs can still be defined the same way.

[5] Any strong solution is also a weak solution. But the reverse is not true.

integrals are added together, they should yield the increment $S_t - S_0$. This would *verify* the solution.

This approach verifies the solution using the corresponding integral equation rather than using the SDE directly. Why is this so? Note that according to the discussion up to this point, we do not have a theory of differentiation in stochastic environments. Hence, if we have a candidate for a solution of an SDE, we *cannot* take derivatives and see if the corresponding derivatives satisfy the SDE. Two alternative routes exist.

The process of verifying solutions to SDEs can best be understood if we start with a deterministic example. Consider the simple ordinary differential equation

$$\frac{dX_t}{dt} = aX_t \tag{11.16}$$

where a is a constant and X_0 is given. There is no random innovation term; this is not a stochastic differential equation. A candidate for the solution can be verified directly. For example, suppose it is suspected that the function

$$X_t = X_0 e^{at} \tag{11.17}$$

is a solution of (11.16). Then, the solution must satisfy two conditions. First, if we take the derivative of X_t with respect to t, this derivative must equal a times the function itself. Second, when evaluated at $t = 0$, the function should give a value equal to X_0, the initial point, which is assumed to be known.

We proceed to verify the solution to Eq. (11.16). Taking the straightforward derivative of X_t,

$$\frac{d}{dt}\left(X_0 e^{at}\right) = a\left[X_0 e^{at}\right] \tag{11.18}$$

which is indeed a times the function itself. The first condition is satisfied.

Letting $t = 0$, we get

$$\left(X_0 e^{a0}\right) = X_0 \tag{11.19}$$

Hence, the candidate solution satisfies the initial condition as well. We thus say that X_t solves the ODE in (11.16). This method verified the solution using the concept of derivative.[6]

If there is no differentiation theory of continuous stochastic processes, a similar approach cannot be utilized in verifying solutions of SDEs. In fact, if one uses the same differentiation methodology, assuming (mistakenly) that it holds in stochastic environments, and tries to "verify" solutions to SDEs by taking derivatives, one would get the *wrong* answer. As seen earlier, the rules of differentiation that hold for deterministic functions are not valid for functions of random variables.

Some further comments on this point might be useful. Note that in an ordinary differential equation,

$$\frac{dX_t}{dt} = aX_t, \tag{11.20}$$

with X_0 given, both sides of the equation contain terms in the unknown function X_t. That is why the ODE is an *equation*. The *solution* of the equation is then a specific *function* that depends on the remaining parameters and known variables in the ODE. The parameters are $\{a, X_0\}$, and the only known variable is the time t. Hence, the solution expresses the unknown function X_t as a function of the known quantities:

$$X_t = X_0 e^{at} \tag{11.21}$$

Verifying the solution involves differentiating this function X_t with respect to the right-hand-side variable t, and then checking to see if the ODE is satisfied.

Now consider the special case of the SDE given by

$$dS_t = a dt + \sigma\, dW_t, \quad t \geq 0 \tag{11.22}$$

[6]Of course, the reader may wonder how the candidate solution was obtained to begin with. This topic belongs to texts on differential equations. Here, we just deal with models routinely used in finance.

with S_0 given.[7] When a strong solution of this SDE is obtained, it will be some function $f(\cdot)$ that depends on the time t, on the parameters $\{a, \sigma, S_0\}$, and on the W_t:

$$S_t = f(a, \sigma, S_0, t, W_t) \tag{11.23}$$

Hence, the solution will be a stochastic process because it depends on the random process W_t[8]:

Using deterministic differentiation formulas to check whether this satisfies the SDE in (11.22) means taking the derivatives of S_t and W_t with respect to t. But these derivatives with respect to t are not well defined. Hence, the solution cannot be verified by using the same methodology as in the deterministic case.

Instead, one should consider a candidate solution, and then, using Ito's Lemma, try to see if this candidate satisfies the SDE or the corresponding integral equation. In the example below, we consider this point in detail.

11.3.5 Verification of Solutions to SDEs

Again consider the special SDE,

$$dS_t = \mu S_t dt + \sigma S_t dW_t \tag{11.24}$$

which was used by Black–Scholes (1973) in pricing call options. Here, S_t represents the price of a security that does not pay any dividends.

Dividing both sides by S_t, we get

$$\frac{1}{S_t} dS_t = \mu\, dt + \sigma\, dW_t \tag{11.25}$$

First, we calculate the implied integral equation:

$$\int_0^t \frac{dS_u}{S_u} = \int_0^t \mu\, du + \int_0^t \sigma\, dW_u \tag{11.26}$$

[7]The drift and diffusion parameters are constant and do not depend on the information available at time t.

[8]It is important to keep in mind that the SDE discussed above is a special case. In general, the strong solution S_t shown in (11.23) will depend on the *integrals* of $a(S_u, u), \sigma(S_u, u)$, and dW_u. Hence, the dependence will be on the whole trajectory of W_t.

Since the first integral on the right-hand side does not contain any random terms, it can be calculated in the standard way:

$$\int_0^t \mu\, du = \mu t \tag{11.27}$$

The second integral does contain a random term, but the coefficient of dW_u is a time-invariant constant. Hence, this integral can also be taken in the usual way

$$\int_0^t \sigma\, dW_u = \sigma\,[W_t - W_0] \tag{11.28}$$

where by definition $W_0 = 0$. Thus, we have

$$\int_0^t \sigma\, dW_u = \mu t + \sigma W_t \tag{11.29}$$

Any solutions of the SDE must satisfy this integral equation. In this particular case, we can show this simply by using Ito's Lemma.

Consider the candidate

$$S_t = S_0 e^{\left\{\left(a - \frac{1}{2}\sigma^2\right)t + \sigma W_t\right\}} \tag{11.30}$$

Note that this solution candidate is indeed a function of the parameters a and of time t, and of the random variable W_t. Clearly, we are dealing with a strong solution, since S_t depends on W_t and is I_t-adapted.

How do we verify that this function is indeed a solution?

Consider calculating the stochastic differential dS_t using Ito's Lemma:

$$\begin{aligned} dS_t &= \left[S_0 e^{\left\{\left(a - \frac{1}{2}\sigma^2\right)t + \sigma W_t\right\}}\right]\left[\left(a - \frac{1}{2}\sigma^2\right) dt \right. \\ &\quad \left. + \sigma\, dW_t + \frac{1}{2}\sigma^2 dt\right] \end{aligned} \tag{11.31}$$

where the very last term on the right-hand side corresponds to the second-order term in Ito's Lemma.

Canceling similar terms and replacing by S_t, we obtain

$$dS_t = S_t\left[adt + \sigma\, dW_t\right] \tag{11.32}$$

which is the original SDE with a equal to μ. It is interesting to note that the terms $\frac{1}{2}\sigma^2 dt$ are eliminated by the application of Ito's Lemma. If the rules of deterministic differentiation were used, these terms would not disappear in Eq. (11.32), and the function in (11.30) would not verify the SDE.

In fact, if we had used ordinary calculus, total differentiation would instead give

$$dS_t = S_t\left[\left(a - \frac{1}{2}\sigma^2\right)dt + \sigma\, dW_t\right] \tag{11.33}$$

and this would not be the same as the original SDE if a equals μ. Hence, if we had used ordinary calculus, we would have mistakenly concluded that the function in (11.30) is not a solution of the SDE in (11.24).

11.3.6 An Important Example

Suppose S_t is some asset price with a *random* rate of appreciation. In other words, we have

$$dS_t = rS_t dt + \sigma S_t dW_t, \quad t \in [0, \infty) \tag{11.34}$$

The previous section discussed a candidate for the (strong) solution of this SDE:

$$S_t = S_0 e^{\left\{\left(r - \frac{1}{2}\sigma^2\right)t + \sigma W_t\right\}} \tag{11.35}$$

Now, suppose S_T is the price at some future time $T > t$. As of time t, this S_T is unknown. But it can be predicted, and the best prediction will be given by the conditional expectation:

$$\mathbb{E}_t\left[S_T\right] = \mathbb{E}\left[S_T | I_t\right] \tag{11.36}$$

In asset pricing theory, one is interested in whether the following equality will hold:

$$S_t = e^{-r(T-t)}\mathbb{E}_t\left[S_T\right] \tag{11.37}$$

This would make the current price equal to the expected price at time T discounted at a rate r. This martingale property is of interest because it can be exploited to calculate the current price S_t.

We now calculate $\mathbb{E}_t\left[S_T\right]$. The first step is to realize the following:

$$S_T = \left[S_0 e^{\left(r - \frac{1}{2}\sigma^2\right)T}\right]\left[e^{\sigma W_T}\right] \tag{11.38}$$

so that expectations of S_T depend on expectations of the term

$$e^{\sigma WT} \tag{11.39}$$

where, for future reference, the expression is a nonlinear function of W_T. Hence, the S_T is a nonlinear function of W_T as well. This means that in taking the expectation $\mathbb{E}_t\left[S_T\right]$, we cannot "move" the $\mathbb{E}_t\left[\cdot\right]$ operator in front of the random term W_T.

We can approach the expectation in two different ways. One method would be to use the density function for the Wiener process W_T and "take" the expectation directly by integrating

$$\mathbb{E}_t\left[e^{\sigma W_t}\right] = \int_{-\infty}^{\infty} e^{\sigma W_T} f(W_T | W_t)\, dW_T \tag{11.40}$$

where the term in brackets inside the integral is the (conditional) density function of W_T. The (conditional) mean is W_t, and the variance is $T - t$.

Calculating this integral is not difficult. But we prefer using a second method, which is specific to "stochastic calculus." This method will illustrate Ito's Lemma once again, and will introduce an important integral equation that sees frequent use in stochastic calculus.

According to Eq. (11.38), S_T is given by the function

$$S_T = \left[S_0 e^{\left(r - \frac{1}{2}\sigma^2\right)T}\right]\left[e^{\sigma W_T}\right] \tag{11.41}$$

The idea behind the second method is to transform this nonlinear expression in W_t into a linear one, and then take the expectations directly without having to use the density function of the Wiener process.

The method is indirect, but fairly simple. First, denote the nonlinear random term in Eq. (11.35) by Z_t[9]:

$$Z_t = e^{\sigma W_t} \tag{11.42}$$

Second, apply Itô's Lemma:

$$dZ_t = \sigma e^{\sigma W_t} dW_t + \frac{1}{2}\sigma^2 e^{\sigma W_t} dt \tag{11.43}$$

Third, consider the corresponding integral equation:

$$\mathbb{Z}_t = Z_0 + \sigma \int_0^t e^{\sigma W_s} dW_s + \int \frac{1}{2}\sigma^2 e^{\sigma W_s} ds \tag{11.44}$$

Finally, take expectations on both sides, and note the following:

$$\mathbb{E}[Z_0] = 1 \tag{11.45}$$

since, by definition, $W_0 = 0$. Also,

$$\mathbb{E}\left[\int_0^t e^{\sigma W_s} dW_s\right] \tag{11.46}$$

since the increments in a Wiener process are independent from the observed past. Consequently,

$$\mathbb{E}[Z_t] = 1 + \int \frac{1}{2}\sigma^2 \mathbb{E}[Z_t]\, ds \tag{11.47}$$

with the substitution $e^{\sigma W_s} = Z_s$, which is true by the definition of Z_s.

Note some interesting characteristics of Eq. (11.47). First of all, this equation does not contain any integrals defined with respect to a random variable. Second, the equation is linear in $\mathbb{E}[Z_t]$. Hence, it can be solved in a standard fashion. For example, we can treat $\mathbb{E}[Z_t]$ as a deterministic variable, call it x_t, and then recognize that

$$x_t = 1 + \int_0^t \frac{1}{2}\sigma^2 x_s ds \tag{11.48}$$

is equivalent to the ordinary differential equation[10]

$$\frac{dx_t}{dt} = \frac{1}{2}\sigma^2 x_t \tag{11.49}$$

with initial condition $x_0 = 1$. The solution of this ordinary differential equation is known to be

$$x_t = \mathbb{E}[Z_t] = e^{\frac{1}{2}\sigma^2 t} \tag{11.50}$$

with $\mathbb{E}[Z_0] = 1$. Going back to $\mathbb{E}_t[S_t]$,

$$\mathbb{E}_t[S_t] = \left[S_0 e^{\left(r - \frac{1}{2}\sigma^2\right)T}\right] \mathbb{E}[Z_t] \tag{11.51}$$

Using the result just derived for $\mathbb{E}[Z_t]$,

$$\mathbb{E}_t[S_t] = \left[S_0 e^{\left(r - \frac{1}{2}\sigma^2\right)T}\right] e^{\sigma W_t} e^{\frac{1}{2}\sigma^2 (T-t)} \tag{11.52}$$

where the W_t term on the right-hand side appears due to conditioning on information at time t. Recognizing that

$$S_t = S_0 e^{\left(r - \frac{1}{2}\sigma^2\right)t + \sigma W_t}$$

we obtain

$$S_t e^{r(T-t)} \tag{11.53}$$

which implies that

$$S_0 = e^{-rT} \mathbb{E}_0[S_T] \tag{11.54}$$

That is, at time $t = 0$, the asset price equals the expected future price discounted at a rate r. For any time t we have, correspondingly,

$$S_t = e^{-r(T-t)} \mathbb{E}_t[S_T] \tag{11.55}$$

It is worthwhile to repeat the way Ito's Lemma is used in these calculations. By using it we were able to obtain an integral Eq. (11.47) linear in Z_t. This way we could move the $\mathbb{E}[\cdot]$ operator in front of Z_t and use the fact that increments in Wiener process have zero expectations. This eliminates the integral with respect to the

[9]The next few derivations use the t subscript instead of T. This does not cause any loss of generality. It simplifies exposition.

[10]By taking the derivatives, with respect to t, of (11.48).

random variable. The second integral was with respect to time, and could be handled using standard calculus.

At this point, if instead of Ito's Lemma we had used the rules of standard calculus, Eq. (11.43) would become

$$dZ_t = \sigma e^{\sigma W_t} dW_t \tag{11.56}$$

and the expected value of the stock price would be written as

$$\mathbb{E}[S_T] = S_t e^{\left(r+\frac{1}{2}\sigma^2\right)(T-t)} \tag{11.57}$$

The use of standard calculus implies that today's stock price is not equal to the expected future value discounted at a rate r. We lose the martingale equality.

11.4 MAJOR MODELS OF SDEs

There are some specific SDEs that are found to be quite useful in practice. In this section, we discuss these cases and show what types of asset prices they could represent and how they could be useful.

11.4.1 Linear Constant Coefficient SDEs

The simplest case of stochastic differential equations is where the drift and diffusion coefficients are independent of the information received over time:

$$dS_t = \mu\, dt + \sigma\, dW_t, \quad t \in [0, \infty) \tag{11.58}$$

where W_t is a standard Wiener process with variance t.

In this SDE, the coefficients μ and σ do not have time subscripts t. This means that they are constant as time passes. Hence, they do not depend on the information sets I_t. The mean of ΔS_t during a small interval of length h is given by

$$\mathbb{E}_t(\Delta S_t) = \mu h \tag{11.59}$$

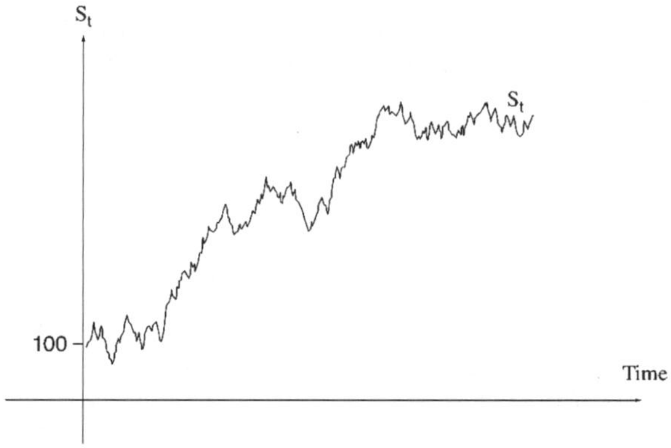

FIGURE 11.2 A sample path of the linear constant coefficient SDE.

The expected variation in ΔS_t will be

$$\mathbb{V}[\Delta S_t] = \sigma^2 h \tag{11.60}$$

An example of the paths that can be described by this SDE is shown in Figure 11.2. Computer simulations were used to obtain this path. First, some desired values for μ and σ were selected:

$$\mu = 0.01 \tag{11.61}$$

$$\sigma = 0.03 \tag{11.62}$$

Then, a small but finite interval size was decided upon:

$$h = 0.001 \tag{11.63}$$

This is assumed to be an approximation to the infinitesimal interval dt. The initial point was selected as

$$S_0 = 100 \tag{11.64}$$

Finally, a random number generator was used to obtain 1000 independent, normally distributed random variables, with mean zero and variance 0.001. The fact that W_t in (11.59) is a martingale permits the use of independent (normally distributed) random variables.

A discrete approximation of Eq. (11.59) was used to obtain the S_t plotted in Figure 11.2. The observations were determined from the iterations

$$S_k = S_{k-1} + 0.01\,(0.001) + 0.03\,(\Delta W_k)\,,$$
$$k = 1, 2, \ldots, 1000 \qquad (11.65)$$

With the initial point S_0 given, one substitutes randomly drawn normal random numbers for ΔW_k and obtains the S_k successively.

As can be seen from this figure, the behavior of S_t seems to fluctuate around a straight line with slope μ. The size of σ determines the extent of the fluctuations around this line. Note that these fluctuations do not become larger as time passes.

This suggests when such a stochastic differential equation is appropriate in practice. In particular, this SDE will be a good approximation if the behavior of the asset prices is stable over time, if the "trend" is linear, and if the "variations" do not get any larger. Finally, it will be a good approximation if there do not appear to be systematic "jumps" in asset prices.

11.4.2 Geometric SDEs

The standard SDE used to model underlying asset prices is not the linear constant coefficient model, but is the geometric process. It is the model exploited by Black and Scholes:

$$dS_t = \mu S_t dt + \sigma S_t dW_t, \quad t \in [0, \infty) \qquad (11.66)$$

This model implies that in terms of the formal notation,

$$a\,(S_t, t) = \mu S_t \qquad (11.67)$$

and

$$\sigma\,(S_t, t) = \sigma S_t \qquad (11.68)$$

Hence, the drift and the diffusion coefficients depend on the information that becomes available at time t. However, this dependence is rather straightforward. The drift and the standard deviation change proportionally with S_t. In fact, dividing both sides by S_t, we obtain

$$\frac{dS_t}{S_t} = \mu S_t + \sigma S_t \qquad (11.69)$$

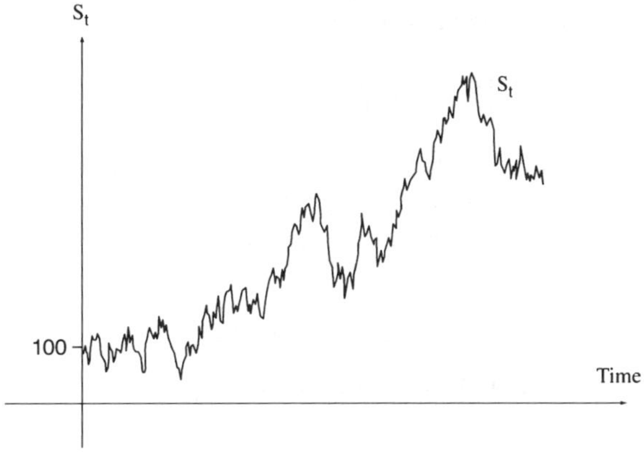

FIGURE 11.3 A sample path of geometric SDE.

This means that although the drift and the diffusion part of the increment in asset price changes, the drift and diffusion of percentage change in S_t still has time-invariant parameters.

Figure 11.3 shows one realization of the S_t obtained from a finite difference approximation of

$$dS_t = 0.15 S_t dt + 0.30 S_t dW_t \qquad (11.70)$$

with the initial point $S_0 = 100$. As can be seen from this graph, the S_t is made of two components. First, there is an *exponential* trend that grows at 15%. Second, there are random fluctuations around this trend. These variations *increase* over time because of higher prices.

What is the empirical relevance of this model when compared with constant coefficient SDEs?

It turns out that the constant coefficient SDE described an asset price that fluctuated around a linear trend, while this model gives prices that fluctuate randomly around an exponential trend. For most asset prices, the exponential trend is somewhat more realistic.

But this says nothing about the assumption concerning the diffusion coefficient. Is a diffusion coefficient proportional to S_t more realistic as well?

To answer this, we note that the "variance" of an incremental change in S_t between times t_k and

t_{k-1} could be approximated by

$$\mathbb{V}\left[S_k - S_{k-1}\right] = \sigma^2 S_{k-1}^2 \tag{11.71}$$

Hence, the variance increases in a way proportional to the square of S_t. In some practical cases, this may add too much variation to S_t.

11.4.3 Square Root Process

A model close to the one just discussed is the square root process,

$$dS_t = \mu S_t dt + \sigma \sqrt{S_t} dW_t, \quad t \in [0, \infty) \tag{11.72}$$

Here, the S_t is made to follow an exponential trend, while the standard deviation is made a function of the square root of S_t, rather than of S_t itself. This makes the "variance" of the error term proportional to S_t.

Hence, if the asset price volatility does not increase "too much" when S_t increases, this model may be more appropriate. This will, of course, be the case if $S_t > 1$.

As an example we provide, in Figure 11.4, the sample path obtained from the same dW_t terms used to generate Figure 11.3. We consider the equation

$$dS_t = 0.15S_t + 0.30\sqrt{S_t} dW_t \tag{11.73}$$

where the drift and diffusion coefficients are as in the case of Figure 11.3, but where the diffusion is now proportional instead of being proportional to S_t. We select the initial point as $S_0 = 100$.

Clearly, the fluctuations in Figure 11.4 are more subdued than the ones in Figure 11.3, yet the sample paths have "similar" trends.

Finally, another characteristic of this process is the meaning of the parameter σ. Note that with this specification of the diffusion component, σ cannot be interpreted as percentage volatility of S_t. Markets, on the other hand, quote by convention the percentage volatility of underlying assets.

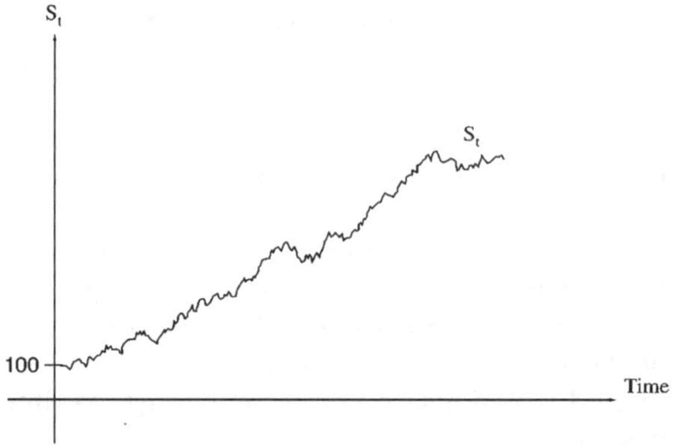

FIGURE 11.4 A sample path of square root process.

11.4.4 Mean Reverting Process

An SDE that has been found useful in modeling asset prices is the *mean reverting* model[11]:

$$dS_t = \lambda (\mu - S_t) dt + \sigma S_t dW_t \tag{11.74}$$

As S_t falls below some "mean value" μ, the term in parentheses, $(\mu - S_t)$, will become positive. This makes dS_t more likely to be positive. *St* will eventually move toward and revert to the value μ.

A related SDE is the one where the drift is of the mean reverting type, but the diffusion is dependent on the square root of S_t:

$$dS_t = \lambda (\mu - S_t) dt + \sigma \sqrt{S_t} dW_t \tag{11.75}$$

There is a significant difference between the mean reverting SDE and the two previous models.

The mean-reverting process has a trend, but the deviations around this trend are not completely random. The process S_t can take an excursion away from the long-run trend. It eventually reverts to that trend, but the excursion may take some time. The average length of these excursions is controlled by the parameter $\lambda > 0$. As this parameter becomes smaller, the excursions

[11]This is often used to model interest rate dynamics.

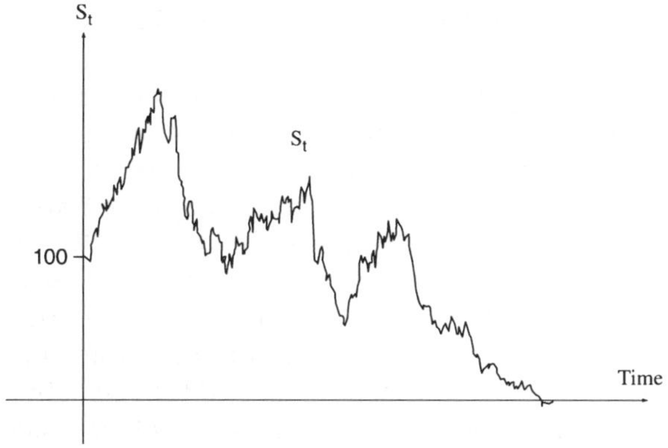

FIGURE 11.5 A sample path of a mean reverting process.

take longer. Thus, asset prices may exhibit some predictable periodicities. This usually makes the model inconsistent with market efficiency.

An example of a sample path of a mean reverting process is shown in Figure 11.5. We selected

$$\mu = 0.05, \quad \lambda = 0.5, \quad \sigma = 0.8 \tag{11.76}$$

This implies a long-run mean of 5% and a volatility of 80% during a time interval of length 1. The λ implies an adjustment of 50%.

We then selected the length of finite subintervals as $h = 0.001$. According to this, during a time interval of length 1, we will observe 1000 S_t's.

Random numbers with mean zero and variance 0.001 were obtained, and the sample path was generated by using the increments

$$\Delta S_k = 0.5\left(0.05 - S_{k-1}\right)(0.001) + 0.8\Delta W_k,$$
$$k = 1, 2, \ldots 1000 \tag{11.77}$$

where the initial point was $S_0 = 100$.

The trajectory is shown in Figure 11.5. Because the diffusion term does not depend on S_t, in this particular case, the process may become negative.

11.4.5 Ornstein–Uhlenbeck Process

Another useful SDE is the Ornstein–Uhlenbeck process,

$$dS_t = -\mu S_t dt + \sigma\, dW_t \tag{11.78}$$

where $\mu > 0$. Here, the drift depends on S_t negatively through the parameter μ, and the diffusion term is of the constant parameter type. Obviously, this is a special case of "mean reverting SDE."

This model can be used to represent asset prices that fluctuate around zero. The fluctuations can be in the form of excursions, which eventually revert to the long-run mean of zero. The parameter μ controls how long excursions away from this mean will take. The larger the μ, the faster the S_t will go back toward the mean.

11.5 STOCHASTIC VOLATILITY

All previous examples of SDEs consisted of modeling the drift and diffusion parameters of SDEs in some convenient fashion. The simplest case showed constant drift and diffusion. The most complicated case was the mean reverting process.

A much more general SDE can be obtained by making the drift and the diffusion parameters random. In the case of financial derivatives, this may have some interesting applications, because it implies that the volatility may be considered not only time-varying, but also random, given the S_t.

For example, consider the SDE for an asset price S_t,

$$dS_t = \mu\, dt + \sigma\, dW_{1t} \tag{11.79}$$

where the drift parameter is constant, while the diffusion parameter is assumed to change over time. More specifically, σ_t is assumed to change according to another SDE,

$$d\sigma_t = \lambda\left(\sigma_0 - \sigma_t\right) dt + \alpha \sigma_t dW_t$$

where the Wiener processes W_{1t}, W_{2t} may very well be dependent.

Note what Eq. (11.80) says about the volatility. The volatility of the asset has a long-run mean of σ_0. But at any time t, the actual volatility may deviate from this long-run mean, the adjustment parameter being λ. The increments dW_{2t} are unpredictable shocks to volatility that are independent of the shocks to asset prices S_t. The $\alpha > 0$ is a parameter.

The market participant has to calculate predictions for asset prices and for volatility. Using such layers of SDEs, one can obtain more complicated models for representing real-life, financial phenomena. On the other hand, stochastic volatility adds additional diffusion components and possibly new risks to be hedged. This may lead to models that are not "complete."

PURE JUMP FRAMEWORK

In Chapter 8, we discussed Levy processes. In the rest of this chapter, we introduce pure jump processes, we discuss how to obtain a pure jump process, the characteristic of pure jump processes, and we later derive a partial-integro differential equation (PIDE) in the value function of the claim. All the previous models we have discussed have concentrated on modifying the volatility of the underlying process in order to better capture a dynamic volatility structure or modifying the drift in order to introduce market observed mean reverting behavior. However, real financial markets contain prices and rates which do not move smoothly through time, but in fact jump to different levels instantaneously. The effects of these types of price movements can be seen in the market prices for options. Indeed, the importance of introducing a jump component in modeling stock price dynamics has been noted by experts in the field, who argue that pure diffusion-based models have difficulties in explaining the very steep smile effect in short-dated option prices. Thus a concerted effort has been made to design models which admit price jumps, and Poisson-type jump components and jump-diffusion models are designed to address these concerns.

Although historically models in mathematical finance were based on Brownian motion and thus are models with continuous price paths, jump processes play now a key role across all areas of finance (see e.g., Carr and Madan, 1999). One reason for this move into a new class of processes is that because of their distributional properties diffusions in many cases cannot provide a realistic picture of empirically observed facts. Another reason is the enormous progress which has been made in understanding and handling jump processes due to the development of semimartingale theory on one side and of computational power on the other side.

The simplest jump process is a process with just one jump. Let T be a random stopping time with respect to an information structure given by a filtration $(\mathcal{F}_t)_{t\geq 0}$, then

$$X_t = \mathbb{1}_{T\leq t} \quad (t \geq 0) \tag{11.80}$$

has the value 0 until a certain event occurs and then 1. As simple as this process looks like, as important it is in modeling credit risk, namely as the process which describes the time of default of a company. We get back to this in Chapter 23.

Now consider the geometric Brownian motion under physical measure

$$\frac{dS_t}{S_t} = \mu\, dt + \sigma\, dW_t \tag{11.81}$$

where μ is the instantaneous expected return per unit time, and σ is the instantaneous volatility per unit time, and W_t is a standard Wiener process under the physical measure. There are two principal methods to add jumps to this framework. The first one is to simply add a jump component to it, the so-called jump-diffusion process (Merton, 1976), and the second one is to perform the time change that lets the Wiener process move according to different times as opposed to calendar time; that way we create jumps into the process, the so-called pure jump processes (Madan et al., 1998; Carr et al., 2002).

To extend it to a jump-diffusion framework, we do as follows. Assume a Poisson process N_t, independent of the jump sizes J and the Wiener process W_t, with arrival intensity λ per unit time under the physical measure $\mathbb{P}$, in order that increments satisfy

$$dN_t = \begin{cases} 1 & \text{with probability } \lambda\, dt \\ 0 & \text{with probability } 1 - \lambda\, dt \end{cases}$$

The expected proportional jump size is

$$\kappa = \mathbb{E}(e^J - 1) \tag{11.82}$$

The jump magnitudes J are independent and identically distributed random variables that are the jump-size J are drawn from a pre-specified probability distribution. The dynamics under consideration are given by

$$\frac{dS_t}{S_t} = (\mu - \lambda\kappa)dt + \sigma\, dW_t + (e^J - 1)dN_t \tag{11.83}$$

This model is called Merton jump-diffusion model (Merton, 1976). Under the risk-neutral measure we can show the dynamics is

$$\frac{dS_t}{S_t} = (\mu - \lambda^*\kappa^*)dt + \sigma\, d\tilde{W}_t + (e^J - 1)dN_t \tag{11.84}$$

11.5.1 Variance Gamma Process

The variance gamma (VG) process is a pure jump process that accounts for high activity, in keeping with the normal distribution, by having an infinite number of jumps in any interval of time. Unlike many other jump models, it is not necessary to introduce a diffusion component for the VG process, as the Black–Scholes model is a parametric special case already and high activity accounted for. Unlike normal diffusion, the sum of absolute log price changes is finite for the VG process. Since VG has finite variation, it can be written as the difference of two increasing processes, the first of which accounts for the price increases, while the second explains the price decreases. In the case of the VG process, the two increasing processes that are subtracted to obtain the VG process are themselves gamma processes.

11.5.1.1 Stochastic Differential Equation

The variance gamma process is a three parameter generalization of a Brownian motion as a model for the dynamics of the logarithm of some underlying market variable. The variance gamma process is obtained by evaluating a Brownian motion with a constant drift and constant volatility at a *random time change* given by a gamma process. That is,

$$\begin{aligned} b(t,\sigma,\theta) &= \theta t + \sigma W_t \\ X(t;\sigma,\nu,\theta) &= b(\gamma(t;1,\nu),\sigma,\theta) \\ &= \theta\gamma(t;1,\nu) + \sigma W(\gamma(t;1,\nu)) \end{aligned}$$

Each unit of calendar time may be viewed as having an economically relevant time length given by an independent random variable that has a gamma density with unit mean and positive variance, which we write as $\gamma(t;1,\nu)$. Thus we can view this model as accounting for different levels of trading activity during different time periods. As stated in Carr et al. (2003), the economic intuition underlying the stochastic time change approach to stochastic volatility arises from the Brownian scaling property. This property relates changes in scale to changes in time, and thus random changes in volatility can alternatively be captured by random changes in time. Thus the stochastic time change of the variance gamma model is an alternative way to represent stochastic volatility in a pure jump process.

Under the variance gamma model the unit period continuously compounded return is normally distributed conditional on the realization of a random process—a random time with a gamma density. The resulting process and associated pricing model provide us with a robust three parameter generalization of the standard Brownian motion model. The log of the asset price process under the variance gamma model is given by

$$\ln S_t = \ln S_0 + (r - q + \omega)t + X(t;\sigma,\nu,\theta)$$

or equivalently

$$S_t = S_0 e^{(r-q+\omega)t+X(t;\sigma,\nu,\theta)}$$

ω is determined so that

$$\mathbb{E}(S_t) = S_0 e^{(r-q)t}$$

The density of the log asset price under the variance gamma model at time t can be expressed conditional on the realization of gamma time change g as a normal density function. The unconditional density may then be obtained on integrating out g.

$$\begin{aligned} f(x;\sigma,\nu,\theta) &= \int_0^\infty \phi(\theta g, \sigma^2 g) \times \text{gamma} \\ &= \left(\frac{t}{\nu}, \nu\right) dg \int_0^\infty \frac{1}{\sigma\sqrt{2\pi g}} \\ &\quad \exp\left(-\frac{(x-\theta g)^2}{2\sigma^2 g}\right) \frac{g^{t/\nu-1} e^{-g/\nu}}{\nu^{t/\nu}\Gamma(t/\nu)} dg \end{aligned}$$

The generalization of this model allows for parameters which control not only the volatility of the Brownian motion, but also (i) *kurtosis* fat tailedness, a symmetric increase in the left and right tail probabilities, relative to the normal for the return distribution and (ii) *skewness* that allows for the asymmetry of the left and right tails of the return density.

An additional attractive feature of VG is that it nests the lognormal density and the Black–Scholes formula as a parametric special case.

11.5.1.2 Characteristic Function

The characteristic function of a VG process can be obtained by first conditioning on the gamma time g.

$$\begin{aligned} \mathbb{E}(e^{iuX_t}|g) &= \mathbb{E}\left(e^{iu(\theta g+\sigma W_g)}\right) \\ &= e^{iu\theta g}\mathbb{E}\left(e^{iu\sigma W_g}\right) \\ &= e^{iu\theta g}\mathbb{E}\left(e^{iu\sigma\sqrt{g}Z}\right) \\ &= e^{iu\theta g} e^{-\frac{(u\sigma\sqrt{g})^2}{2}} \\ &= e^{iu\theta g} e^{\frac{-u^2\sigma^2 g}{2}} \\ &= e^{i\left(u\theta + i\frac{u^2\sigma^2}{2}\right)g} \end{aligned}$$

Now to calculate the characteristic function of a VG process, we have to integrate over g.

$$\begin{aligned} \mathbb{E}(e^{iuX_t}) &= \mathbb{E}_g\left(e^{i\left(u\theta+i\frac{u^2\sigma^2}{2}\right)g}\right) \\ &= \int_0^\infty e^{iu\theta g} e^{\frac{-u^2\sigma^2 g}{2}} \frac{g^{t/\nu-1}e^{-g/\nu}}{\nu^{t/\nu}\Gamma(t/\nu)} dg \end{aligned}$$

which is the characteristic function of a gamma process with shape parameter $\frac{t}{\nu}$ and scale parameter ν evaluated at $u\theta + i\frac{u^2\sigma^2}{2}$. Following expression in Eq. (5.43) we obtain

$$\begin{aligned} \mathbb{E}_g\left(e^{i\left(u\theta+i\frac{u^2\sigma^2}{2}\right)g}\right) &= \left(\frac{\frac{1}{\nu}}{\frac{1}{\nu} - i(u\theta + i\frac{u^2\sigma^2}{2})}\right)^{\frac{t}{\nu}} \\ &= \left(\frac{1}{1 - iu\theta\nu + \frac{u^2\sigma^2\nu}{2}}\right)^{\frac{t}{\nu}} \end{aligned} \tag{11.85}$$

Therefore the characteristic function of the VG process with parameters σ, ν, and θ at time t is

$$\mathbb{E}(e^{iuX(t)}) = \left(\frac{1}{1 - iu\theta\nu + \sigma^2 u^2\nu/2}\right)^{\frac{t}{\nu}} \tag{11.86}$$

11.6 CONCLUSIONS

This chapter introduced the notion of solutions for SDEs. We distinguished between two types of solutions. The strong solution is similar to the case of ordinary differential equations. The weak solution is novel.

We did not discuss the weak solution in detail here. An important example will be discussed in later chapters.

This chapter also discussed major types of stochastic differential equations used to model asset prices.

11.7 REFERENCES

In this chapter, we followed the treatment of Oksendal (2000), which has several other examples of SDEs. An applications-minded reader will also benefit from having access to the literature on the numerical solution of SDEs. The book by Kloeden et al. (1994) is both very accessible and comprehensive. It may very well be said that the best way to understand SDEs is to work with their numerical solutions.

11.8 EXERCISES

1. Consider the following SDE:

$$d\left(W_t^3\right) = 3\left[W_t dt + W_t^2 dW_t\right]$$

(a) Write the above SDE in the integral form.
(b) What is the value of the integral

$$\int_0^t W_s^2 ds$$

2. Consider the geometric SDE:

$$dS_t = \mu S_t dt + \sigma S_t dW_t$$

where S_t is assumed to represent an equity index. The current value of the index is

$$S_0 = 940$$

It is known that the annual percentage volatility is 0.15. The risk-free interest rate is constant at 5%. Also, as is the case in practice, the effect of dividends is eliminated in calculating this index. Your interest is confined to an eight-day period. You do not see any harm in dividing this horizon into four consecutive two-day intervals denoted by Δ.

(a) Use coin tossing to generate random errors that will approximate the term dW_t, with

$$H = +1$$
$$T = -1$$

(b) How can you make sure that the limiting mean and variance of the random process generated by coin tossing matches that of dW_t, as $\Delta \to 0$?
(c) Generate *three* approximate random paths for S_t over this 8-day period.

3. Consider the linear SDE that represents the dynamics of a security price:

$$dS_t = 0.01 S_t dt + 0.05\sigma S_t dW_t$$

with $S_0 = 1$ given. Suppose a European call option with expiration $T = 1$ and strike $K = 1.5$ is written on this security. Assume that the risk-free interest rate is 3%.

(a) Using your computer, generate five normally distributed random variables with mean zero and variance $\sqrt{2}$.
(b) Obtain one simulated trajectory for the S_t. Choose $\Delta t = 0.2$.
(c) Determine the value of the call at expiration.
(d) Now repeat the same experiment with five uniformly distributed random numbers, with appropriate mean and variances.
(e) If we conducted the same experiment 1000 times, would the calculated price differ significantly in two cases? Why?
(f) Can we combine the two Monte–Carlo samples and calculate the option price using 2000 paths?

4. Consider the SDE:

$$dS_t = 0.05\,dt + 0.1\,dW_t$$

Suppose dW_t is approximated by the following process:

$$\Delta W_t = \begin{cases} +\Delta & \text{with probability } 0.5 \\ -\Delta & \text{with probability } 0.5 \end{cases}$$

   (a) Consider intervals of size $\Delta = 1$. Calculate the values of S_t beginning from $t = 0$ to $t = 3$. Note that you need $S_0 = 100$
   (b) Let $\Delta = 0.5$ and repeat the same calculations.
   (c) Plot these two realizations.
   (d) How would these graphs look if $\Delta = 0.01$?
   (e) Now multiply the variance of S_t by 3, let $dt = 1$, and obtain a new realization for S_t. (To generate any needed random variables, you can toss a coin.)

5. Consider the mean-reverting process $dS_t = \lambda(\mu - S_t)dt + \sigma St\,dW_t$. Create a time discretization of $\Delta = \frac{1}{252}$ and assume that the appropriate monthly model parameters are:

$$\lambda = 0.75$$
$$\sigma = 0.35$$
$$S_0 = 40$$

Consider the cases where $\mu = 30$ and $\mu = 50$. Create a Monte–Carlo algorithm for pricing call options under this model. In order to do this, generate a random normal sample for each timestep on each path in order to create an estimate for dS_t at each timestep. The stock price can be computed along each step as:

$$S_{t+1} = S_t + dS_t \tag{11.87}$$

$$dS_t = \lambda(\mu - S_t)\Delta + \sigma StW \tag{11.88}$$

Here W is a normally distributed random variable with mean 0 and variance Δ.
Generate 1000 paths of 3 years by sampling daily. Using the Monte–Carlo paths generates calls prices for options at the following strikes: $30, 40, 50, 60, 70$. Comment on the difference in prices caused by changing μ.

6. Write a simulation program to price a European call option, whose underlying stock price follows a geometric Brownian motion with volatility $\sigma = 0.1$. Other parameters are $r = 0.05, q = 0, S = K = 10, T = 1$.

CHAPTER

12

Pricing Derivative Products: Partial Differential Equations

OUTLINE

12.1 INTRODUCTION

Thus far we have learned about major tools for modeling the dynamic behavior of a random process in continuous time, and how one can (and cannot) take derivatives and integrals under these circumstances.

These tools were not discussed for their own sake. Rather, they were discussed because of their usefulness in pricing various derivative

Financial Derivatives, *Third Edition*.
http://dx.doi.org/10.1016/B978-0-12-384682-2.00012-8

instruments in financial markets. Far from being mere theoretical developments, these tools are practical methods that can be used by market professionals. In fact, because of some special characteristics of derivative products, abstract theoretical models in this area are much more amenable to practical applications than in other areas of finance.

Modern finance has developed two major methods of pricing derivative products. The first of these leads to the utilization of partial differential equations, which are the subject of this chapter. The second requires transforming underlying processes into martingales. This necessitates utilization of equivalent martingale measures, which is the topic of Chapter 14. In principle, both methods should give the same answer. However, depending on the problem at hand, one method may be more convenient or cheaper to use than the other. The mathematical tools behind these two pricing methods are, however, very different.

First, we will briefly discuss the logic behind the method of pricing securities that leads to the use of PDEs. These results will be utilized in Chapter 13.

12.2 FORMING RISK-FREE PORTFOLIOS

Derivative instruments are contracts written on other securities, and these contracts have finite maturities. At the time of maturity denoted by T, the price F_T of the derivative contract should depend solely on the value of the underlying security S_T, the time T, and nothing else:

$$F_T = F(S_T, T) \tag{12.1}$$

This implies that at expiration we know the exact form of the function $F(S_T, T)$. We assume that the same relationship is true for times other than T and that the price of the derivative product can be written as

$$F(S_t, t) \tag{12.2}$$

The increments in this price will be denoted by dF_t. At the outset, a market participant will not know the functional form of $F(S_t, t)$ at times other than expiration. This function needs to be found.

This suggests that if we have a law of motion for the S_t process—i.e., if we have an equation describing the way dS_t is determined—then we can use Ito's Lemma to obtain dF_t. But this means that dF_t and dS_t would be increments that have the same source of underlying uncertainty, namely the innovation part in dS_t. In other words, at least in the present example, we have two increments, dF_t and dS_t, that depend on one *innovation* term. Such dependence makes it possible to form *risk-free portfolios* in continuous time.

Let P_t dollars be invested in a combination of $F(S_t, t)$ and S_t:

$$P_t = \theta_1 F(S_t, t) + \theta_2 S_t \tag{12.3}$$

where θ_1, θ_2 are the quantities of the derivative instrument and the underlying security purchased. They represent portfolio weights.

The value of this portfolio changes as time t passes because of changes in $F(S_t, t)$ and S_t. Taking θ_1, θ_2 as constant, we can write this change as[1]

$$dP_t = \theta_1 dF_t + \theta_2 dS_t \tag{12.4}$$

In general, θ_1, θ_2 will vary over time and hence will carry a time subscript as well. At this point we ignore such dependence. In this equation, both dF_t and dS_t are increments that have an unpredictable component due to the innovation term dW_t in dS_t.

[An important remark about notation. dF_t should again be read as the total change in the derivative price $F(S_t, t)$ during an interval dt. This should not be confused with F_t, which we reserve for the partial derivative of $F(S_t, t)$ with respect to t].

[1] Strictly speaking, this stochastic differential is correct only when the portfolio weights do not depend on S_t. Otherwise, there will be further terms on the right. This point will be quite relevant when we discuss the Black–Scholes framework below.

Our main interest is in the price of the derivative product, and how this price changes. Thus, we begin by positing a model that determines the dynamics of the underlying asset S_t, and from there we try to determine how $F(S_t, t)$ behaves. Accordingly, we assume that the stochastic differential dS_t obeys the SDE

$$dS_t = a(S_t, t)\,dt + \sigma(S_t, t)\,dW_t \tag{12.5}$$

Using this, we can apply Ito's Lemma to find dF_t:

$$dF_t = F_t dt + \frac{1}{2}F_{ss}\sigma_t^2 dt + F_s dS_t \tag{12.6}$$

We substitute for dS_t using Eq. (12.5), and obtain the SDE for the derivative asset price:

$$dF_t = \left[F_s a_t + \frac{1}{2}F_{ss}\sigma_t^2 + F_t\right]dt + F_s\sigma_t dW_t \tag{12.7}$$

Note that we simplified the notation by writing a_t for the drift and σ_t for the diffusion parameter. If we knew the form of the function $F(S_t, t)$, we could calculate the corresponding partial derivatives, F_s, F_{ss}, F_t, and then obtain explicitly this SDE that governs the dynamics of the financial derivative. The functional form of $F(S_t, t)$, however, is not known. We can use the following steps to determine it.

We first see that the SDE in (12.7), describing the dynamics of dF_t, is driven by the same Wiener increment dW_t that drives the S_t. One should, in principle, be able to use one of these SDEs to eliminate the randomness in the other. In forming risk-free portfolios, this is in fact what is done.

We now show how this is accomplished. First, note that it is the market participant who selects the portfolio weights θ_1, θ_2.

Second, the latter can always be set such that the dP_t is independent of the innovation term dW_t and hence is completely predictable. The reason is as follows. Given that dF_t and dS_t have the same unpredictable component, and given that θ_1, θ_2 can be set as desired, one can always eliminate the dW_t component from Eq. (12.4). To do this, consider again

$$dP_t = \theta_1 dF_t + \theta_2 dS_t \tag{12.8}$$

and substitute for dF_t using (12.6)[2]:

$$dP_t = \theta_1\left[F_s dS_t + \frac{1}{2}F_{ss}\sigma_t^2 dt + F_t dt\right] + \theta_2 dS_t \tag{12.9}$$

In this equation we are free to set θ_1, θ_2 the way we wish. Suppose we ignore for a minute that F_s depends on S_t and select

$$\theta_1 = 1 \tag{12.10}$$

and

$$\theta_2 = -F_s \tag{12.11}$$

These particular values for portfolio weights will lead to cancellation of the terms involving dS_t in (12.9) and reduces it to

$$dP_t = F_t dt + \frac{1}{2}F_{ss}\sigma_t^2 dt \tag{12.12}$$

Clearly, given the information set I_t, in this expression there is no random term. The dP_t is a completely predictable, deterministic increment for all times t. This means that the portfolio P_t is risk-free.[3]

Since there is no risk in P_t, its appreciation must equal the earnings of a risk-free investment during an interval dt in order to avoid arbitrage. Assuming that the (constant) risk-free interest rate is given by r, the expected capital gains must equal

$$rP_t dt \tag{12.13}$$

in the case where S_t pays no "dividends," and must equal

$$rP_t dt - \delta\, dt \tag{12.14}$$

[2]Recall that this will be correct mathematically if θ_1, θ_2 do not depend on S_t.

[3]Note this important point: The value of θ_2 set at $-F_s$ will vary over time. For nonlinear products such as options, or structures containing options, the F_s will be a function of S_t. This means that the risk-free portfolio method is not satisfactory mathematically, yet it will give the "correct" PDE.

in the case where S_t pays dividends of δ per unit time. In the latter case, the capital gains in (12.14) *plus* the dividends earned will equal the risk-free rate.[4]

Utilizing the case with no dividends, Eqs. (12.12) and (12.13) yield

$$rP_t dt = F_t dt + \frac{1}{2} F_{ss} \sigma_t^2 dt \tag{12.15}$$

Since the dt terms are common to all factors, they can be "eliminated" to obtain a partial differential equation:

$$r\left(F\left(S_t, t\right) - F_s S_t\right) = F_t + \frac{1}{2} F_{ss} \sigma_t^2 \tag{12.16}$$

[We replace $P(t)$ in (12.15) by its components.] We rewrite Eq. (12.16) as

$$-rF + rF_s S_t + F_t + \frac{1}{2} F_{ss} \sigma_t^2 = 0, \quad 0 \leq S_t,\ 0 \leq t \leq T \tag{12.17}$$

where the derivative asset price $F(S_t, t)$ is denoted simply by the letter F for notational convenience.

We have an additional piece of information. The derivative product will have an expiration date T, and the relationship between the price of the underlying asset and that of the derivative asset will, in general, be known exactly at expiration. That is, we know at expiration that the price of the derivative product is given by

$$F\left(S_T, T\right) = G\left(S_T, T\right) \tag{12.18}$$

where $G(\cdot)$ is a known function of S_T and T. For example, in the case of a call option, $G(\cdot)$, the expiration price of the call with a strike price K is

$$G\left(S_T, T\right) = \max\left[S_T - K, 0\right] \tag{12.19}$$

According to this equation, if at expiration the stock price is below the strike price, $S_T - K$ will be negative and the call option will not be exercised. It will be worthless. Otherwise, the option will have a price equal to the differential between the stock and the strike price.

[4]Note the role of dt. Some infinitesimal time must pass in order to earn interest or receive dividends. If no time passes, regardless of the level of interest rates r, the interest earnings will be zero. The same is true for dividend earnings.

Equation (12.17) is known as a *partial differential equation* (PDE). Equation (12.18) is an associated *boundary condition*.

The reason this method "works" and eliminates the innovation term from Eq. (12.4) is that $F(\cdot)$ represents a price of a derivative instrument, and hence has the same inherent unpredictable component dW_t as S_t. Thus, by combining these two assets, it becomes possible to eliminate their common unpredictable movements. As a result, P_t becomes a risk-free investment, since its future path will be known with certainty.

This construction of a risk-free portfolio is heuristic. From a mathematical point of view, it is not satisfactory. In a formal approach, one should form self-financing portfolios using completeness of markets with respect to a class of *trading strategies* and using the implied *"synthetic"* equivalents of the assets under consideration. Jarrow (1996) is an excellent source on these concepts. The next section discusses this point in more detail.

12.3 ACCURACY OF THE METHOD

The previous section illustrated the method of risk-free portfolios in obtaining the PDE's corresponding to the arbitrage-free price $F(S_t, t)$ of a derivative asset written on S_t.

Recall that the idea was to form a risk-free portfolio by combining the underlying asset and, say, a call option written on it:

$$P_t = \theta_1 F\left(S_t, t\right) + \theta_2 S_t \tag{12.20}$$

where θ_1, θ_2 are the portfolio weights. Then we took the differential during an infinitesimal time period dt by letting:

$$dP_t = \theta_1 dF\left(S_t, t\right) + \theta_2 dS_t \tag{12.21}$$

Mathematically speaking, this equation treated the θ_1, θ_2 as if they are constants, because they were not differentiated. Up to this point, there is really nothing wrong with the risk-free portfolio method. But consider what happens when we select the portfolio weights.

We selected the portfolio weights as:

$$\theta_1 = 1, \quad \theta_2 = -F_s \tag{12.22}$$

This selection "works" in the sense that it eliminates the "unpredictable" random component and makes the portfolio risk-free, but unfortunately it also violates the assumption that θ_1, θ_2 are constant. In fact, the θ_2 is now dependent on S_t because, in general, F_s is a function of S_t and t. Thus, first replacing the θ_1, θ_2 with their selected values, and then taking the differential should give a very different result.

Writing the dependence of F_s on S_t explicitly:

$$P_t = F(S_t, t) - F_s(S_t, t)\, S_t \tag{12.23}$$

Then, differentiating yields:

$$dP_t = dF(S_t, t) - F_s dS_t - S_t dF_s - dF_s(S_t, t) dS_t \tag{12.24}$$

Note that we now have a third term since the F_s is dependent on S_t, and, hence, is time dependent and stochastic. In general, this term will not vanish. In fact, we can use Ito's Lemma and calculate the dF_s, which is a function of S_t and t. This is equivalent to taking the stochastic differential of the derivative's DELTA:

$$dF_s(S_t, t) = F_{st} dt + F_{ss} dS_t + \frac{1}{2} F_{sss} \sigma^2 S_t^2 dt$$

where the third derivative of F is there because we are applying Ito's Lemma to the F already differentiated with respect to S_t. After replacing the differential dS_t, and arranging:

$$\begin{aligned} dF_s(S_t, t) &= F_{st} dt + F_{ss}\left(\mu S_t dt + \sigma dW_t\right) \\ &\quad + \frac{1}{2} F_{sss} \sigma_t^2 S_t^2 dt \\ &= \left[F_{st} + F_{ss}\mu S_t + \frac{1}{2} F_{sss} \sigma_t^2 S_t^2\right] dt \\ &\quad + F_{ss} \sigma S_t dW_t \end{aligned}$$

Thus, the formal differential of

$$P_t = \theta_1 F(S_t, t) + \theta_t S_t \tag{12.25}$$

when θ_2 is equal to $-F_s$ will be given by:

$$\begin{aligned} dP_t &= dF(S_t, t) - F_s dS_t \\ &\quad - S_t \left[\left[F_{st} + F_{ss}\mu S_t + \frac{1}{2} F_{sss}\sigma_t^2 S_t^2\right] dt \right. \\ &\qquad \left. + F_{ss}\sigma S_t dW_t\right] - F_{ss}\sigma_t^2 S_t^2 dt \end{aligned} \tag{12.26}$$

Clearly, this portfolio will not be self-financing in general, since we do not have:

$$dP_t = dF(S_t, t) - F_s dS_t \tag{12.27}$$

On the right-hand side there are extra terms, and these extra terms will not equal zero unless we have:

$$S_t^2 F_{ss}\left(\sigma dW_t + (\mu - r)\, dt\right) = 0$$

which will, in general, not be the case. In order to see this, note that differentiating the Black–Scholes PDE in (12.17) with respect to S_t again, we can write

$$F_{st} + F_{ss} r S_t + \frac{1}{2} F_{sss}\sigma^2 S_t^2 + \sigma^2 F_{ss} S_t = 0$$

Using this equation eliminates most of the unwanted terms in (12.26). But we are still left with:

$$\begin{aligned} dP_t &= dF(S_t, t) - F_s dS_t - S_t\left[F_{ss}(\mu - r) S_t dt\right] \\ &\quad - F_{ss}\sigma S_t^2 dW_t \end{aligned} \tag{12.28}$$

Thus, in order to make the portfolio P_t self-financing, we need

$$S_t^2 F_{ss}\left(\sigma dW_t + (\mu - r)\, dt\right) = 0$$

which will not hold in general.

12.3.1 An Interpretation

Although, formally speaking, the risk-free portfolio method is not satisfactory and, in general, makes one work with portfolios that require infusions of cash or leave some capital gains, the method still gives us the correct PDE. How can we interpret this result?

The answer is in the additional term. This term has nonzero expectation under the true probability $\mathbb{P}$. But once we switch to a risk-free measure $\mathbb{Q}$ and define a new Wiener process W_t^* under this probability, we can write:

$$dW_t^* = \left(\sigma dW_t + (\mu - r)\, dt\right)$$

We will have:

$$\mathbb{E}^{\mathbb{P}}\left[S_t^2 F_{ss}\left(\sigma dW_t + (\mu - r)\, \Delta\right)\right] \approx 0$$

Thus, in small intervals, the extra cost (gain) associated with the portfolio P_t has zero expectation. It is as if, on the average, it is self-financing. But, it is interesting that this "average" is taken with respect to the synthetic risk-neutral measure and not with respect to real-life probability. See Musiela and Rutkowski (1997) for more details.

12.4 PARTIAL DIFFERENTIAL EQUATIONS

We rewrite the partial differential equation (12.17) in a general form, using the shorthand notation $F(S_t, t) = F$,

$$a_0 F + a_1 F_s S_t + a_2 F_t + a_3 F_{ss} = 0, \quad 0 \le S_t,\ 0 \le t \le T \tag{12.29}$$

with the boundary condition

$$F(S_T, T) = G(S_T, T) \tag{12.30}$$

$G(\cdot)$ being a known function.[5]

[5]In the literature, the PDE notation is different than what is adopted in this section. For example, the PDE in (12.29) would be written as

$$a_0 F(X,t) + a_1 F_s(X,t) + a_2 F_t(X,t) + a_3 F_{ss}(X,t) = 0,$$
$$0 \le X,\ 0 \le t \le T \tag{12.31}$$

with the boundary condition

$$F(X,t) = G(X,T)$$

In this section, we keep using S_t instead of switching to a generic variable X, as is usually done.

The method of forming such risk-free portfolios in order to obtain arbitrage-free prices for derivative instruments will always lead to PDEs.

Since derivative securities are always "derived" from some underlying asset(s), the formation of such arbitrage-free portfolios is in general quite straightforward. On the other hand, the boundary conditions as well as the implied PDEs may get more complicated depending on the derivative product one is working with. But, overall, the method will center on the solution to a PDE. This concept should be discussed in detail.

We discuss partial differential equations in several steps.

12.4.1 Why is the PDE an "Equation?"

In what sense is the PDE in (12.29) an equation? With respect to what "unknown" is this equation to be solved?

Unlike the usual cases in algebra where equations are solved with respect to some variable or vector x, the unknown in Eq. (12.29) is in the form of a function. It is not known what type of function $F(S_t, t)$ represents. What is known is that if one takes various partial derivatives of $F(S_t, t)$ and combines them by multiplying by coefficients a_i, the result will equal zero. Also, at time $t = T$, this function must equal the (known) $G(S_T, T)$—i.e., it must satisfy the boundary condition.

Hence, in solving PDEs, one tries to find a function whose partial derivatives satisfy Eqs. (12.29) and (12.30).

12.4.2 What is the Boundary Condition?

Partial differential equations are obtained by combining various partial derivatives of a function and then setting the combination equal to zero. The boundary conditions are an integral part of such equations. In physics, boundary conditions are initial or terminal states of some physical phenomenon that evolve over time according to the PDE.

In finance, boundary conditions play a similar role. They represent some contractual clauses of various derivative products. Depending on the product and the problem at hand, boundary conditions may change. The most obvious boundary values are initial or terminal values of derivative contracts. Often, finance theory tells us some plausible conditions that prices of derivative contracts must satisfy at maturity. For example, futures prices and cash prices cannot be (very) different at the delivery date. In the case of options, option prices must satisfy an equation such as (12.19). In case of a discount bond, the asset price equals 100 at maturity.

If there are no boundary conditions, then finding price functions $F(S_t, t)$ that satisfy a given PDE will, in general, not be possible. Further, the fact that derivative products are known functions of the underlying asset at expiration will always yield a boundary condition to a market participant.

To see the role of boundary conditions and to consider some simple PDEs, we look at some examples.

12.5 CLASSIFICATION OF PDEs

One can classify PDEs in several different ways. First of all, PDEs can be *linear* or *nonlinear*. This refers to the coefficients applied to partial derivatives in the equation. If an equation is a linear combination of F and its partial derivatives, it is called a linear PDE.[6]

The second type of classification has to do with the *order* of differentiation. If all partial derivatives in the equation are first order, then the PDE will also be first order. If there are cross-partials, or second partials, then the PDE becomes second order. For nonlinear financial derivatives such as options, or instruments containing options, the resulting PDE will always be second order.

Thus far, these classifications are similar to the case of ordinary differential equations. The third type of classification is specific to PDEs. The latter can also be classified as *elliptic, parabolic,* or *hyperbolic*. The PDEs we encounter in finance are similar to parabolic PDEs.

We first consider examples of linear first- and second-order PDEs. These examples have no direct relevance in finance. Yet they may help establish an intuitive understanding of what PDEs are, and why boundary conditions are important.

12.5.1 Example 1: Linear, First-Order PDE

Consider the PDE for a function $F(S_t, t)$:

$$F_t + F_s = 0, \quad 0 \le S_t,\ 0 \le t \le T \tag{12.32}$$

According to this PDE, the negative of the partial of $F(\cdot)$ with respect to t is equal to its partial with respect to S_t. If t were to represent time, and S_t were to represent the price of the underlying security, then (12.32) would mean that the negative of the price change during a small time interval with S_t fixed, equals the price change due to a small movement in the price of the underlying asset when t is fixed.

In a financial market, there is no compelling reason why such a relationship should exist between the two partial derivatives. But suppose (12.32) is nevertheless written down and a solution $F(S_t, t)$ is sought. What would this function $F(S_t, t)$ look like?

We can immediately guess a solution:

$$F(S_t, t) = \alpha S_t - \alpha t + \beta \tag{12.33}$$

where α, β are any constants. With such a function, the partials will be given by

$$\frac{\partial F}{\partial t} = -\alpha \tag{12.34}$$

and

$$\frac{\partial F}{\partial S} = \alpha \tag{12.35}$$

[6]This means that the coefficients of the partial derivatives are not functions of F.

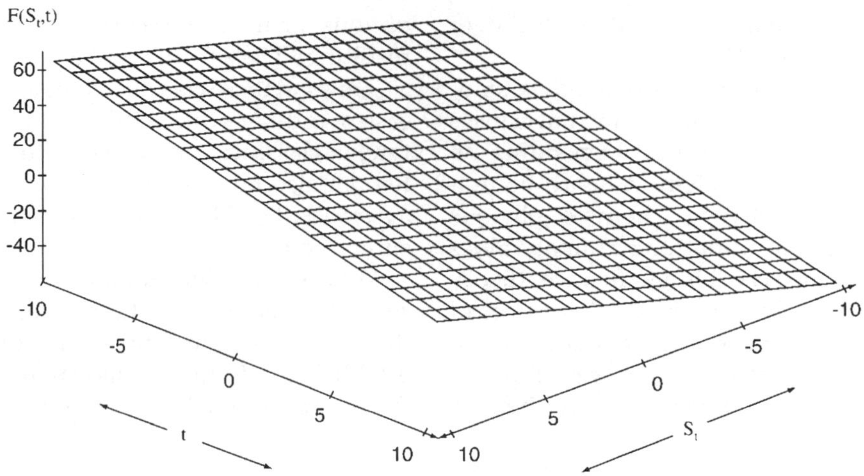

FIGURE 12.1 Plot of the plane $F(S_t, t) = 3S_t - 3t + 4$.

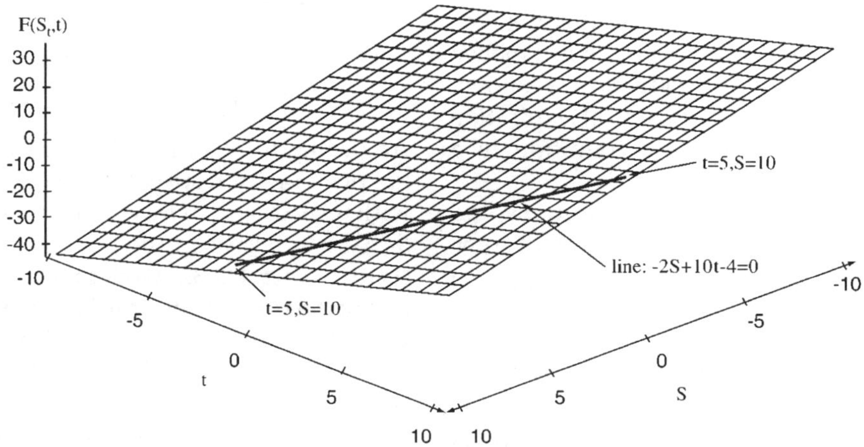

FIGURE 12.2 Plot of the plane $F(S_t, t) = \alpha S_t - \alpha t + \beta$ for $\alpha = 2$ and $\beta = 4$.

Their sum will equal zero, and this is exactly what the PDE in (12.32) implies.

The solution suggested by the function (12.33) is a plane in a three-dimensional space. If no boundary conditions are given, this is all we know. We would not be able to determine exactly which plane $F(S_t, t)$ would represent, since we would not be able to pinpoint the values of α, β given the information in (12.33). All we can say is the following: at $t = 0, S_0 = 0$ the intercept will equal β. For a fixed S_t, the $F(S_t, t)$ has contours that are straight lines with slope $-\alpha$. For fixed t, the contours are straight lines with slope α.

Figures 12.1 and 12.2 show two examples of $F(S_t, t)$ that "solve" the PDE in (12.32). Figure 12.1 is the plot of the plane:

$$F(S_t, t) = 3S_t - 3t + 4, \quad -10 \le t \le 10, \quad -10 \le S_t \le 10 \tag{12.36}$$

Note that in this case $F_s = 3$ and $F_t = -3$. Hence, this function satisfies the PDE in (12.32). This solution is a plane that increases with respect to S_t, but decreases with respect to t.

Figure 12.2 shows another example where

$$F(S_t, t) = -2S_t + 2t - 4, \quad -10 \leq t \leq 10, \quad -10 \leq S_t \leq 10 \tag{12.37}$$

We again see that $F(Sl, t)$ is a plane. But in this case it increases with respect to t, and decreases with respect to S_t; the contours are again straight lines.

The examples of $F(S_t, t)$ given in (12.36) and (12.37) are very different-looking functions. Yet they both solve the PDE in (12.32). This is because Eq. (12.32) does not contain sufficient information to allow the function $F(S_t, t)$ to be determined precisely. There are uncountably many functions $F(S_t, t)$ whose first partials with respect to S_t and t are equal.

Now, if in addition to (12.33) we are given some boundary conditions as well, then we can determine the $F(S_t, t)$ precisely. For example, suppose we know that at expiration time $t = 5$ (the boundary for t) we have

$$F(S_5, 5) = 6 - 3S_5 \tag{12.38}$$

We can now determine the unknowns α and β in Eq. (12.33):

$$\alpha = 2 \tag{12.39}$$

$$\beta = 4 \tag{12.40}$$

This is the plane shown in Figure 12.2.

On the other hand, if we had a second boundary condition, say, at $S_t = 100$,

$$F(100, t) = 5 + 0.3t \tag{12.41}$$

then there will be no meaningful solution because Eqs. (12.41) and (12.42) overdetermine the constants α and β.

Thus, when $F(S_t, t)$ is a plane, we need a single boundary condition to exactly pinpoint the function that solves the PDE.

This is easy to see geometrically, since a boundary condition corresponds to first selecting the "endpoint" for t (or S_t) and then obtaining the intersection of the plane with a surface orthogonal to the time axis and passing from that t. In Figure 12.2, the boundary condition at $t = 5$,

$$F(S_5, 5) = 6 - 3S_5 \tag{12.42}$$

is shown explicitly. Note that the other candidate for $F(S_t, t)$ shown in Figure 12.1 will not pass from this line at $t = 5$. Hence, it cannot be a solution.

Also, when $F(S_t, t)$ is a plane, the terminal conditions with respect to t or S_t will be straight lines.

12.5.1.1 *Remark*

The solutions to the class of PDEs

$$F_t + F_s = 0 \tag{12.43}$$

are not restricted to planes. In fact, consider the function

$$F(S_t, t) = e^{\alpha S_t - \alpha t} \tag{12.44}$$

This function will also satisfy the equality (12.43). It is the boundary condition that will determine the unique solution.

12.5.2 Example 2: Linear, Second-Order PDE

It was easy to guess the solution of the first-order PDE discussed in Example 1. Now consider a second-order PDE

$$\frac{\partial^2 F}{\partial t^2} = 0.3\frac{\partial^2 F}{\partial S_t^2} \tag{12.45}$$

or, more succinctly,

$$-0.3F_{ss} + F_{tt} = 0 \tag{12.46}$$

First, note that we are again dealing with a linear PDE, since the partials in question are combined by using constant coefficients.

Again, ignore the boundary conditions for the moment. We can try to guess a solution to (12.46). It is clear that the function $F(\cdot)$ has to be such that the second partials of $F(S_t, t)$ with respect to S_t

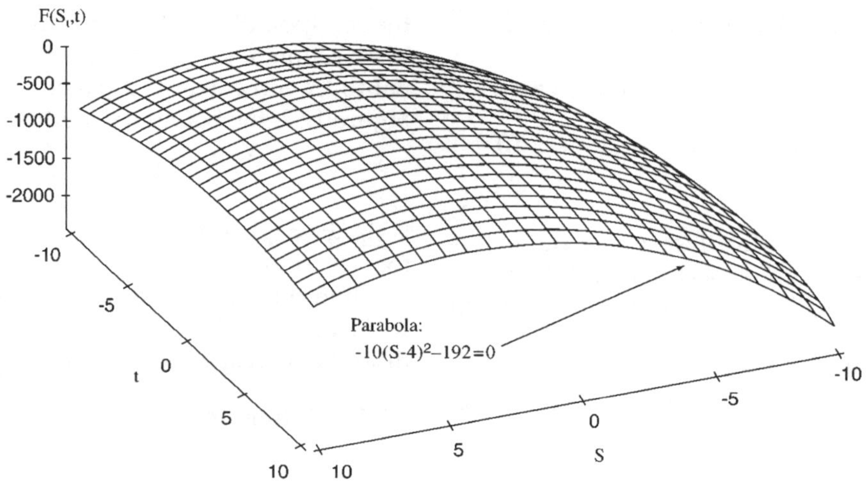

FIGURE 12.3

and t are proportional with a factor of proportionality equal to 0.3. This relationship between F_{ss} and F_{tt} should be true at any S_t and t. What could this function be?

Consider the formula

$$F(S_t, t) = \frac{1}{2}\alpha (S_t - S_0)^2 + \frac{0.3}{2}\alpha (t - t_0)^2 + \beta (t - t_0) \tag{12.47}$$

where S_0, t_0 are unknown constants and where the parameters α and β are again unknown.

Now, if we take the second partials of $F(S_t, t)$:

$$\frac{\partial^2 F}{\partial t^2} = 0.3\alpha \tag{12.48}$$

$$\frac{\partial^2 F}{\partial S^2} = 1\alpha \tag{12.49}$$

Hence the second partials F_{tt}, F_{ss} of the $F(S_t, t)$ in (12.48) and (12.49) will satisfy Eq. (12.45). Thus, the $F(S_t, t)$ given in (12.47) is a solution of the partial differential equation (12.45).

Note that for fixed $F(S_t, t)$,

$$F(S_t, t) = F_0 \tag{12.50}$$

the contours of this function are ellipses.[7]

[7]See the next section.

Again, the solution of (12.45) is not unique, since the $F(S_t, t)$ with any α, β, S_0, t_0 could be a solution, as long as it is of the form (12.47). To obtain a unique solution we need boundary conditions.

One boundary condition could be at $S_t = 10$:

$$F(10, t) = 100 + t^2 \tag{12.51}$$

This is a function that traces a parabola in the F, t plane.

Yet such a boundary condition is not sufficient to determine all the parameters α, β, S_0, t_0. One would need a second boundary condition, say, at $t = 0$:

$$F(S_0, 0) = 50 + S_0^2 \tag{12.52}$$

This equation is another parabola. But the relevant plane is F, S_t.

We give an example of such an $F(S_t, t)$ in Figure 12.3. The figure displays the three-dimensional plot of the function

$$F(S_t, t) = -10(S_t - 4)^2 - 3(t - 2)^2, \quad -10 \le t \le 10, \quad -10 \le S_t \le 10 \tag{12.53}$$

The surface has contours as ellipses. In terms of boundary conditions, we can pick $t = 10$ as the terminal value for t and get a boundary condition that has the form of a parabola:

$$F(S_{10}, 10) = -10(S_{10} - 4)^2 - 192 \quad (12.54)$$

The boundary condition for $S_t = 0$ will be another parabola:

$$F(0, t) = -160 - 3(t - 2)^2 \quad (12.55)$$

These two boundary conditions are satisfied for $\alpha = -20, \beta = 0, S_0 = 4, t_0 = 2$.

12.6 A REMINDER: BIVARIATE, SECOND-DEGREE EQUATIONS

It turns out that frequently encountered graphs such as circles, ellipses, parabolas, or hyperbolas can all be represented by a second-degree equation. In this section we briefly review this aspect of analytical geometry, since it relates to the terminology concerning PDEs.

For the time being, let x, y denote two deterministic variables. We can define an equation of the second degree as

$$Ax^2 + Bxy + Cy^2 + Dx + Ey + F = 0 \quad (12.56)$$

Here A, B, C, D, E, F represent various constants. The equation is of the second degree, because the highest power of x or of y is a square.

By choosing different values for A, B, C, D, E, F, the locus of the equation can be in the form of an ellipse, a parabola, a hyperbola, or a circle. It is worth discussing these briefly.

12.6.1 Circle

Consider the case where

$$A = C \quad \text{and} \quad B = 0 \quad (12.57)$$

The second-degree equation reduces to

$$Ax^2 + Ay^2 + Dx + Ey + F = 0 \quad (12.58)$$

After completing the square, this can always be written as

$$(x - x_0)^2 + (y - y_0)^2 = R \quad (12.59)$$

which most readers will recognize as the equation of a circle with radius R and center at (x_0, y_0). To see why, expand (12.59):

$$x^2 + y^2 - 2x_0x - 2y_0y + x_0^2 + y_0^2 = R \quad (12.60)$$

In this equation, we can always let

$$A = \frac{1}{R} \quad (12.61)$$

$$\frac{-2x_0}{R} = D \quad (12.62)$$

$$\frac{-2y_0}{R} = E \quad (12.63)$$

and

$$\frac{x_0^2 + y_0^2}{R} = F \quad (12.64)$$

Hence, with $A = C, B = 0$, the x and the y that satisfy the second-degree equation will always trace a circle in the x, y plane.

In the special case when $R = 0$, the circle reduces to a point. Another degenerate case can be obtained when $A = C = 0$. Then the circle has degenerated into a straight line, but the equation is not second degree.

12.6.2 Ellipse

The second case of interest is when

$$B^2 - 4AC < 0 \quad (12.65)$$

This is similar to the case of a circle, except B is not zero, and the coefficients of x^2 and y^2 are different. We can again rewrite the second-degree equation in a different form,

$$\alpha(x - x_0)^2 + \beta(y - y_0)^2 + \gamma(x - x_0)(y - y_0) = R \quad (12.66)$$

which will be recognized as the equation of an ellipse, where the center is at x_0, y_0.

Given values for A, B, C, D, E, F, we can always determine the values of the parameters $x_0, y_0, \alpha, \beta, \gamma, R$, since by equating the coefficients of the expanded form of (12.66) with those of (12.56), we will have six equations in six unknowns.

12.6.2.1 *Example*

The method of completing the square is useful for differentiating among ellipses, circles, parabolas, and hyperbolas. We illustrate this with a simple example.[8]

Consider the second-degree equation

$$9x^2 + 16y^2 - 54x - 64y + 3455 = 0 \tag{12.67}$$

Note that

$$B^2 - 4AC = -576 \tag{12.68}$$

so we must be dealing with an ellipse. We directly show this by "completing the squares":

$$9\left(x^2 - 6x + ?\right) + 16\left(y^2 - 6y + ?\right) = 3455 \tag{12.69}$$

By filling in for the question marks, we can make the two terms in parentheses become squares. We replace the first question mark with 9 for the first parenthesis. This requires adding 81 to the right-hand side. The second question mark needs to be replaced by 4. This requires adding 64 to the right-hand side. We obtain

$$9(x-3)^2 + 16(y-2)^2 = 3600 \tag{12.70}$$

or

$$\frac{(x-3)^2}{400} + \frac{(y-2)^2}{225} = 1 \tag{12.71}$$

This is the formula of an ellipse with center at $x = 3, y = 2$.

[8]The method of "completing the square" is used frequently in calculations involving geometric SDEs.

12.6.3 Parabola

The second-degree equation in (12.56) reduces to a parabola when we have

$$B^2 - 4AC = 0 \tag{12.72}$$

The easiest way to see this is to note that $B = 0$ and either $A = 0$ or $C = 0$ satisfies the required condition. But, when this happens, the second-degree equation reduces to

$$Ax^2 + Dx + Ey + F = 0 \tag{12.73}$$

which is the general equation for a parabola.

12.6.4 Hyperbola

The general second-degree equation in (12.56) represents a hyperbola if the condition

$$B^2 - 4AC > 0 \tag{12.74}$$

is satisfied. This case will have limited use for us, so we will skip the details.

12.7 TYPES OF PDEs

Example 2 suggests that the contours of $F(S_t, t)$ would in general be nonlinear equations. In case of Example 2, they were ellipses. In fact, partial differential equations of the form

$$a_0 + a_1F_t + a_2F_s + a_3F_{ss} + a_4F_{tt} + a_5F_{st} = 0 \tag{12.75}$$

are called *elliptic* PDEs if we have

$$a_5^2 - 4a_3a_4 < 0 \tag{12.76}$$

The PDE in (12.75) is called *parabolic* if

$$a_5^2 - 4a_3a_4 = 0 \tag{12.77}$$

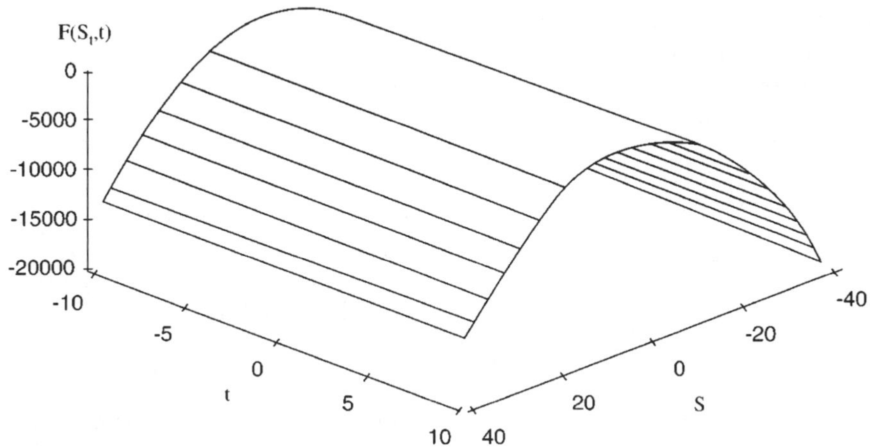

FIGURE 12.4

Finally, the PDE is called hyperbolic if

$$a_5^2 - 4a_3a_4 > 0 \tag{12.78}$$

Clearly, $F(S_t, t)$ graphed in Figure 12.3 is a solution to a PDE that satisfies the condition of an elliptic PDE, since $a_4 = 0$ and both a_3 and a_4 are of the same sign. As a result, the condition

$$a_5^2 - 4a_3a_4 < 0 \tag{12.79}$$

is satisfied.

12.7.1 Example: Parabolic PDE

Figure 12.4 gives the graph of the function $F(S_t, t)$ defined as

$$F(S_t, t) = -10(S_t - 4)^2 - 3(t - 2) \tag{12.80}$$

Note that the contours of this function are parabolas. This $F(S_t, t)$ will have boundary conditions as parabolas with respect to t, and as straight lines with respect to S_t.

Such an $F(S_t, t)$ is one of the solutions of the PDE

$$-\frac{1}{4}F_{ss} + \frac{5}{3}F_t = 0 \tag{12.81}$$

The coefficients of the PDE are such that

$$a_5^2 - 4a_3a_4 = 0 \tag{12.82}$$

since $a_4 = 0$ and $a_5 = 0$. Hence, this PDE is parabolic.

12.8 PRICING UNDER VARIANCE GAMMA MODEL

We will cover derivatives pricing under the variance gamma model analytically (via a transform method) and numerically (by solving the associated partial integro-differential equation) depending on the type of the option under consideration. At this point it should be obvious that pricing European options under the variance gamma model involves first conditioning on the random time g and then simply using a Black–Scholes type formula to solve for the conditional option value. Thus, the VG European option price, $C(S_0, K, T)$, is obtained on integrating with respect to the gamma density.

$$C(S_0, K, T) = \int_0^\infty \text{Black–Scholes}(S_0, K, g) \frac{g^{t/\nu-1}e^{-g/\nu}}{\nu^{t/\nu}\Gamma(t/\nu)}dg$$

Also, by applying Eqs. (12.85) and (12.86) from the Lévy–Khintchine theorem (Sato, 2001) it can be shown that the Lévy measure for the variance gamma process can be written as $d\nu(x) = k(x)dx$ where $k(x)$ is given[9] by

$$\begin{aligned} d\nu(x) &= k(x)dx \\ k(x) &= \frac{e^{-\lambda_p x}}{\nu x}\mathbb{1}_{x>0} + \frac{e^{-\lambda_n |x|}}{\nu |x|}\mathbb{1}_{x<0} \\ \lambda_p &= \left(\frac{\theta^2}{\sigma^4} + \frac{2}{\sigma^2 \nu}\right)^{\frac{1}{2}} - \frac{\theta}{\sigma^2} \\ \lambda_n &= \left(\frac{\theta^2}{\sigma^4} + \frac{2}{\sigma^2 \nu}\right)^{\frac{1}{2}} + \frac{\theta}{\sigma^2} \end{aligned}$$

Derivation of the (PIDE) Partial-integro Differential Equation

Before we dive straight into the derivation of the PIDE for the VG process, we define Lévy processes in general and the Lévy–Khintchine representation. Lévy processes seem to be relatively new to financial engineering and mathematics of finance students and not used extensively in practice. We refer readers to Bertoin (1996) and Papapantoleon (2008) for further reading on Lévy processes and infinitely divisible distributions. They seem to really be essential in terms of understanding pure jump and general Lévy models and the decomposition of any Lévy process into a Brownian, small Poisson jump and large Poisson jump component, as well as a derivation of the form of the Lévy density. We refer the reader to Cont and Tankov (2003) for financial modeling with jump processes, which is a good introduction to general Lévy models in finance.

A Lévy process is a stochastic process with stationary independent increments. The Lévy–Khintchine theorem provides a characterization of a Lévy process in terms of the characterization function of the process; that is, there exists a measure ν such that for all $u \in \mathbb{R}$ and t non-negative

$$\mathbb{E}(e^{iuX_t}) = \exp(t\phi(u)) \tag{12.83}$$

[9] $d\nu(x)$ is the Lévy measure under risk-neutral measure $\mathbb{Q}$.

where

$$\begin{aligned} \phi(u) &= i\gamma u - \frac{1}{2}\sigma^2 u^2 \\ &\quad + \int_{-\infty}^{+\infty} (e^{iuy} - 1 - iuy\mathbb{1}_{\{|y|\le 1\}})d\nu(y) \end{aligned} \tag{12.84}$$

Here γ and σ are real numbers, ν is a measure on $\mathbb{R}$ such that $\nu(\{0\}) = 0$, and $\int_{-\infty}^{+\infty} \min(1, x^2)d\nu(x)$ is bounded. Assume a Lévy process, $\{X_t\}_{t\ge 0}$, of the following form:

$$X_t = (r - q + \mu)t + Z_t \tag{12.85}$$

This process has a drift term controlled by μ and a pure jump component $\{Z_t\}_{t\ge 0}$. Here we focus on the case that the Lévy measure associated to the pure jump component can be written as $d\nu(y) = k(y)dy$, where $k(y)$ is defined as

$$\begin{aligned} d\nu(y) &= k(y)dy \\ k(y) &= \frac{e^{-\lambda_p y}}{\nu y^{1+Y}}\mathbb{1}_{y>0} + \frac{e^{-\lambda_n |y|}}{\nu |y|^{1+Y}}\mathbb{1}_{y<0} \\ \lambda_p &= \left(\frac{\theta^2}{\sigma^4} + \frac{2}{\sigma^2 \nu}\right)^{\frac{1}{2}} - \frac{\theta}{\sigma^2} \\ \lambda_n &= \left(\frac{\theta^2}{\sigma^4} + \frac{2}{\sigma^2 \nu}\right)^{\frac{1}{2}} + \frac{\theta}{\sigma^2} \end{aligned} \tag{12.86}$$

The variable Y allows for control of the sign of large and small jumps. By raising Y above zero, one may induce greater activity near zero and less activity further away from zero. There are also some critical values of Y of interest.

- $Y = 1$ separates finite variation $Y < 1$ from $Y > 1$ infinite variation
- $Y = 0$ separates finite arrival rate $Y < 0$ from $Y > 0$ infinite arrival rate
- $Y = -1$ separates activity concentrated away from zero $Y < -1$ from $Y > -1$ activity concentrated at zero

Let function $F(S_t, t)$ be the value of a derivative security. Applying Itô's lemma for semimartingales (Øksendal, 2000) on $e^{r(T-t)}F(S_t, t)$, we get

$$
\begin{aligned}
F(S_T,T) &= e^{rT}F(S_0,0) \\
&\quad + \int_0^T e^{r(T-t)}\frac{\partial V}{\partial S}(S_t,t)dS_t \\
&\quad + \int_0^T e^{r(T-t)}\int_{-\infty}^{+\infty} \\
&\quad \times\Big[F(S_{t-}e^x,t) - F(S_{t-},t) \\
&\quad -\frac{\partial V}{\partial S}(S_{t-},t)S_{t-}(e^x-1)\Big]\mu(dx,dt) \\
&\quad + \int_0^T e^{r(T-t)}\left[\frac{\partial V}{\partial t}(S_t,t) - rF(S_t,t)\right]dt \\
&= e^{rT}F(S_0,0) + \int_0^T e^{r(T-t)}\frac{\partial V}{\partial S}(S_t,t) \\
&\quad \times\left[dS_t-(r-q)S_tdt\right] \\
&\quad + \int_0^T e^{r(T-t)}\int_{-\infty}^{+\infty} \\
&\quad \times\Big[F(S_{t-}e^x,t) - F(S_{t-},t) \\
&\quad -\frac{\partial V}{\partial S}(S_{t-},t)S_{t-}(e^x-1)\Big]\mu(dx,dt) \\
&\quad + \int_0^T e^{r(T-t)}\Big[\frac{\partial V}{\partial t}(S_t,t) + (r-q)S_t \\
&\quad \frac{\partial V}{\partial S}(S_t,t) - rF(S_t,t)\Big]dt
\end{aligned}
$$

where $\mu(dx,dt)$ is the integer valued random measure which counts the number of jumps in any region of space-time. The density $\nu(dy)dt$ is the compensator of $\mu(dx,dt)$ (Jacod and Shiryaev, 1987). Add and subtract the following term to the above equation:

$$
\begin{aligned}
&\int_0^T e^{r(T-t)}\int_{-\infty}^{+\infty}\Big[F(S_{t-}e^y,t) - F(S_{t-},t) \\
&\quad -\frac{\partial V}{\partial S}(S_{t-},t)S_{t-}(e^y-1)\Big]\nu(dy)dt
\end{aligned}
$$

to get

$$
\begin{aligned}
F(S_T,T) &= F(S_0,0)e^{rT} \\
&\quad + \int_0^T e^{r(T-t)}\frac{\partial V}{\partial S}(S_t,t)\left[dS_t-(r-q)S_tdt\right] \\
&\quad + \int_0^T e^{r(T-t)}\int_{-\infty}^{+\infty} \\
&\quad \times\left[F(S_te^y,t) - F(S_t,t) - \frac{\partial V}{\partial S}(S_t,t)S_t(e^x-1)\right] \\
&\quad \times\left[\mu(dx,dt)-\nu(dy)dt\right] \\
&\quad + \int_0^T e^{r(T-t)}\int_{-\infty}^{+\infty} \\
&\quad \times\Big[F(S_te^x,t) - F(S_{t-},t) \\
&\quad -\frac{\partial V}{\partial S}(S_t,t)S_{t-}(e^x-1)\Big]\nu(du)dt \\
&\quad + \int_0^T e^{r(T-t)} \\
&\quad \times\Big[\frac{\partial V}{\partial t}(S_t,t) + (r-q)S_t\frac{\partial V}{\partial S}(S_t,t) \\
&\quad -rF(S_t,t)\Big]dt
\end{aligned}
$$

Now, taking the expectation under $\mathbb{Q}$, we will get

$$
\begin{aligned}
\mathbb{E}^{\mathbb{Q}}F(S_T,T) &= F(S_0,0)e^{rT} \\
&\quad + \int_0^T e^{r(T-t)}\Big\{\int_{-\infty}^{+\infty}\Big[F(S_te^y,t) - F(S_t,t) \\
&\qquad -\frac{\partial V}{\partial S}(S_t,t)S_{t-}(e^y-1)\Big]\nu(dy) \\
&\qquad +\frac{\partial V}{\partial t}(S_t,t) + (r-q)S_t\frac{\partial V}{\partial S}(S_t,t) \\
&\qquad -rF(S_t,t)\Big\}dt
\end{aligned}
$$

We know that

$$
\mathbb{E}^{\mathbb{Q}}\left(F(S_T,T)\right) = F(S_0,0)e^{rT}
$$

Therefore

$$
\begin{aligned}
&\int_0^T e^{r(T-t)}\Big\{\int_{-\infty}^{+\infty}\Big[F(S_{t-}e^y,t) - F(S_{t-},t) \\
&\qquad -\frac{\partial V}{\partial S}(S_{t-},t)S_{t-}(e^y-1)\Big]\nu(dy)
\end{aligned}
$$

$$+\frac{\partial V}{\partial t}(S_t, t) + (r-q)S_t$$
$$\left.\frac{\partial V}{\partial S}(S_t, t) - rF(S_t, t)\right\} dt = 0$$

Since the integrand is non-negative, that implies

$$\int_{-\infty}^{+\infty} \Bigg[F(S_{t-}e^y, t) - F(S_{t-}, t)$$
$$\left. - \frac{\partial V}{\partial S}(S_{t-}, t)S_{t-}(e^y - 1)\right] \nu(dy)$$
$$+\frac{\partial V}{\partial t}(S_t, t) + (r-q)S_t\frac{\partial V}{\partial S}(S_t, t)$$
$$-rF(S_t, t) = 0$$

Note that the PIDE is pretty generic for any Lévy density $\nu(dy)$. Writing $\nu(dy) = k(y)dy$ we get

$$\int_{-\infty}^{\infty} \Bigg[F(S_{t-}e^y, t) - F(S_{t-}, t)$$
$$\left. - \frac{\partial V}{\partial S}(S_{t-}, t)S_{t-}(e^y - 1)\right] k(y)dy$$
$$+\frac{\partial V}{\partial t}(S_t, t) + (r-q)S_t\frac{\partial V}{\partial S}(S_t, t)$$
$$-rF(S_t, t) = 0 \tag{12.87}$$

which is the partial-integro differential equation (PIDE).

12.9 CONCLUSIONS

In this chapter, we introduced the notion of a partial differential equation (PDE). These are functional equations, whose solutions are functions of the underlying variables. We briefly discussed various forms of PDEs and introduced the related terminology.

This chapter also showed that the relationship between financial derivatives and the underlying assets can be exploited to obtain PDEs that derivative asset prices must satisfy.

12.10 REFERENCES

Most of our readers are interested in PDEs because, at one point, they will be applying them in practical derivative asset pricing. Thus, rather than books on the theory of PDEs, sources dealing with the numerical solution of PDEs will be more useful. In most cases, these sources contain a brief summary of the underlying theory as well. We recommend two books on PDEs. Smith (1985) is easy to read. Thomas (1995) is a more comprehensive and recent treatment.

12.11 EXERCISES

1. You are given a function $f(x,z,y)$ of three variables, x, z, y. The following PDE is called Laplace's equation:

$$f_{xx} + f_{yy} + f_{zz} = 0$$

According to this, in Laplace's equation, the sum of second partials with respect to the variables in the function must equal zero.

(a) Do the following equations satisfy Laplace's equation?

i. $f(x,z,y) = 4z^2y - x^2y - y^3$
ii. $f(x,y) = x^2 - y^2$
iii. $f(x,y) = x^3 - 3xy$
iv. $f(x,y,z) = \frac{x}{y+z}$

(b) Why is it that more than one function satisfies Laplace's equation? Is it "good" to have many solutions to an equation in general?

2. A function $f(x,z,y,t)$ of four variables, x, z, y, t, that satisfy the following PDE is called the heat equation:

$$f_t = a^2\left(f_{xx} + f_{yy} + f_{zz}\right)$$

where a is a constant. According to the heat equation, first partial with respect to t is proportional to the sum of second partials with respect to the variables in the function. Do the following functions satisfy the heat equation?

(a) $f(x,y,z) = e^{[29a^2\pi^2 t+\pi(3x+2y+4z)]}$
(b) $f(x,z,y) = 3x^2 + 3y^2 - 6z^2 + x + y - 9z - 3$

3. Consider the PDE:

$$f_x + 0.2f_y = 0$$

with $X \in [0,1]$ and $Y \in [0,1]$.

(a) What is the unknown in this equation?
(b) Explain this equation using plain English.
(c) How many functions $f(x,y)$ can you find that will satisfy such an equation?
(d) Now suppose you know the boundary condition:

$$f(0,Y) = 1$$

Can you find a solution to the PDE? Is the solution unique?

4. Consider the PDE:

$$f_{xx} + 0.2f_t = 0$$

with the boundary condition

$$f(x,1) = \max[x-6, 0]$$

Let

$$0 \le x \le 12$$

and

$$0 \le t \le 1$$

(a) Is the single boundary condition sufficient for calculating a numerical approximation to $f(x,t)$?
(b) Impose additional boundary conditions of your choice on $f(0,t)$ and $f(12,t)$.
(c) Choose grid sizes of $\Delta x = 3$ and $\Delta t = 0.25$ and calculate a numerical approximation to $f(x,t)$ under the boundary conditions you have imposed.

5. Consider the problem of pricing the following options via PDE:

(a) European Call
(b) European Put
(c) Up-and-Out Call
(d) Down-and-Out Put

In each case explain reasonable boundary conditions at $f(S_{min},t)$ and $f(S_{max},t)$ that could be used when solving the PDE.

6. Write a simulation program to price a European digital option whose underlying stock price follows a geometric Brownian motion with volatility $\sigma = 0.1$. Other parameters are $r = 0.05, q = 0, S = K = 10, T = 1$.

CHAPTER

13

PDEs and PIDEs—An Application

OUTLINE

13.1 INTRODUCTION

In this chapter, we provide some examples of partial differential equation methods using derivative asset pricing.

One purpose of this is to have a geometric look at the function that solves the PDE obtained by Black and Scholes (1973). The geometry of the Black–Scholes formula helps with the understanding of PDEs. In particular, we show geometrically the implications of having a single random factor in pricing call options.

Next, we complicate the original Black–Scholes framework by introducing a second

Financial Derivatives, Third Edition.
http://dx.doi.org/10.1016/B978-0-12-384682-2.00013-X

factor. This leads to some major difficulties, which we will discuss briefly.

Finally, we compare closed-form solutions for PDEs with numerical approaches. We conclude with an example of a numerical asset price calculation.

13.2 THE BLACK–SCHOLES PDE

In Chapter 12 we obtained the PDE that the price of a derivative written on the underlying asset S_t must satisfy under some conditions. The underlying security did not pay a dividend, and the risk-free interest rate was assumed to be constant at r.

Now, suppose we consider the special SDE where

$$a\left(S_t, t\right) = \mu S_t \tag{13.1}$$

and, more importantly,

$$\sigma\left(S_t, t\right) = \sigma S_t, \quad t \in [0, \infty) \tag{13.2}$$

We occasionally write σ_t to denote σS_t. Under these conditions the fundamental PDE of Black and Scholes and the associated boundary condition are given by

$$-rF + rF_t S_t + F_t + \frac{1}{2}\sigma^2 F_{ss} S_t^2 = 0,$$
$$0 \le S_t, \quad 0 \le t \le T \tag{13.3}$$

$$F(T) = \max\left[S_T - K, 0\right] \tag{13.4}$$

Equations (13.3) and (13.4) were first used in finance by Black and Scholes (1973). Hence we call these equations "the fundamental PDE of Black and Scholes."[1]

[1]Only one of the second partials, namely the one with respect to S_t, is present in this PDE. Also, note that there is no constant term. Under these conditions, we can easily calculate the value of the expression from Chapter 12,

$$a_5^2 - 4a_3a_4 \tag{13.5}$$

as zero. This means that leaving aside the presence of S_t and S_t^2, which are always positive, the Black–Scholes PDE is of the parabolic form.

Black and Scholes solve this PDE and obtain the form of the function $F(S_t, t)$ explicitly:

$$F\left(S_t, t\right) = S_t N\left(d_1\right) + K e^{-r(T-t)} N\left(d_2\right) \tag{13.6}$$

where

$$d_1 = \frac{\log\left(\frac{S_t}{K}\right) + \left(r + \frac{1}{2}\sigma^2\right)(T - t)}{\sigma\sqrt{T - t}} \tag{13.7}$$

$$d_2 = d_1 - \sigma\sqrt{T - t} \tag{13.8}$$

$N(d_i), i = 1, 2,$ are two integrals of the standard normal density:

$$N\left(d_i\right) = \int_{-\infty}^{d_i} \frac{1}{\sqrt{2\pi}} e^{-\frac{1}{2}x^2} dx \tag{13.9}$$

To show that this function satisfies the Black–Scholes PDE and the corresponding boundary condition, we have to take the first and second partials of (13.6) with respect to S_t, and plug these in (13.3) with the $F(S_t, t)$ and its partial with respect to t. The result should equal zero. As t approaches T, the function should equal (13.4).

13.2.1 A Geometric Look at the Black–Scholes Formula

We saw in Chapter 12 that functions $F(S_t, t)$ satisfying various PDEs could be represented in three-dimensional space. We can do the same for the Black–Scholes PDE. The solution of this PDE was given by (13.6). We would like to pick numerical values for the parameters K, r, σ, T and represent this formula in the three-dimensional space $F \times S \times t$.

We pick

$$r = 0.065, \quad K = 100, \quad \sigma = 0.080, \quad T = 1 \tag{13.10}$$

and substitute these in formula (13.6). These numbers imply a 6.5% risk-free borrowing cost, and an 80% volatility during the interval $t \in [0, 1]$. This type of volatility is high for most mature financial markets. But it makes the graphics easier to read. The life of the call option is normalized to 1, with $T = 1$ implying one year, and the initial time is set at $t_0 = 0$. Finally, the strike price is set at 100.[2]

To plot the Black–Scholes formula with these particular parameters, we must select a range for the two variables S_t and t. We let S_t range from 50 to 140, and let t range from 0 to 1. The resulting surface is shown in Figure 13.1.

In Figure 13.1 we have a "horizontal surface" defined by the axes labeled S_t and $T - t$, where the latter represents "time to expiration." These two axes form a plane. For example, point A represents an underlying asset price of 130, and a "time to expiration" equal to 0.80. By going up vertically toward the surface we reach point B, which is in fact the value of the Black–Scholes formula evaluated at A:

$$B(130, 0.2) \tag{13.11}$$

We display two types of contours on the surface. First, we fix S_t at a particular level and vary t. This gives lines such as $aa\prime$, which show how the call price will change as t goes from 0 to 1 when S_t is fixed at 100.

The second contour is shown as $bb\prime$ and represents $F(S_t, t)$ when we fix t at 0.4 and move S_t from 60 to 140. It is interesting to see that as t goes toward 1, this contour goes toward the limit shown as $cc\prime$. The latter is the usual graph with a kink at K, which shows the option payoff at expiration.

We would like to emphasize a potentially confusing point using Figure 13.1. The Black–Scholes formula gives a surface, once we fix K, r, and σ. This surface will *not* move as random events occur and realized values of dW_t become known.

[2]If $T = 1$ means "one year," the interest rate and the volatility will be yearly rates. But $T = 1$ may very well mean six months, three months, or any time interval during which the financial instrument will exist. We merely used $T = 1$ as a normalization. Under such conditions, the interest rate or the volatility numbers must be adjusted to the relevant time period.

Realization of the Wiener increments would only cause random movements on the surface. One such example is the trajectory denoted by C_0, C_T in Figure 8. Because the increments of the Wiener process are unpredictable, the movement of the stock price along the t direction will proceed in "random steps." Over infinitesimal intervals these steps are also infinitesimal, yet still unpredictable.

The trajectory C_0, C_T is interesting from another angle as well. As time passes, S_t will trace the trajectory shown on the $S_t \times t$ plane. Going vertically to the surface, we obtain the trajectory C_0, C_T. Note that there is a *deterministic* correspondence between the two trajectories. Given the trajectory of S_t on the horizontal plane, there is only one trajectory for $F(S_t, t)$ to follow on the surface. This is the consequence of having the same randomness in S_t and in $F(S_t, t)$.

13.3 LOCAL VOLATILITY MODEL

Yet, we consider another special SDE where as before

$$a\,(S_t, t) = \mu S_t \tag{13.12}$$

and, more importantly,

$$\sigma\,(S_t, t) = \sigma(S_t, t) S_t, \quad t \in [0, \infty) \tag{13.13}$$

As we see, unlike Black–Scholes, where σ assumed constant, here it is level and time dependent. This model is called local volatility model. Under these conditions the fundamental PDE of local volatility model and the associated boundary condition are given by

$$-rF + rF_t S_t + F_t + \frac{1}{2}\sigma(S_t, t)^2 F_{ss} S_t^2 = 0,$$
$$0 \le S_t, \quad 0 \le t \le T \tag{13.14}$$

$$F(T) = \max\,[S_T - K, 0] \tag{13.15}$$

Unlike in the case of the Black–Scholes model there is no closed-form solution for the local volatility model and we have to utilize some

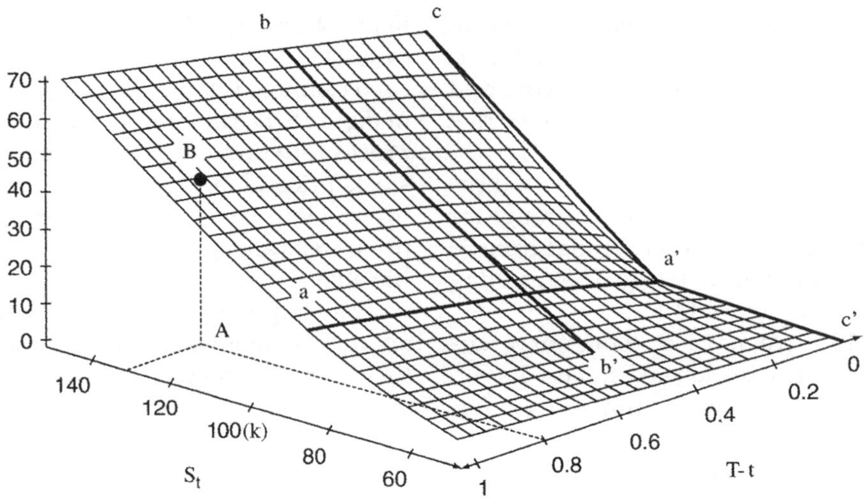

FIGURE 13.1 Black–Scholes call price surface.

numerical method to solve the equation to find an approximation to its solution.

Due to volatility dependence on the stock level and time, this model can capture skew and kurtosis, something that the Black–Scholes model lacks.

13.4 PARTIAL INTEGRO-DIFFERENTIAL EQUATIONS (PIDEs)

13.4.1 Rewriting the Black–Scholes Partial Differential Equation

Before giving any formal introduction, we start from Black–Scholes PDE and rewrite it as a partial integro-differential equation and investigate what to expect when working with pure jump processes. This is a complete reverse engineering problem.

Consider the Black–Scholes partial differential equation

$$\frac{\partial F}{\partial t} + \frac{\sigma^2 S^2}{2}\frac{\partial^2 F}{\partial S^2} + (r-q)S\frac{\partial F}{\partial S} = rF(S,t) \quad (13.16)$$

We do not bring in terminal or boundary conditions which are dependent on the type of the option we wish to price. Make the change of variable $x = \ln S$ and call $\bar{F}(x,t) = F(S,t)$; note that by chain rule we have

$$\frac{\partial \bar{F}}{\partial x}(x,t) = S\frac{\partial F}{\partial S}(S,t) \quad (13.17)$$

$$\frac{\partial^2 \bar{F}}{\partial x^2}(x,t) - \frac{\partial \bar{F}}{\partial x}(x,t) = S^2\frac{\partial^2 F}{\partial S^2}(S,t) \quad (13.18)$$

Rewriting (13.16) in terms of $\bar{F}(x,t)$ by substituting (13.17) and (13.18), we obtain

$$\frac{\partial \bar{F}}{\partial t} + (r-q)\frac{\partial \bar{F}}{\partial x} + \frac{\sigma^2}{2}\left(\frac{\partial^2 \bar{F}}{\partial x^2} - \frac{\partial \bar{F}}{\partial x}\right) = r\bar{F}(S,t) \quad (13.19)$$

As we see later it is intentional that we do not gather terms together. If $k(y)$ is defined as

$$k(y) = \begin{cases} \frac{e^{-\lambda|y|}}{\nu|y|} & \text{for } y < 0 \\ \frac{e^{-\bar{\lambda} y}}{\nu y} & \text{for } y > 0 \end{cases}$$

with $\lambda = \sqrt{\frac{2}{\sigma^2 \nu}}$, then it can be shown that as ν approaches zero we get

$$\int_{|y|\leq\epsilon} y^2 k(y)dy = \sigma^2 \tag{13.20}$$

It begs the question: Where is Lévy density $k(y)$ coming from? What are the parameters ν and λ? And is it independent of ϵ? Having that, we can rewrite (13.19) as

$$\frac{\partial \bar{F}}{\partial t} + (r-q)\frac{\partial \bar{F}}{\partial x} + \int_{|y|\leq\epsilon} \frac{1}{2}\left(\frac{\partial^2 \bar{F}}{\partial x^2} - \frac{\partial \bar{F}}{\partial x}\right) \times y^2 k(y)dy = r\bar{F}(x,t) \tag{13.21}$$

Note that we can move the partial derivatives inside the integral because they are independent of y. Now, we want to rewrite the integrand differently and make it dependent on y. Thinking of jumps, we would think of terms like $\bar{F}(x+y,t)$, where y is the jump size, and as we see in the Merton jump-diffusion model we think of the term $(e^y - 1)$. Applying the Taylor expansion to $\bar{F}(x+y,t)$ we see that

$$\bar{F}(x+y,t) = \bar{F}(x) + y\frac{\partial \bar{F}}{\partial x} + \frac{y^2}{2}\frac{\partial^2 \bar{F}}{\partial x^2} + O(y^3) \tag{13.22}$$

Now apply Taylor expansion to e^y to see

$$e^y = 1 + y + \frac{y^2}{2} + O(y^3) \tag{13.23}$$

In both expansions, the remainder term is of order y^3. From (13.22) we have

$$\frac{y^2}{2}\frac{\partial^2 \bar{F}}{\partial x^2} = \bar{F}(x+y,t) - \bar{F}(x) - y\frac{\partial \bar{F}}{\partial x} + O(y^3) \tag{13.24}$$

From (13.23) we have

$$\frac{y^2}{2} = e^y - 1 - y + O(y^3) \tag{13.25}$$

Substituting (13.24) and (13.25) in (13.21), one gets

$$\frac{\partial \bar{F}}{\partial t} + (r-q)\frac{\partial \bar{F}}{\partial x} + \int_{|y|\leq\epsilon} \left(\bar{F}(x+y) - \bar{F}(x) - y\frac{\partial \bar{F}}{\partial x} - (e^y - 1 - y)\frac{\partial \bar{F}}{\partial x}\right) k(y)dy = r\bar{F}(x,t) \tag{13.26}$$

or, simplifying it further,

$$\frac{\partial \bar{F}}{\partial t} + (r-q)\frac{\partial \bar{F}}{\partial x} + \int_{|y|\leq\epsilon} \left(\bar{F}(x+y) - \bar{F}(x) - (e^y - 1)\frac{\partial \bar{F}}{\partial x}\right) k(y)dy = r\bar{F}(x,t) \tag{13.27}$$

Here we drop the remainder $O(y^3)$, knowing $|y| \leq \epsilon$ and assuming ϵ is small enough, one can neglect the term $O(y^3)$. Rewriting back in $S = e^x$ and noting $\bar{F}(x+y,t) = F(Se^y,t)$, we get

$$\frac{\partial F}{\partial t} + (r-q)S\frac{\partial F}{\partial S} + \int_{|y|\leq\epsilon} \left(F(Se^y,t) - F(S,t) - (e^y - 1)\frac{\partial F(S,t)}{\partial S}\right) k(y)dy = rF(S,t) \tag{13.28}$$

Here are some remarks on what we have written so far:

- We see that the diffusion term is replaced by an integral in a narrow range, considering that, after all, Black–Scholes is a pure diffusion case that is what one would expect to get; nonetheless, it can be written as an integral with no second derivative.
- The $k(y)$ is a measure-type function which it counts the number of jumps and in this case just infinitely many small jumps.
- One can conjecture to move a pure jump process we need an integral which can resume large jumps no matter how infrequent those jumps could be.
- We see this as a good start, knowing that Black–Scholes in some special case can be written as an integral and that means pure jump processes that we consider should always resume Black–Scholes in a specific case, e.g., if the market is Black–Scholes we should get Black–Scholes back from it.

- How to drive partial integro-differential equation for a pure jump process.
- How to arrive at a pure jump process in a rhetoric fashion.
- What is $k(y)$, how to interpret it and how to obtain it.
- This seems very appealing, knowing that Black–Scholes is linked to a pure jump process and can be written this way.
- Better able to fit smile.
- There exists a consistent theoretical framework.

13.5 PDEs/PIDEs IN ASSET PRICING

The partial differential equation obtained by Black and Scholes is relevant under some specific assumptions. These are (1) the underlying asset is a stock, (2) the stock does not pay any dividends, (3) the derivative asset is a European-style call option that cannot be exercised before the expiration date, (4) the risk-free rate is constant, and (5) there are no indivisibilities or transaction costs such as commissions and bid-ask spreads.

In most applications of pricing, one or more of these assumptions will be violated. If so, in general, the Black–Scholes PDE will not apply and a new PDE should be found. One exception is the violation of assumption (13.3). If the option is American style, the PDE will remain the same.

The relevant PDEs under these more complicated circumstances fall into one of a few general classes of applications. We discuss a simple case next.

13.5.1 Constant Dividends

If one is trying to price a call option, and if the option is written on a stock that pays dividends at a constant rate of δ units per time, the resulting PDE will change only slightly.

Suppose we change one of the Black—Scholes assumptions and introduce a constant rate of dividends, δ, paid by the underlying asset S_t.

Again, we can try to form the same "approximately"risk-free portfolio by combining the underlying asset and the call option written on it:

$$P_t = \theta_1 F(S_t, t) + \theta_2 S_t \tag{13.29}$$

The portfolio weights θ_1, θ_2 can be selected as

$$\theta_1 = 1, \quad \theta_2 = -F_s \tag{13.30}$$

so that the "unpredictable"random component is eliminated and a hedge is formed:

$$dP_t = F_t dt + \frac{1}{2} F_{ss} \sigma_t^2 dt \tag{13.31}$$

Up to this point there is no difference from the original Black–Scholes approach discussed in Chapter 12. The time path of the P_t will again be completely predictable.

The difference occurs in deciding how much this portfolio should appreciate in value. Before, the (completely predictable) capital gains were exactly equal to earnings of a risk-free investment. But now, the underlying stock pays a dividend that is predictable at a rate of δ. Hence, the capital gains *plus* the dividends received must equal the earnings of a risk-free portfolio:

$$dP_t + \delta dt = rP_t dt \tag{13.32}$$

or

$$dP_t = -\delta dt + rP_t dt \tag{13.33}$$

Putting this together with (13.14) we get a slightly different PDE:

$$rF - rF_s S_t - \delta - F_t - \frac{1}{2} F_{ss} \sigma_t^2 = 0 \tag{13.34}$$

There is now a constant term δ. Hence stocks paying dividends at a constant rate δ do not present a major problem.

13.6 EXOTIC OPTIONS

In the previous section, a complication to the Black–Scholes framework was discussed. The PDE satisfied by the arbitrage-free price of the derivative asset did change as the assumptions concerning dividend payments changed. This section discusses another complication.

Suppose the derivative asset is an option with a possibly *random* expiration date. For example, there are some "down-and-out" and "up-and-out" options that are known as *barrier* derivatives.[3] Unlike "standard" options, the payoff of these instruments *also* depends on whether or not the spot price of the underlying asset crossed a certain barrier during the life of the option. If such a crossing has occurred, the payoff of the option changes. We briefly review some of these "exotic"options.

13.6.1 Lookback Options

In the standard Black–Scholes case, the call option payoff is equal to $S_T - K$, if the option expires in the money. In this payoff S_T is the price of the underlying asset at expiration and K is the constant strike price.

In the case of a *floating* lookback call option, the payoff is the difference $S_T - S_{\min}$, where $S_{\min}$ is the minimum price of the underlying asset observed during the life of the option.

A *fixed* lookback call option, on the other hand, pays the difference (if positive) between a fixed strike price K and $S_{\max}$, where the latter is the maximum reached by the underlying asset price during the life of the option.[4] These options have the characteristic that some positive payoff is guaranteed if the option is in the money during some time over its life. Hence, everything else being the same, they are more expensive.

[3]These are also known as "knock-out" and "knock-in" options.

[4]The lookback option is *floating* because the strike price is not fixed.

13.6.2 Ladder Options

A ladder option has several *thresholds*, such that if the underlying price reaches these thresholds, the return of the option is "locked in."

13.6.3 Trigger or Knock-in Options

A down-and-in option gives its holder a European option if the spot price falls below a *barrier* during the life of the option. If the barrier is not reached, the option expires with some *rebate* as a payoff.[5]

13.6.4 Knock-out Options

Knock-out options are European options that expire immediately if, for example, the underlying asset price falls below a barrier during the life of the option. The option pays a rebate if the barrier is reached. Otherwise, it is a "standard" European option. Such an option is called "down-and-out."[6]

13.6.5 Other Exotics

There are obviously many different ways one can structure an exotic option. Some common cases include the following:

- *Basket options*, which are derivatives where the underlying asset is a basket of various financial instruments. Such baskets dampen the volatility of the individual securities. Basket options become more affordable in the case of *emerging market* derivatives.
- *Multi-asset options* have payoffs depending on the underlying price of more than one asset.

[5]Similarly, there are up-and-in options that come into effect if the underlying asset price has an upcrossing of a certain barrier.

[6]The up-and-out option expires immediately if the underlying asset price has an upcrossing of a certain barrier.

For example, the payoff of such a call may be

$$F(S_{1T}, S_{2T}, T) = \max[0, \max(S_{1T}, S_{2T}) - K] \quad (13.35)$$

Another possibility is the *spread call*

$$F(S_{1T}, S_{2T}, T) = \max[0, (S_{1T} - S_{2T}) - K] \quad (13.36)$$

or the *portfolio call*

$$F(S_{1T}, S_{2T}, T) = \max[0, (\theta_1 S_{1T} + \theta_2 S_{2T}) - K] \quad (13.37)$$

where θ_1, θ_2 are known portfolio weights. As a final example, one may have a *dual strike call option*:

$$F(S_{1T}, S_{2T}, T) = \max[0, S_{1T} - K, S_{2T} - K] \quad (13.38)$$

- *Average or Asian options* are quite common and have payoffs depending on the average price of the underlying asset over the lifetime of the option.[7]

13.6.6 The Relevant PDEs

It is clear from this brief list of exotic options that there are three major differences between exotics and the standard Black–Scholes case.

First, the *expiration value* of the option may depend on some event happening over the life of the option (e.g., it may be a function of the maximum of the underlying asset price). Clearly, these make the boundary conditions much more complicated than the Black–Scholes case.

Second, derivative instruments may have random expiration *dates*. Third, the derivative may be written on more than one asset. All these may lead to changes in the basic PDE that we derived in the Black–Scholes case. Not all examples can be discussed here. But consider the case of *knock-out options*. We discuss the case of a "down-and-out" call.

[7]Often arithmetic averages are used, and the average can be computed on a daily, weekly, or monthly basis.

Let the K_t be the *barrier* at time t. Let S_t and $F(S_t, t, K_t)$, respectively, be the price of the underlying asset and the price of the knock-out option. If the S_t reaches the K_t during the life of the option, the option holder receives a rebate R_t and the option suddenly expires. Otherwise, it is a standard European option.

In deriving the relevant PDE, the main difference from the standard case is in the boundary conditions. As long as the underlying asset price is above the *barrier* K_t during the life of the option, $t \in [0, T]$, the same PDE as in the standard case prevails:

$$\frac{1}{2}\sigma_t^2 F_{ss} + rF_s S_t - rF + F_t = 0, \quad \text{if } S_t > K_t \quad (13.39)$$

and

$$F(S_T, T, K_T) = \max[0, S_T - K_T] \quad (13.40)$$

But if the S_t falls below K_t during the life of the option, we have

$$F(S_t, t, K_t) = R, \quad \text{if } S_t \leq K_t \quad (13.41)$$

The form of the PDE is the same, but the boundary is different. This will result in a different solution for $F(S_t, t, K_t)$, as was discussed earlier.

13.7 SOLVING PDEs/PIDEs IN PRACTICE

Once a trader obtains a PDE representing the behavior over time of a derivative price $F(S_t, t)$ there will be two ways to proceed in calculating this value in practice.

13.7.1 Closed-Form Solutions

The first method is similar to the one used by Black and Scholes, which involves solving the PDE for a closed-form formula. It turns out that the PDEs describing the behavior of derivative prices cannot in every case be solved for closed forms. In general, either such PDEs are not easy to solve, or they do not have solutions that one can express as closed-form formulas.

First, let us discuss the difference between closed forms and numerical solutions of a PDE. The function $F(S_t, t)$ solves a PDE if the appropriate partial derivatives satisfy an equality such as

$$\frac{1}{2}\sigma_t^2 F_{ss} + rF_s S_t - rF + F_t = 0, \quad 0 \leq S_t, 0 \leq t \leq T \tag{13.42}$$

Now, it is possible that one can find a continuous surface such that the partial derivatives do indeed satisfy the PDE. But it may still be impossible to represent this surface in terms of an easy and convenient formula, as in the case of Black–Scholes. In other words, although a solution may exist, this solution may not be representable as a convenient function of S_t and t.

We will discuss this by using an analogy. Consider the function of time $F(t)$ shown in Figure 13.2.

The way it is drawn, $F(t)$ is clearly continuous and smooth. So, in the region shown, $F(t)$ has derivatives with respect to time. But $F(t)$ was drawn in some *arbitrary* fashion, and there is no reason to expect that this curve can be represented by a compact formula involving a few terms in t. For example, an exponential formula

$$F(t) = a_2 e^{a_1 t} + a_3 \tag{13.43}$$

where $a_i, i = 1, 2, 3$ are constants, cannot represent this curve. In fact, for a general continuous and smooth function, such closed-form formulas will not exist.[8,9]

[8]On the other hand, if the curve is of a "special" type, one may be able to identify it as a simple polynomial and represent it with a formula. For example, the curve in Figure 13.3 looks like a parabola and has a simple closed-form representation as $F = a_0 + a_1 t + a_2 t^2$.

[9]If a curve is smooth and continuous, it may, however, be expanded as an infinite Taylor series expansion. Yet Taylor series expansions are not closed-form formulas. They are representations of such $F(\cdot)$.

The solutions of PDEs in the simple Black–Scholes case are surfaces in the three-dimensional space generated by S_t, t and $F(S_t, t)$. Given a smooth and continuous curve in three-dimensional space $F \times t \times S_t$, the partial derivatives may be well defined and may satisfy a certain PDE, but the surface may not be representable by a compact formula.

Hence, a solution to a PDE may exist, but a closed-form expression for the formula may not. In fact, given that such formulas are very constrained in representing smooth surfaces in three (or higher) dimensions, this may often be the case rather than the exception.

13.7.2 Numerical Solutions

When a closed-form solution does not exist, a market participant is forced to obtain numerical solutions to PDEs. A *numerical solution* is like calculating the surface represented by $F(S_t, t)$ directly, without first obtaining a closed-form formula for $F(S_t, t)$. Consider again the PDE obtained from the Black–Scholes framework:

$$\frac{1}{2}\sigma_t^2 F_{ss} + rF_s S_t - rF + F_t = 0, \quad 0 \leq S_t, 0 \leq t \leq T \tag{13.44}$$

To solve this PDE *numerically*, one assumes that the PDE is valid for finite increments in S_t and t. Two "partitions" are needed.

1. A grid size for ΔS must be selected as a minimum increment in the price of the underlying security.
2. Time t is the second variable in $F(S_t, t)$. Hence, a grid size for Δt is needed as well. Needless to say, $\Delta t, \Delta S$ must be "small." How small is "small," can be decided by trial and error.
3. Next, one has to decide on the range of possible values for S_t. To be more precise, one selects, a priori, the minimum $S_{\min}$ and the maximum $S_{\max}$ as possible values of S_t. These extreme values should be selected so that observed prices remain within the range

$$S_{\min} \leq S_t \leq S_{\max} \tag{13.45}$$

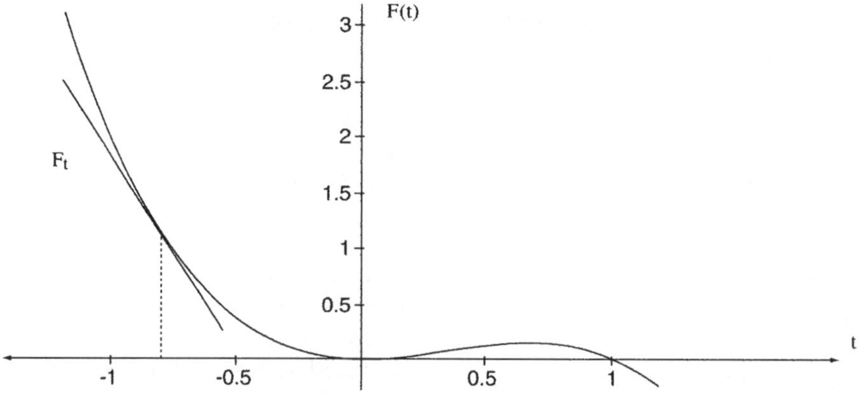

FIGURE 13.2 A smooth function with no closed-form representation.

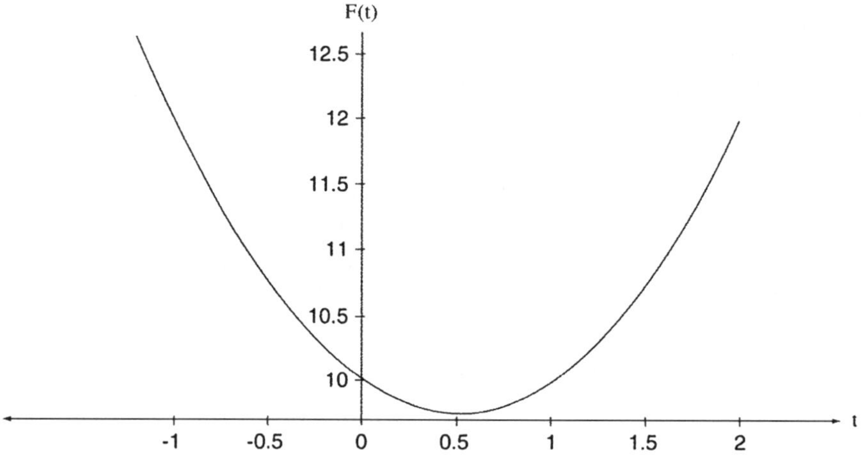

FIGURE 13.3 A function with a simple closed-form representation.

4. The boundary conditions must be determined.
5. Assuming that for small but noninfinitesimal ΔS_t and Δt the same PDE is valid, the values of $F(S_t, t)$ at the grid points should be determined.

To illustrate the last step, let

$$F_{ij} = F(S_i, t_j) \tag{13.46}$$

where F_{ij} is the value at time t_j if the price of the underlying asset is at S_i. The limits of i, j will be determined by the choice of $\Delta S, \Delta t$ and of $S_{\min}, S_{\max}$.

We want to approximate $F(S_t, t)$ at a finite number of points F_{ij}. This is shown in Figure 13.4 for an arbitrary surface and in Figure 13.5 for the Black–Scholes surface. In either case, the dots represent the points at which $F(S_t, t)$ will be evaluated. The sizes of the grids ΔS and Δt determine how "close" these dots will be on the surface. Obviously, the closer these dots are, the better the approximation of the surface.

We let F_{ij} denote the "dot" that represents the ith value for S_t and the jth value for t. These values for S_t and t will be selected from their respective axes and then "plugged in" to $F(S_t, t)$. The result is written as F_{ij}.

To carry on this calculation, we need to change the partial differential equation to a difference equation by replacing all differentials with

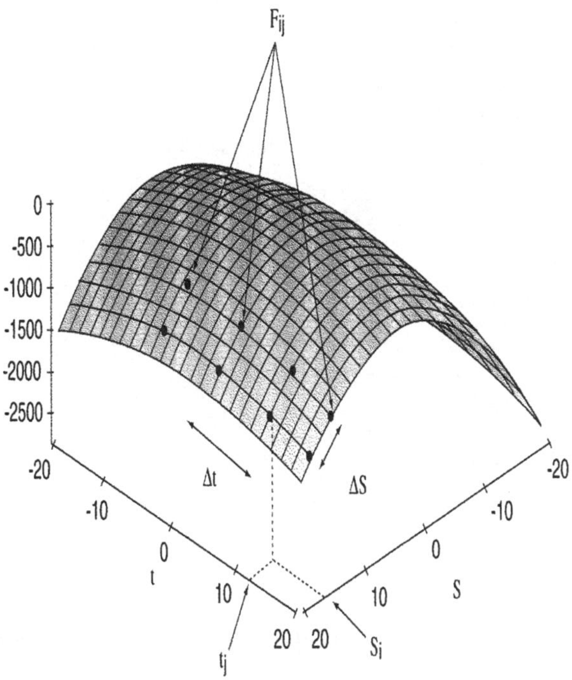

FIGURE 13.4 An arbitrary surface.

appropriate differences. There are various methods of doing this, each with a different degree of accuracy. Here, we use the simplest method[10]

$$\frac{\Delta F}{\Delta t} + rS\frac{\Delta F}{\Delta S} + \frac{1}{2}\sigma^2 S^2 \frac{\Delta^2 F}{\Delta S^2} \approx rF \qquad (13.47)$$

where the first-order partial derivatives are approximated by the corresponding differences. For first partials we can use the backward differences

$$\frac{\Delta F}{\Delta t} \approx \frac{F_{ij} - F_{i,j-1}}{\Delta t} \qquad (13.48)$$

$$\frac{\Delta F}{\Delta S} \approx \frac{F_{ij} - F_{i-1,j}}{\Delta S} \qquad (13.49)$$

[10]We are ignoring i, j subscripts for notational convenience. As will be seen below, elements of this difference equation depend on i, j. For each i, j there exists one equation such as (13.30).

or we can use forward differences,[11] an example of which is

$$rS\frac{\Delta F}{\Delta S} \approx rS_j \frac{F_{i+1,j} - F_{i,j}}{\Delta S} \qquad (13.50)$$

For the second-order partials, we use the approximations

$$\frac{\Delta^2 F}{\Delta S^2} \approx \left[\frac{F_{i+1,j} - F_{i,j}}{\Delta S} - \frac{F_{i,j} - F_{i-1,j}}{\Delta S}\right] \frac{1}{\Delta S} \qquad (13.51)$$

where $i = 1, \ldots, n$ and $j = 1, \ldots, N$. The parameters N and n determine the number of points at which we decided to calculate the surface $F(S_t, t)$.

For example, in Figure 13.5 we can let $n = 5$ and $N = 22$. Hence, excluding the points on the boundary values, we have a total of 80 dots to calculate on the surface. These values can be calculated by solving recursively the (system of) equations in (13.30).

The recursive nature of the problem is due to the existence of boundary conditions. The next section deals with these.

13.7.3 Boundary Conditions

Now, some of the F_{ij} are known because of endpoint conditions. For example, we always know the value of the option as a function of S_t at expiration. For extreme values of S_t, we can use some approximations that are valid in the limit. In particular:

- For S_t that is very high, we let $S_t = S_{\max}$ and

$$F(S_{\max}, t) \approx S_{\max} - Ke^{-r(T-t)} \qquad (13.52)$$

  Here, $S_{\max}$ is a price chosen so that the call premium is very close to the expiration date payoff.
- For S_t that is very low, we let $S_t = S_{\min}$ and

$$F(S_{\min}, t) \approx 0 \qquad (13.53)$$

  In this case, $S_{\min}$ is an extremely low price. There is almost no chance that the option will

[11]We can also use *centered* differences.

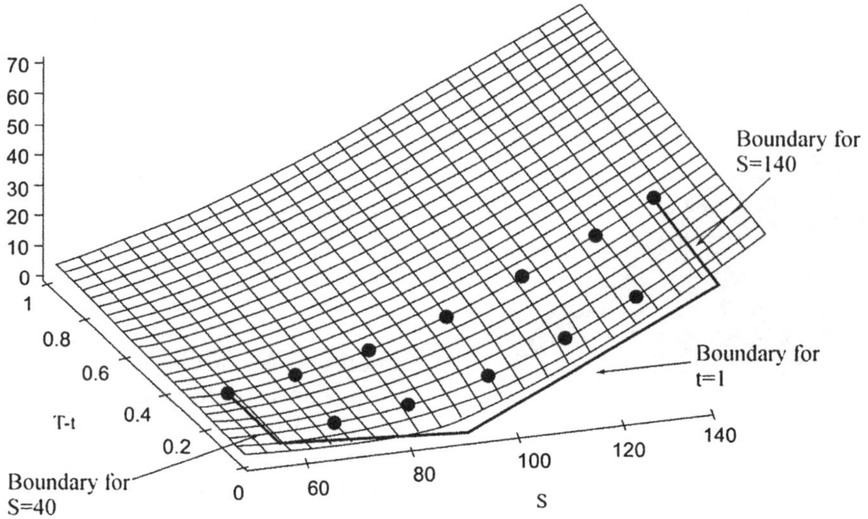

FIGURE 13.5 A Black–Scholes surface.

expire in the money. The resulting call premium is close to zero.

- For $t = T$, we know exactly that

$$F(S_T, T) \max[S_T - K, 0] \tag{13.54}$$

These give the boundary values for F_{ij}. In Figure 13.5, these boundary regions are shown explicitly. Using these boundary values in Eq. (13.30), we can solve for the remaining unknown F_{ij}.

13.7.4 Tips on Numerical Solution of Partial Integro-Differential Equations

We just focus on the integral part of PIDE and give some tips on numerical solution of how to numerically solve it. Here we consider

$$\int_{-\infty}^{\infty} (F(S_{t-}e^y, t) - F(S_{t-}, t))k(y)dy \tag{13.55}$$

We are going to solve numerically for the given Lévy density

$$k(y) = \frac{e^{-\lambda_p y}}{\nu y} \mathbb{1}_{y>0} + \frac{e^{-\lambda_n |y|}}{\nu |y|} \mathbb{1}_{y<0} \tag{13.56}$$

Without loss of generality we just focus on positive y.

$$\int_{0}^{\infty} (F(S_{t-}e^y, t) - F(S_{t-}, t))k(y)dy \tag{13.57}$$

By making the change of variables $x = \ln S$ and calling $\bar{F}(x, t) = F(S, t)$ we have

$$\int_{0}^{\infty} (\bar{F}(x + y, t) - \bar{F}(x, t))k(y)dy \tag{13.58}$$

By making the change of variable $x = \ln S$.

There is a region of interest and that is the region around zero. It is natural to look at the following interval first

$$\int_{0}^{\Delta x} (\bar{F}(x + y, t) - \bar{F}(x, t))k(y)dy \tag{13.59}$$

for some small $\Delta x > 0$. Using Taylor expansion, we can write $\bar{F}(x + y, t)$ as follows:

$$\bar{F}(x + y, t) = \bar{F}(x, t) + y\frac{\partial \bar{F}}{\partial x}(x, t) + \frac{1}{2}y^2\frac{\partial^2 \bar{F}}{\partial x^2}(\xi, t) \tag{13.60}$$

or equivalently

$$\bar{F}(x+y,t)-\bar{F}(x,t)=y\frac{\partial\bar{F}}{\partial x}(x,t)+O(y^2) \quad (13.61)$$

Substituting and we get

$$\int_0^{\Delta x}(\bar{F}(x+y,t)-\bar{F}(x,t))k(y)dy$$

$$=\int_0^{\Delta x}y\frac{\partial\bar{F}}{\partial x}(x,t)k(y)dy \quad (13.62)$$

$$=\frac{\partial\bar{F}}{\partial x}(x,t)\int_0^{\Delta x}y\frac{e^{-\lambda_p y}}{\nu y}dy \quad (13.63)$$

$$=\frac{1}{\nu}\frac{\partial\bar{F}}{\partial x}(x,t)\int_0^{\Delta x}e^{-\lambda_p y}dy \quad (13.64)$$

$$=\frac{1}{\nu\lambda_p}\frac{\partial\bar{F}}{\partial x}(x,t)(1-e^{-\lambda_p\Delta x}) \quad (13.65)$$

Away from zero, we can look at some arbitrary small subinterval. For $y\in(k\Delta x,(k+1)\Delta x)$, for some integer k we are interested in evaluating the following integral

$$\int_{k\Delta x}^{(k+1)\Delta x}(\bar{F}(x+y,t)-\bar{F}(x,t))k(y)dy \quad (13.66)$$

Using linear approximation yields we can write $\bar{F}(x+y,t)$ on $(k\Delta x,(k+1)\Delta x)$ as follows

$$\bar{F}(x+y,t)=\bar{F}(x+k\Delta x,t)+\frac{\bar{F}(x+(k+1)\Delta x,t)-\bar{F}(x+k\Delta x,t)}{\Delta x}\times(y-k\Delta x)+O((\Delta x)^2) \quad (13.67)$$

Substituting it would yield

$$\int_{k\Delta x}^{(k+1)\Delta x}\left(\bar{F}(x+k\Delta x,t)+\frac{\bar{F}(x+(k+1)\Delta x,t)-\bar{F}(x+k\Delta x,t)}{\Delta x}\times(y-k\Delta x)-\bar{F}(x,t)\right)\frac{e^{-\lambda_p y}}{\nu y}dy$$

$$=(\bar{F}(x+k\Delta x,t)-\bar{F}(x,t)-k(\bar{F}(x+(k+1)\Delta x,t)-\bar{F}(x+k\Delta x,t)))\left(\int_{k\Delta x}^{(k+1)\Delta x}\frac{e^{-\lambda_p y}}{\nu y}dy\right)+\frac{\bar{F}(x+(k+1)\Delta x,t)-\bar{F}(x+k\Delta x,t)}{\nu\Delta x}\left(\int_{k\Delta x}^{(k+1)\Delta x}e^{-\lambda_p y}dy\right)$$

$$=\frac{(\bar{F}(x+k\Delta x,t)-\bar{F}(x,t)-k(\bar{F}(x+(k+1)\Delta x,t)-\bar{F}(x+k\Delta x,t)))}{\nu}\times\left(\mathrm{ei}(k\Delta x\lambda_p)-\mathrm{ei}((k+1)\Delta x\lambda_p)\right)+\frac{\bar{F}(x+(k+1)\Delta x,t)-\bar{F}(x+k\Delta x,t)}{\lambda_p\nu\Delta x}\left(e^{-\lambda_p k\Delta x}-e^{-\lambda_p(k+1)\Delta x}\right) \quad (13.68)$$

where

$$\mathrm{ei}(\xi)=\int_{\xi}^{\infty}\frac{e^{-t}}{t}dt \quad (13.69)$$

which is called *exponential integral* and there are programming routines for fast calculation of it.

13.8 CONCLUSIONS

This chapter discussed some examples of PDEs/PIDEs faced in pricing derivative assets. We illustrated the difficulties of introducing a second random element in pricing call options. We also discussed some exotic derivatives and the way PDEs/PIDEs would change.

One important point was the geometry of the Black–Scholes surfaces. We saw that random trajectories for the underlying assets price led to random paths on this surface. This is shown in Figure 13.6.

13.9 REFERENCES

Ingersoll (1987) provides several examples of PDEs from asset pricing. Our treatment of this topic is clearly intended to provide examples for a simple introduction to PDEs. An interested reader should consult other sources if information beyond a simple introduction is needed. One good introduction to PDEs is Betounes (1998). This book illustrates the basics using MAPLE.

13.10 EXERCISES

The exercises in this section prepare the reader for the next three chapters instead of dealing with

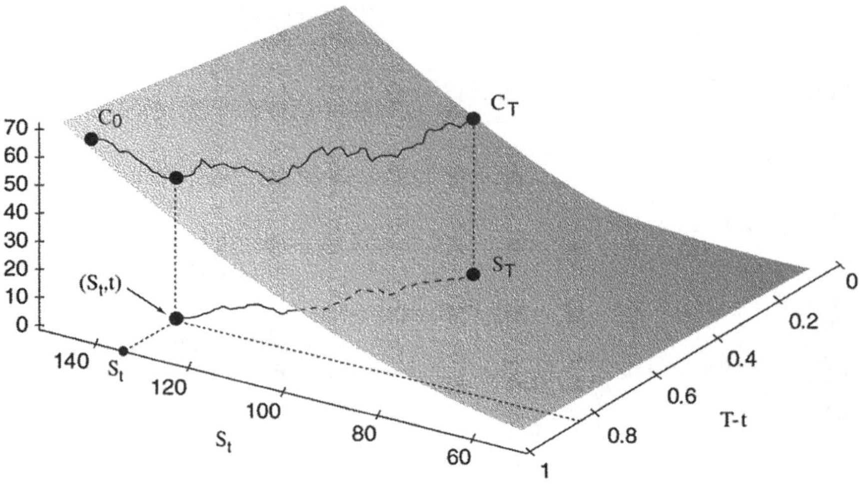

FIGURE 13.6 Random trajectory for the underlying asset price leads to a random path on Black—Scholes surface.

the PDEs. An interested reader will find several useful problems in Betounes (1998).

1. Let X_t be a geometric Wiener process,

$$X_t e^{Y_t}$$

where

$$Y_t \sim \mathcal{N}\left(\mu t, \sigma^2 t\right)$$

   (a) Consider the definition

$$\mathbb{E}\left[X_t \middle| X_s, s < t\right] = \mathbb{E}\left[e^{Y_t} \middle| Y_s \quad, s < t\right]$$

   And the trivial equality

$$\mathbb{E}\left[e^{Y_t} \middle| Y_s, s < t\right] = \mathbb{E}\left[e^{(Y_t - Y_s) + Y_s} \middle| Y_s \,, s < t\right]$$

   Using these, calculate the expectation:

$$\mathbb{E}\left[X_t | X_s, s < t\right]$$

2. This exercise deals with obtaining martingales. Suppose X_t is a geometric process with drift μ and diffusion parameter σ.

   (a) When would the e^{-rtX_t} be a martingale? That is, when would the following equality hold:

$$\mathbb{E}\left[X_t \middle| X_s, s < t\right] = e^{-rt} X_s$$

   (b) More precisely, remember from the previous derivation that

$$\begin{aligned}\mathbb{E}\left[X_t \middle| X_s, s < t\right] &= X_s e^{-rt} e^{\left(\mu + \frac{1}{2}\sigma^2\right)\,(t-s)} \\ &= X_s e^{-rs} e^{-r(t-s)} e^{(\mu + \frac{1}{2}\sigma^2)\,(t-s)}\end{aligned}$$

   or, again,

$$\mathbb{E}\left[X_t \middle|\ X_s, s < t\right] = X_s e^{-rs} e^{\left(\mu + \frac{1}{2}\sigma^2 - r\right)\,(t-s)}$$

   Which selection of μ would make e^{-rtX_t} a martingale? Would μ work?

   (c) How about

$$\mu = r + \sigma^2$$

   (d) Now try:

$$\mu = r - \frac{1}{2}\sigma^2$$

   Note that each one of these selections defines a different distribution for the e^{-rtX_t}.

3. Consider

$$Z_t = e^{-rt} X_t$$

where X_t is an exponential Wiener process:

$$X_t = e^{W_t}$$

(a) Calculate the expected value of the increment $dZ(t)$.
(b) Is Z_t a martingale?
(c) Calculate $\mathbb{E}[Z_t]$. How would you change the definition of X_t to make Z_t a martingale?
(d) How would $\mathbb{E}[Z_t]$ then change?

4. Consider the portfolio $C(s,t) - \Delta S_t$. Assume that the dynamics of $S(t)$ are:

$$dS_t = \mu S_t + \sigma dW_t \tag{13.70}$$

Apply Ito's Lemma to $C(s,t) - \Delta S_t$ to obtain the PDE of this portfolio as a function of dS and dt. Let Δ be equal to the delta of the option, that is: $\Delta = \partial C / \partial S$. Comment on the relationship between the derivative with respect to time, theta, and the second derivative with respect to the underlying, gamma, from the perspective of an investor.

5. Write a Matlab function to price an European call.

CHAPTER

14

Pricing Derivative Products: Equivalent Martingale Measures

OUTLINE

14.1 TRANSLATIONS OF PROBABILITIES

Recent methods of derivative asset pricing do not necessarily exploit PDEs implied by arbitrage-free portfolios. They rest on converting prices of such assets into martingales. This is done through transforming the underlying probability distributions using the tools provided by the Girsanov theorem.

This approach is quite different from the method of PDEs. The tools involved exploit the existence of arbitrage-free portfolios indirectly, and hence are more difficult to visualize.

Financial Derivatives, Third Edition.
http://dx.doi.org/10.1016/B978-0-12-384682-2.00014-1

A student of finance or economics is likely to be even less familiar with this new set of tools than with, say, the PDEs.

This chapter discusses these tools. We adopt a step-by-step approach. First, we review some simple concepts and set the notation. As motivation, we show some simple examples of the way the Girsanov theorem is used. The full theorem is stated next. We follow this with a section dealing with the intuitive explanation of various concepts utilized in the theorem. Finally, the theorem is applied in examples of increasing complexity. Overall, few examples are provided from financial markets. The next chapter deals with that. The purpose of the present chapter is to clarify the notion of transforming underlying probability distributions.

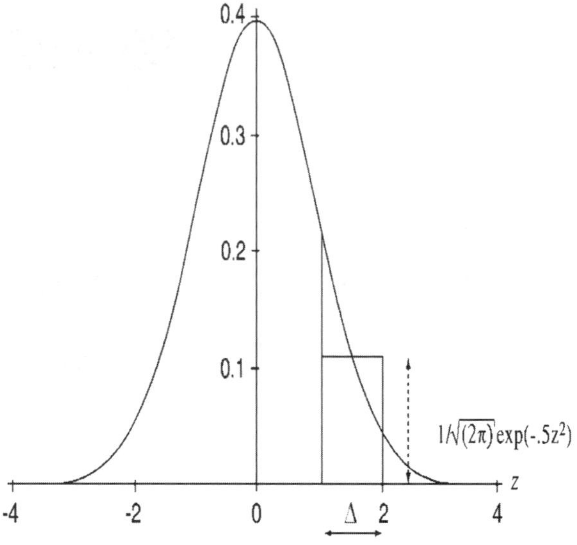

FIGURE 14.1 Calculating the integral of the normal density over a small region using a rectangle.

14.1.1 Probability as "Measure"

Consider a normally distributed random variable z_t at a fixed time t, with zero mean and unit variance. Formally,

$$z_t \sim \mathcal{N}(0,1) \tag{14.1}$$

The probability density $f(z_t)$ of this random variable is given by the well-known expression

$$f(z_t) = \frac{1}{\sqrt{2\pi}} e^{-\frac{z_t^2}{2}} \tag{14.2}$$

Suppose we are interested in the probability that z_t falls *near* a specific value $\bar{z}$. Then, this probability can be expressed by first choosing a small interval $\Delta > 0$, and next by calculating the integral of the normal density over the region in question:

$$\mathbb{P}\left(\bar{z} - \frac{1}{2}\Delta < z_t < \bar{z} + \frac{1}{2}\Delta\right) = \int_{\bar{z}-\frac{1}{2}\Delta}^{\bar{z}+\frac{1}{2}\Delta} \frac{1}{\sqrt{2\pi}} e^{-\frac{z_t^2}{2}} dz_t \tag{14.3}$$

Now, if the region around is $\bar{z}$ small, then $f(z_t)$ will not change very much as z_t varies from $\bar{z} - \frac{1}{2}\Delta$ to $\bar{z} + \frac{1}{2}\Delta$. This means we can approximate $f(z_t)$ by $f(\bar{z})$ during this interval and write the integral on the right-hand side of (14.3) as

$$\int_{\bar{z}-\frac{1}{2}\Delta}^{\bar{z}+\frac{1}{2}\Delta} \frac{1}{\sqrt{2\pi}} e^{-\frac{z_t^2}{2}} dz_t \approx \frac{1}{\sqrt{2\pi}} e^{-\frac{\bar{z}^2}{2}} \int_{\bar{z}-\frac{1}{2}\Delta}^{\bar{z}+\frac{1}{2}\Delta} dz_t \tag{14.4}$$

$$= \frac{1}{\sqrt{2\pi}} e^{-\frac{\bar{z}^2}{2}} \Delta \tag{14.5}$$

This construction is shown in Figure 14.1. The probability in (14.5) is a "mass" represented (approximately) by a rectangle with base Δ and height $f(\bar{z})$.

Visualized this way, probability corresponds to a "measure" that is associated with possible values of z_t in small intervals. Probabilities are called measures because they are mappings from arbitrary sets to nonnegative real numbers $\mathbb{R}^+$. For infinitesimal Δ, which we write as dz_t, these measures are denoted by the symbol $d\mathbb{P}(z_t)$, or simply $d\mathbb{P}$ when there is no confusion about the underlying random variable:

$$d\mathbb{P}(\bar{z}) = \mathbb{P}\left(\bar{z} - \frac{1}{2}dz_t < z < \bar{z} + \frac{1}{2}dz_t\right) \tag{14.6}$$

This can be read as the probability that the random variable z_t will fall within a small interval centered on $\bar{z}$ and of infinitesimal length dz_t. The sum of all such probabilities will then be given by adding these for various values of $\bar{z}$. Formally, this is expressed by the use of the integral

$$\int_{-\infty}^{\infty} d\mathbb{P}(z_t) = 1 \tag{14.7}$$

A similar approach is used for calculating the expected value of z_t,

$$\mathbb{E}[z_t] = \int_{-\infty}^{\infty} z_t d\mathbb{P}(z_t) \tag{14.8}$$

which can be seen as an "average" value of z_t. Geometrically, this determines the *center* of the probability mass. The variance is another weighted average:

$$\mathbb{E}[z_t - \mathbb{E}[z_t]]^2 = \int_{-\infty}^{\infty} [z_t - \mathbb{E}[z_t]]^2 d\mathbb{P}(z_t) \tag{14.9}$$

The variance has a geometric interpretation as well. It gives an indication of how the probability mass spreads around the center.

Accordingly, when we talk about a certain probability measure, $d\mathbb{P}$, we always have in mind a *shape* and a *location* for the density of the random variable.[1]

Under these conditions, we can subject a probability distribution to two types of transformations:

- We can leave the shape of the distribution the same, but move the density to a different location. Figure 14.2 illustrates a case where the normal density that was centered at

$$\mu = -5 \tag{14.10}$$

  is transformed into another normal density, this time centered at zero:

$$\mu = 0 \tag{14.11}$$

- We can also change the shape of the distribution. One way to do this is to increase or decrease the variance of the distribution. This can be accomplished by scaling the original random variable. Figure 14.3 displays a case where the variance of the random variable z_t is reduced from 4 to 1.

Modern methods for pricing derivative assets utilize a novel way of transforming the probability measure $d\mathbb{P}$ so that the mean of a random process z_t changes. The transformation permits treating an asset that carries a positive "risk premium" as if it were risk-free. This chapter deals with this complicated idea.

In the following section, we discuss two different methods of switching means of random variables.

14.2 CHANGING MEANS

Now, fix t and let z_t be a univariate random variable. There are two ways one can change the mean of z_t. In the first case, we operate on the values assumed by z_t. In the second, and counterintuitive, case, we leave the values assumed by z_t unchanged, but instead operate on the probabilities associated with z_t.

Both operations lead to a change in the original mean, while preserving other characteristics of the original random variable. However, while the first method cannot, in general, be used in asset pricing, the second method becomes a very useful tool.

We discuss these methods in detail next. The discussion proceeds within the context of a single random variable, rather than a stochastic process. The more complicated case of a continuous-time process is treated in the section on the Girsanov theorem.

14.2.1 Method 1: Operating on Possible Values

The first and standard method for changing the mean of a random variable is used routinely

[1] In this book we always assume that this density exists. In other settings, the density function of the underlying random variables may not exist.

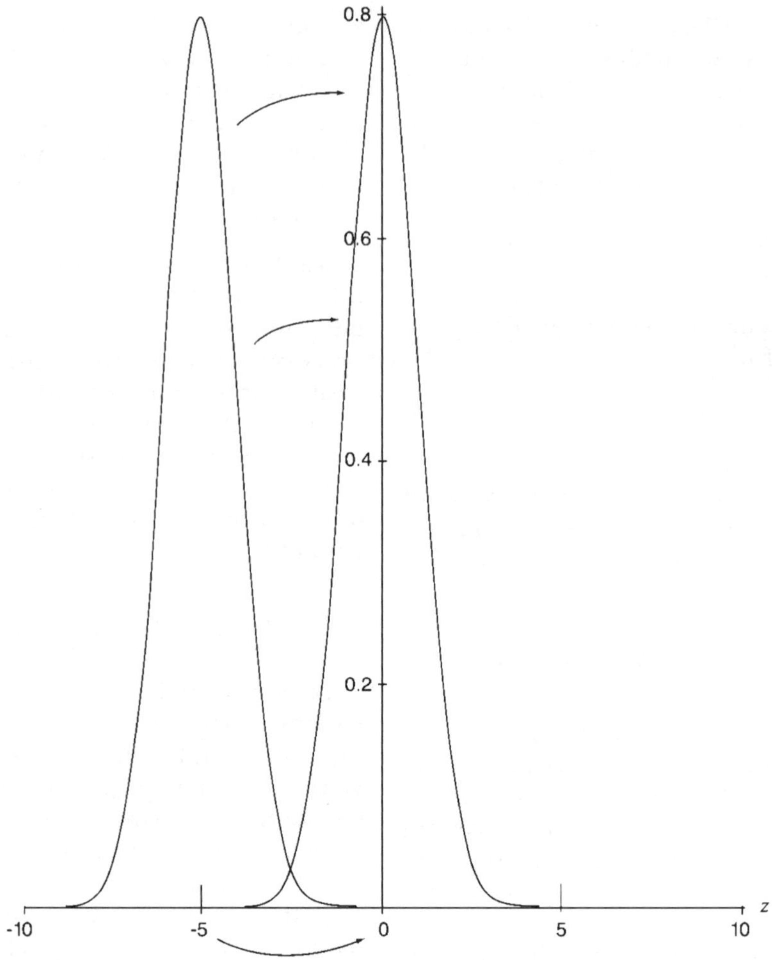

FIGURE 14.2 Transferring a normal density centered at $\mu = -5$ to a normal density centered at $\mu = 0$.

in econometrics and statistics. One simply adds a constant μ to z_t in order to obtain a new random variable $\widetilde{z}_t = z_t + \mu$.[2]

The defined $\widetilde{z}_t$ this way will have a new mean. For example, if originally

$$\mathbb{E}[z_t] = 0 \tag{14.12}$$

then the new random variable will be such that

$$\mathbb{E}[\widetilde{z}_t] = \mathbb{E}[z_t] + \mu = \mu \tag{14.13}$$

[2] μ can be negative.

14.2.1.1 *Example 1*

In spite of the simplicity of this transformation, it is important for later discussions to look at a precise example.

Suppose the random variable Z is defined as follows. A die is rolled and the values of Z are set according to the rule

$$Z = \begin{cases} 10 \text{ roll of 1 or 2} \\ -3 \text{ roll of 3 or 4} \\ -1 \text{ roll of 5 or 6} \end{cases} \tag{14.14}$$

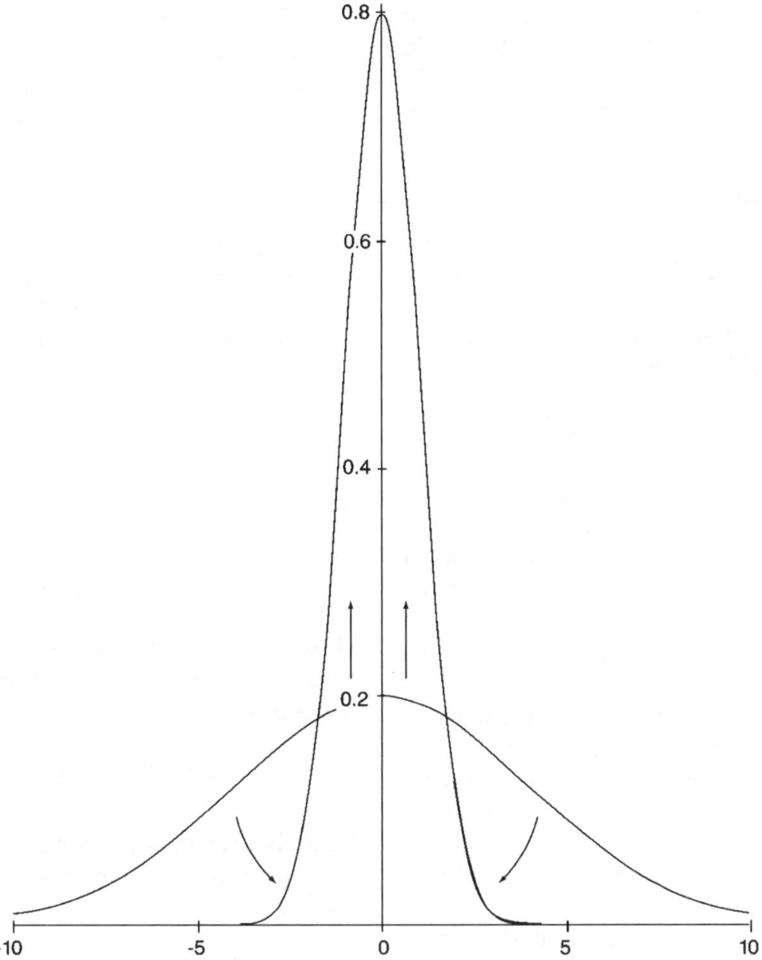

FIGURE 14.3 Changing the shape of a normal distribution by scaling the original one, here the variance of the random variable is reduced from 4 to 1.

Assuming that the probability of getting a particular number is 1/6, we can easily calculate the mean of Z as a weighted average of its possible values:

$$\mathbb{E}[Z] = \frac{1}{3}[10] + \frac{1}{3}[-3] + \frac{1}{3}[-1] \quad (14.15)$$

$$2 \quad (14.16)$$

Now, suppose we would like to change the mean of Z using the method outlined earlier. More precisely, suppose we would like to calculate a new random variable with the same variance but with a mean of one. We call this new random variable $\widetilde{Z}$ and let

$$\widetilde{Z} = Z - 1 \quad (14.17)$$

Using the formula for the mean, we calculate the $\mathbb{E}[\widetilde{Z}]$

$$\mathbb{E}[\widetilde{Z}] = \frac{1}{3}[10-1] + \frac{1}{3}[-3-1] + \frac{1}{3}[-1-1] \quad (14.18)$$

$$1 \quad (14.19)$$

As can be seen from this transformation, in order to change the mean of Z, we operated on the values assumed by Z. Namely, we subtracted 1 from each possible value. The probabilities were not changed.

14.2.1.2 Example 2

We can illustrate this method for changing the means of random variables using a more relevant example from finance.

The yield of a triple-A-rated corporate bond R_t with fixed t will have the expected value

$$\mathbb{E}[R_t] = r_t + \mathbb{E}\left[\text{risk premium}\right] \tag{14.20}$$

where r_t is the known risk-free rate of treasury bonds with comparable maturity, and where $\mathbb{E}[\cdot]$ expresses the expectation over possible states of the world. Let α be the (constant) expected risk premium:

$$\mathbb{E}[R_t] = r_t + \alpha \tag{14.21}$$

Then R_t is a random variable with mean $r_t+\alpha$.

The first method to change the mean of R_t is to add a constant and obtain a new random variable. This random variable will have the mean

$$\mathbb{E}[R_t + \mu] = r_t + \mu + \alpha \tag{14.22}$$

In the case of normally distributed random variables, this is equivalent to preserving the shape of the density, while sliding the center of the distribution to a new location. Figure 14.2 displays an example.

If μ is selected as $-\alpha$, then such a transformation would eliminate the risk premium from R_t. Note that in order to use this method for changing means, we need to know the risk premium α. Only under these conditions could we arrive at subtracting the "right" quantity from R_t and obtain the equivalent risk-free yield.

This example is simple and does not illustrate why somebody might want to go through such a transformation of means to begin with. The next example is more illustrative in this respect.

14.2.1.3 Example 3

The example is discussed in discrete time first. Let $St, t = 1, 2, \ldots$ be the price of some financial asset that pays no dividends. The S_t is observed over discrete times $t, t+1, \ldots$.

Let r_t be the rate of risk-free return. A typical risky asset S_t must offer a rate of return R_t greater than r_t, since otherwise there will be no reason to hold it. This means that, using $\mathbb{E}[\cdot]$, the expectation operator conditional on information available as of time t satisfies

$$\mathbb{E}_t\left[S_{t+1}\right] > (1 + r_t)\, S_t \tag{14.23}$$

That is, on the average, the risky asset will appreciate faster than the growth of a risk-free investment. This equality can be rewritten as

$$\frac{1}{1+r_t}\mathbb{E}_t\left[S_{t+1}\right] > S_t \tag{14.24}$$

Here, the left-hand side represents the expected future price discounted at the risk-free rate. For some $\mu > 0$,

$$\frac{1}{1+r_t}\mathbb{E}_t\left[S_{t+1}\right] = S_t\,(1 + \mu) \tag{14.25}$$

Note that the positive constants μ or $\mu + \mu r_t$ can be interpreted as a risk premium. Transforming (14.25),

$$\frac{\mathbb{E}_t\left[S_{t+1}\right]}{S_t} = (1 + r_t)(1 + \mu) \tag{14.26}$$

The term on the left-hand side of this equation, $\mathbb{E}_t\left[S_{t+1}/S_t\right]$, represents expected gross return, $\mathbb{E}_t[1 + R_t]$. This means that

$$\mathbb{E}_t[1 + R_t] = (1 + r_t)(1 + \mu) \tag{14.27}$$

which says that the expected return of the risky asset must exceed the risk-free return approximately by μ:

$$\mathbb{E}_t[R_t] \approx 1 + r_t \tag{14.28}$$

in the case where r_t and μ are small enough that the cross-product term can be ignored.

Under these conditions, μ is the risk premium for holding the asset for one period, and is the *risk-free discount* factor.

Now consider the problem of a financial analyst who wants to obtain the fair market value of this asset today. That is, the analyst would like to calculate S_t. One way to do this is to exploit the relation

$$\mathbb{E}_t\left[\frac{1}{1+R_t}S_{t+1}\right] = S_t \tag{14.29}$$

by calculating the expectation on the left-hand side.[3]

But doing this requires a knowledge of the distribution of R_t, which requires knowing the risk premium μ.[4] Yet knowing the risk premium before knowing the fair market value S_t is rare. Utilization of (14.29) will go nowhere in terms of calculating S_t.[5]

On the other hand, if one could "transform" the mean of R_t without having to use the μ, the method might work.[6] Another way of transforming the distribution of R_t must be found.

If a new expectation using a different probability distribution $\mathbb{Q}$ yields an expression such as

$$\mathbb{E}_t^{\mathbb{Q}}\left[\frac{1}{1+r_t}S_{t+1}\right] = S_t \tag{14.31}$$

this can be very useful for calculating S_t. In fact, one could exploit this equality by "forecasting" S_{t+1}, using a model that describes the dynamics of S_t, and then discounting the "average forecast" by the (known) r_t. This would provide an estimate of S_t.

What would $\mathbb{E}_t^{\mathbb{Q}}[\cdot]$ and r_t represent in this particular case? r_t will be the risk-free rate. The expectation operator would be given by the risk-neutral probabilities. By making these transformations, we would be eliminating the risk premium from R_t:

$$R_t - \mu = r_t \tag{14.32}$$

The trick here is to accomplish this transformation in the mean without having to use the value of μ explicitly. Even though this seems an impossible task at the outset, the second method for changing means does precisely this.

14.2.2 Method 2: Operating on Probabilities

The second way of changing the mean of a random variable is to leave the random variable "intact," but transform the corresponding *probability measure* that governs z_t. We introduce this method using a series of examples that get more and more complicated. At the end we provide the Girsanov theorem, which extends the method to continuous-time stochastic processes. The idea may be counterintuitive, but is in fact quite simple, as Example 1 will show.

14.2.2.1 Example 1

Consider the first example of the previous section, with Z defined as a function of rolling the die:

$$Z = \begin{cases} 10 & \text{roll of 1 or 2} \\ -3 & \text{roll of 3 or 4} \\ -1 & \text{roll of 5 or 6} \end{cases} \tag{14.33}$$

with a previously calculated mean of

$$\mathbb{E}[Z] = 2 \tag{14.34}$$

[3] This relation is just the definition of the yield R_t. If we discount the next period's price by $1 + R_t$, we naturally recover today's value.

[4] Only by knowing μ can the mean of R_t be calculated, and the distribution of R_t be pinned down.

[5] There is an additional difficulty. The term on the left-hand side of (14.29) is a nonlinear function of R_t. Hence, we cannot simply move the expectation operator in front of R_t:

$$\mathbb{E}\left[\frac{1}{1+R_t}S_{t+1}\right] \neq \left[\frac{1}{1+\mathbb{E}[R]_t}S_{t+1}\right] \tag{14.30}$$

This further complicates the calculations.

[6] Because the mean of the distribution of R_t could be made equal to r_t.

and a variance

$$\mathbb{V}[Z] = \mathbb{E}[Z - \mathbb{E}[Z]]^2 = \frac{1}{3}[10-2]^2 + \frac{1}{3}[-3-2]^2 + \frac{1}{3}[-1-2]^2 = \frac{98}{3} \quad (14.35)$$

Suppose we want to transform this random variable so that its mean becomes one, while leaving the variance unchanged.

Consider the following transformation of the original probabilities associated with rolling the die:

$$\mathbb{P}\left(\text{getting a 1 or 2}\right) = \frac{1}{3} \to \mathbb{Q}\left(\text{getting a 1 or 2}\right) = \frac{122}{429} \quad (14.36)$$

$$\mathbb{P}\left(\text{getting a 3 or 4}\right) = \frac{1}{3} \to \mathbb{Q}\left(\text{getting a 3 or 4}\right) = \frac{22}{39} \quad (14.37)$$

$$\mathbb{P}\left(\text{getting a 5 or 6}\right) = \frac{1}{3} \to \mathbb{Q}\left(\text{getting a 5 or 6}\right) = \frac{5}{33} \quad (14.38)$$

Note that the new probabilities are designated by $\mathbb{Q}$.

Now calculate the mean under these new probabilities:

$$\mathbb{E}^{\mathbb{Q}}[Z] = \left[\frac{122}{429}\right][10] + \frac{22}{39}[-3] + \frac{5}{33}[-1] = 1 \quad (14.39)$$

The mean is indeed one. Calculate the variance:

$$\mathbb{E}^{\mathbb{Q}}[Z]^2 = \left[\frac{122}{429}\right][10-1]^2 + \frac{22}{39}[-3-1]^2 + \frac{5}{33}[-1-1]^2 = \frac{98}{3} \quad (14.40)$$

The variance has not changed. The transformation of probabilities shown in (14.38) accomplishes exactly what the first method did. Yet this second method operated on the probability measure $\mathbb{P}(Z)$, rather than on the values of Z itself.

It is worth emphasizing that these new probabilities do not relate to the "true" odds of the experiment. The "true" probabilities associated with rolling the die are still given by the original numbers, $\mathbb{P}$.

The reader may have noticed the notation we adopted. In fact, we need to write the new expectation operator as $\mathbb{E}^{\mathbb{Q}}[\cdot]$ rather than $\mathbb{E}[\cdot]$. The probabilities used in calculating the averages are no longer the same as $\mathbb{P}$, and the use of $\mathbb{E}[\cdot]$ will be misleading. When this method is used, special care should be given to designating the probability distribution utilized in calculating expectations under consideration. [7]

14.3 THE GIRSANOV THEOREM

The examples just discussed were clearly simplified. First, we dealt with random variables that were allowed to assume a finite number of values—the state space was finite. Second, we dealt with a single random variable instead of using a random process.

The Girsanov theorem provides the general framework for transforming one probability measure into another "equivalent" measure in more complicated cases. The theorem covers the case of Brownian motion. Hence, the state space is continuous, and the transformations are extended to continuous-time stochastic processes.

The probabilities so transformed are called "equivalent" because, as we will see in more detail later in this chapter, they assign positive probabilities to the same domains. Thus, although the two probability distributions are

[7] Some readers may wonder how we found the new probabilities. In this particular case, it was easy. We considered the probabilities as unknowns and used three conditions to solve them. The first condition is that the probabilities sum to one. The second is that the new mean is one. The third is that the variance equals 98/3.

different, with appropriate transformations one can always recover one measure from the other. Since such recoveries are always possible, we may want to use the "convenient" distribution for our calculations, and then, if desired, switch back to the original distribution.

Accordingly, if we have to calculate an expectation, and if this expectation is easier to calculate with an equivalent measure, then it may be worth switching probabilities, although the new measure may not be the one that governs the true states of nature. After all, the purpose is not to make a statement about the odds of various states of nature. The purpose is to calculate a quantity in a convenient fashion.

The general method can be summarized as follows: (1) We have an expectation to calculate. (2) We transform the original probability measure so that the expectation becomes easier to calculate. (3) We calculate the expectation under the new probability. (4) Once the result is calculated and, if desired, we transform this probability back to the original distribution.

We now discuss such probability transformations in more complex settings. The Girsanov theorem will be introduced using special cases with growing complexity. Then we provide the general theorem and discuss its assumptions and implications.

14.3.1 A Normally Distributed Random Variable

Fix t and consider a normally distributed random variable z_t:

$$z_t \sim \mathcal{N}(0,1) \tag{14.41}$$

Denote the density function of z_t by $f(z_t)$ and the implied probability measure by $\mathbb{P}$ such that

$$d\mathbb{P}(z_t) = \frac{1}{\sqrt{2\pi}} e^{-\frac{1}{2}(z_t)^2} dz_t \tag{14.42}$$

In this example, the state space is continuous, although we are still working with a single random variable, instead of a random process.

Next, define the function

$$\xi(z_t) = e^{z_t \mu - \frac{1}{2}\mu^2} \tag{14.43}$$

When we multiply $\xi(z_t)$ by $d\mathbb{P}(z_t)$, we obtain a new probability. This can be seen from the following:

$$\left[d\mathbb{P}(z_t)\right][\xi(z_t)] = \frac{1}{\sqrt{2\pi}} e^{-\frac{1}{2}(z_t)^2 + \mu z_t - \frac{1}{2}(\mu)^2} dz_t \tag{14.44}$$

After grouping the terms in the exponent, we obtain the expression

$$d\mathbb{Q}(z_t) = \frac{1}{\sqrt{2\pi}} e^{-\frac{1}{2}(z_t - \mu)^2} dz_t \tag{14.45}$$

Clearly $d\mathbb{Q}(z_t)$ is a *new* probability measure, defined by

$$d\mathbb{Q}(z_t) = d\mathbb{P}(z_t)\,\xi(z_t) \tag{14.46}$$

By simply reading from the density in (14.45), we see that is the probability associated with a normally distributed random variable mean μ and variance 1.

It turns out that by multiplying $d\mathbb{P}(z_t)$ by the function $\xi(z_t)$, and then switching to $\mathbb{Q}$, we succeeded in changing the mean of z_t. Note that in this particular case, the multiplication by $\xi(z_t)$ preserved the shape of the probability measure. In fact, (14.45) is still a bell-shaped, Gaussian curve with the same variance. But $\mathbb{P}(z_t)$ and $\mathbb{Q}(z_t)$ are different measures. They have different means and they assign different weights to intervals on the z-axis.

Under the measure $\mathbb{P}(z_t)$, the random variable z_t has mean zero, $\mathbb{E}^{\mathbb{P}}[z_t] = 0$, and variance $\mathbb{E}^{\mathbb{P}}\left[z_t^2\right] = 1$. However, under the new probability measure $\mathbb{Q}(z_t), z_t$ has mean $\mathbb{E}^{\mathbb{Q}}[z_t] = \mu$. The variance is unchanged.

What we have just shown is that there exists a function $\xi(z_t)$ such that if we multiply a probability measure by this function, we get a new probability. The resulting random variable is again normal but has a different mean.

Finally, the transformation of measures,

$$d\mathbb{Q}(z_t) = \xi(z_t)\, d\mathbb{P}(z_t) \tag{14.47}$$

which changed the mean of the random variable z_t, is reversible:

$$\xi(z_t)^{-1}\, d\mathbb{Q}(z_t) = d\mathbb{P}(z_t) \tag{14.48}$$

The transformation leaves the variance of z_t unchanged, and is unique, given μ and σ.

We can now summarize the two methods of changing means[8]:

- Method 1: Subtraction of means. Given a random variable

$$Z \sim \mathcal{N}(\mu, 1) \tag{14.49}$$

define a new random variable $\widetilde{Z}$ by transforming Z:

$$\widetilde{Z} = \frac{Z - \mu}{1} \sim \mathcal{N}(0, 1) \tag{14.50}$$

Then $\widetilde{Z}$ will have a zero mean.
- Method 2: Using equivalent measures. Given a random variable Z with probability $\mathbb{P}$,

$$Z \sim \mathbb{P} = \mathcal{N}(\mu, 1) \tag{14.51}$$

transform the probabilities $d\mathbb{P}$ through multiplication by $\xi(Z)$ and obtain a new probability $\mathbb{Q}$ such that

$$Z \sim \mathbb{Q} = \mathcal{N}(0, 1) \tag{14.52}$$

The next question is whether we can accomplish the same transformations if we are given a sequence of normally distributed random variables, $z_1, z_2, \ldots, z_t$.

14.3.2 A Normally Distributed Vector

The previous example showed how the mean of a normally distributed random variable could be changed by multiplying the corresponding probability measure by a function $\xi(z_t)$. The transformed measure was shown to be another probability that assigned a different mean to z_t, although the variance remained the same.

Can we proceed in a similar way if we are given a vector of normally distributed variables?

The answer is yes. For simplicity, we show the bivariate case. Extension to an n-variate Gaussian vector is analogous.

With fixed t, suppose we are given the random variables z_{1t}, z_{2t}, *jointly* distributed as normal. The corresponding density will be

$$f(z_{1t}, z_{2t}) = \frac{1}{2\pi\sqrt{|\Omega|}} e^{-\frac{1}{2}[(z_{1t}-\mu),(z_{2t}-\mu)]'\Omega^{-1}[(z_{1t}-\mu),(z_{2t}-\mu)]} \tag{14.53}$$

where Ω is the variance covariance matrix of $[(z_{1t} - \mu), (z_{2t} - \mu)]$,

$$\Omega = \begin{bmatrix} \sigma_1^2 & \sigma_{12} \\ \sigma_{12} & \sigma_2^2 \end{bmatrix} \tag{14.54}$$

with $\sigma_i^2, i = 1, 2$, denoting the variances and σ_{12} the covariance between z_{1t}, z_{2t}. The $|\cdot|$ represents the determinant:

$$|\Omega| = \sigma_1^2\sigma_2^2 - \sigma_{12}^2 \tag{14.55}$$

Finally, μ_1, μ_2 are the means corresponding to z_{1t} and z_{2t}.

The joint probability measure can be defined using

$$d\mathbb{P}(z_{1t}, z_{2t}) = f(z_{1t}, z_{2t})\, dz_{1t} dz_{2t} \tag{14.56}$$

This expression is the probability mass associated with a small rectangle dz_{1t}, dz_{2t} centered at a particular value for the pair z_{1t}, z_{2t}. It gives the probability that z_{1t}, z_{2t} will fall in that particular rectangle jointly. Hence the term joint density function.

Suppose we want to change the means of z_{1t}, z_{2t} from μ_1, μ_2 to zero, while leaving the variances unchanged. Can we accomplish this by transforming the probability $d\mathbb{P}(z_{1t}, z_{2t})$

[8]We simplify the notation slightly.

just as in the previous example, namely, by multiplying by a function $\xi(z_{1t}, z_{2t})$?

The answer is yes. Consider the function defined by

$$\xi(z_{1t}, z_{2t}) = e^{-\frac{1}{2}[z_{1t}, z_{2t}]'\Omega^{-1}[\mu_1, \mu_1] + \frac{1}{2}[\mu_1, \mu_1]'\Omega^{-1}[\mu_1, \mu_1]} \tag{14.57}$$

Using this, we can define a new probability measure $\mathbb{Q}(z_{1t}, z_{2t})$ by

$$d\mathbb{Q}(z_{1t}, z_{2t}) = \boldsymbol{\xi}(z_{1t}, z_{2t})\, d\mathbb{P}(z_{1t}, z_{2t}) \tag{14.58}$$

(z_{1t}, z_{2t}) can be obtained by multiplying expression (14.53) by $\boldsymbol{\xi}(z_{1t}, z_{2t})$, shown in (14.57). The product of these two expressions gives

$$d\mathbb{Q}(z_{1t}, z_{2t}) = \frac{1}{2\pi\sqrt{|\Omega|}} e^{-\frac{1}{2}[z_{1t}, z_{2t}]'\Omega^{-1}[z_{1t}, z_{2t}]} dz_{1t} dz_{2t} \tag{14.59}$$

We recognize this as the bivariate normal probability distribution for a random vector $[z_{1t}, z_{2t}]$ with mean zero and variance–covariance matrix Ω. The multiplication by $\xi(z_{1t}, z_{2t})$ accomplished the stated objective. The nonzero mean of the bivariate vector was eliminated through a transformation of the underlying probabilities.

This example dealt with a bivariate random vector. Exactly the same transformation can be applied if instead we have a random sequence of k normally distributed random variables, $[z_{1t}, z_{2t}, \ldots z_{kt}]$. Only the orders of the corresponding vectors and matrices in (14.53) need to be changed, with similar adjustments in (14.57).

14.3.2.1 A Note

With future discussion in mind, we would like to emphasize one regularity that the reader may already have observed.

Think of z_t as representing a vector of length k, or simply as a univariate random variable. In transforming the probability measures $\mathbb{P}(z_t)$ into $\mathbb{Q}(z_t)$, the function $\xi(z_t)$ was utilized. This function had the following structure,

$$\xi(z_t) = e^{-\frac{1}{2}z_t'\Omega^{-1}\mu + \frac{1}{2}\mu'\Omega^{-1}\mu} \tag{14.60}$$

which in the scalar case became

$$\xi(z_t) = e^{-\frac{1}{2}\frac{z_t\mu}{\sigma^2} + \frac{1}{2}\frac{\mu^2}{\sigma^2}} \tag{14.61}$$

We will now discuss where this functional form comes from. In normal distributions, the parameter μ, which represents the mean, shows up only as an exponent of e. What is more, this exponent is in the form of a square:

$$-\frac{1}{2}\frac{(z_t - \mu)^2}{\sigma^2} \tag{14.62}$$

In order to convert this expression into

$$-\frac{1}{2}\frac{z_t^2}{\sigma^2} \tag{14.63}$$

we need to add

$$-\frac{z_t\mu + 1/2\mu^2}{\sigma^2} \tag{14.64}$$

This is what determines the functional form of $\xi(z_t)$. Multiplying the original probability measure by $\xi(z_t)$ accomplishes this transformation in the exponent of e.

Given this, a reader may wonder if we could attach a deeper interpretation to what the $\xi(z_t)$ really represents. The next section discusses this point.

14.3.3 The Radon–Nikodym Derivative

Consider again the function $\xi(z_t)$ with $\sigma = 1$[9]:

$$\xi(z_t) = e^{-\mu z_t + \frac{1}{2}\mu^2} \tag{14.67}$$

[9]Incidentally, the function

$$\xi(z_t) = e^{-\mu z_t + \frac{1}{2}\mu^2} \tag{14.65}$$

subtracts a mean from z_t, whereas the function

$$\xi(z_t)^{-1} = e^{\mu z_t - \frac{1}{2}\mu^2} \tag{14.66}$$

would *add* a mean μ to a z with an original mean of zero.

We used the $\xi(z_t)$ in obtaining the new probability measure $d\mathbb{Q}(z_t)$ from $d\mathbb{P}(z_t)$:

$$d\mathbb{Q}(z_t) = d\xi(z_t)d\mathbb{P}(z_t) \tag{14.68}$$

Or, dividing both sides by $d\mathbb{P}(z_t)$,

$$\frac{d\mathbb{Q}(z_t)}{d\mathbb{P}(z_t)} = d\xi(z_t) \tag{14.69}$$

This expression can be regarded as a derivative. It reads as if the "derivative" of the measure with respect to $\mathbb{P}$ is given by $\xi(z_t)$. Such derivatives are called Radon–Nikodym derivatives, and $\xi(z_t)$ can be regarded as the density of the probability measure $\mathbb{Q}$ with respect to the measure $\mathbb{P}$.

According to this, if the Radon–Nikodym derivative of with respect to $\mathbb{P}$ exists, then we can use the resulting density $\xi(z_t)$ to transform the mean of z_t by leaving its variance structure unchanged.

Clearly, such a transformation is very useful for a financial market participant, because the risk premiums of asset prices can be "eliminated" while leaving the volatility structure intact. In the case of options, for example, the option price does not depend on the mean growth of the underlying asset price, whereas the volatility of the latter is a fundamental determinant. In such circumstances, transforming original probability distributions using $\xi(z_t)$ would be very convenient.

In Figure 14.4, we show one example of this function $\xi(z_t)$.

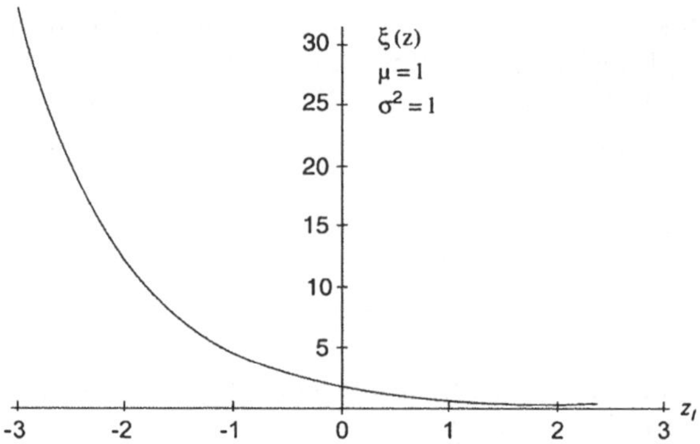

FIGURE 14.4 Plot of a Randon–Nikodym derivative.

14.3.4 Equivalent Measures

When would the Radon–Nikodym derivative,

$$\frac{d\mathbb{Q}(z_t)}{d\mathbb{P}(z_t)} = d\xi(z_t) \tag{14.70}$$

exist? That is, when would we be able to perform transformations such as

$$d\mathbb{Q}(z_t) = d\xi(z_t)d\mathbb{P}(z_t) \tag{14.71}$$

In heuristic terms, note that in order to write the ratio

$$\frac{d\mathbb{Q}(z_t)}{d\mathbb{P}(z_t)} \tag{14.72}$$

meaningfully, we need the probability mass in the denominator to be different from zero. To perform the inverse transformation, we need the numerator to be different from zero. But the numerator and the denominator are probabilities assigned to infinitesimal intervals dz. Hence, in order for the Radon–Nikodym derivative to exist, when $\mathbb{Q}$ assigns a nonzero probability to dz, so must $\mathbb{P}$, and vice versa. In other words:

Condition: Given an interval dz_t, the probabilities $\mathbb{P}$ and $\mathbb{Q}$ satisfy

$$\mathbb{Q}(dz_t) > 0 \quad \text{if and only if} \quad \mathbb{P}(dz_t) > 0 \tag{14.73}$$

If this condition is satisfied, then $\xi(z_t)$ would exist, and we can always go back and forth between the two measures; and $\mathbb{P}$ using the relations

$$d\mathbb{Q}(z_t) = \xi(z_t)\, d\mathbb{P}(z_t) \tag{14.74}$$

and

$$d\mathbb{P}(z_t) = \xi(z_t)^{-1}\, d\mathbb{Q}(z_t) \tag{14.75}$$

This means that for all practical purposes, the two measures are equivalent.

Hence, they are called equivalent probability measures.

14.4 STATEMENT OF THE GIRSANOV THEOREM

In applications of continuous-time finance, the examples provided thus far will be of limited use. Continuous-time finance deals with continuous or right continuous stochastic processes, whereas the transformations thus far involved only a *finite* sequence of random variables. The Girsanov theorem provides the conditions under which the Radon–Nikodym derivative $\xi(z_t)$ exists for cases where z_t is a continuous stochastic process. Transformations of probability measures in continuous finance use this theorem.

We first state the formal version of the Girsanov theorem. A motivating discussion follows afterwards.

The setting of the Girsanov theorem is the following. We are given a family of information sets $\{I_t\}$ over a period $[0,T]$. T is finite.[10]

Over this interval, we define a random process ξ_t:

$$\xi_t = e^{\left(\int_0^t X_u dW_u - \frac{1}{2}\int_0^t X_u^2 du\right)} \quad t \in [0,\infty) \tag{14.76}$$

where X_t is an I_t-measurable process.[11] The W_t is a Wiener process with probability distribution $\mathbb{P}$.

We impose an additional condition on X_t; X_t should not vary "too much":

$$\mathbb{E}\left[e^{\int_0^t X_u^2 du}\right] < \infty \quad t \in [0,\infty) \tag{14.77}$$

This means that X_t cannot increase (or decrease) rapidly over time. Equation (14.77) is known as the Novikov condition.

In continuous time, the density ξ_t has a "new" property that turns out to be very important. It turns out that if the Novikov condition is satisfied, then ξ_t will be a square integrable martingale. We first show this explicitly.

[10]Note that this is not a very serious restriction in the case of financial derivatives. Almost all financial derivatives have finite expiration dates. Often, the maturity of the derivative instrument is less than 1 year.

[11]That is, given the information set I_t, the value of X_t is known exactly.

Using Ito's lemma, calculate the differential

$$d\xi_t = \left[e^{\left(\int_0^t X_u dW_u - \frac{1}{2}\int_0^t X_u^2 du\right)}\right]\left[X_t dW_t\right] \tag{14.78}$$

which reduces to

$$d\xi_t = \xi_t X_t dW_t \tag{14.79}$$

Also, we see by simple substitution of $t = 0$ in (14.76),

$$\xi_0 = 1 \tag{14.80}$$

Thus, by taking the stochastic integral of (14.79), we obtain

$$\xi_t = 1 + \int_0^t \xi_s X_s dW_s \tag{14.81}$$

But the term

$$\int_0^t \xi_s X_s dW_s \tag{14.82}$$

is a stochastic integral with respect to a Wiener process. Also, the term $\xi_s X_s$ is I_t-adapted and does not move rapidly. All these imply, as shown in Chapter 6, that the integral is a (square integrable) martingale,

$$\mathbb{E}\left[\int_0^t \xi_s X_s dW_s \,\middle|\, I_u\right] = \int_0^u \xi_s X_s dW_s \tag{14.83}$$

where $u < t$. Due to (14.81), this implies that ξ_t is a (square integrable) martingale.

We are now ready to state the Girsanov theorem.

Theorem 6. *If the process ξ_t defined by (14.76) is a martingale with respect to information sets I_t, and the probability $\mathbb{P}$, then W_t^*, defined by*

$$W_t^* = W_t - \int_0^t X_u dW_u \quad t \in [0,\infty) \tag{14.84}$$

is a Wiener process with respect to I_t and with respect to the probability measure $\mathbb{Q}$ given by

$$\mathbb{Q}(A) = \mathbb{E}^{\mathbb{P}}\left(1_A \xi_T\right) \tag{14.85}$$

with A being an event determined by I_T and 1_A being the indicator function of the event.

In heuristic terms, this theorem states that if we are given a Wiener process W_t, then, multiplying the probability distribution of this process by ξ_t, we can obtain a new Wiener process with probability distribution $\mathbb{Q}$. The two processes are related to each other through

$$dW_t^* = dW_t - X_t dt \tag{14.86}$$

That is, it is obtained by subtracting an I_t-adapted drift from W_t.

The main condition for performing such transformations is that ξ_t is a martingale with $\mathbb{E}[\xi_T] = 1$.

We now discuss the notation and the assumptions of the Girsanov theorem in detail. The proof of the theorem can be found in Liptser and Shiryayev (1977).

14.5 A DISCUSSION OF THE GIRSANOV THEOREM

In this section, we go over the notation and assumptions used in the Girsanov theorem systematically, and relate them to previously discussed examples. We also show their relevance to concepts in financial models.

We begin with the function ξ_t:

$$\xi_t = e^{\frac{1}{\sigma^2}\left[\int_0^t X_u dW_u - \frac{1}{2}\int_0^t X_u^2 du\right]} \tag{14.87}$$

where we explicitly factored out the (constant) σ^2 term from the integrals. Alternatively, this term can be incorporated in X_u.

Suppose the X_u was constant and equaled μ:

$$X_u = \mu \tag{14.88}$$

Then, taking the integrals in the exponent in a straightforward fashion, and remembering that $W_0 = 0$,

$$\xi_t = e^{\frac{1}{\sigma^2}\left[\mu W_t - \frac{1}{2}\mu^2 t\right]} \tag{14.89}$$

which is similar to the $\xi(z_t)$ discussed earlier. This shows the following important points:

1. The symbol X_t used in the Girsanov theorem plays the same role μ played in simpler settings. It measures how much the original "mean" will be changed.
2. In earlier examples, μ was time independent. Here, X_t may depend on any random quantity, as long as this random quantity is known by time t. That is the meaning of making X_t I_t-adapted. Hence, much more complicated drift transformations are allowed for.
3. The ξ_t is a martingale with $\mathbb{E}[\xi_t] = 1$.

Next, consider the Wiener process. There is something counterintuitive about this process. It turns out that *both* W_t^* and W_t are standard Wiener processes. Thus, they do not have any drift. Yet they relate to each other by

$$dW_t^* = dW_t - X_t dt \tag{14.90}$$

which means that at least *one* of these processes must have nonzero drift if X_t is not identical to zero. How can we explain this seemingly contradictory point?

The point is, W_t^* has zero drift under $\mathbb{Q}$, whereas W_t has zero drift under $\mathbb{P}$, hence can be used to represent unpredictable errors in dynamic systems *given that* we switch the probability measures from $\mathbb{P}$ to $\mathbb{Q}$.

Also, because contains a term $-X_t dt$, using it as an error term in lieu of W_t would reduce the drift of the original SDE under consideration exactly by $-X_t dt$. If the X_t is interpreted as the time-dependent risk premium, the transformation would make all risky assets grow at a risk-free rate.

Finally, consider the relation

$$\mathbb{Q}(A) = \mathbb{E}^{\mathbb{P}}(1_A \xi_T) \tag{14.91}$$

What is the meaning of 1_A? How can we motivate this relation?

1_A is simply a function that has value 1 if A occurs. In fact, we can rewrite the preceding equation as

$$\mathbb{Q}(A) = \mathbb{E}^{\mathbb{P}}(1_A \xi_T) = \int_A \xi_T d\mathbb{P} \tag{14.92}$$

In the case where A is an infinitesimal interval, this means

$$d\mathbb{Q} = \xi_T d\mathbb{P} \tag{14.93}$$

which is similar to the probability transformations seen earlier in much simpler settings.

14.5.1 Application to SDEs

We give a heuristic example.

Let dS_t denote incremental changes in a stock price. Assume that these changes are driven by infinitesimal shocks that have a normal distribution, so that we can represent S_t using the stochastic differential equation driven by the Wiener process W_t

$$dS_t = \mu dt + \sigma dW_t \quad t \in [0, \infty) \tag{14.94}$$

with $W_0 = 0$.[12] The W_t is assumed to have the probability distribution $\mathbb{P}$, with

$$d\mathbb{P}(W_t) = \frac{1}{\sqrt{2\pi t}} e^{-\frac{1}{2t}(W_t)^2} dW_t \tag{14.95}$$

Clearly, S_t cannot be a martingale if the drift term μdt is nonzero. Recall that

$$S_t = \mu \int_0^t ds + \sigma \int_0^t dW_s \quad t \in [0, \infty) \tag{14.96}$$

with $S_0 = 0$. Or,

$$S_t = \mu t + \sigma W_t \tag{14.97}$$

We can write

$$\begin{aligned} \mathbb{E}[S_{t+s} | S_t] &= \mu (t+s) + \sigma \mathbb{E}[W_{t+s} - E_t | S_t] \\ &\quad + \sigma W_t \qquad (14.98) \\ &= S_t + \mu s \qquad (14.99) \end{aligned}$$

since $[W_{t+s} - W_t]$ is unpredictable, given S_t. Thus, for $\mu > 0, s > 0$:

$$\mathbb{E}[S_{t+s} | S_t] > S_t \tag{14.100}$$

S_t is not a martingale.

[12]This formulation again permits negative prices at positive probability. We use it because it is notationally convenient. In any case, the geometric SDE will be dealt with in the next chapter.

Yet we can *easily* convert S_t into a martingale by eliminating its drift.

One method, discussed earlier, was to subtract an appropriate mean from S_t and define

$$S_t^* = S_t - \mu t \tag{14.101}$$

Then S_t^* will be a martingale.

One disadvantage of this transformation is that in order to obtain S_t^*, one would need to know μ. But μ incorporates any risk premium that the risky stock return has. In general, such risk premiums are not known before one finds the fair market value of the asset.

The second method to convert S_t into a martingale is much more promising. Using the Girsanov theorem, we could easily switch to an equivalent measure $\mathbb{Q}$, so that the drift of S_t is zero.

To do this, we have to come up with a function $\xi(S_t)$, and multiply it by the original probability measure associated with S_t. S_t may be a submartingale under $\mathbb{P}$,

$$\mathbb{E}^{\mathbb{P}}[S_{t+s} | S_t] > S \tag{14.102}$$

but it will be a martingale under $\mathbb{Q}$:

$$\mathbb{E}^{\mathbb{Q}}[S_{t+s} | S_t] = S \tag{14.103}$$

As usual, the superscript of the $\mathbb{E}[\cdot|\cdot]$ operator represents the probability measure used to evaluate the expectation.

In order to perform this transformation, a $\xi(S_t)$ function needs to be calculated. First, recall that the density of S_t is given by

$$f_s = \frac{1}{\sqrt{2\pi\sigma^2 t}} e^{-\frac{1}{2t\sigma^2}[S_t - \mu t]^2} \tag{14.104}$$

This defines the probability measure $\mathbb{P}$.

We would like to switch to a new probability $\mathbb{Q}$ such that under $\mathbb{Q}$, S_t becomes a martingale.

Define

$$\xi(S_t) = e^{-\frac{1}{\sigma_2}\left[\mu S_t - \frac{1}{2}\mu^2 t\right]} \tag{14.105}$$

Multiply f_s by this $\xi(S_t)$ to get

$$\begin{aligned} d\mathbb{Q}(S_t) &= \xi(S_t)\,d\mathbb{P}(S_t) \\ &= e^{-\frac{1}{\sigma^2}\left[\mu S_t - \frac{1}{2}\mu^2 t\right]} \frac{1}{\sqrt{2\pi\sigma^2 t}} e^{-\frac{1}{2t\sigma^2}[S_t-\mu t]^2} dS_t \end{aligned} \tag{14.106}$$

Or, rearranging the exponents

$$= \frac{1}{\sqrt{2\pi\sigma^2 t}} e^{-\frac{S_t^2}{2t\sigma^2}} dS_t \tag{14.107}$$

But this is a probability measure associated with a normally distributed process with zero drift and diffusion σ. That means we can write the increments of S_t in terms of a new driving term:

$$dS_t = \sigma dW_t^* \tag{14.108}$$

Such an S_t process was shown to be a martingale. The Weiner process is defined with respect to probability $\mathbb{Q}$.

14.6 WHICH PROBABILITIES?

The role played by the synthetic probabilities $\mathbb{Q}$ appears central to pricing of financial securities even at this level of discussion. According to the discussion in Chapter 2, under the condition of no-arbitrage and in a discrete time setting, the "fair" price of any security that trades in liquid markets will be given by the martingale equality:

$$S_t = \mathbb{E}^{\mathbb{Q}}[D_t S_T] \tag{14.109}$$

where $t < T$ and the D_t is a discount factor, known or random, depending on the normalization adopted. In case there are no foreign currencies or payouts, and in case savings account normalization is utilized, D_t will be a function of the risk-free rate r_t. If $r_s = r$ is constant, the D_t will be known and will factor out of the expectation operator.

The fact that the D_t and the probability $\mathbb{Q}$ are known makes Eq. (14.109) a very useful analytical tool, because for all derivative assets there will exist an expiration time T, such that the dependence of the derivative asset's price, C_T, on S_T is contractually specified. Hence, using $C_t = F(S_t, t)$ we can write:

$$\begin{aligned} C_t &= \mathbb{E}^{\mathbb{Q}}[D_t C_T] \\ &= \mathbb{E}^{\mathbb{Q}}[D_t F(S_T, T)] \end{aligned}$$

with *known* $F(\cdot)$. A market practitioner would then need to take the following "straightforward" steps in order to *price* the derivative contract:

- First, the probability distribution $\mathbb{Q}$ needs to be selected. This is, in general, done indirectly by selecting the first- and second-order moments of the underlying processes, as implied by the fundamental theorem of finance. For example, in case the security does not have any payouts and there is no foreign currency involved, we let for a small $\Delta > 0$:

$$\frac{\mathbb{E}^{\mathbb{Q}}[S_{t+\Delta}]}{S_t} \approx r_t \Delta \tag{14.110}$$

  This determines the arbitrage-free dynamics of the postulated stochastic differential equation.
- Second, the market practitioner needs to calibrate the SDE's volatility parameter(s). This nontrival task is often based on the existence of liquid options, or caps/floors markets, that provide direct volatility quotes. But, even then calibration needs to be done carefully.
- Once the underlying synthetic probability and the dynamics are determined, the task reduces to one of calculating in (14.109) the expectation itself. This can be done either by calculating the implied closed-form solution, or by numerical evaluation of the expectation. In case of closed-form solution, one would "take" the integral, which gives the expectation $\mathbb{E}^{\mathbb{Q}}[F(S_T, T)]$

$$\int_{S_{\min}}^{S_{\max}} F(S_T, T)\, d\mathbb{Q}(S_T) \tag{14.111}$$

where the $d\mathbb{Q}(S_T)$ is the martingale probability associated with that particular infinitesimal variation in S_T. The $S_{\min}, S_{\max}$ is the range of possible movements in S_T. In case of "Monte Carlo" evaluation, one would use the approximation:

$$\mathbb{E}^{\mathbb{Q}}[F(S_T, T)] \approx \frac{1}{N}\sum_{i=1}^{N}\left[F\left(S_T^j, T\right)\right] \quad (14.112)$$

where the $j = 1, \ldots, N$ is an index that represents trajectories of S_T randomly selected from the arbitrage-free distribution. This and similar procedures are called Monte Carlo methods. The law of large numbers guarantees that, if the randomness is correctly modeled in the selection of $\mathbb{Q}$, and if the number of paths N goes to infinity, the above average will converge to the true expectation. Hence, the approximation can be made arbitrarily good.[13]

- The last step is simple. In case the discount factor D_t is known, one divides by D_t to express the value in current dollars. If the D_t is itself random, then its random behavior needs to be taken into account jointly, with the S_T within the expectation operator.

The role played by the $\mathbb{Q}$ in these calculations is clearly very important. Thanks to the use of the martingale probability, the pricing can proceed without having to model the *true* probability distribution $\mathbb{P}$ of the process S_t, or for that matter without having to model the risk premium. Both of which require difficult and *subjective* modeling decisions.[14]

This brings us to the main question that we want to discuss. Martingale probability appears to be an important tool to a market practitioner. Is it also as important to, say, an econometrician?

In general, not at all. Suppose the econometrician's objective is to obtain the best prediction of S_T. Then, the use of $\mathbb{Q}$ would yield miserable results. In order to see this, suppose the world at time T has M possible states. The "best" forecast of S_T, denoted by $\widetilde{S}_T$, will then be given by:

$$\widetilde{S}_T = \mathbb{P}_1 S_T^1 + \cdots + \mathbb{P}_M S_T^M \quad (14.113)$$

$$= \sum_{i=1}^{M} \mathbb{P}_j S_T^j \quad (14.114)$$

That is, $\widetilde{S}_T$ will be obtained by multiplying the possible values by the true probabilities $\mathbb{P}_j$ that correspond to the possible states.[15] Clearly, if one used $\mathbb{Q}$ in the place of $\mathbb{P}_j$, the resulting forecast

$$\widetilde{S}_T = \mathbb{Q}_1 S_T^1 + \cdots + \mathbb{Q}_M S_T^M \quad (14.115)$$

$$= \sum_{i=1}^{M} \mathbb{Q}_j S_T^j \quad (14.116)$$

would be quite an inaccurate reflection of where S_t would be within an interval Δ, because under $\mathbb{Q}$ the S_t would grow at the (inaccurate) rate $r_t\Delta$ rather than the true expected growth $((r_t + \mu)\,\Delta)$ that incorporates the risk premium μ. Having misrepresented the possible growth in S_t, the martingale probability $\mathbb{Q}$ could certainly not generate satisfactory forecasts. Yet, the $\mathbb{Q}$ is useful in the process of pricing. For forecasting exercises, a decision maker should clearly use the real-world probability $\mathbb{P}$ and apply the operator $\mathbb{E}^{\mathbb{P}}[\cdot]$.

14.7 A METHOD FOR GENERATING EQUIVALENT PROBABILITIES

As seen in Girsanov theorem, there is an interesting way one can use martingales to generate probabilities. For example, assume that we define a random process Z_t that assumes only

[13] It may, however, take a significant effort in time and technology to obtain the desired numbers.

[14] In contrast, the $\mathbb{Q}$ is unique and "objective." All practitioners will have to agree on it.

[15] To be more exact, here the $\mathbb{P}_j$ would be conditional probabilities.

nonnegative values. Suppose we select a random process Z that has the following properties:

$$\mathbb{E}^{\mathbb{P}}[Z_t] = 1 \tag{14.117}$$

and

$$0 \leq Z_t \tag{14.118}$$

for all t, under a probability $\mathbb{P}$. We show that such Z can be very useful in generating new probabilities.

Consider a set A in the real line $\mathbb{R}$, and define its indicator function as 1_A:

$$1_A = \begin{cases} 1 \text{ if } Z_t \in A \\ 0 \text{ otherwise} \end{cases} \tag{14.119}$$

That is, 1_A is one if Z_t assumes a value that falls in A, otherwise it is zero.

We would like to investigate the meaning of the expression:

$$\mathbb{E}^{\mathbb{P}}[Z_t 1_A] \tag{14.120}$$

where A represents a set of possible values that Z_t can adopt. In particular, we would like to show how this expression defines a new probability for the Z_t process.

First, some heuristics. The expected value of Z_t is one. By multiplying this process by the indicator function 1_A, we are in fact "zeroing out" the values assumed by Z_t, other than those that fall in the set A. Also recall that Z_t cannot be negative. Thus we must have:

$$0 \leq \mathbb{E}^{\mathbb{P}}[Z_t 1_A] \tag{14.121}$$

Second, suppose Ω represents all possible values of Z_t[16] and that we split this set into n mutually exclusive sets, A_i, such that

$$A_1, \ldots, A_n = \Omega \tag{14.122}$$

[16]If Z_t represents the price of a financial asset, then the Ω will be all positive real numbers. In case there is "minimum tick," the Ω will be a countable set of positive rational numbers.

Then

$$1_{A_1} + \cdots + 1_{A_n} = 1_\Omega \tag{14.123}$$

regardless of the value assumed by Z_t. Thus we can write

$$\mathbb{E}^{\mathbb{P}}[Z_t] = \mathbb{E}^{\mathbb{P}}[Z_t 1_\Omega] \tag{14.124}$$

or, after replacing,

$$\mathbb{E}^{\mathbb{P}}[Z_t] = \mathbb{E}^{\mathbb{P}}\left[Z_t 1_{A_1}\right] + \mathbb{E}^{\mathbb{P}}\left[Z_t 1_{A_2}\right] + \cdots + \mathbb{E}^{\mathbb{P}}\left[Z_t 1_{A_n}\right] \tag{14.125}$$

$$= 1 \tag{14.126}$$

Thus each $\mathbb{E}^{\mathbb{P}}\left[Z_t 1_{A_i}\right]$ is positive and together they sum to one.

If we denote these terms by

$$\mathbb{E}^{\mathbb{P}}[Z_t 1_A] = \mathbb{Q}(A_i) \tag{14.127}$$

we can claim to have obtained a new probability associated with Z_t for sets A_i. That is, we have obtained:

$$\mathbb{Q}(A_i) \geq 0 \tag{14.128}$$

and

$$\sum_{i=1}^{n} \mathbb{Q}(A_i) = 1 \tag{14.129}$$

Note that the values of $\mathbb{Q}(A_i)$ may be quite different from the original, "true" probabilities, $\mathbb{P}(A_i)$,

$$\Pr(Z_t \in A_i) = \mathbb{P}(A_i) \tag{14.130}$$

associated with the Z_t.

Thus, in this special case, starting with the true probability, $\mathbb{P}$, and the expectation,

$$\mathbb{E}^{\mathbb{P}}[Z_t 1_A] \tag{14.131}$$

we could generate a new probability distribution, if

$$\mathbb{E}^{\mathbb{P}}[Z_t] \tag{14.132}$$

and

$$Z_t \geq 0 \tag{14.133}$$

If, in addition, the Z_t process is a martingale, then the consistency conditions for the new set of probabilities over different time periods will also be satisfied.

Without discussing this technical point we instead look at another way of expressing these concepts. Suppose the "true" probability $\mathbb{P}$ has a density function $f(z)$:

$$\Pr(Z_t \in A_i) = \int_{A_i} f(z)\,dz \tag{14.134}$$

Then, by definition, we have

$$\mathbb{E}^{\mathbb{P}}[Z_t] = \int_{\Omega} zf(z)\,dz = 1 \tag{14.135}$$

and

$$\mathbb{E}^{\mathbb{P}}[Z_t 1_A] = \int_{\Omega} 1_{A_i} zf(z)\,dz \tag{14.136}$$

$$= \int_{A_i} zf(z)\,dz$$

$$= \mathbb{Q}(A_i) \tag{14.137}$$

as before.

Now, suppose we need to calculate an expectation of some function $g(X_t)$ under the probability $\mathbb{P}$:

$$\mathbb{E}^{\mathbb{P}}[g(X_t)] = \int_{\Omega} g(x)f(x)\,dx \tag{14.138}$$

Suppose also that we found a way of writing this $g(x)$ using a Z_t (depending on X_t) as above:

$$g(X_t) = Z_t h(X_t)$$

Then note the following useful transformations:

$$\mathbb{E}^{\mathbb{P}}[g(X_t)] = \int_{\Omega} g(x)f(x)\,dx \tag{14.139}$$

$$= \int_{\Omega} zh(x)f(x)\,dx$$

$$= \int_{\Omega} h(x)\widetilde{f}(x)\,dx = \mathbb{E}^{\mathbb{Q}}[h(X_t)] \tag{14.140}$$

It turns out that this last integral could be easier to deal with than the original one in (14.139). We now see an application of these concepts.

14.7.1 An Example

Consider the random variable Z_t defined by

$$Z_t = e^{\left[\sigma W_t - \frac{1}{2}\sigma^2 t\right]} \tag{14.141}$$

where W_t is a Wiener process with respect to a probability, $\mathbb{P}$, having zero mean and variance t. The $0 < \sigma$ is a known constant.

Note that by definition $0 \le Zt$. Now consider its first moment. Taking the expectation directly:

$$\mathbb{E}^{\mathbb{P}}[Z_t] = \int_{-\infty}^{\infty} e^{\left[\sigma W_t - \frac{1}{2}\sigma^2 t\right]} \frac{1}{\sqrt{2\pi t}} e^{-\frac{1}{2t}W_t^2} dW_t \tag{14.142}$$

This simplifies to

$$\mathbb{E}^{\mathbb{P}}[Z_t] = \int_{-\infty}^{\infty} \frac{1}{\sqrt{2\pi t}} e^{-\frac{1}{2t}\left(W_t^2 - 2\sigma t W_t + \sigma^2 t^2\right)} dW_t \tag{14.143}$$

$$= \int_{-\infty}^{\infty} \frac{1}{\sqrt{2\pi t}} e^{-\frac{1}{2t}(W_t - \sigma t)^2 dW_t} \tag{14.144}$$

But the function under the integral sign is the density of a normally distributed random variable with mean σt and variance t. Consequently, when it is integrated from minus to plus infinity, we should obtain:

$$\int_{-\infty}^{\infty} \frac{1}{\sqrt{2\pi t}} e^{-\frac{1}{2t}(W_t - \sigma t)^2} dW_t = 1 \tag{14.145}$$

This means that

$$\mathbb{E}^{\mathbb{P}}[Z_t] = 1 \tag{14.146}$$

Hence, $e^{\left[\sigma W_t - \frac{1}{2}\sigma^2 t\right]}$ is one convenient candidate to the nonnegative process Z_t that we discussed in the previous section. It can be used to generate equivalent probabilities because it is positive and its expectation is equal to one.

In fact, given a set $A \subseteq \mathbb{R}$, we can define a new probability starting from the original probability $\mathbb{P}$ by calculating the expectation:

$$\mathbb{Q}(A) = \mathbb{E}^{\mathbb{P}}\left[e^{\left[\sigma W_t - \frac{1}{2}\sigma^2 t\right]} 1_A\right] \tag{14.147}$$

How could this function be used in pricing financial securities?

Consider the arbitrage-free price of a call option C_t with strike price K, written under the Black–Scholes assumptions:

$$C_t = e^{-r(T-t)}\mathbb{E}^{\mathbb{P}}\left[\max\{S_T - K, 0\}\right] \quad (14.148)$$

where, according to Black–Scholes framework, the S_t is a geometric process obeying, under the risk-neutral probability $\mathbb{Q}$, the SDE:

$$dS_t = rS_t dt + \sigma S_t dW_t \quad (14.149)$$

Now, we know that the solution of this SDE will give an S_t such as:

$$S_t = S_0 e^{\left[rt+\sigma W_t - \frac{1}{2}\sigma^2 t\right]} \quad (14.150)$$

Thus, substituting for S_T, in (14.148) the option price will be given by:

$$C_t = e^{-r(T-t)}\mathbb{E}^{\mathbb{P}} \left[\max\left[S_t e^{r(T-t)+\sigma(W_T-W_t)-\frac{1}{2}\sigma^2(T-t)} - K, 0\right]\right] \quad (14.151)$$

Note an interesting occurrence. A version of the variable Z_t introduced in (14.141) is embedded in this expression. In fact, splitting the exponential term into two we can write:

$$\mathbb{E}^{\mathbb{P}}\left[\max\left[S_t e^{r(T-t)+\sigma(W_T-W_t)-\frac{1}{2}\sigma^2(T-t)} - K, 0\right]\right] \quad (14.152)$$

$$= \mathbb{E}^{\mathbb{P}}\left[\max\left[S_t e^{\sigma(W_T-W_t)-\frac{1}{2}\sigma^2(T-t)} e^{r(T-t)} - K, 0\right]\right] \quad (14.153)$$

Or, after factoring out:

$$\mathbb{E}^{\mathbb{P}}\left[e^{\sigma(W_T-W_t)-\frac{1}{2}\sigma^2(T-t)} \max\left[S_t e^{r(T-t)} - e^{-\sigma(W_T-W_t)+\frac{1}{2}\sigma^2(T-t)}K, 0\right]\right] \quad (14.154)$$

Now, as before, let

$$Z_T = e^{\sigma(W_T-W_t)-\frac{1}{2}\sigma^2(T-t)}$$

We obtain:

$$\mathbb{E}^{\mathbb{P}}\left[e^{\sigma(W_T-W_t)-\frac{1}{2}\sigma^2(T-t)} \max\left[S_t e^{r(T-t)} - e^{-\sigma(W_T-W_t)+\frac{1}{2}\sigma^2(T-t)}K, 0\right]\right]$$

$$= \mathbb{E}^{\mathbb{P}}\left[Z_T \max\left[S_t e^{r(T-t)} - e^{-\sigma(W_T-W_t)+\frac{1}{2}\sigma^2(T-t)}K, 0\right]\right] \quad (14.155)$$

$$= \mathbb{E}^{\mathbb{Q}}\left[\max\left[S_t e^{r(T-t)} - e^{-\sigma(W_T-W_t)+\frac{1}{2}\sigma^2(T-t)}K, 0\right]\right] \quad (14.156)$$

for some probability defined by:

$$\mathbb{Q}(A) = \mathbb{E}^{\mathbb{P}}\left[Z 1_A\right]$$

Note that, by switching to the probability $\mathbb{Q}$, the term represented by Z_T has simply "disappeared" and the expectation is easier to calculate. In the case of pricing exotic options, transformations that use this method turn out to be convenient ways of obtaining pricing formulas. Essentially, we see that expectations involving geometric processes will contain implicitly terms that can be represented by such Z_T. It then immediately becomes possible to change measures using the trick discussed in this section. The resulting expectations may be easier to evaluate.

14.8 CONCLUSION

As conclusions, we review some of the important steps of transforming the S_t into a martingale process.

- The transformation was done by switching the distribution of S_t from $\mathbb{P}$ to $\mathbb{Q}$. This was accomplished by using a new error term, $\widetilde{W}_t$.
- This new error term, $\widetilde{W}_t$, still had the same variance.
- What distinguishes representation (14.108) from (14.94) is that the mean of S_t is altered, while preserving the zero mean property of the error terms. This was accomplished by

changing the distributions, rather than subtracting a constant from the underlying random variable.

- More importantly, in this example, the transformation was used to convert S_t into a martingale. In financial models, one may want to apply the transformation to $e^{-rt}S_t$ rather than S_t. $e^{-rt}S_t$ would represent the discounted value of the asset price, where the discount is done with respect to the (risk-free) rate r. The $\xi(S_t)$ function has to be redefined in order to accomplish this.

14.9 REFERENCES

Transforming stochastic processes into martingales through the use of the Girsanov theorem is a deeper topic in stochastic calculus. The sources that provide the technical background of this method will all be at an advanced level. Karatzas and Shreve (1991) provides one of the more intuitive discussions. Liptser and Shiryayev (1977) is a comprehensive reference.

14.10 EXERCISES

1. Consider a random variable Δx with the following values and the corresponding probabilities:

$$\{\Delta x = 1, p(\Delta x = 1) = 0.3\}$$
$$\{\Delta x = -0.5, p(\Delta x = -0.5) = 0.2\}$$
$$\{\Delta x = 0.2, p(\Delta x = 0.2) = 0.5\}$$

(a) Calculate the mean and the variance of this random variable.

(b) Change the mean of this random variable to 0.05 by subtracting an appropriate constant from Δx. That is, calculate

$$\Delta y = \Delta x - \mu$$

such that Δy has mean 0.05.

(c) Has the variance changed?

(d) Now do the same transformation using a change in probabilities, so that again the variance remains constant.

(e) Have the values of Δx changed?

2. Assume that the return R_t of a stock has the following log-normal distribution for fixed t:

$$\log(R_t) \sim \mathcal{N}\left(\mu, \sigma^2\right)$$

Suppose we let the density of $\log(R_t)$ be denoted by $f(R_t)$ and hypothesize that $\mu = 0.17$. We further estimate the variance as $\sigma^2 = 0.09$.

(a) Find a function $\xi(R_t)$ such that under the density, $\xi(R_t)f(R_t)$, R_t has a mean equal to the risk-free rate $r = 0.05$.

(b) Find a $\xi(R_t)$ such that R_t has mean zero.

(c) Under which probability is it "easier" to calculate

$$\mathbb{E}\left[R_t^2\right]$$

(d) Is the variance different under these probabilities?

3. The long rate R and the short rate r are known to have a jointly normal distribution with variance–covariance matrix Σ and mean μ. These moments are given by

$$\Sigma = \begin{bmatrix} 0.5 & 0.1 \\ 0.1 & 0.9 \end{bmatrix}$$

and

$$\mu = \begin{bmatrix} 0.07 \\ 0.05 \end{bmatrix}$$

Let the corresponding joint density be denoted by $f(R, r)$.

(a) Using Mathematica or Maple plot this joint density.

(b) Find a function $\xi(R, r)$ such that the interest rates have zero mean under the probability:

$$d\mathbb{Q} = \xi(R, r) f(R, r)\, dR dr$$

(c) Plot the $\xi(R, r)$ and the new density.

(d) Has the variance–covariance matrix of interest rate vector changed?

4. Generate a Monte Carlo algorithm for estimating the probability of a number less than -4 from a standard normal density. To do this, generate $N = 10,000$ random normal variables with mean 0 and variance 1. Run this algorithm several times and see how much the estimate changes. Comment on the efficiency and stability of this estimate. Next, consider the use of Girsanov theorem to improve our estimate. Describe the transformation used. Have we improved our estimate?

5. Estimate the price of a European barrier option that expires in 30 days and that allows the owner to purchase a share of a given stock for \$100 provided its price exceeds \$110 some time in the next 30 days. The present price of the share is \$100. Assume the stock price follows geometric Brownian motion with $r = 0.05$ and $\sigma = 0.2$.

CHAPTER

15

Equivalent Martingale Measures

OUTLINE

15.1 INTRODUCTION

In this chapter, we show how the method of equivalent martingale measures can be applied. We use option pricing to do this. We know that there are two ways of calculating the arbitrage-free price of a European call option C_t written on a stock S_t that does not pay any dividends.

1. The original Black–Scholes approach, where: (1) a riskless portfolio is formed, (2) a partial differential equation in $F(S_t, t)$ is obtained, and (3) the PDE is solved either directly or numerically.
2. The martingale methods, where one finds a "synthetic"probability $\mathbb{Q}$ under which St becomes a martingale. One then calculates

$$C_t = \mathbb{E}^{\mathbb{Q}} e^{r(T-t)} \left[\max\{S_T - K, 0\}\right] \qquad (15.1)$$

again, either analytically or numerically.

The first major topic of this chapter is a step-by-step treatment of the martingale approach. We begin with the assumptions set by Black

Financial Derivatives, Third Edition.
http://dx.doi.org/10.1016/B978-0-12-384682-2.00015-3

and Scholes and show how to convert the (discounted) asset prices into martingales. This is done by finding an equivalent martingale measure $\mathbb{Q}$. This application does not use the Girsanov theorem directly.

The Girsanov theorem is applied explicitly in the second half of the chapter, where the correspondence between two approaches to asset pricing is also discussed. In particular, we show that converting (discounted) call prices into martingales is equivalent to forcing the $F(S_t, t)$ to satisfy a particular partial differential equation, which turns out to be the Black–Scholes PDE introduced earlier. We conclude that the PDE and the martingale approaches are closely related.

15.2 A MARTINGALE MEASURE

The method of forming risk-free portfolios and using the resulting PDEs was discussed in Chapter 12, although a step-by-step derivation of the Black–Scholes formula was not provided there.

The method of equivalent martingale measures adopts a different way of obtaining the same formula. The derivation is tedious at points, but rests on straightforward mathematics and consequently is conceptually very simple. We will provide a step-by-step derivation of the Black–Scholes formula using this approach.

First, some intermediate results need to be discussed. These results are important in their own right, since they occur routinely in asset pricing formulas.

15.2.1 The Moment-Generating Function

Now let Y_t be a continuous-time process,[1]

$$Y_t \sim \mathcal{N}(\mu t, \sigma t) \tag{15.2}$$

with Y_0 given.

[1] Y_t is sometimes called a generalized Wiener process, because it obeys a normal distribution, has a nonzero mean, and has a variance not necessarily equal to one.

We define S_t as the geometric process

$$S_t = S_0 e^{Y_t} \tag{15.3}$$

S_0 is the initial point of S_t and is given exogenously.[2] We would like to obtain the moment-generating function of Y_t.

The moment-generating function denoted by $M(\lambda)$ is a specific expectation involving Y_t,

$$M(\lambda) = \mathbb{E}\left[e^{\lambda Y_t}\right] \tag{15.4}$$

where λ is an arbitrary parameter. The explicit form of this moment-generating function is useful in asset pricing formulas. More importantly, the types of calculations one has to go over to obtain the moment-generating function illustrate some standard operations in stochastic calculus. The following section is useful in this respect as well.

15.2.1.1 Calculation

Using the distribution in (15.2), $\mathbb{E}\left[e^{\lambda Y_t}\right]$ can be calculated explicitly. Substituting from the definition in (15.4), we can write

$$\mathbb{E}\left[e^{\lambda Y_t}\right] = \int_{-\infty}^{\infty} e^{\lambda Y_t} \frac{1}{\sqrt{2\pi\sigma^2 t}} e^{-\frac{1}{2}\frac{(Y_t-\mu t)^2}{\sigma^2 t}}\, dY_t \tag{15.5}$$

The expression inside the integral can be simplified by grouping together the exponents:

$$\mathbb{E}\left[e^{\lambda Y_t}\right] = \int_{-\infty}^{\infty} \frac{1}{\sqrt{2\pi\sigma^2 t}} e^{-\frac{1}{2}\frac{(Y_t-\mu t)^2}{\sigma^2 t}+\lambda Y_t} dY_t \tag{15.6}$$

In this expression, the exponent is not a perfect square, but can be completed into one by multiplying the right-hand side by

$$e^{-\left(\lambda\mu t+\frac{1}{2}\sigma^2 t\lambda^2\right)} e^{\left(\lambda\mu t+\frac{1}{2}\sigma^2 t\lambda^2\right)} = 1 \tag{15.7}$$

[2] S_0 may be random as long as it is independent of Y_t.

Then the equality in (15.6) becomes

$$\mathbb{E}\left[e^{\lambda Y_t}\right] = \int_{-\infty}^{\infty} \frac{1}{\sqrt{2\pi\sigma^2 t}} e^{\left(\lambda\mu t + \frac{1}{2}\sigma^2 t\lambda^2\right)} \times e^{-\frac{1}{2}\frac{(Y_t-\mu t)^2}{\sigma^2 t} + \lambda Y_t - \left(\lambda\mu t + \frac{1}{2}\sigma^2 t \lambda^2\right)} dY_t \tag{15.8}$$

The exponent of the second exponential function can now be completed into a square. The terms that do not depend on Y_t can be factored out. Doing this, we get

$$\mathbb{E}\left[e^{\lambda Y_t}\right] = e^{\left(\lambda\mu t + \frac{1}{2}\sigma^2\lambda^2 t\right)} \int_{-\infty}^{\infty} \frac{1}{\sqrt{2\pi\sigma^2 t}} e^{-\frac{1}{2}\frac{\left(Y_t - \left(\mu t + \sigma^2 t\lambda\right)\right)^2}{\sigma^2 t}} dY_t \tag{15.9}$$

But the integral on the right-hand side of this expression is the area under the density of a normally distributed random variable. Hence, it sums to one. We obtain

$$M(\lambda) = e^{\left(\lambda\mu t + \frac{1}{2}\sigma^2\lambda^2 t\right)} \tag{15.10}$$

The moment-generating function is a useful tool in statistics. If its kth derivative with respect to λ is calculated and evaluated at $\lambda = 0$, one finds the kth moment of the random variable in question.

For example, the first moment of Y_t can be calculated by taking the derivative of (15.10) with respect to λ:

$$\frac{\partial M}{\partial \lambda} = \left(\mu t + \sigma^2 t\lambda\right) e^{\left(\lambda\mu t + \frac{1}{2}\sigma^2\lambda^2 t\right)} \tag{15.11}$$

Now substitute 0 for λ in this formula to get

$$\left.\frac{\partial M}{\partial \lambda}\right|_{\lambda=0} = \mu t \tag{15.12}$$

For the second moment, we take the second derivative and set λ equal to zero:

$$\left.\frac{\partial^2 M}{\partial \lambda^2}\right|_{\lambda=0} = \sigma^2 t \tag{15.13}$$

These are useful properties. But they are of secondary importance in asset pricing. The usefulness of the moment-generating formula in asset pricing is tied to Eq. (15.10). We exploit the relationship

$$\mathbb{E}\left[e^{\lambda Y_t}\right] = e^{\left(\lambda\mu t + \frac{1}{2}\sigma^2\lambda^2 t\right)} \tag{15.14}$$

as a result by itself. At several points later, we have to take expectations of geometric processes. The foregoing result is very convenient, in that it gives an explicit formula for expectations involving geometric processes.

15.2.2 Conditional Expectation of Geometric Processes

In pricing financial derivatives using martingale methods, one expression that needs to be evaluated is the conditional expectation $\mathbb{E}\left[S_t | S_u, u < t\right]$, where S_t is the geometric process discussed earlier. This is the second intermediate result that we need before proceeding with martingale methods. We use the same assumptions as in the previous section and assume that

$$S_t = S_0 e^{Y_t}, \quad t \in [0, \infty) \tag{15.15}$$

where Y_t again had the distribution

$$Y_t \sim \mathcal{N}\left(\mu t, \sigma^2 t\right) \tag{15.16}$$

By definition, it is always true that

$$Y_t = Y_s + \int_s^t dY_u \tag{15.17}$$

Define ΔY_t by

$$\Delta Y_t = \int_s^t dY_u \tag{15.18}$$

Note that, by the definition of generalized Wiener processes,

$$\Delta Y_t \sim \mathcal{N}\left(\mu(t-s), \sigma^2(t-s)\right) \tag{15.19}$$

Thus, ΔY_t is a normally distributed random variable as well. According to calculations of the previous section, its moment-generating function is given by

$$M(\lambda) = e^{\lambda\mu(t-s)+\frac{1}{2}\sigma^2\lambda^2(t-s)} \tag{15.20}$$

Using these, we can calculate the conditional expectation of a geometric Brownian motion. Begin with

$$\mathbb{E}\left[\frac{S_t}{S_u}\middle| S_u, u < t\right] = \mathbb{E}\left[e^{\Delta Y_t}\middle| S_u\right] \tag{15.21}$$

because S_u can be treated as nonrandom at time u. Recall that ΔY_t is independent of $Y_u, u < t$. This means that

$$\mathbb{E}\left[e^{\Delta Y_t}\middle| S_u\right] = \mathbb{E}\left[e^{\Delta Y_t}\right] \tag{15.22}$$

But $\mathbb{E}\left[e^{\Delta Y_t}\right]$ is the moment-generating function in (15.10) evaluated at $\lambda = 1$. Substituting this value of λ in (15.10), we get

$$\mathbb{E}\left[e^{\Delta Y_t}\right] = e^{\mu(t-s)+\frac{1}{2}\sigma^2(t-s)} \tag{15.23}$$

$$= \mathbb{E}\left[\frac{S_t}{S_u}\middle| S_u\right] \tag{15.24}$$

Or, multiplying both sides by S_u,

$$\mathbb{E}\left[S_t | S_u\right] = S_u e^{\mu(t-s)+\frac{1}{2}\sigma^2(t-s)} \tag{15.25}$$

This formula gives the conditional expectation of a geometric process. It is routinely used in asset pricing theory and will be utilized during the following discussion.

15.3 CONVERTING ASSET PRICES INTO MARTINGALES

Suppose we have as before

$$S_t = S_0 e^{Y_t}, \quad t \in [0, \infty) \tag{15.26}$$

where Y_t is a Wiener process whose distribution we label with $\mathbb{P}$. Here, $\mathbb{P}$ is the "true" probability measure that is behind the infinitesimal shocks affecting the asset price S_t.

Observed values of S_t will occur according to the probabilities given by $\mathbb{P}$. But this does not mean that a financial analyst would find this distribution most convenient to work with. In fact, according to the discussion in Chapter 14, one may be able to obtain an equivalent probability $\mathbb{Q}$ under which pricing assets becomes much easier. This will especially be the case if we work with probability measures that convert asset prices into martingales.

In this section we discuss an example of how to find such a probability measure.

Recall that the "true" distribution of S_t is determined by the distribution of Y_t. Hence, the probability $\mathbb{P}$ is given by

$$Y_t \sim \mathcal{N}\left(\mu t, \sigma^2 t\right) \tag{15.27}$$

Now, assume that S_t represents the value of an underlying asset at time t, and let $S_u, u < t$ be a price observed at an earlier date u.

First of all, we know that because the asset S_t is risky when discounted by the risk-free rate, it cannot be a martingale. In other words, under the true probability measure $\mathbb{P}$, we *cannot* have

$$\mathbb{E}^{\mathbb{P}}\left[e^{-rt}S_t \middle| S_u, u < t\right] = e^{-ru}S_u \tag{15.28}$$

In fact, because of the existence of a risk premium, in general, we have

$$\mathbb{E}^{\mathbb{P}}\left[e^{-rt}S_t \middle| S_u, u < t\right] > e^{-ru}S_u \tag{15.29}$$

Under the "true" probability measure $\mathbb{P}$, the discounted process Z_t, defined by

$$Z_t = e^{-rt}S_t \tag{15.30}$$

cannot be a martingale.

Yet, the ideas introduced in Chapter 14 can be used to change the drift of Z_t and convert it into a martingale. Under some conditions, we might be able to find an equivalent probability measure $\mathbb{Q}$, such that the equality

$$\mathbb{E}^{\mathbb{Q}}\left[e^{-rt}S_t \middle| S_u, u < t\right] = e^{-ru}S_u \tag{15.31}$$

is satisfied. This can also be expressed using Zt:

$$\mathbb{E}^{\mathbb{Q}}\left[Z_t | Z_u, u < t\right] = Z_u \quad (15.32)$$

The drift in dZ_t will be zero as one switches the driving error term from the Wiener process, W_t, to a new process, W_t^*, with distribution $\mathbb{Q}$.

The question is how to find such a probability measure. We do this explicitly in the next section.

15.3.1 Determining $\mathbb{Q}$

Our problem is the following. We need to find a probability measure, $\mathbb{Q}$, such that expectations calculated with it have the property

$$\mathbb{E}^{\mathbb{Q}}\left[e^{-rt}S_t \middle| S_u, u < t\right] = e^{-ru}S_u \quad (15.33)$$

That is, S_t becomes a martingale.[3]

How can we find such a $\mathbb{Q}$? What is its form?

The step-by-step derivation that follows will answer this question. We know that

$$S_t = S_0 e^{Y_t} \quad (15.34)$$

where Y_t has the distribution denoted by $\mathbb{P}$:

$$Y_t \sim \mathcal{N}\left(\mu t, \sigma^2 t\right) \quad (15.35)$$

Now, define a new probability $\mathbb{Q}$ by

$$Y_t \sim \mathcal{N}\left(\rho t, \sigma^2 t\right) \quad (15.36)$$

where the drift parameter ρ is arbitrary and is the only difference between the two measures $\mathbb{P}$ and $\mathbb{Q}$. Both probabilities have the same variance parameter.

Now we can evaluate the conditional expectation

$$\mathbb{E}^{\mathbb{Q}}\left[e^{-rt}S_t \middle| S_u, u < t\right] \quad (15.37)$$

[3]As usual, we assume that S_t will satisfy other regularity conditions for being a martingale.

using the probability given in (15.36). In fact, the formula for such a conditional expectation was derived earlier in Eq. (15.25). We have

$$\mathbb{E}^{\mathbb{Q}}\left[e^{-rt}S_t \middle| S_u, u < t\right] = S_u e^{-r(t-u)} e^{\rho(t-u)+\frac{1}{2}\sigma^2(t-s)} \quad (15.38)$$

Note that because the expectation is taken with respect to the probability $\mathbb{Q}$, the right-hand side of the formula depends on ρ instead of μ.

Recall that the parameter ρ in (15.36) is *arbitrary*. We can select it as desired, as long as the expectation under $\mathbb{Q}$ satisfies the martingale condition. Define ρ as

$$\rho = r - \frac{1}{2}\sigma^2 \quad (15.39)$$

The parameter ρ is now fixed in terms of the volatility σ and the risk-free interest rate r. The important aspect of this choice for ρ is that the exponential on the right-hand side of (15.38) will equal one, since with this value of ρ,

$$-r(t-u) + \rho(t-u) + \frac{1}{2}\sigma^2(t-s) = 0 \quad (15.40)$$

Substituting this in (15.38):

$$\mathbb{E}^{\mathbb{Q}}\left[e^{-r(t-u)}S_t \middle| S_u, u < t\right] = S_u \quad (15.41)$$

Transferring e^{ru} to the right,

$$\mathbb{E}^{\mathbb{Q}}\left[e^{-rt}S_t \middle| S_u, u < t\right] = S_u e^{-ru} \quad (15.42)$$

This is the martingale condition. It implies that $e^{-rt}S_t$ has become a martingale under $\mathbb{Q}$.

By determining a particular value for ρ, we were able to find a probability distribution under which expectations of asset prices had the martingale property. This distribution is normal in this particular case, and its form is given by

$$\mathcal{N}\left(\left(r - \frac{1}{2}\sigma^2\right)t, \sigma^2 t\right) \quad (15.43)$$

This probability is different from the "true" probability measure $\mathbb{P}$ given in (15.35). The difference is in the mean.

15.3.2 The Implied SDEs

The previous section discussed how to determine an equivalent martingale measure $\mathbb{Q}$, when the "true" distribution of asset prices was governed by the probability measure $\mathbb{P}$. It is instructive to compare the implied stochastic differential equations (SDE) under the two probability measures.

S_t was given by

$$S_t = S_0 e^{Y_t}, \quad t \in [0, \infty) \tag{15.44}$$

where Y_t was normally distributed with mean μt and variance $\sigma^2 t$. In other words, the increments dY_t have the representation

$$dY_t = \mu\, dt + \sigma\, dW_t \tag{15.45}$$

To get the SDE satisfied by S_t, we need to obtain the expression for stochastic differentials dS_t. Because S_t is a function of Y_t, and because we have an SDE for the latter, Ito's lemma can be used:

$$dS_t = S_0 e^{Y_t} \left[\mu\, dt + \sigma\, dW_t\right] + \left[S_0 e^{Y_t}\right] \frac{1}{2}\sigma^2\, dt \tag{15.46}$$

or, after substituting S_t and grouping,

$$dS_t = \left[\mu S_t + \frac{1}{2}\sigma^2 S_t\right] dt + \sigma S_t\, dW_t \tag{15.47}$$

Under the "true" probability $\mathbb{P}$, the asset price S_t satisfies an SDE with

1. a drift coefficient $\left(\mu + \frac{1}{2}\sigma^2\right) S_t$,
2. a diffusion coefficient σS_t,
3. and a driving Wiener process W_t.

The SDE under the martingale measure $\mathbb{Q}$ is calculated in a similar fashion, but the drift coefficient is now different. To get this SDE, we simply replace μ with ρ and W_t with in (15.47). By following the same steps, we obtain

$$dS_t = \left[\rho S_t + \frac{1}{2}\sigma^2 S_t\right] dt + \sigma S_t\, dW_t^* \tag{15.48}$$

Here we emphasize, in passing, a critical step that may have gone unnoticed. By substituting in place of W_t, we are implicitly switching the underlying probability measures from $\mathbb{P}$ to $\mathbb{Q}$. This is the case because only under $\mathbb{Q}$ will the error term in Eq. (15.48) be a *standard* Wiener process. If we continue to use $\mathbb{P}$, the error terms will have a nonzero drift.

In Eq. (15.48), ρ can now be replaced by its value

$$\rho = r - \frac{1}{2}\sigma^2 \tag{15.49}$$

Substituting this in (15.48),

$$dS_t = \left[\left(r - \frac{1}{2}\sigma^2\right) S_t + \frac{1}{2}\sigma^2 S_t\right] S_t\, dt + \sigma S_t\, dW_t^* \tag{15.50}$$

The terms involving $\frac{1}{2}\sigma^2$ cancel out and we obtain the SDE:

$$dS_t = r S_t\, dt + \sigma S_t\, dW_t^* \tag{15.51}$$

This is an interesting result. *The probability that makes S_t a martingale switches the drift parameter of the original SDE to the risk-free interest rate r.* The μ contained a risk premium that is in general not known before S_t is calculated. The r, on the other hand, is the risk-free rate and is known by assumption.

Note the second difference between the two SDEs. The SDE in (15.51) is driven by a new Wiener process W_t^*, which has the distribution $\mathbb{Q}$. This has nothing to do with the actual occurrence of various states of the world. The probability measure $\mathbb{P}$ determines that. On the other hand, it is a very convenient measure to work with. Under this measure, (discounted) asset prices are martingales, and this is a very handy property to have in valuing derivative assets. Also, we know from finance theory that under appropriate conditions, the existence of such a "synthetic" probability $\mathbb{Q}$ under which asset prices are martingales is guaranteed if there is no arbitrage.

15.4 APPLICATION: THE BLACK–SCHOLES FORMULA

The Black–Scholes formula gives the price of a call option, $F(S_t, t)$, when the following conditions apply:

1. The risk-free interest rate is constant over the option's life.
2. The underlying security pays no dividends before the option matures.
3. The call option is of the European type, and thus cannot be exercised before the expiration date.
4. The price S_t of the underlying security is a geometric Brownian motion with drift and diffusion terms proportional to S_t.
5. Finally, there are no transaction costs, and assets are infinitely divisible.

Under these conditions, the Black–Scholes formula can be obtained by solving the following PDE analytically:

$$0 = -rF + F_t + rF_s S_t + \frac{1}{2}\sigma^2 F_{ss} S_t^2, \quad 0 \leq S_t, \quad 0 \leq t \leq T \tag{15.52}$$

where the boundary condition is $F(S_T, T) = \max[S_T - K, 0]$.

The resulting formula is given by

$$F(S_t, t) = S_t N(d_1) - Ke^{-r(T-t)} N\left(d_1 - \sigma\sqrt{T-t}\right) \tag{15.53}$$

with

$$d_1 = \frac{\log\left(\frac{S_t}{K}\right) + \left(r + \frac{1}{2}\sigma^2\right)(T-t)}{\sigma\sqrt{T-t}} \tag{15.54}$$

In these expressions, T is the expiration date of the call option, r is the risk-free interest rate, K is the strike price, and σ is the volatility. The function $N(x)$ is the probability that a standard normal random variable is less than x. For example, $N(d_1)$ is given by

$$N(d_1) = \int_{-\infty}^{d_1} \frac{1}{\sqrt{2\pi}} e^{-\frac{x^2}{2}} dx \tag{15.55}$$

Let S_t be an underlying asset, and C_t the price of a European call option written on this asset. Assume the standard Black–Scholes framework, with no dividends, a constant risk-free rate, and no transaction costs. Our objective in this section is to derive the Black–Scholes formula directly by using the equivalent martingale measure $\mathbb{Q}$.

The basic relation is the martingale property that the $e^{-rt}C_t$ must satisfy under the probability $\mathbb{Q}$,

$$C_t = \mathbb{E}^{\mathbb{Q}}\left[e^{-r(T-t)} C_T\right] \tag{15.56}$$

where $T > t$ is the expiration date of the call option.

We know that at expiration, the option's payoff will be $S_T - K$ if $S_T > K$. Otherwise, the call option expires with zero value. This permits one to write the boundary condition

$$C_T = \max[S_T - K, 0] \tag{15.57}$$

and the martingale property for $e^{-rt}C_t$ implies

$$C_t = \mathbb{E}_t^{\mathbb{Q}}\left[e^{-r(T-t)} \max\{S_T - K, 0\}\right] \tag{15.58}$$

In order to derive the Black–Scholes formula, this expectation will be calculated explicitly. The derivation is straightforward, yet involves lengthy expressions. It is best to simplify the notation. We make the following simplifications:

- Let $t = 0$ and calculate the option price as of time zero.
- Accordingly, the current information set I_t becomes I_0. This way, instead of using conditional expectations, we can use the unconditional expectation operator $\mathbb{E}^{\mathbb{Q}}[\cdot]$.

We now proceed with the step-by-step derivation of the Black–Scholes formula by directly evaluating

$$C_0 = \mathbb{E}^{\mathbb{Q}}\left[e^{-rT} \max\{S_T - K, 0\}\right] \tag{15.59}$$

using the probability measure $\mathbb{Q}$.

The probability $\mathbb{Q}$ is the equivalent martingale measure and was derived in the previous section,

$$d\mathbb{Q} = \frac{1}{\sqrt{2\pi\sigma^2 T}} e^{-\frac{1}{2\pi\sigma^2 T}\left(Y_T - \left(r - \frac{1}{2}\sigma^2\right)T\right)^2} dY_T \tag{15.60}$$

with

$$S_T = S_0 e^{Y_T} \tag{15.61}$$

Using this density, we can directly evaluate the expression

$$C_0 = \mathbb{E}^{\mathbb{Q}}\left[e^{-rT} \max\{S_T - K, 0\}\right] \tag{15.62}$$

which can be written as

$$C_0 = \int_{-\infty}^{\infty} e^{-rT} \max[S_T - K, 0]\, d\mathbb{Q} \tag{15.63}$$

where we also have

$$S_T = S_0 e^{Y_T} \tag{15.64}$$

Substituting these in (15.63),

$$C_0 = \int_{-\infty}^{\infty} e^{-rT} \max\left[S_0 e^{Y_T} - K, 0\right] \times \frac{1}{\sqrt{2\pi\sigma^2 T}} e^{-\frac{1}{2\pi\sigma^2 T}\left(Y_T - \left(r - \frac{1}{2}\sigma^2\right)T\right)^2} dY_T \tag{15.65}$$

To eliminate the max function from inside the integral, we change the limits of integration. We note that, after taking logarithms, the condition

$$S_0 e^{Y_T} \geq K \tag{15.66}$$

is equivalent to

$$Y_T \geq \log\left(\frac{K}{S_0}\right) \tag{15.67}$$

Using this in (15.65),

$$C_0 = \int_{\log\left(\frac{K}{S_0}\right)}^{\infty} e^{-rT}\left(S_0 e^{Y_T} - K\right) \frac{1}{\sqrt{2\pi\sigma^2 T}} e^{-\frac{1}{2\pi\sigma^2 T}\left(Y_T - \left(r - \frac{1}{2}\sigma^2\right)T\right)^2} dY_T \tag{15.68}$$

The integral can be split into two pieces:

$$C_0 = S_0 \int_{\log\left(\frac{K}{S_0}\right)}^{\infty} e^{-rT} e^{Y_T} \frac{1}{\sqrt{2\pi\sigma^2 T}} e^{-\frac{1}{2\pi\sigma^2 T}\left(Y_T - \left(r - \frac{1}{2}\sigma^2\right)T\right)^2} dY_T + Ke^{-rT} \int_{\log\left(\frac{K}{S_0}\right)}^{\infty} \frac{1}{\sqrt{2\pi\sigma^2 T}} e^{-\frac{1}{2\pi\sigma^2 T}\left(Y_T - \left(r - \frac{1}{2}\sigma^2\right)T\right)^2} dY_T \tag{15.69}$$

We can now evaluate the two integrals on the right-hand side of this expression separately.

15.4.1 Calculation

First, we apply a transformation that simplifies the notation further. We define a new variable Z by

$$Z = \frac{Y_T - \left(r - \frac{1}{2}\sigma^2\right)T}{\sigma\sqrt{T}} \tag{15.70}$$

This requires an adjustment of the lower integration limit, and the second integral on the right-hand side of (15.69) becomes

$$Ke^{-rT} \int_{\log\left(\frac{K}{S_0}\right)}^{\infty} \frac{1}{\sqrt{2\pi\sigma^2 T}} e^{-\frac{1}{2\pi\sigma^2 T}\left(Y_T - \left(r - \frac{1}{2}\sigma^2\right)T\right)^2} dY_T = Ke^{-rT} \int_{\frac{\log\left(\frac{K}{S_0}\right) - \left(r - \frac{1}{2}\sigma^2\right)T}{\sigma\sqrt{T}}}^{\infty} \frac{1}{\sqrt{2\pi}} e^{-\frac{1}{2}Z^2}\, dY_T \tag{15.71}$$

But the lower limit of the integral is closely related to the parameter d_2 in the Black–Scholes formula.[4] Letting

$$-\log\left(\frac{K}{S_0}\right) = \log\left(\frac{S_0}{K}\right) \tag{15.73}$$

[4]To see why the limits of the integration change, note that when Y_T goes from $\log\left(\frac{K}{S_0}\right)$ to ∞, the transformed Z defined by (15.70) will be between

$$\frac{\log\left(\frac{S_0}{K}\right) - \left(r - \frac{1}{2}\sigma^2\right)T}{\sigma\sqrt{T}} \tag{15.72}$$

and infinity.

we obtain the d_2 parameter of the Black–Scholes formula:

$$-\frac{\log\left(\frac{S_0}{K}\right)+\left(r-\frac{1}{2}\sigma^2\right)T}{\sigma\sqrt{T}}=-d_2 \qquad (15.74)$$

We recall that the normal distribution has various symmetry properties. One of these states that with $f(x)$ standard normal density, we can write

$$\int_L^{\infty} f(x)dx=\int_{-\infty}^{-L} f(x)dx \qquad (15.75)$$

Using the transformations in (15.74) and (15.75), we write

$$Ke^{-rT}\int_{-d_2}^{\infty}\frac{1}{\sqrt{2\pi}}e^{-\frac{1}{2}Z^2}\,dY_T$$

$$=Ke^{-rT}\int_{-\infty}^{d_2}\frac{1}{\sqrt{2\pi}}e^{-\frac{1}{2}Z^2}\,dY_T \qquad (15.76)$$

$$=Ke^{-rT}N\left(d_2\right) \qquad (15.77)$$

Hence, we derived the second part of the Black–Scholes formula, as well as the value of the parameter d_2.

We are left to derive the first part, $S_0N(d_1)$, and show the connection between d_1 and d_2. This requires manipulating the first integral on the right-hand side of (15.69). As a first step, we again use the variable Z defined in (15.70):

$$\int_{\log\left(\frac{K}{S_0}\right)}^{\infty} e^{-rT}S_0e^{Y_T}\frac{1}{\sqrt{2\pi\sigma^2T}}$$

$$e^{-\frac{1}{2\sigma^2T}\left(Y_T-\left(r-\frac{1}{2}\sigma^2\right)T\right)^2}dY_T$$

$$=e^{\left(r-\frac{1}{2}\sigma^2\right)T}e^{-rT}S_0\int_{\log\left(\frac{K}{S_0}\right)}^{\infty}$$

$$e^{\sigma Z\sqrt{T}}\frac{1}{\sqrt{2\pi}}e^{-\frac{1}{2}Z^2}\,dZ \qquad (15.78)$$

We transform the integral on the right-hand side, using properties of the normal density:

$$=e^{-rT}S_0e^{\left(r-\frac{1}{2}\sigma^2\right)T}\int_{-\infty}^{d_2}\frac{1}{\sqrt{2\pi}}e^{-\frac{1}{2}(Z^2+2\sigma Z\sqrt{T})}dZ \qquad (15.79)$$

Next, we complete the square in the exponent by adding and subtracting

$$\frac{\sigma^2T}{2} \qquad (15.80)$$

This gives:

$$=e^{-rT}S_0e^{\frac{T\sigma^2}{2}}e^{\left(r-\frac{1}{2}\sigma^2\right)T}\int_{-\infty}^{d_2}\frac{1}{\sqrt{2\pi}}e^{-\frac{1}{2}(Z-\sigma\sqrt{T})^2}dZ \qquad (15.81)$$

The terms in front of the integral cancel out, except for S_0.

Finally, we make the substitution

$$H=Z+\sigma\sqrt{T} \qquad (15.82)$$

to obtain

$$=S_0\int_{-\infty}^{d_2+\sigma\sqrt{T}}\frac{1}{\sqrt{2\pi}}e^{-\frac{1}{2}H^2}\,dH \qquad (15.83)$$

where

$$d_1=d_2+\sigma\sqrt{T} \qquad (15.84)$$

This gives the first part of the Black–Scholes formula and completes the derivation. We emphasize that during this derivation, no PDE was solved.

15.5 COMPARING MARTINGALE AND PDE APPROACHES

We have seen two contrasting approaches that can be used to calculate the fair market value of a derivative asset price. The first approach obtained the price of the derivative instrument by forming risk-free portfolios. Infinitesimal adjustments in portfolio weights and changes in the option price were used to replicate unexpected movements in the underlying asset, S_t. This eliminated all the risk from the portfolio, at the same time imposing restrictions on the way $F(S_t, t), S_t$, and the risk-free asset could jointly move over time. The assumption that we could make infinitesimal changes in positions played an important role here and showed the advantage of continuous-time asset pricing models.

The second method for pricing a derivative asset rested on the claim that we could find a probability measure $\mathbb{Q}$, such that under this probability, $e^{-rt}F(S_t,t)$ becomes a martingale. This means that

$$e^{-rt}F(S_t,t) = \mathbb{E}^{\mathbb{Q}}\left[e^{-rT}F(s_T,T)\Big| I_t\right], \quad t<T \tag{15.85}$$

or, heuristically, that the drift of the stochastic differential

$$d\left[e^{-rt}F(S_t,t)\right], \quad 0 \le t \tag{15.86}$$

was zero.

The Black–Scholes formula can be obtained from either approach. One could either solve the fundamental PDE of Black and Scholes or, as we did earlier, one could calculate the expectation $\mathbb{E}^{\mathbb{Q}}\left[e^{-rT}F(s_T,T)\big| I_t\right]$ explicitly using the equivalent measure $\mathbb{Q}$. In their original article, Black and Scholes chose the first path. The previous section derived the same formula using the martingale approach. This involved somewhat tedious manipulations, but was straightforward in terms of mathematical operations concerned.

Obviously, these two methods should be related in some way. In this section, we show the correspondence between the two approaches.

The discussion is a good opportunity to apply some of the more advanced mathematical tools introduced thus far. In particular, the discussion will be another example of the following:

- application of differential and integral forms of Ito's lemma,
- the martingale property of Ito integrals,
- an important use of the Girsanov theorem.

We show the correspondence between the PDE and martingale approaches in two stages. The first stage uses the symbolic form of Ito's lemma. It is concise and intuitive, yet many important mathematical questions are not explicitly dealt with. The emphasis is put on the application of the Girsanov theorem. In the second stage, the integral form of Ito's lemma is used.

In the following, Ito's lemma will be applied to processes of the form $e^{-rt}F(S_t,t)$. This requires that $F(\cdot)$ be twice differentiable with respect to S_t, and once differentiable with respect to t. These assumptions will not be repeated in the following discussion.

15.5.1 Equivalence of the Two Approaches

In order to show how the two approaches are related, we proceed in steps. In the first step, we show how $e^{-rt}S_t$ can be converted into a martingale by switching the driving Wiener process, and the associated probability measure. In the second step, we do the same for the derivative asset $e^{-rt}F(S_t,t)$.

These conversions are done by a direct application of the Girsanov theorem. (The switching of probabilities from $\mathbb{P}$ to $\mathbb{Q}$ during the derivation of the Black–Scholes formula did not use the Girsanov theorem explicitly).

15.5.1.1 *Converting $e^{-rt}S_t$ into a Martingale*

We begin with the basic model that determines the dynamics of the underlying asset price S_t. Suppose the underlying asset price follows the stochastic differential equation

$$dS_t = \mu(S_t)\,dt + \sigma(S_t)\,dW_t, \quad t \in [0,\infty) \tag{15.87}$$

where the drift and the diffusion terms only depend on the observed underlying asset price S_t. It is assumed that these coefficients satisfy the usual regularity conditions. W_t is the usual Wiener process with probability measure $\mathbb{P}$.

We simplify this SDE to keep the notation clear. We write it as

$$dS_t = \mu_t\,dt + \sigma_t\,dW_t \tag{15.88}$$

In the first section of this chapter, $e^{-rt}S_t$ was converted into a martingale by directly finding a probability measure $\mathbb{Q}$. Next, we do the same using the Girsanov theorem.

We can calculate the SDE followed by $e^{-rt}S_t$, the price discounted by the risk-free rate. Applying Ito's lemma to $e^{-rt}S_t$, we obtain

$$d\left[e^{-rt}S_t\right] = S_t d\left[e^{-rt}\right] + e^{-rt}\,dS_t \tag{15.89}$$

Substituting for dS_t and grouping similar terms,

$$d\left[e^{-rt}S_t\right] = e^{-rt}\left[\mu_t - r\right]S_t\,dt + e^{-rt}\sigma_t\,dW_t \tag{15.90}$$

In general, this equation will not have a zero drift, and $e^{-rt}S_t$ will not be a martingale,

$$[\mu_t - r] > 0 \tag{15.91}$$

since S_t is a risky asset.[5]

But, we can use the Girsanov theorem to convert $e^{-rt}S_t$ into a martingale. We go over various steps in detail, because this is a fundamental application of the Girsanov theorem in finance.

The Girsanov theorem says that we can find an I_t-adapted process X_t and W_t^* a new Wiener process such that

$$dW_t^* = dX_t + dW_t \tag{15.92}$$

The probability measure associated with W_t^* is given by

$$d\mathbb{P} = \xi_t\,d\mathbb{Q} \tag{15.93}$$

where the ξ_t is defined as

$$\xi_t = e^{\int_0^t X_u dW_u - \frac{1}{2}\int_0^t X_u^2 du} \tag{15.94}$$

We assume that the process X_t satisfies the remaining integrability conditions of the Girsanov theorem.[6]

The important equation for our purposes is the one in (15.92). We use this to eliminate the dW_t in (15.90). Rewriting, after substitution of dW_t^*:

$$d\left[e^{-rt}S_t\right] = e^{-rt}\left[\mu_t - rS_t\right]dt + e^{-rt}\sigma_t\left[dW_t^* - dX_t\right] \tag{15.95}$$

[5]Here $rS_t dt$ is an incremental earning if S_t dollars were kept in the risk-free asset, and $\mu\,tdt$ is an actual expected earning on the asset during an infinitesimal period, dt.

[6]This means, among other things, that the drift and diffusion parameters of the original system are well-behaved.

Grouping the terms:

$$d\left[e^{-rt}S_t\right] = e^{-rt}\left[\mu_t - rS_t\right] dt - e^{-rt}\sigma_t\,dX_t + e^{-rt}\sigma_t\,dW_t^* \tag{15.96}$$

According to the Girsanov theorem, if we define this SDE under the new probability $\mathbb{Q}$, W_t^* will be a standard Wiener process. In addition, $\mathbb{Q}$ will be a martingale measure, if we equate the drift term to zero. This can be accomplished by picking the value of dX_t as

$$dX_t = \left[\frac{\mu_t - rS_t}{\sigma_t}\right]dt \tag{15.97}$$

We assume that the integrability conditions required by the Girsanov theorem are satisfied by this dX_t equaling the term in the brackets.

This concludes the first step of our derivation. We now have a martingale measure $\mathbb{Q}$, a new Wiener process dW_t^*, and the corresponding drift adjustment X_t, such that $e^{-rt}S_t$ is a martingale and obeys the SDE

$$d\left[e^{-rt}S_t\right] = e^{-rt}\sigma_t\,dW_t^* \tag{15.98}$$

We use these in converting $[e^{-rt}F(S_t,t)]$ into a martingale.

15.5.1.2 *Converting $e^{-rt}F(S_t,t)$ into a Martingale*

The derivation of the previous section gave the precise form of the process X_t needed to apply the Girsanov theorem to derivative assets. To price a derivative asset, we need to show that $e^{-rt}F(S_t,t)$ has the martingale property under $\mathbb{Q}$. In this section, the Girsanov theorem will be used to do this.

We go through similar steps. First, we use the differential form of Ito's lemma to obtain a stochastic differential equation for $e^{-rt}F(S_t,t)$, and then apply the Girsanov transformation to the driving Wiener process.

Taking derivatives in a straightforward manner, we obtain

$$d\left[e^{-rt}F(S_t,t)\right] = d\left[e^{-rt}\right]F + e^{-rt}\,dF \tag{15.99}$$

Note that on the right-hand side we abbreviated $F(S_t, t)$ as F. Substituting for dF using Ito's lemma gives the SDE that governs the differential $d[e^{-rt}F(S_t, t)]$:

$$d\left[e^{-rt}F(S_t, t)\right] = e^{-rt}[-rF\,dt] + e^{-rt}\left[F_t\,dt + F_s\,dS_t + \frac{1}{2}F_{ss}\sigma_t^2\,dt\right] \quad (15.100)$$

The important question now is what to substitute for dS_t. We have two choices. Under W_t^* and $\mathbb{Q}$, $e^{-rt}S_t$ is a martingale. We can use

$$d\left[e^{-rt}S_t\right] = e^{-rt}\sigma_t\,dW_t^* \quad (15.101)$$

Or we can use the original SDE in (15.87):

$$dS_t = \mu_t\,dt + \sigma_t\,dW_t \quad (15.102)$$

We choose the second step to illustrate once again at what point the Girsanov theorem is exploited. Eliminating the dS_t from (15.100) using (15.102),

$$d\left[e^{-rt}F(S_t, t)\right] = e^{-rt}[-rF\,dt] + e^{-rt}\left[F_t\,dt + F_s\left[\mu_t\,dt + \sigma_t\,dW_t\right] + \frac{1}{2}F_{ss}\sigma_t^2\,dt\right] \quad (15.103)$$

Rearranging,

$$d\left[e^{-rt}F(S_t, t)\right] = e^{-rt}\left[-rF + F_t + F_s\mu_t + \frac{1}{2}F_{ss}\sigma_t^2\,dt\right]dt + e^{-rt}F_s\sigma_t\,dW_t \quad (15.104)$$

Now we apply the Girsanov theorem for a second time. We again consider the Wiener process, W_t^*, defined by:

$$dW_t^* = dW_t + dX_t \quad (15.105)$$

and transform the SDE in (15.104) using the Girsanov transformation:

$$d\left[e^{-rt}F(S_t, t)\right] = e^{-rt}\left[-rF + F_t + F_s\mu_t + \frac{1}{2}F_{ss}\sigma_t^2\,dt\right]dt - e^{-rt}F_s\sigma_t\,dX_t + e^{-rt}F_s\sigma_t\,dW_t^* \quad (15.106)$$

Again, note the critical argument here. We know that the error term dW_t^* that drives Eq. (15.106) is a standard Wiener process only under the probability measure $\mathbb{Q}$. Hence, $\mathbb{Q}$ becomes the relevant probability.

The value of dX_t has already been derived in Eq. (15.97):

$$dX_t = \frac{\mu_t - rS_t}{\sigma_t}\,dt \quad (15.107)$$

We substitute this in (15.106):

$$d\left[e^{-rt}F(S_t, t)\right] = e^{-rt}\left[-rF + F_t + F_s\mu + \frac{1}{2}F_{ss}\sigma_t^2\,dt - \frac{\mu_t - rS_t}{\sigma_t}\right]dt + e^{-rt}F_s\sigma_t\,dW_t^* \quad (15.108)$$

Simplifying,

$$d\left[e^{-rt}F(S_t, t)\right] = e^{-rt}\left[-rF + F_t + F_s rS_t + \frac{1}{2}F_{ss}\sigma_t^2\,dt\right]dt + e^{-rt}F_s\sigma_t\,dW_t^* \quad (15.109)$$

But, in order for $e^{-rt}F(S_t, t)$ to be a martingale under the pair $W_t^*, \mathbb{Q}$, the drift term of this SDE must be zero.[7] This is the desired result:

$$-rF + F_t + F_s rS_t + \frac{1}{2}F_{ss}\sigma_t^2\,dt = 0 \quad (15.110)$$

This expression is identical to the fundamental PDE of Black and Scholes. With this choice of dX_t, the derivative price discounted at the risk-free rate obeys the SDE

$$d\left[e^{-rt}F(S_t, t)\right] = e^{-rt}\sigma_t F_s\,dW_t^* \quad (15.111)$$

The drift parameter is zero.

15.5.2 Critical Steps of the Derivation

There were some critical steps in this derivation that are worth further discussion.

[7] We know that if there are no-arbitrage possibilities, the same $\mathbb{Q}$ will convert all asset prices into martingales.

First, note the way the Girsanov theorem was used. We are given a Wiener process-driven SDE for the price of a financial asset discounted by the risk-free rate. Initially, the process is not a martingale. The objective is to convert it into one.

To do this, we use the Girsanov theorem and find a new Wiener process *and* a new probability, $\mathbb{Q}$, such that the discounted asset price becomes a martingale. The probability measure $\mathbb{Q}$ is called an *equivalent martingale measure*. This operation gives the drift adjustment term X_t, required by the Girsanov theorem. In the preceeding derivation this was used twice, in (15.95) and in (15.106).

This brings us to the second critical point of the derivation. We go back to Eq. (15.106):

$$d\left[e^{-rt}F(S_t,t)\right] = e^{-rt}\left[-rF + F_t + F_s\mu_t + \frac{1}{2}F_{ss}\sigma_t^2\, dt\right]dt - e^{-rt}F_s\sigma_t\, dX_t + e^{-rt}F_s\sigma_t\, dW_t^* \tag{15.112}$$

Here, substituting the value of dX_t means adding

$$dX_t = \frac{\mu_t - rS_t}{\sigma_t} \tag{15.113}$$

to the drift term. Note the subtle role played by this transformation. The dX_t is defined such that the term $F_s\mu_t\, dt$ in Eq. (15.104) will be eliminated and will be replaced by $F_s r\, dt$.

In other words, the application of the Girsanov theorem amounts to transforming the drift term μt into rS_t, the risk-free rate. Often, books on derivatives do this mechanically, by replacing all drift parameters with the risk-free rate. The Girsanov theorem is provided as the basis for such transformations. Here, we see this explicitly.

Finally, a third point. How do we know that the pair $W_t^*, \mathbb{Q}$ that converts $e^{-rt}S_t$ into a martingale will also convert $e^{-rt}F(S_t,t)$ into a martingale? This question is important, because a function of a martingale need not itself be a martingale.

This step is related to equilibrium and arbitrage valuation of financial assets. It is in the domain of dynamic asset pricing theory. We briefly mention a rationale. As was discussed heuristically in Chapter 2, under proper conditions, arbitrage relations among asset prices will yield a unique martingale measure that will convert all asset prices, discounted by the risk-free rate, into martingales.

Hence, the use of the same pair $W_t^*, \mathbb{Q}$ in Girsanov transformations is a consequence of asset pricing theory. If arbitrage opportunities existed, we could not have done this.

15.5.3 Integral Form of the Ito Formula

The relationship between the PDE and martingale approaches was discussed using the symbolic form of Ito's lemma, which deals with stochastic differentials.

As emphasized several times earlier, the stochastic differentials under consideration are symbolic terms, which stand for integral equations in the background. The basic concept behind all SDEs is the Ito integral. We used stochastic differentials because they are convenient, and because the calculations already involved tedious equations.

The same analysis can be done using the integral form of Ito's lemma. Without going over all the details, we repeat the basic steps.

The value of a call option discounted by the risk-free rate is represented as usual by $e^{-rt}F(S_t,t)$. Applying the integral form of Ito's lemma,

$$e^{-rt}F(S_t,t) = F(S_0,0) + \int_0^t e^{-ru}\left[-rF + F_t + \frac{1}{2}F_{ss}\sigma_u^2 + F_s rS_u\right]du + \int_0^t e^{-ru}\sigma_u F_s\, dW_t^* \tag{15.114}$$

Note that we use W_t^* in place of W_t, and consequently "replace" μt by r, the risk-free interest rate.

We assume σ_t is such that

$$\mathbb{E}^{\mathbb{Q}}\left[e^{\int_0^t (F_s e^{-ru}\sigma_u)^2 du}\right] \tag{15.115}$$

This is the Novikov condition of the Girsanov theorem and implies that the integral

$$\int_0^t F_s e^{-ru}\sigma_u du \tag{15.116}$$

is a martingale under $\mathbb{Q}$.

But the derivative asset price discounted by e^{-rt} is also a martingale. This makes the first integral on the right-hand side of (15.114),

$$\int_0^t e^{-ru}\left[-rF + F_t + \frac{1}{2}F_{ss}\sigma_u^2 + F_s r S_u\right] du \tag{15.117}$$

a (trivial) martingale as well. But this is an integral taken with respect to time, and martingales are not supposed to have nonzero drift coefficients. Thus, the integral must equal zero. This gives the partial differential equation

$$-rF+F_t+\frac{1}{2}F_{ss}\sigma_u^2+F_s r S_u = 0 \quad 0 \le S_t, \quad 0 \le t \le T \tag{15.118}$$

This is again the fundamental PDE of Black and Scholes.

15.6 CONCLUSIONS

This chapter dealt with applications of the Girsanov theorem. We discussed several important technical points. In terms of broad conclusions, we retain the following.

There is a certain equivalence between the martingale approach to pricing derivative assets and the one that uses PDEs.

In the martingale approach, we work with conditional expectations taken with respect to an equivalent martingale measure that converts all assets discounted by the risk-free rate into martingales. These expectations are very easy to conceptualize once the deep ideas involving the Girsanov theorem are understood. Also, in the case where the derivative asset is of the European type, these expectations provide an easy way of numerically obtaining arbitrage-free asset prices.

It was shown that the martingale approach implies the same PDEs utilized by the PDE methodology. The difference is that, in the martingale approach, the PDE is a consequence of risk-neutral asset pricing, whereas in the PDE method, one begins with the PDEs to obtain risk-free prices.

15.7 REFERENCES

The section where we obtain the Black–Scholes formula follows the treatment of Ross (1993). Cox and Huang (1989) is an excellent summary of the main martingale results. The same is true, of course, of the treatment of Duffie (1996).

15.8 EXERCISES

1. In this exercise we use the Girsanov theorem to price the *chooser* option. The chooser option is an exotic option that gives the holder the right to choose, at some future date, between a call and a put written on the same underlying asset. Let the T be the expiration date, S_t be the stock price, K the strike price. If we buy the chooser option at time t, we can choose between call or put with strike K, written on S_t. At time t the value of the call is

$$C(S_t,t) = e^{-r(T-t)}\mathbb{E}\left[\max(S_T - K, 0)|\, I_t\right]$$

whereas the value of the put is:

$$P(S_t,t) = e^{-r(T-t)}\mathbb{E}\left[\max(K - S_T, 0)|\, I_t\right]$$

and thus, at time t, the chooser option is worth:

$$H(S_t,t) = \max\left[C(S_t,t), P(S_t,t)\right]$$

(a) Using these, show that:

$$C(S_t,t) - P(S_t,t) = S_t - Ke^{-r(T-t)}$$

Does this remind you of a well-known parity condition?

(b) Next, show that the value of the chooser option at time t is given by

$$H(S_t,t) = \max\Big[C(S_t,t), C(S_t,t) - S_t + Ke^{-r(T-t)}\Big]$$

(c) Consequently, show that the option price at time zero will be given by

$$H(S_0,0) = C(S_0,0) + e^{-rT}\mathbb{E}\Big[\max\Big[K - S_0 e^{rT} e^{\sigma W_t - \frac{1}{2}\sigma^2 t}, 0\Big]\Big]$$

where S_0 is the underlying price observed at time zero.

(d) Now comes the point where you use the Girsanov theorem. How can you exploit the Girsanov theorem and evaluate the expectation in the above formula *easily*?

(e) Write the final formula for the chooser option.

2. In this exercise we work with the Black–Scholes setting applied to foreign currency denominated assets. We will see a different use of Girsanov theorem. [For more details see Musiela and Rutkowski (1997).] Let r, f denote the domestic and the foreign risk-free rates. Let S_t be the exchange rate, that is, the price of one unit of foreign currency in terms of domestic currency. Assume a geometric process for the dynamics of S_t:

$$dS_t = (r - f)S_t dt + \sigma S_t dW_t$$

(a) Show that

$$S_t = S_0 e^{\left(r - f - \frac{1}{2}\sigma^2\right)t + \sigma W_t}$$

where W_t is a Wiener process under probability $\mathbb{P}$.

(b) Is the process

$$\frac{S_t e^{ft}}{S_0 e^{rt}} = e^{\sigma W_t - \frac{1}{2}\sigma^2 t}$$

a martingale under measure $\mathbb{P}$?

(c) Let $\mathbb{Q}$ be the probability

$$\mathbb{Q}(A) = \int_A e^{\sigma W_T - \frac{1}{2}\sigma^2 T} dP$$

What does Girsanov theorem imply about the process, $W_t - \sigma t$, under $\mathbb{Q}$?

(d) Show, using Ito formula, that

$$dZ_t = Z_t\Big[\left(f - r + \sigma^2\right)dt - \sigma\, dW_t\Big]$$

where $Z_t = 1/S_t$.

(e) Under which probability is the process $Z_t e^{rt}/e^{ft}$ a martingale?

(f) Can we say that $\mathbb{Q}$ is the arbitrage-free measure of the foreign economy?

3. Consider the following SDE for S_t under the physical measure: $dS_t = \mu\, dt + \sigma\, dW_t$. Write the corresponding dynamics under the risk-neutral measure, such that the discounted stock price is a martingale. Under the risk-neutral measure, use the methodology detailed in the book for the log-normal model to derive the closed-form value of a call.

4. Write a Matlab program to compute the European digit option without using simulation method.

CHAPTER

16

New Results and Tools for Interest-Sensitive Securities

OUTLINE

16.1 INTRODUCTION

The first part of this book dealt with an introduction to quantitative tools that are useful for the *Classical Black–Scholes approach*, where underlying security S_t was a nondividend-paying stock, the risk-free interest rate r and the underlying volatility σ were constant, the option was European, and where there were no transactions costs or indivisibilities.

The types of derivative securities traded in financial markets are much more complicated than such "plain vanilla" call or put options that may fit this simplified framework reasonably well. In fact, some of the assumptions used by Black–Scholes, although often quite robust, may fall significantly short in the case of interest-sensitive securities.[1] New assumptions introduced in their place require more complicated tools.

These new instruments may be similar in some ways to the plain-vanilla derivatives already discussed. Yet, there are some nontrivial complications. More importantly, some new results have

[1]Robustness of Black–Scholes assumptions is one reason why the formula continues to be very popular with market professionals. For example, one still obtains reasonably accurate prices when volatility is stochastic, or when interest rates move randomly. A comprehensive source on this aspect of Black–Scholes formula is El Karoui et al. (1998).

Financial Derivatives, Third Edition.
http://dx.doi.org/10.1016/B978-0-12-384682-2.00016-5

recently been obtained in dealing with interest-sensitive instruments and term structure of interest rates. These powerful results require a different set of quantitative tools in their own respect.

Recall that the examples discussed in previous chapters were by and large in line with the basic Black–Scholes assumptions. In particular, two aspects of Black–Scholes framework were always preserved.

1. Early exercise possibilities of American-style derivative securities were not dealt with.
2. The risk-free interest rate r was always kept constant.

These are serious restrictions for pricing a large majority of financial derivatives.[2]

First, a majority of financial derivatives are American style, containing early exercise clauses. A purchaser of financial derivatives often does not have to wait until the expiration date to exercise options that he or she has purchased. This complicates derivative asset pricing significantly. New mathematical tools need to be introduced.

Second, it is obvious that risk-free interest rates are not constant. They are subject to unpredictable, infinitesimal shocks just like any other price. For some financial derivatives, such as options on stocks, the assumption of constant risk-free rate may be incorrect, but still is a reasonable approximation.

However, especially for interest rate derivatives, such an assumption cannot be maintained. It is precisely the risk associated with the interest rate movements that makes these derivatives so popular. Introducing unpredictable Wiener components into risk-free interest rate models leads to some further complications in terms of mathematical tools.

[2]Merton (1973) was an early attempt to introduce stochastic interest rates. Yet, this was in a world where the underlying asset was again a stock. Such a complication can, by and large, still be handled by using classical tools. New tools start being more practical when the derivative is interest sensitive, in the sense that the payoff depends on the value and/or path followed by interest rates.

Finally, notice that Black–Scholes assumptions can be maintained as long as derivatives are short-dated, whereas the consideration of longer dated instruments may, by itself, be a sufficient reason for relaxing assumptions on constant interest rates and volatility.

This second part of the book discusses new tools required by such modifications and introduces the important new results applicable to term structure models.

16.2 A SUMMARY

In this chapter we briefly outline the basic ideas behind the new tools. The issues discussed in the following chapters are somewhat more advanced, but they all have practical implications in terms of pricing highly liquid derivative structures.

Chapter 17 will reintroduce the simple two-state framework that motivated the first part of this book. But, in the new version of models used in Chapter 2, we will complicate the simple setup by allowing for stochastic short rates and by considering interest-sensitive instruments. This way, we can motivate important concepts such as *normalization* and tools such as the *forward measure*.

The major topic of Chapter 18 is the foundations for modeling the term structure of interest rates. The definitions of a forward rate, spot rate, and term structure are given here formally. More important, Chapter 18 introduces the two broad approaches to modeling term structure of interest rates, namely, the *classical* and the *Heath-Jarrow-Morton* approach. Learning the differences between the assumptions, the basic philosophies, and the practical implementations that one can adopt in each case, is an important step for understanding the valuation of interest-sensitive instruments.

Chapter 19 discusses classical PDE analysis for interest-sensitive securities. This approach can be regarded as an attempt to follow steps similar to those used with Black–Scholes PDE, and then obtaining PDEs satisfied by default-free zero-coupon bond prices and derivatives written on them. The main difficulty is to find ways of adjusting the drift of the short-rate process. Short-rate is not an *asset*, so this drift cannot be replaced with the risk-free spot rate, r, as in the case of Black–Scholes. A more complicated operation is needed. This leads to the introduction of the notion of a *market price of interest rate risk*. The corresponding PDEs will now incorporate this additional (unobserved) variable.

Chapter 20 is a discussion of the so-called classical PDE approach to fixed income. Chapter 21 deals with the recent tools that are utilized in pricing, hedging, and arbitraging interest rate sensitive securities. The first topic here consists of the fundamental relationship that exists between a class of conditional expectations of stochastic processes and some partial differential equations. Once this correspondence is established, financial market participants gain a very important tool with practical implications. This tool is related to the Feynman–Kac formula and it is dealt with in this chapter. Using this "correspondence," one can work either with conditional expectations taken with respect to martingale measures, or with the corresponding PDEs. The analyst could take the direction which promises simpler (or cheaper) numerical calculations.

Some of the other concepts introduced in Chapter 21 are the *generator* of a stochastic process, Kolmogorov's backward equation, and the implications of the so-called Markov property. The latter is especially important for models of short rate, because the latter is shown *not* to behave as a Markov process, a property which complicates the utilization of Feynman–Kac type correspondences.

Finally, Chapter 22 discusses *stopping times*, which are essential in dealing with American-style derivatives. This concept is introduced along with a certain algorithm called *dynamic programming* that is very important in its own right. In this chapter we also show the correspondence between using binomial trees for American-style securities and stopping times. We see that the pricing is based on applications of dynamic programming.

Stopping times are random variables whose outcomes are some particular points in time where a certain process is being "stopped." For example, an American-style call option can be exercised before the expiration date. Initially, such execution times are unknown. Hence, the execution *date* of an option can be regarded as a random variable. Stopping times provide the mathematical tools to incorporate in pricing the effects of such random variables.

These mathematical tools are particularly useful in case of interest-sensitive derivatives. Hence, before we proceed with the discussion of the tools, we need to discuss briefly some of these instruments. This is done in the following section.

16.3 INTEREST RATE DERIVATIVES

One of the most important classes of derivative instruments that violate the assumptions of Black–Scholes environment are derivatives written on interest-sensitive securities.

Some well-known interest rate derivatives are the following[3]:

- **Interest rate futures and forwards**. Let L_{t_i} represent the annualized simple interest rate on a loan that begins at time t_i and ends at time t_{i+1}. Suppose there are no bid-ask spreads or default risks involved. Then, at time t, where $t < t_i < t_{i+1}$, we can write futures and forward contracts on these "Libor rates," L_{t_i}.[4] For

[3]In the following, the reader will notice a slight change in notation. In particular, the time subscript will be denoted by t_i. This is required by the new instruments.

[4]Libor is the London Interbank Offered Rate. It is an interbank rate asked by sellers of funds. It is obtained by polling selected banks in London and then averaging the quotes. Hence, depending on the selection of banks, there may be several Libor rates on the same maturity. The British Bankers Association calculates an "official" Libor that forms the basis of most of these Libor Instruments.

example, *forward loans* for the period $[t_i, t_{i+1}]$ can be contacted at time t, with an interest rate F_t. The buyer of the forward will receive, as a loan, a certain sum N at time t_i and will pay back at time t_{i+1} the sum $N(1+F_t\delta)$, where the δ is the day's adjustment factor.[5]

- **Forward rate agreements** (FRA). Already discussed in Chapter 1, these instruments provide a more convenient way of hedging interest rate risk. Depending on the outcome of $F_t > L_{t_i}$, or $F_t < L_{t_i}$, the buyer of a FRA *paid in-arrears* receives, at time t_{i+1}, the sum

$$N\left[F_t - L_{t_i}\right]\delta$$

if it is positive, or pays

$$N\left[F_t - L_{t_i}\right]\delta$$

if it is negative. The FRA rate F_t is selected so that the time t price of the FRA contract equals zero. This situation is shown in Figure 16.1. In case of FRAs traded in actual markets, often the payment is made at the same time the L_{t_i} is observed. Hence, it has to be discounted by $(1 + L_{t_i}\delta)$. This is also shown in Figure 16.1.

- **Caps and floors**. Caps and floors are among some of the most liquid interest rate derivatives. Caps can be used to hedge the risk of increasing interest rates. Floors do the same for decreasing rates. They are essentially baskets of options written on Libor rates. Suppose t denotes the *present* and let $t_0, t \leq t_0$ be the *starting date* of an interest rate cap. Let t_n be the ending date of the cap for some fixed $n, t < t_0 < t_n$. Let the $t_1, t_2, \dots t_{n-1}$ be reset dates. Then for every *caplet* that applies to the period t_i, t_{i+1}, the buyer of the cap will receive, at time t_{i+1}, the sum

$$N \max\left[\delta\left(L_{t_i} - R_{cap}\right), 0\right] \qquad (16.1)$$

where L_{t_i} is the underlying Libor rate observed at time t_i, the δ is the day's adjustment, and the N is a notional amount to be decided at time t. The R_{cap} is the cap rate, which plays the role of a strike price. In a sense, a caplet will compensate the buyer for any increase in the future Libor rates beyond the level R_{cap}. Thus, it is equivalent to a put option with expiration date t_i, written on a default-free discount bond with maturity date t_{i+1}, with a strike price obtained from R_{cap}. In particular, the strike price that applies to this option is the $100/(1 + R_{cap}\delta)$, where R_{cap} is the cap rate, and the δ is, as usual, the day's adjustment.

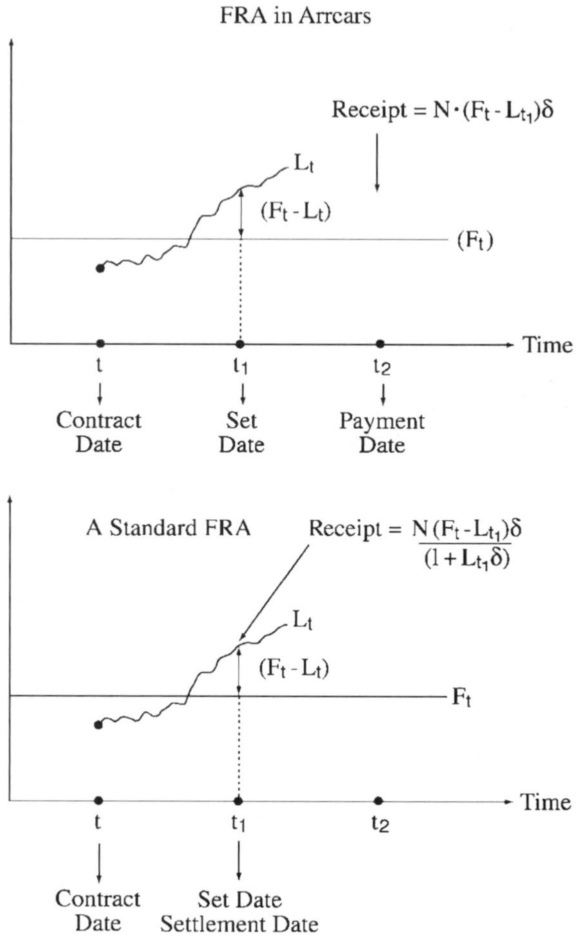

FIGURE 16.1 Forward rate agreements.

[5] For example, it is equal to the number of days during $[t_i, t_{i+1}]$ divided by 365.

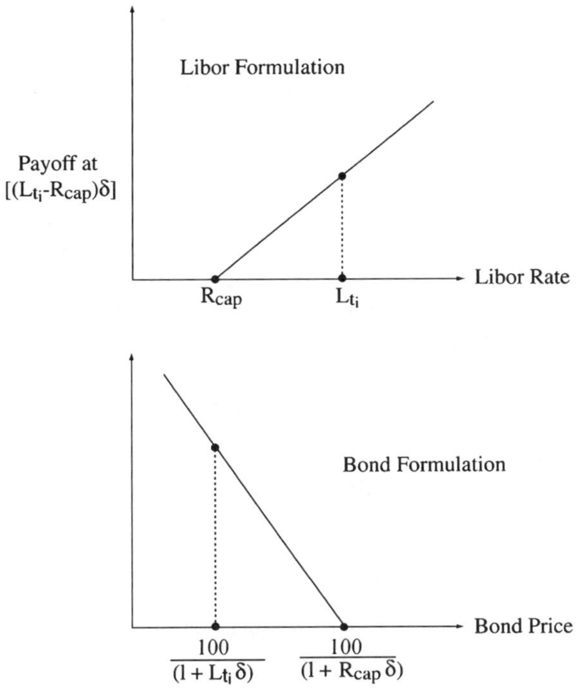

FIGURE 16.2 Interest rate caps.

This formulation is shown in Figure 16.2. At time t_i the option expires. If the price of the bond is lower than $100/(1 + R_{cap}\delta)$, then the holder receives the difference:

$$Payoff = \frac{100}{(1 + R_{cap}\delta)} - \frac{100}{(1 + L_{t_i}\delta)}$$

Otherwise, the holder receives nothing. That the caplet is equivalent to a call option on L_{t_i} with a strike R_{cap} is also seen in Figure 16.2. Here, if we view the caplet as a call with expiration date t_{i+1}, written on the Libor rate L_{t_i}, then it should be kept in mind that the settlement will be done at time t_{i+1}, rather than at time t_i. An interest rate floorlet can similarly be shown to be equivalent to a call option with expiration t_i, on a discount bond with maturity t_{i+1}. Equivalently, it can be viewed as a put option on the Libor rate that expires at time t_{i+1}.

- **Interest rate swaps**. These instruments were also discussed in Chapter 1. Plain vanilla interest rate swaps paid in-arrears involve an exchange of cash flows generated by a fixed pre-set *swap rate* κ against cash flows generated by floating Libor rates L_{t_i}. The cash flows are based on a notional amount N and are settled at times t_{i+1}. Clearly, a swap is a more complex form of a sequence of FRAs. The swap rate κ is set so that the time t price of the swap contract equals zero.
- **Bond options**. A call option written on a bond gives its holder the right to buy a bond with price B_t at the strike price K. Since the price of a bond depends on the current and future spot rates, bond options will be sensitive to movements in r_t or, for that matter, to movements in Libor rates L_{t_i}.
- **Swaptions**. Swaptions are options written on swap contracts. Depending on maturity, they are very liquid. At time t a practitioner may buy an option on a swap contract with strike price κ and notional amount N. The option expires at time T and the swap will start at some date $T_1, T \leq T_1$ and end at time T_2, $T_1 < T_2$. The buyer of the swaption contract will, at expiration, have the right to get in a fixed-payer swap contract with swap rate κ, notional amount N, start date T_1, and end date T_2. Hence, the value of the swaption will be positive if actual swap rates at time T_1, namely the R_{t_i}, have moved above κ.

These are some of the basic interest rate derivatives. The scope of this book prevents us from going into more exotic products. Instead, we now would like to summarize some key elements of these instruments and see what types of new tools would be required.

16.4 COMPLICATIONS

Introducing interest rate derivatives leads to several complications. This is best seen by looking at bond options, and then comparing these with the Black–Scholes framework.

The price of a bond B_t depends on the stochastic behavior of the current and future spot rates in the economy. Hence, at the outset, two new assumptions are required: (1) Bond price B_t must be a function of the current and future spot rates, and (2) the spot rate r_t cannot be assumed constant, because this would amount to saying that B_t would be completely predictable, which in turn would mean that the volatility of the underlying security is zero. Hence, there would be no demand for any call or put options written on the bond.

Thus, the very first requirement is that we work with stochastic interest rates. But then, the resulting discount factors *and* the implied payoffs would be dependent on interest rates. Clearly this would make arbitrage-free pricing more complicated.

The second complication is that most interest rate derivatives may be American style and any explicit or implicit options may be exercised before their respective expiration date, if desired.

Third, the payouts of the underlying security may be different for interest rate derivatives. For example, in case of mark-to-market adjustments, the fact that spot rates are stochastic will, in general, make a difference in evaluating an arbitrage-free futures price compared to forward prices. This is the case, because with mark-to-market adjustments the holder of the contracts makes/receives periodic payments that fluctuate as interest rates change. But these mark-to-market cash flows will also be discounted by stochastic discount factors that are affected by the same interest rate movements. The resulting futures prices may be different from the price of a forward contract that has no mark-to-market requirement.

Similarly, if a bond makes coupon payments, the underlying security, B_t, will also be different than a no-dividend paying stock, S_t.

These are some of the obvious modifications that are required to deal with interest rate derivatives. There are also some more technical implications that may not be as obvious at the outset. One of these was mentioned above.

16.4.1 Drift Adjustment

Interest rates are not assets, they are more like "returns" on assets. This means that the arbitrage-free restriction that consists of removing the unknown drift, μ , of an asset price, St, in the dynamics:

$$dS_t = \mu S_t\, dt + \sigma S_t\, dW_t$$

and then replacing it with the risk-free rate, r, is not a valid procedure anymore. In the dynamics of r_t written as,

$$dr_t = a\,(r_t, t)\, dt + \sigma\,(r_t, t)\, dW_t$$

the drift $a(r_t, t)$ has to be risk-adjusted by other means. This makes the practical application of Girsanov theorem much more complicated. In fact, the switch from W_t to a new Wiener process, W_t^*, defined under the risk-free measure $\mathbb{Q}$, cannot be done in a straightforward way. Given the Girsanov correspondence between the two Wiener processes:

$$dW_t = dW_t^* - \lambda_t\, dt$$

The modifications of interest rate dynamics require:

$$dr_t = \big(a\,(r_t, t) - \lambda_t \sigma\,(r_t, t)\, dt\big) + \sigma\,(r_t, t)\, dW_t^*$$

And, it is not clear at the outset how λ_t can be determined.

This is a significant complication compared to the Black–Scholes substitution:

$$r = (a\mu - \lambda_t \sigma) \tag{16.2}$$

which is possible when S_t is a traded asset. Fundamental theorem of finance would then imply this equality. Yet r_t not being an "asset," a similar substitution is not valid.

16.4.2 Term Structure

Another complication is the coexistence of *many* interest rates. Note that within the simple Black–Scholes world, there is one underlying asset S_t. Yet within the fixed-income sector, there are many interest rates implied by different maturities. Moreover, these interest rates cannot follow very different dynamics from each other because they relate, after all, to similar instruments.

Thus, in contrast to the Black–Scholes case for interest rates, one would deal with a vector of random processes that must obey complex interrelations due to arbitrage possibilities. The resulting k-dimensional dynamics are bound to be more complicated.

Note that in case of a classical Black–Scholes environment, modeling the risk-free dynamics of the underlying asset meant modeling a single SDE, where overtime arbitrage restrictions on a single variable had to be taken into account. But in the case of interest rates, the same overtime restrictions need to be modeled for k-variables. There is more. Now, arbitrage restrictions *across* variables need to be specified as well.

Last but not least, there is the modeling of volatilities. The volatility of a bond has to vary over time. After all, the bond matures at some specific date. Hence, these volatilities cannot be assumed constant, as in the case of stocks.

Clearly, this very broad class of interest rate derivatives cannot be treated using the assumptions of the Black–Scholes environment.

Some of these complications can be handled within a Black–Scholes framework by either making small modifications in the assumptions or by "tricking" them in some ingenious way. But the early exercise possibility of interest rate derivatives and stochastic interest rates are two modifications that have to be incorporated in derivative asset pricing using new mathematical tools. The following chapters are intended to do this.

16.5 CONCLUSIONS

This chapter is simply a brief summary and cannot be considered an introduction to interest-sensitive securities. It has, however, the bare minimum necessary for understanding the tools discussed in the remaining chapters.

16.6 REFERENCES

The book of readings published by Risk, "Vasicek and Beyond," is highly recommended as an excellent collection of readings concerning interest-rate derivatives and their pricing. The reader should also consult Hull (2009) and the extensive treatment in Rebonato (1998).

16.7 EXERCISES

1. Plot the payoff diagrams for the following instruments:

- **(a)** A caplet with cap rate $R_{cap} = 6.75\%$ written on 3-month Libor L_t that is about to expire.
- **(b)** A forward contract written on a default-free discount bond with maturity 2 years. The forward contract expires in 3 months. The contracted price is 89.5.
- **(c)** A 3 by 6 FRA contract that pays the fixed 3-month rate, F, against Libor.
- **(d)** A fixed payer interest rate swap with swap rate $\kappa = 7.5\%$. The swap has maturity 2 years and receives 6-month Libor. Start date was exactly 6 months ago.
- **(e)** A swaption that expires in 6 months on a 2-year fixed-payer swap with swap rate $\kappa = 0.6\%$.

2. Which one(s) of the following are assets traded in financial markets:

- **(a)** 6-month Libor
- **(b)** A 5-year Treasury bond
- **(c)** A FRA contract

(d) A caplet
(e) Returns on 30-year German Bonds
(f) Volatility of Federal Funds rate
(g) An interest rate swap

3. Explain why it might be reasonable to model interest rates by a mean-reverting process. What is the fundamental difference that makes this (potentially) appropriate for interest rates but not for stocks?

4. Consider the Vasicek Model under risk neutral probability measure:

$$dr_t = \alpha(\mu - r_t)dt + \sigma\, dW_t \tag{16.3}$$

with $r_0 = 0.01, \alpha = 0.2, \mu = 0.01, \sigma = 0.05$. Write a Matlab program to simulate a sample path of r_t from $t = 0$ to $t = 1$.

CHAPTER

17

Arbitrage Theorem in a New Setting

OUTLINE

17.1 INTRODUCTION

The motivation for the main tools in derivatives pricing was introduced in the simple model of Chapter 2. There we discussed a simple construction of synthetic (martingale) probabilities that played an essential role in the first part of this book. Because the setting was very simple, it was well-suited for motivating complex notions such as risk-neutral probabilities and the crucial role played by martingale tools.

Chapter 2 considered a model where lending and borrowing at a *constant* risk-free rate was one of the three possible ways of investing, the other two being stocks and options written on these stocks. Throughout Chapter 2 interest rates were assumed to be constant and a discussion of interest-sensitive financial derivatives was deliberately omitted. Yet, in financial markets a large majority of the instruments that trade are interest-sensitive products. These are used to hedge, to arbitrage the interest rate risk, and to speculate on it. Relaxing the assumption of constant interest rates is, thus, essential. As mentioned in Chapter 16, relaxing the assumption of constant interest rates and then introducing complex interest rate derivatives creates a need for new

Financial Derivatives, *Third Edition*.
http://dx.doi.org/10.1016/B978-0-12-384682-2.00017-7

mathematical tools, most of which were discovered only lately.

This chapter attempts to *motivate* these notions and introduces the new tools by using a simple discrete-state approach similar to the one utilized in Chapter 2. However, the model is extended in new directions so that these new tools and concepts can also be easily understood. By expanding the simplified framework of Chapter 2, one can discuss at least three major additional results.

The first set of issues can be grouped under the concept of *normalization*. This is the technique of obtaining pricing equations for *ratios* of asset prices instead of prices themselves. A ratio has a numerator and a denominator. In a dynamic setting, both of these change. The expected rate of change of each element may be unknown, but under some conditions, the expected rate of change of the *ratio* of the *two* may be a *known* number. For example, the numerator and denominator of a deterministic ratio may grow at the same unknown rate. But the ratio itself will stay the same. Thus, if the numerators and denominators in pricing formulas are carefully selected, and if the Girsanov theorem is skillfully exploited, modeling of asset price dynamics can be greatly simplified.

In order to start discussing the issue of normalization, we first let the short rate fluctuate randomly from one period to another and then try to see whether basic results obtained in Chapter 2 remain the same. Clearly, this makes the discussion directly applicable to interest-sensitive derivatives, given that pricing of such securities needs to assume stochastic interest rates. But this is not the main point.

It turns out that once interest rates become stochastic, we have new ways of searching for synthetic probabilities, especially when we deal with interest-sensitive instruments. Although the general philosophy of the approach introduced in Chapter 2 remains the same, the mechanics change in a dramatic way. In fact, one can show that using *different* synthetic probabilities will be more practical for different classes of financial derivatives. Obviously, the final arbitrage-free price that one obtains will be identical in each case. After all, what matters is not the synthetic probability, but the underlying unique state-price vector. Yet some synthetic probabilities may be more practical than others.

This simple step, which appears at the outset inconsequential, turns out to be very important for the practical utilization of synthetic probabilities, or "measures," as a pricing tool in finance. In fact, we discover that choosing one measure over another equally "correct" probability can simplify the pricing effort dramatically. The second objective of this chapter is to explain this complex idea in a simple setting.[1]

It is also the case that earlier chapters dealt with a very limited number of derivative instruments. Most discussion centered on plain vanilla options within the Black–Scholes environment. Occasionally, some forward contract was discussed. The present chapter is a new step in this respect as well. Forward contracts and options written on Libor rates or bonds are the most liquid of all derivative instruments, yet their treatment within the simple setting of Chapter 2 was not possible with constant spot rates. In this chapter, we incorporate these important instruments in the context of the Fundamental Theorem of Finance and show that their treatment requires additional tools.

17.2 A MODEL FOR NEW INSTRUMENTS

We need to remember first the simplified setting of Chapter 2. A non-dividend paying stock S_t, a European call option C_t, and risk-free borrowing and lending were considered in a two-state, one-period setting. The Fundamental

[1]One can also ask the following question. Given that we want to convert asset prices into martingales by modifying the true probability distribution, is there a way we can choose this synthetic measure in some "best" way?

Theorem of Finance then gives the following linear relation between the possible future values and the current arbitrage-free prices of the three assets under consideration:

$$\begin{bmatrix} 1 \\ S_t \\ C_t \end{bmatrix} = \begin{bmatrix} (1+r\Delta) & (1+r\Delta) \\ S^u_{t+\Delta} & S^d_{t+\Delta} \\ C^u_{t+\Delta} & C^d_{d+\Delta} \end{bmatrix} \begin{bmatrix} \psi^u \\ \psi^d \end{bmatrix} \tag{17.1}$$

where Δ is the time that elapses between the two time periods, the u and the d represent the two-states under consideration, and the $\left\{\psi^u > 0, \psi^d > 0\right\}$ are state prices. The first row represents the payoffs of risk-free lending and borrowing, the second row represents the payoffs of the stock S_t, and the third row represents payoffs of the option C_t.[2]

According to the Fundamental Theorem of Finance, the $\left\{\psi^u, \psi^d\right\}$ will exist and will be positive if there are no arbitrage possibilities given $\{r, S_t, C_t\}$. The reverse is also true. If the $\left\{\psi^u, \psi^d\right\}$ exist and are positive, then there will be no arbitrage opportunity at the prices shown on the left-hand side.

The risk-free probability $\mathbb{Q}$ was obtained from the first row of this matrix,

$$1 = (1+r\Delta)\,\psi^u + (1+r\Delta)\,\psi^d$$

which by defining

$$\mathbb{Q}_u = (1+r\Delta)\,\psi^u$$

gave

$$\mathbb{Q}_d = (1+r\Delta)\,\psi^d$$

The conditions $0 \leq \mathbb{Q}_u, 0 \leq \mathbb{Q}_d$ are satisfied, given the positiveness of state prices ψ^u, ψ^d.

Thus with $\mathbb{Q}_u, \mathbb{Q}_d$ we had two numbers that were positive and that summed to one. These satisfy the requirements of a probability distribution within this simple setting, and hence, we called the $\mathbb{Q}_u, \mathbb{Q}_d$ synthetic, or more precisely, risk-neutral probabilities. These probabilities, which said nothing about the real-world odds of the states u, d, were called "risk-neutral" due to the following.

Consider the second and third rows of the system above in isolation:

$$S_t = S^u_{t+\Delta}\psi^u + S^d_{t+\Delta}\psi^d \tag{17.2}$$

$$C_t = C^u_{t+\Delta}\psi^u + C^d_{t+\Delta}\psi^d \tag{17.3}$$

Multiply the ψ^u, ψ^d by $(1+r\Delta)/(1+r\Delta)$ and introduce the $\mathbb{Q}_u, \mathbb{Q}_d$ to obtain the pricing equations:

$$\begin{aligned} S_t &= S^u_{t+\Delta}\frac{\mathbb{Q}_u}{1+r\Delta} + S^d_{t+\Delta}\frac{\mathbb{Q}_d}{1+r\Delta} \\ &= \frac{1}{1+r\Delta}\mathbb{E}^{\mathbb{Q}}\left[S_{t+\Delta}\right] \end{aligned} \tag{17.4}$$

and

$$\begin{aligned} C_t &= C^u_{t+\Delta}\frac{\mathbb{Q}_u}{1+r\Delta} + C^d_{t+\Delta}\frac{\mathbb{Q}_u}{1+r\Delta} \\ &= \frac{1}{1+r\Delta}\mathbb{E}^{\mathbb{Q}}\left[C_{t+\Delta}\right] \end{aligned} \tag{17.5}$$

where the $\mathbb{E}^{\mathbb{Q}}[\cdot]$ denotes, as usual, the (conditional) expectation operator that uses the probabilities $\mathbb{Q}_u, \mathbb{Q}_d$. Note that we are omitting the t subscript in $\mathbb{E}^{\mathbb{Q}}[\cdot]$ to simplify the notation in this chapter.

According to these pricing equations, expected future payoffs of the risky assets discounted by the *risk-free* rate give the current arbitrage-free price. It is in this sense that $\mathbb{Q}_u, \mathbb{Q}_d$ are "risk-neutral." Even though market prices S_t, C_t contain risk premia, they are nevertheless obtained using the $\mathbb{Q}_u, \mathbb{Q}_d$, as if they come from a risk-neutral world.

There was a second important result that was obtained from these pricing equations. Rearranging (17.4) and (17.5), we get

$$1 + r\Delta = \mathbb{E}^{\mathbb{Q}}\left[\frac{S_{t+\Delta}}{S_t}\right]$$

$$1 + r\Delta = \mathbb{E}^{\mathbb{Q}}\left[\frac{C_{t+\Delta}}{C_t}\right]$$

[2]We make slight modifications in the notation compared to the simple model used in Chapter 2. In particular we introduce indexing by u and d, which stand for the two-states.

Thus, the probability $\mathbb{Q}$ "modified" expected returns of the risky assets so that all expected returns became equal to the risk-free rate r. Hence, the term risk-neutral measure or probability.

In the next section we extend this framework in two ways. First, we add another time period so that the effects of random fluctuations in the spot rate can be taken into account. Second, we change the types of instruments considered and introduce interest-sensitive securities.

17.2.1 The New Environment

We consider two periods described by dates $t_1 < t_2 < t_3$, but keep the assumption of two possible states in each time period the same. Adding one more time period still increases the number of possibilities. This way, in looking at time $t_3 = 1+2\Delta$ from time $t_1 = 1$ there will be *four* possible states, $\{\omega_i, i = 1, \ldots, 4\}$, describing the possible paths the prices can follow at time-nodes $\{t_1, t_2, t_3\}$:

$$\left\{\begin{array}{ll} \omega_1 = \text{down, down} & \omega_2 = \text{up, down} \\ \omega_3 = \text{down, up} & \omega_4 = \text{up, up} \end{array}\right\}$$

It turns out that a minimum of two time periods is necessary to factor in the effects of the random spot rate r_t. The situation is shown in Figures 17.1 and 17.2. An investor who would like to lend his or her money between t_1 and t_2 does this at the risk-free rate contracted at time t_1. Then, no matter which state occurs in the immediate future, his or her return is not risky, because the payoff is known.[3]

Regardless of the state *up* or *down* that may occur at t_2, the investor will receive the same income $\{1 + r_{t_1}\Delta\}$. Because the riskless borrowing and lending yields the same return whether the *up* and *down* state occurs, in a model where there are two time-nodes, t_1, t_2, it will be *as if* the spot rate does not fluctuate. So the effect of any randomness in r_t cannot be analyzed.

[3]It is assumed that there is no default risk in this time setting.

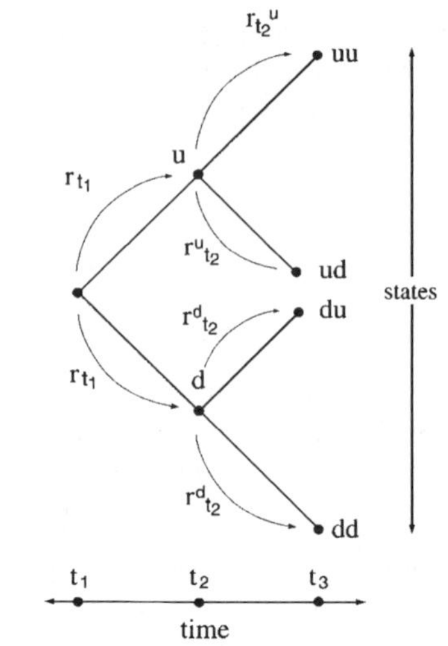

FIGURE 17.1 Illustration of randomness of risk-free rate.

But as we add one more time period this changes. Looked at from time $t_1 = 1$, the spot rate that the investor will be offered at time $t_2 = 1+\Delta$ will present some risks. By staying with this type of investment, the investor may end up lending his or her money either at a "high" rate of interest $r^d_{t_2}$ or at a "lower" rate $r^u_{t_2}$. Which of these spot rates will be available at t_2 is *not* known at time t_1. Thus, with three time-nodes t_1, t_2, t_3,[4] the randomness of the "risk-free" rates will be an important factor, although with the one-period framework of Chapter 2 this randomness was not relevant. This situation was presented in Figure 17.1. For time-node t_2 we have two possible spot rates, $r^u_{t_2}$ and $r^d_{t_2}$, and hence the value of r_{t_2} is random.[5]

[4]This means two time periods.

[5]This shows that the term "risk-free" investment is not entirely appropriate for a savings (or money market) account. The investment is *risk-free* in the sense that what will be received at the end of the contract is known. The contract has no "market" risk. Its price will not fluctuate during the contract period because the r_t is constant. In addition, we assume that there is no default-risk either. So the payoff at the end of the contract period is constant. Yet, an investor who keeps his or her funds in the savings account will experience random fluctuations in the payoffs if this investment is rolled-over.

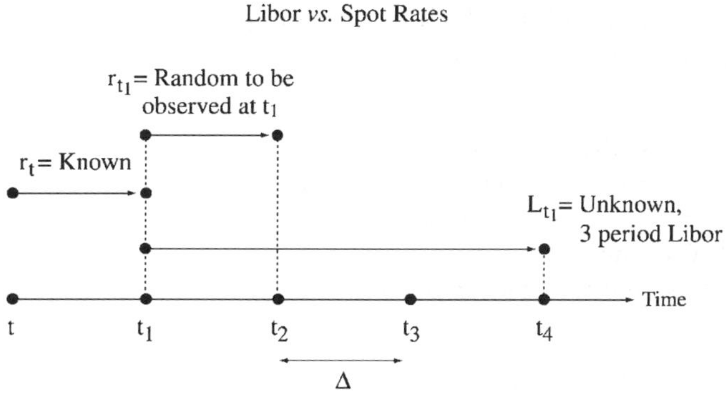

FIGURE 17.2 Effects of the random spot rate r_t

The second modification that we introduce to Chapter 2 is in the selection of instruments. Instead of dealing with stocks and options written on these stocks, we consider interest-sensitive securities and forwards. This extends arbitrage pricing to more interesting assets and, at the same time, gives us an easy way of showing why changing normalization is a useful tool for the financial market practitioner. In particular, within this framework we will be able to introduce the so-called *forward* measure and compare its properties with the *risk-neutral* measure seen earlier.

Hence, we assume that there are liquid markets for the following instruments[6]:

- A *savings account* with no default risk. At time t one can contract the rate r_t, and after an interval Δ, one is paid $(1 + r\Delta)$. If the investor wants to stay with this short-term investment, he or she will have to contract a new spot rate $r_t + \Delta$ at time $t + \Delta$.
- A *forward contract* on an interest rate L_t. This interest rate is default-free. It is also a spot rate that can be contracted for more than just one period. For example, it could represent the *simple* interest rate at which a business borrows for 6 months, such as the 6-month Libor. Hence the choice of L_t as the symbol.[7]
- A *short*-maturity default-free discount bond with time-t price $B(t, t_3)$, $t < t_3$. This bond pays one dollar at maturity t_3, and nothing else at other times.
- A *long*-maturity default-free discount bond with time-t price $B(t, T)$. This bond pays 1 at maturity date T. Because we want this bond to have longer maturity, we let $t < t_3 < T$. Choosing a numerical value for T is not necessary in our model.
- A FRA contract written on the L_{t_2} that results in payoff at time t_3. Here the F_{t_2} is a forward rate contracted at time t_1. If $F_{t_2} > L_{t_2}$, the buyer of the FRA (Forward Rate Agreement) pays the net amount. If $F_{t_2} < L_{t_2}$, the buyer receives the net amount. Note that the payment depends on the rate that becomes known at time t_2, but the proceeds from the FRA are paid (received) at time t_3. This is what makes the FRA *in-arrears*. Δ is the day's adjustment.
- And, finally, we consider an interest rate *derivative*, say a call option written on

[6]In this context, the liquidity of markets means that the assets can be instantaneously bought and sold at the quoted prices.

[7]As defined in the previous chapter, Libor is the London Interbank Offered Rate, an interest rate at which banks can borrow money in London. Libor rates are used as benchmarks and Libor-based instruments form an important proportion of assets in bank balance sheets.

$B(t_1, t_3)$ or a caplet that involves the L_{t_2}. The derivative expires at time $t = t_3$ and has the current price C_{t_1}.

We now need to stack these assets and the corresponding payoffs in a matrix equation similar to the one used in Chapter 2. But, first we make some notational simplifications.

We define the *gross* risk-free returns for periods t_1 and t_2 as follows:

$$R_{t_1} = (1 + r_{t_1}\Delta)$$
$$R_{t_2} = (1 + r_{t_2}\Delta)$$

Although we will revert back to the original notation in later chapters, for the sake of simplifying the matrix equation discussed below, we simplify the notation for bonds as well. We let

$$B^s_{t_1} = B(t_1, t_3)$$

which represents the price of the "short" bond at current period t_1, and let the

$$B_{t_1} = B(t_1, T)$$
$$B_{t_3} = B(t_3, T)$$

represent the price of the "long" bond at times t_1 and t_3, respectively. We do not need to consider the price of the long bond for the interim period t_2.

Then, we set the notional amount of the FRA contract denoted by N equal to 1, because this parameter plays an inconsequential role in our model.

Finally, we assume that all interest rates are expressed as rates over periods of length Δ. This way we do not need to multiply an "annual" rate r_t by a factor of Δ to obtain corresponding returns. This is also intended to simplify the notation. Alternatively one can take $\Delta = 1$ as equaling one year.

Now we can write the matrix equation implied by the Fundamental Theorem of Finance. Stacking the current prices of the five instruments discussed above in a (5×1) vector on the left-hand side, we obtain the relation:

$$\begin{bmatrix} 1 \\ 0 \\ B^s_{t_1} \\ B_{t_1} \\ C_{t_1} \end{bmatrix} = \begin{bmatrix} R_{t_1}R^u_{t_2} & R_{t_1}R^u_{t_2} & R_{t_1}R^d_{t_2} & R_{t_1}R^d_{t_2} \\ \left(F_{t_1} - L^u_{t_2}\right) & \left(F_{t_1} - L^u_{t_2}\right) & \left(F_{t_1} - L^d_{t_2}\right) & \left(F_{t_1} - L^d_{t_2}\right) \\ 1 & 1 & 1 & 1 \\ B^{uu}_{t_3} & B^{ud}_{t_3} & B^{du}_{t_3} & B^{dd}_{t_3} \\ C^{uu}_{t_3} & C^{ud}_{t_3} & C^{du}_{t_3} & C^{dd}_{t_3} \end{bmatrix} \times \begin{bmatrix} \psi^{uu} \\ \psi^{ud} \\ \psi^{du} \\ \psi^{dd} \end{bmatrix} \tag{17.6}$$

where the right-hand side consists of the product of possible payoffs at time t_3, multiplied by the states prices ψ^{ij}. This matrix equation is similar to the one used in Chapter 2, yet, given the new complications, several comments are in order.

The first row of this system describes what happens to an investment in the "risk-free" savings account. If one dollar is invested here, it will return a *known* $R_{t_1} = (1 + r_{t_1})$ at time t_2 and an *unknown* $R_{t_2} = (1 + r_{t_2})$ at time t_3. Time t_3 return is random because, in contrast to R_{t_1} the R_{t_2} is unknown at time t_1. At time t_2, there are two possibilities, and this is indicated by the superscripts u, d on the $R^d_{t_2}$, $R^u_{t_2}$. This row is similar to the first row in Eq. (17.1), except here the elements are not constant.

Next, consider the second row of this matrix equation. The F_{t_1} is a forward rate contracted at time t_1 on the random Libor rate L_{t_2} which will be observed at time t_2. Hence, we have here a Forward Rate Agreement. It turns out that FRAs have the arbitrage-free value *zero* at *contract*-time, because no up-front payment is required for signing these contracts. This explains the second element of the vector on the left-hand side. Also, according to this FRA contract, the difference between the known F_{t_1} and the unknown L_{t_3} will be paid (received) at time t_3, and this explains the second row of the matrix. Clearly, there are four possibilities here.

The third and fourth rows of the matrix equation deal with the two bonds we included in the system. The $B^s_{t_1}$ *and* B_{t_1} denote the time-t_1 arbitrage-free prices of the two zero-coupon bonds, the first maturing at time t_3, the second at some future date T, respectively. Note that the value of the short bond is constant and equal to one at t_3 because this happens to be the maturity date. On the other hand, the price of the long bond does not have this property. There are four possible values that B_{t_3} can assume.

The last row of the matrix equation represents the price, C_{t_1}, of a derivative security written on one or more of these assets.

Finally, the $\left\{\psi^{ij}, i,j = u,d\right\}$ are the 4 state prices for time t_3. They exist and they are positive if and only if there are no arbitrage opportunities. As was the case in Chapter 2, it is important that for all i,j:

$$\psi^{ud} > 0$$

Consider now how this setup differs from that of Chapter 2. First, in this matrix equation, the risks of investing in a "risk-free" savings account can be seen explicitly. The investment is risk-free only for one period. An investor may be certain about the payoff at time $t_2 = t+\Delta$, the immediate future. But, one period down the line, spot rates may yield a higher or lower return depending on the state of the world $\omega_{i,j}$. Hence, in general terms, the current spot rate r_t is known, but $r_{t+\Delta}$ is still random. It may be "low" (the up state), or "high" (the down state). Consequently it carries a superscript indicating this dependence on the realized state at time t_2.

Second, note that the securities included in this model are quite different from those of Chapter 2. The forward contract and the bonds considered here are interest-sensitive instruments and the pricing of them is likely to be more delicate than the assets selected for the simpler model of Chapter 2. The same is true for the option C_t. The option is written on interest-sensitive securities.

Finally, note a straightforward aspect of the model. Because one of the bonds matures at time t_3, its payoff is known and is *constant* at that time. This simple point will have important implications for *choosing* a synthetic probability (martingale measure) that is more convenient for pricing the new assets introduced here.

We can now consider the important issue of normalization that determines the choice of measure for the instruments under consideration. But first we need the following caveat.

17.2.1.1 A Remark

In this chapter the R_t and the L_t denote the short rate and the Libor process, respectively. In principle, these two *are* different processes, with L_t having a different maturity than the spot rate, which is by definition the rate on the shortest possible tenor. But, because we want to keep the instruments and the model at a minimum level of complication, we assume that the L_t is the one-period Libor. This would make R_t and L_t the same, but even with this consideration all results of this chapter still hold. The chapter treats these as if they are different notationally because with longer maturity Libor rates this equivalence will disappear. Using different notation for L_t and R_t will help us better understand the Libor instruments and their relationship to spot rates in these more general cases.

An alternative is to consider a two-period Libor. But this would require a three-period model, which will lead to a much more complicated matrix equation than the one considered here. In this sense, the compromise of two periods and four states of the world is the smallest system in which the issue of normalization can be discussed.

17.2.2 Normalization

Again we begin with a review of the framework in Chapter 2. Consider how the risk-neutral probabilities were obtained given the Fundamental Theorem of Finance. More precisely, take the equation for S_t given by Eq. (17.1) and earlier in Chapter 2:

$$S_t = S^u_{t+\Delta}\psi^u + S^d_{t+\Delta}\psi^d \tag{17.7}$$

In order to introduce the risk-neutral probability in this equation, we multiplied each ψ^i by the ratio $(1 + r\Delta) / (1 + r\Delta)$:

$$S_t = S^u_{t+\Delta}\psi^u \frac{(1+r\Delta)}{(1+r\Delta)} + S^d_{t+\Delta}\psi^d \frac{(1+r\Delta)}{(1+r\Delta)} \tag{17.8}$$

and then recognized that the $\psi^i (1 + r\Delta)$ are in fact the $\mathbb{Q}_i$. This resulted in

$$S_t = \frac{1}{(1+r\Delta)} S^u_{t+\Delta}\mathbb{Q}_u + \frac{1}{(1+r\Delta)} S^d_{t+\Delta}\mathbb{Q}_d \tag{17.9}$$

Now, during this operation, when the $\mathbb{Q}_i$ are substituted for the $\psi^i (1 + r\Delta)$, an extra factor is left in the denominator of each term. This factor is $(1 + r\Delta)$, and represents the return of the risk-free investment in that particular state. But, recall that one-period-ahead risk-free return is constant, and hence this factor was successfully factored out to give:

$$S_t = \frac{1}{(1+r\Delta)} \left[S^u_{t+\Delta}\mathbb{Q}_u + S^d_{t+\Delta}\mathbb{Q}_d \right] \tag{17.10}$$

However, note that in the new model with two periods, this return will be a random variable and a similar operation will *not* be possible.

Nevertheless, the point to remember here is that the process of introducing the risk-neutral probabilities in the pricing equations resulted in the "normalization" of each state's return by the corresponding return of risk-free lending. In fact, the substitution of $\mathbb{Q}_i$ for ψ^i is equivalent to dividing every possible entry in the matrix Eq. (17.1) by the corresponding entry of the first row, which represents the payoff of the risk-free investment.

As mentioned earlier, under the risk-neutral measure $\mathbb{Q}$ asset prices will have trends. Indeed, under $\mathbb{Q}$ all expected returns are converted to r and this means that they will drift upwards. Thus, the prices *themselves* will not be martingales under $\mathbb{Q}$. But, by "normalizing" with risk-free lending, the expected return (i.e., the trend) of the *ratio* becomes zero. In other words, the numerator and the denominator will "trend" upwards by the *same* expected drift r and the normalized variable becomes a martingale. It will have no discernible trend.

Note one additional characteristic of this normalization. The division by $(1 + r\Delta)$ amounts to discounting a future cash flow to present. But, in normalizing by risk-free lending, one *first* discounts, and *then* averages using the probability to get:

$$S_t = \mathbb{E}^{\mathbb{Q}} \left[\frac{S_{t+\Delta}}{(1+r\Delta)} \right] \tag{17.11}$$

The r being constant, this simplifies to

$$S_t = \frac{1}{(1+r\Delta)} \mathbb{E}^{\mathbb{Q}} \left[S_{t+\Delta} \right] \tag{17.12}$$

This is the pricing equation used several times in the first part of this book. As with all expectation operators used in this chapter, $\mathbb{E}^{\mathbb{Q}}$ is the expectation conditional on time t information unless indicated otherwise.

Would the same steps work in the two-period model that incorporates interest rate derivatives? The answer is no.

Consider trying the same strategy in the new model shown in (17.6). Suppose we decide to determine the risk-neutral probabilities $\mathbb{Q}^{ij}$ by starting again with the first row of the system which corresponds to the savings account:

$$1 = R_{t_1}R^u_{t_2}\psi^{uu} + R_{t_1}R^u_{t_2}\psi^{ud} + R_{t_1}R^d_{t_2}\psi^{du} + R_{t_1}R^d_{t_2}\psi^{dd} \tag{17.13}$$

where the state subscript is applied only to R_{t_2} because at time t_1, the R_{t_1} is known with certainty.

We can define the four risk-neutral probabilities in a similar fashion:

$$\mathbb{Q}_{ij} = \left(1 + r_{t_1}\right)\left(1 + r^i_{t_2}\right)\psi^{ij} \tag{17.14}$$

with $i, j = u, d$. Then Eq. (17.13) becomes:

$$1 = \mathbb{Q}_{uu} + \mathbb{Q}_{ud} + \mathbb{Q}_{du} + \mathbb{Q}_{dd} \tag{17.15}$$

If there are no arbitrage opportunities, the state prices $\left\{\psi^{ij}\right\}$ will be positive and we will have

$$\mathbb{Q}_{ij} > 0 \tag{17.16}$$

Clearly, as in Chapter 2, we can use the $\mathbb{Q}_{ij}$ *as if* they are probabilities associated with the states of-the-world, even though they do not have any probabilistic implications concerning the actual realization of any of the four states. Thus, proceeding in a way similar to Chapter 2, we can exploit the remaining equations of system (17.6) in order to obtain the corresponding martingale equalities.

For example, the third row of the system gives the short bond's arbitrage-free price under the measure $\mathbb{Q}$:

$$\begin{aligned} B^s_{t_1} = & \frac{1}{\left(1+r_{t_1}\right)\left(1+r^u_{t_2}\right)}\mathbb{Q}_{uu} \\ & + \frac{1}{\left(1+r_{t_1}\right)\left(1+r^u_{t_2}\right)}\mathbb{Q}_{ud} \\ & + \frac{1}{\left(1+r_{t_1}\right)\left(1+r^d_{t_2}\right)}\mathbb{Q}_{du} \\ & + \frac{1}{\left(1+r_{t_1}\right)\left(1+r^d_{t_2}\right)}\mathbb{Q}_{dd} \end{aligned} \tag{17.17}$$

or

$$B^s_{t_1} = \mathbb{E}^{\mathbb{Q}}\left[\frac{1}{\left(1+r_{t_1}\right)\left(1+r_{t_2}\right)}\right] \tag{17.18}$$

where the r_{t_2} is random and hence cannot be moved out of the (conditional) expectation sign.

By moving to continuous time and then assuming a continuum of states, we can generalize this formula for an arbitrary maturity $T, t < T$. The arbitrage-free price of a default-free zero-coupon bond will then be given by:

$$B(t,T) = \mathbb{E}^{\mathbb{Q}}\left[e^{-\int_t^T r_u du}\right] \tag{17.19}$$

This equation will be used extensively in later chapters.

We now get similar pricing formulas for the long bond in system (17.6) by going over the same steps. The fourth row of (17.6) gives:

$$\begin{aligned} B_{t_1} = & \frac{B^{uu}_{t_3}}{\left(1+r_{t_1}\right)\left(1+r^u_{t_2}\right)}\mathbb{Q}_{uu} \\ & + \frac{B^{ud}_{t_3}}{\left(1+r_{t_1}\right)\left(1+r^u_{t_2}\right)}\mathbb{Q}_{ud} \\ & + \frac{B^{du}_{t_3}}{\left(1+r_{t_1}\right)\left(1+r^d_{t_2}\right)}\mathbb{Q}_{du} \\ & + \frac{B^{dd}_{t_3}}{\left(1+r_{t_1}\right)\left(1+r^d_{t_2}\right)}\mathbb{Q}_{dd} \end{aligned} \tag{17.20}$$

or,

$$B^s_{t_1} = \mathbb{E}^{\mathbb{Q}}\left[\frac{B_{t_3}}{\left(1+r_{t_1}\right)\left(1+r_{t_2}\right)}\right] \tag{17.21}$$

Here, r_{t_2} is again a random variable. But so is B_{t_3} because the time t_3 is not a maturity date for this bond. For this reason, the equation in this form will not be very useful in practical pricing situations.

Finally, using the second and the fifth rows of (17.6) yields the pricing equations for the two Libor instruments, the FRA and the caplet derivative C_t, respectively:

$$0 = \mathbb{E}^{\mathbb{Q}}\left[\frac{1}{\left(1+r_{t_1}\right)\left(1+r_{t_2}\right)}\left[F_{t_1} - L_{t_2}\right]\right] \tag{17.22}$$

$$C_{t_1} = \mathbb{E}^{\mathbb{Q}}\left[\frac{1}{\left(1+r_{t_1}\right)\left(1+r_{t_2}\right)}C_{t_3}\right] \tag{17.23}$$

Thus, proceeding in a way similar to that in Chapter 2, and using the savings account to determine the $\mathbb{Q}$ *does* lead to pricing formulas similar to the ones in (17.18) or (17.19). Yet, within the context of the instruments considered here, and with stochastic spot rates, the use of risk-neutral probabilities $\mathbb{Q}$ turns out to be less convenient and, at times, even inappropriate. It forces a market practitioner into handling unnecessary difficulties. The next section illustrates some of these.

17.2.3 Some Undesirable Properties

The probabilities $\mathbb{Q}^{ij}$ were generated using the savings account equation:

$$1 = R_{t_1} R^u_{t_2} \psi^{uu} + R_{t_1} R^u_{t_2} \psi^{ud} + R_{t_1} R^d_{t_2} \psi^{du} + R_{t_1} R^d_{t_2} \psi^{dd} \tag{17.24}$$

which after relabeling gave:

$$1 = \mathbb{Q}_{uu} + \mathbb{Q}_{ud} + \mathbb{Q}_{du} + \mathbb{Q}_{dd} \tag{17.25}$$

Now consider the details of how these probabilities were used in pricing the FRA contract. First, note that pricing the FRA means determining an F_{t_i} such that the time t_1 value of the contract is zero. This is the case because all FRAs are traded at a price of zero and this we consider as the arbitrage-free price. The task is to determine the arbitrage-free F_{t_i} implied by this price. From the second row of the system in (17.6), we have

$$0 = \left(F_{t_1} - L^u_{t_2}\right) \psi^{uu} + \left(F_{t_1} - L^u_{t_2}\right) \psi^{ud} + \left(F_{t_1} - L^d_{t_2}\right) \psi^{du} + \left(F_{t_1} - L^d_{t_2}\right) \psi^{dd} \tag{17.26}$$

Multiply and divide each term on the right-hand side by the corresponding $(1 + r_{t_1})\left(1 + r^i_{t_2}\right)$ and relabel using

$$(1 + r_{t_1})\left(1 + r^i_{t_2}\right) \psi^{ij} = \mathbb{Q}^{ij} \tag{17.27}$$

to obtain

$$0 = \frac{\left(F_{t_1} - L^u_{t_2}\right)}{(1 + r_{t_1})\left(1 + r^u_{t_2}\right)} \mathbb{Q}^{uu} + \frac{\left(F_{t_1} - L^u_{t_2}\right)}{(1 + r_{t_1})\left(1 + r^u_{t_2}\right)} \mathbb{Q}^{ud} + \frac{\left(F_{t_1} - L^d_{t_2}\right)}{(1 + r_{t_1})\left(1 + r^d_{t_2}\right)} \mathbb{Q}^{du} + \frac{\left(F_{t_1} - L^d_{t_2}\right)}{(1 + r_{t_1})\left(1 + r^d_{t_2}\right)} \mathbb{Q}^{dd} \tag{17.28}$$

Factoring out the F_{t_i}, which is independent of the realization of any future state:

$$F_{t_i} \left[\frac{\mathbb{Q}^{uu}}{(1 + r_{t_1})\left(1 + r^u_{t_2}\right)} + \frac{\mathbb{Q}^{ud}}{(1 + r_{t_1})\left(1 + r^u_{t_2}\right)} + \frac{\mathbb{Q}^{du}}{(1 + r_{t_1})\left(1 + r^d_{t_2}\right)} + \frac{\mathbb{Q}^{dd}}{(1 + r_{t_1})\left(1 + r^d_{t_2}\right)} \right]$$
$$= \frac{L^u_{t_2}\mathbb{Q}^{uu}}{(1 + r_{t_1})\left(1 + r^u_{t_2}\right)} + \frac{L^u_{t_2}\mathbb{Q}^{ud}}{(1 + r_{t_1})\left(1 + r^u_{t_2}\right)} + \frac{L^d_{t_2}\mathbb{Q}^{du}}{(1 + r_{t_1})\left(1 + r^d_{t_2}\right)} + \frac{L^d_{t_2}\mathbb{Q}^{dd}}{(1 + r_{t_1})\left(1 + r^d_{t_2}\right)} \tag{17.29}$$

This we can write as:

$$F_{t_1} \mathbb{E}^{\mathbb{Q}} \left[\frac{1}{(1 + r_{t_1})(1 + r_{t_2})} \right] = \mathbb{E}^{\mathbb{Q}} \left[\frac{L_{t_2}}{(1 + r_{t_1})(1 + r_{t_2})} \right] \tag{17.30}$$

Rearranging, we obtain a pricing formula which gives the arbitrage-free FRA rate:

$$F_{t_1} = \frac{1}{\mathbb{E}^{\mathbb{Q}} \left[\frac{1}{(1+r_{t_1})(1+r_{t_2})} \right]} \mathbb{E}^{\mathbb{Q}} \left[\frac{L_{t_2}}{(1 + r_{t_1})(1 + r_{t_2})} \right] \tag{17.31}$$

This expression yields a formula to determine the contractual rate F_t using the risk-free probability $\mathbb{Q}$. But, unlike the case of option valuation with constant interest rates, we immediately see some undesirable properties of the representation.

First, in general, F_t is not an unbiased estimate of L_{t_2}:

$$F_{t_1} \neq \mathbb{E}^{\mathbb{Q}}\left[L_{t_2}\right] \tag{17.32}$$

The only time this will be the case is when the r_t and the L_t are *statistically independent*. Then, the expectations can be taken separately:

$$F_{t_1} = \frac{1}{\mathbb{E}^{\mathbb{Q}}\left[\frac{1}{(1+r_{t_1})(1+r_{t_2})}\right]} \times \mathbb{E}^{\mathbb{Q}}\left[\frac{1}{\left(1+r_{t_1}\right)\left(1+r_{t_2}\right)}\right]\mathbb{E}^{\mathbb{Q}}\left[L_{t_2}\right] \tag{17.33}$$

After canceling, we get

$$F_{t_1} = \mathbb{E}^{\mathbb{Q}}\left[L_{t_2}\right] \tag{17.34}$$

Under this extreme assumption the forward rate becomes an unbiased estimator of the corresponding Libor process. But, in practice, can we really say that the short rates and the longer maturity Libor rates are *statistically independent*? This will be a difficult assumption to maintain.

Consider the second drawback of using the risk-neutral measure $\mathbb{Q}$. As we noticed earlier, the spot-rate terms inside the expectations taken with respect to $\mathbb{Q}$ do not factor out. In contrast to the simple model of Chapter 2, where r was constant across states, we now have an r_{t_2} that depends on the state u, d. Hence, the denominator terms in Eq. (17.31) are stochastic, and stay *inside* the expectation.

Third, the pricing formula for the FRA in (17.31) is *not* linear. This property, although harmless at first sight, can be quite a damaging aspect of the use of risk-neutral measure. It creates major inconveniences for the market practitioner. In fact, when we try to determine the FRA rate F_{t_2} or the price of the derivative C_t, we now need to model two processes, namely the r_t and L_t, instead of one, the L_t. Worse, these two processes are correlated with each other in some complicated way. The task of evaluating the corresponding expectations can be arduous with nonlinear expressions.

A final comment. Note that, by definition, the F_{t_1} is denominated in a currency value that will be settled in period t_3. Now consider how the risk-neutral measure operates in pricing Eq. (17.31). The pricing formula with $\mathbb{Q}$ works by first discounting to present a value that belongs to time t_3. Then, after taking the average via the expectation operator, the formula tries to reexpress this discounted term in time t_3 dollars, simply because that is eventually how the contract is settled.

Clearly, this is not a very efficient way of calculating the arbitrage-free forward rate. In fact, one can dispense with the discounting altogether, because both the F_{t_1} and L_{t_2} are measured in time t_3 dollars!

Proceeding in a similar fashion for the Libor derivative C_t, we make the same argument. The pricing equation will be given by:

$$C_{t_1} = \mathbb{E}^{\mathbb{Q}}\left[\frac{1}{\left(1+r_{t_1}\right)\left(1+r_{t_2}\right)}C_{t_3}\right] \tag{17.35}$$

Again, if the C_t is an interest-sensitive derivative, the same problems with $\mathbb{Q}$ will be present. The random discount factor cannot be factored out of the expectation and the spot rate will in all likelihood be correlated with the option payoff C_{t_3} if the latter is written on interest-sensitive securities.

Clearly, using the money market account to define the probabilities $\mathbb{Q}$ as done in Eq. (17.24) creates complications which were not present in Chapter 2. Below we will see that a judicious choice of synthetic probabilities can get around these problems in a very convenient and elegant way.

17.2.4 A New Normalization

We now consider an alternative way of obtaining martingale probabilities. Within the same setup as in (17.6) and with the same ψ^{ij}, we can utilize the third equation to write:

$$B^s_{t_1} = \psi^{uu} + \psi^{ud} + \psi^{du} + \psi^{dd} \tag{17.36}$$

Dividing by $B^s_{t_1}$

$$1 = \frac{1}{B^s_{t_1}}\psi^{uu} + \frac{1}{B^s_{t_1}}\psi^{ud} + \frac{1}{B^s_{t_1}}\psi^{du} + \frac{1}{B^s_{t_1}}\psi^{dd} \tag{17.37}$$

and labeling,

$$\pi^{ij} = \frac{1}{B^s_{t_1}}\psi^{ij} \tag{17.38}$$

this equation becomes

$$1 = \pi_{uu} + \pi_{ud} + \pi_{du} + \pi_{dd} \tag{17.39}$$

Because the ψ^{ij} are positive under the condition of no-arbitrage, we have

$$\pi_{ij} > 0, \quad i,j = u,d \tag{17.40}$$

This means that the π_{ij} could be used as a new set of synthetic martingale probabilities. They yield a new set of martingale relationships. We call the π_{ij} the forward measure. Before we consider the advantages of the forward measure, π_{ij}, over the risk-neutral measure, $\mathbb{Q}_{ij}$, we make a few comments on the new normalization.

First, to move from equations written in terms of state prices ψ^{ij}, to those expressed in terms of π, we need to multiply all state-dependent values by $B^s_{t_1}$, which is a value determined at time t_1. Hence, this term is *independent* of the states at future dates and will not carry a state superscript. This means that it will factor out of expectations evaluated under π.

Second, note that we can define a new forward measure for every default-free zero-coupon bond with different maturity. Thus, it may be more appropriate to put a time subscript on the measure, say, π_T, indicating the maturity, T, associated with that particular bond. Given a derivative written on interest-sensitive securities, it is clearly more appropriate to work with a forward measure that is obtained from a bond that matures at the same time that the derivative expires.

Finally, note how normalization is done here. To introduce the probabilities in pricing equations, we multiply and divide each ψ^{ij} by the $B^s_{t_1}$. After relabeling the $\psi^{ij}/B^s_{t_1}$ and π_{ij}, this amounts to multiplying, in the matrix Eq. (17.6), each asset price by the corresponding entry of the short bond $B^s_{t_1}$. Hence, we say that we are "normalizing" by the $B^s_{t_1}$.

These and related issues will be discussed in more detail below.

17.2.4.1 Properties of the Normalization

We now discuss some of the important results of using the new probability measure π instead of $\mathbb{Q}$.

We proceed in steps. First, recall that within the setup in this chapter, the use of the risk-neutral measure leads to an equation where the F_t is a biased estimator of the Libor process L_t. In fact, we had

$$F_{t_1} \neq \mathbb{E}^{\mathbb{Q}}\left[L_{t_2}\right] \tag{17.41}$$

Now consider evaluating the similar expectation under the measure π. To do this, we take the second row in system (17.6) and multiply every element by the ratio $B^s_{t_1}/B^s_{t_1}$, which obviously equals one:

$$0 = \left(F_{t_1} - L^u_{t_2}\right)\frac{B^s_{t_1}}{B^s_{t_1}}\psi^{uu} + \left(F_{t_1} - L^u_{t_2}\right)\frac{B^s_{t_1}}{B^s_{t_1}}\psi^{ud} + \left(F_{t_1} - L^d_{t_2}\right)\frac{B^s_{t_1}}{B^s_{t_1}}\psi^{du} + \left(F_{t_1} - L^d_{t_2}\right)\frac{B^s_{t_1}}{B^s_{t_1}}\psi^{dd} \tag{17.42}$$

Recognizing that the ratios

$$\frac{\psi^{ij}}{B^s_{t_1}} \tag{17.43}$$

are in fact the corresponding elements of π^{ij}, and that they sum to one, we obtain, after factoring out the F_{t_1}:

$$0 = B^s_{t_1}\left[F_{t_1} - \left[L^u_{t_2}\pi_{uu} + L^u_{t_2}\pi_{ud} + L^d_{t_2}\pi_{du} + L^d_{t_2}\pi_{dd}\right]\right] \tag{17.44}$$

Note that here the $B^s_{t_1}$ has conveniently factored out because it is constant, given the observed, arbitrage-free price $B^s_{t_1}$. Canceling and rearranging:

$$F_{t_1} = \left[L^u_{t_2}\pi_{uu} + L^u_{t_2}\pi_{ud} + L^d_{t_2}\pi_{du} + L^d_{t_2}\pi_{dd}\right] \tag{17.45}$$

where the right-hand side is clearly the expectation of the Libor process evaluated using the new martingale probabilities π_{ij}. This means that we now have:

$$F_{t_1} = \mathbb{E}^{\pi}\left[L_{t_2}\right] \tag{17.46}$$

Thus we obtained an important result. Although the F_{t_1} is, in general, a biased estimator of L_{t_2} under the classical *risk-neutral* measure, it becomes an unbiased estimator of L_{t_2} under the new *forward* measure π.

Why is this relevant? How can it be used in practice?

Consider the following general case and revert back to using the Δ instead of the t_i notation. Let the Libor rate for time $t + 2\Delta$ be given by $L_{t+2\Delta}$, the current forward rate be F_t, and consider its future value $F_{t+\Delta}$ with $\Delta > 0$.[8] We can utilize the measure π and write:

$$F_{t+\Delta} = \mathbb{E}^{\pi}_{t+\Delta}\left[L_{t+2\Delta}\right] \tag{17.47}$$

where the subscript of the $\mathbb{E}^{\pi}_t[\cdot]$ operator indicates that the expectation is now taken with respect to information available at time $t + \Delta$. That is,

$$\mathbb{E}^{\pi}_t[\cdot] = \mathbb{E}^{\pi}[\cdot \mid I_t]$$

with the I_t being the information set available at time t. In this particular case, it consists of the current and past prices of all assets under consideration.

[8]Thus F_t is the FRA rate observed "now," whereas the $F_{t+\Delta}$ is the FRA rate that will be observed within a short interval of time Δ.

Next, we recall the recursive property of conditional expectation operators that was used earlier:

$$\mathbb{E}^{\pi}_t\left[\mathbb{E}^{\pi}_{t+\Delta}[\cdot]\right] = \mathbb{E}^{\pi}_t[\cdot] \tag{17.48}$$

which says that the "best" forecasts of future forecasts are simply the forecasts now.[9]

Now, because F_t is an unbiased estimator of $L_{t+2\Delta}$, under π we can write:

$$F_t = \mathbb{E}^{\pi}_t\left[L_{t+2\Delta}\right] \tag{17.49}$$

and use the recursive property of conditional expectations to introduce an $\mathbb{E}^{\pi}_{t+\Delta}$ operator at the "right place":

$$F_{t+\Delta} = \mathbb{E}^{\pi}_t\left[\mathbb{E}^{\pi}_{t+\Delta}\left[L_{t+2\Delta}\right]\right] \tag{17.50}$$

Now, substituting from relation (17.47), this becomes:

$$F_t = \mathbb{E}^{\pi}_t\left[F_{t+\Delta}\right] \tag{17.51}$$

which says that the process $\{F_t\}$ is a *martingale* under the *forward* measure π. As we will see later, this property of forward prices will be very convenient when pricing some interest rate sensitive instruments. A preliminary example of this can already be seen by looking at the similar conditional expectation for the derivative C_t.

Suppose the C_t is the price of a *caplet*. At expiration, the caplet pays the sum:

$$C_{t_3} = N \max\left[L_{t_2} - K, 0\right] \tag{17.52}$$

where N is a notional amount that we set equal to one, the K is the cap-rate selected at time t_1, and the K is the Libor rate realized at time t_2. The payment is made in-arrears at time t_3 and, hence, provides the purchaser of the caplet some sort of insurance against increases in borrowing costs beyond the level K.

How should one price such an instrument? Consider the use of the classical risk-neutral

[9]Again, the "best" is used here in the sense of mean square error.

measure $\mathbb{Q}$. Using standard arguments and the risk-neutral probability $\mathbb{Q}$, we have

$$C_{t_1} = \mathbb{E}^{\mathbb{Q}}\left[\frac{1}{(1+r_{t_1})(1+r_{t_2})}\max\left[L_{t_2}-K,0\right]\right] \tag{17.53}$$

As discussed earlier, in this pricing equation, the (random) spot rate r_{t_2} is likely to be correlated with the (random) Libor rate L_{t_2}, hence the market practitioner will be forced to model *and* calibrate a bivariate process r_t, L_t, in order to price the caplet.

Yet, using the forward measure in the last equation of system (17.6) gives

$$C_{t_1} = B^s_{t_1}\mathbb{E}^{\pi}\left[C_{t_3}\right] \tag{17.54}$$

which means

$$C_{t_1} = B^s_{t_1}\mathbb{E}^{\pi}\left[\max\left[L_{t_2}-K,0\right]\right] \tag{17.55}$$

According to this last equality, the conditional expectation of a function of L_{t_2} is multiplied by the arbitrage-free price of the short bond. That is, the problem of modeling and calibrating a bivariate process has completely disappeared. Inside the expectation sign there is a single random variable, L_{t_2}.

Here, we see the following convenient property of the new measure. The forward measure, π, first calculates the expectation in time t_2 (i.e., forward) dollars and *then* does the discounting using an observed arbitrage-free price, $B^s_{t_2}$. In contrast, the risk-neutral measure first applies a random discount factor to a random payoff, and then does the averaging. Note that in proceeding this way, the risk-neutral measure misses the opportunity of using the discount factor implied by the markets, i.e., the $B^s_{t_2}$ during the pricing process. Instead, the risk-neutral measure is trying to recalculate the discount factor from scratch, as if it is part of the pricing problem, leading to the complicated bivariate dynamics. We will see another example of this in the next section.

17.3 OTHER EQUIVALENT MARTINGALE MEASURES

The forward measure is already discussed. There are other measures that we will cover in this section which are:

share measure
swap measure
spot measure

17.3.1 Share Measure

For the measure that the numeraire is the underlier price S_t is called the share measure, $\mathbb{S}$. Therefore any tradable security deflated by S_t is a martingale under the share measure. This implies that

$$\frac{C_t}{S_t} = \mathbb{E}^{\mathbb{S}}_t\left(\frac{C_T}{S_T}\right) \tag{17.56}$$

For a call option with strike K, we can write

$$\frac{C(K)}{S_0} = \mathbb{E}^{\mathbb{S}}\left(\frac{(S_T-K)^+}{S_T}\right) \tag{17.57}$$

and if we assume the security price S_T is always positive, we get

$$\frac{C(K)}{S_0} = \mathbb{E}^{\mathbb{S}}\left(\left(1-\frac{K}{S_T}\right)^+\right) \tag{17.58}$$

We define $y = \log\left(\frac{S_T}{K}\right)$, which implies $\frac{K}{S_T} = e^{-y}$, and using this definition we can rewrite the normalized call price as follows:

$$\frac{C(K)}{S_0} = \int_0^{\infty}(1-e^{-y})f(y)dy \tag{17.59}$$

where $f(y)$ is the probability density function of $y = \ln(S/K)$ under the share measure.

We perform integration by parts to get

$$\begin{aligned}\frac{C(K)}{S_0} &= \int_0^\infty (1-e^{-y})f(y)dy \\ &= (1-e^{-y})F(y)\Big|_0^\infty - \int_0^\infty e^{-y}F(y)dy \\ &= (F(y)-F(y)e^{-y})\Big|_0^\infty - \int_0^\infty e^{-y}F(y)dy \\ &= 1 - \int_0^\infty e^{-y}F(y)dy \\ &= \int_0^\infty e^{-y}dy - \int_0^\infty e^{-y}F(y)dy \\ &= \int_0^\infty (1-F(y))e^{-y}dy\end{aligned}$$

and thus we have

$$\frac{C(K)}{S_0} = \int_0^\infty (1-F(y))e^{-y}dy$$

For a given y, the expression $1-F(y)$ is the probability that $\ln(S/K)$ is greater than y. Considering that e^{-y} is the probability density function of a positive exponential random variable with $\lambda = 1$, the normalized call price is the probability that under the share measure the logarithm of the stock price exceeds the logarithm of strike by an independent exponential variable (Carr and Madan, 2009) or equivalently

$$\frac{C(K)}{S_0} = P(\ln(S/K) > Y) \tag{17.60}$$

$$= P(\ln S - \ln K > Y) \tag{17.61}$$

$$= P(X - Y > \ln K) \tag{17.62}$$

where X is the logarithm of the stock under the share measure, Y is an independent exponential, and K is the strike.

17.3.2 Spot Measure and Market Models

A Forward Rate Agreement (FRA) is a tradable contract that can be used to directly trade simple forward rates. The contract involves three time instants: (a) the current time t, (b) the expiry time T, where $T > t$, and (c) the maturity time S, with $S > T$. The payoff of the contract at time S is $1+(S-T)F(t;T,S)$, which results in a forward investment of one dollar at time T. However, this investment can be replicated by using the following strategy:

At time t we sell one bond with maturity T, T-bond, and buy $\frac{P(t,T)}{P(t,S)}$ bonds with maturity S, S-bonds, this investment costs zero.
At time T, the T-bond matures and we have to pay one dollar to the holder/buyer of the bond.
At time S, the S-bond matures and we receive $\frac{P(t,T)}{P(t,S)}$ dollars.

The net effect of these transactions is a forward investment of one dollar at time T yielding $\frac{P(t,T)}{P(t,S)}$ dollars at S with certainty. No-arbitrage condition leads to the following equality, $1+(S-T)F(t;T,S) = \frac{P(t,T)}{P(t,S)}$, which implies

$$F(t;T,S) = \frac{1}{S-T}\left(\frac{P(t,T)}{P(t,S)} - 1\right) \tag{17.63}$$

This is called the simple (simply compounded) forward rate for $[T,S]$ prevailing at t. The simple forward rates with a fixed tenor, Δ, are called LIBOR rates, $L(t,T) = F(t,T,T+\Delta)$, where Δ are usually nominally equal, typically $\frac{1}{4}$ or $\frac{1}{2}$.

In market models, LIBOR or swap rates are considered the fundamental interest rates. Considering one of those rates would be our choice; it would be very inconvenient to use risk-neutral measure in pricing, which requires determining the instantaneous short rate at each point in time. As a result we will generally work with other measures. One of those measures is the spot measure. But first we will fix the maturities or tenor dates to which our market models will apply. Considering that prices of liquid securities, such as caps/floors, swaps, and swaptions are determined by the rates (LIBOR or swap) applying to only a finite set of maturities. For that reason, we set in advance a set of tenor dates

$$0 = T_0 < T_1 < T_2 < \cdots < T_M \tag{17.64}$$

with

$$\Delta_i = T_{i+1} - T_i, \quad i = 0, 1, \ldots, M-1 \tag{17.65}$$

Having in mind that the day-count conventions result in slightly different values for each Δ_i. We let $P(t, T_n)$ denote the time t price of a zero-coupon bond maturing at time $T_n > t$ for $n = 1, \ldots, M$. Similarly, we use $L_n(t) = L(t, T_n)$ to denote the time t forward rate applying to the period $[T_n, T_{n+1}]$ for $n = 0, 1, \ldots, M-1$. From Eq. (17.63) we can get the following equation for LIBOR rates

$$L_n(t) = \frac{P(t, T_n) - P(t, T_{n+1})}{\Delta_n P(t, T_{n+1})} \tag{17.66}$$

for $0 \le t \le T_n$ with $n = 0, \ldots, M-1$. The LIBOR rate at time T_i would be

$$L_n(T_i) = \frac{P(T_i, T_n) - P(T_i, T_{n+1})}{\Delta_n P(T_i, T_{n+1})} \tag{17.67}$$

solving for $P(T_i, T_{n+1})$ to get

$$P(T_i, T_{n+1}) = \frac{P(T_i, T_n)}{1 + \Delta_n L_n(T_i)} \tag{17.68}$$

Using (17.68) we can write for $P(T_i, T_n)$

$$P(T_i, T_n) = \frac{P(T_i, T_{n-1})}{1 + \Delta_{n-1} L_{n-1}(T_i)} \tag{17.69}$$

By substituting (17.69) into (17.68), we get

$$P(T_i, T_{n+1}) = \frac{P(T_i, T_{n-1})}{(1 + \Delta_{n-1} L_{n-1}(T_i))(1 + \Delta_n L_n(T_i))} \tag{17.70}$$

By repeating it, we can see that $P(T_i, T_{n+1})$ can be written in terms of LIBOR rates as

$$P(T_i, T_{n+1}) = \Pi_{j=i}^{n} \frac{1}{1 + \Delta_j L_j(T_i)} \tag{17.71}$$

From Eq. (17.71) we can see that we can only determine the bond prices at the fixed maturity dates. What happens if we are interested in finding the price of a zero-coupon bond an arbitrary date t for $0 \le t \le T_n$ that is $p(t, T_N)$. Define index ℓ where T_ℓ is very first tenor date after time, t i.e., $T_{\ell-1} < t \le T_\ell$. It is easy to show that

$$P(t, T_n) = P(t, T_\ell) \Pi_{j=\ell}^{n-1} \frac{1}{1 + \Delta_j L_j(t)} \tag{17.72}$$

Let us define a numeraire B_t^* called rolling over bank account which is formed as follows: at time $T_0 = 0$ we start with \$1 to purchase $\frac{1}{P(0,T_1)}$ of the zero-coupon bonds maturing at time T_1. At time T_1 we have $\$\frac{1}{P(0,T_1)}$. We now reinvest the proceeds to purchase $\frac{1}{P(0,T_1)P(T_1,T_2)}$ of the zero-coupon bond maturing at time T_2; at time T_2 we own $\$\frac{1}{P(0,T_1)P(T_1,T_2)}$. We can reinvest it to purchase $\frac{1}{P(0,T_1)P(T_1,T_2)P(T_2,T_3)}$ of the zero-coupon bond maturing at time T_3. By continuing in this fashion, we see that at time T_ℓ it would be worth

$$\frac{1}{P(0, T_1)P(T_1, T_2)P(T_2, T_3) \cdots P(T_{\ell-1}, T_\ell)} \tag{17.73}$$

Writing it with respect to LIBOR rates by using its definition, we get

$$\Pi_{j=0}^{\ell-1}(1 + \Delta_j L_j(T_j)) \tag{17.74}$$

To find its value at time t we just discount it by $P(t, T_\ell)$ to get

$$B_t^* = P(t, T_\ell)\Pi_{j=0}^{\ell-1}(1 + \Delta_j L_j(T_j)) \tag{17.75}$$

The measure that assumes B_t^* is the numeraire is called spot measure, $\mathbb{P}^*$. Deflating any tradable instrument by B_t^* would be a martingale under spot measure $\mathbb{P}^*$. Here we deflate zero-coupon bond prices by the spot numeraire and calling it $D_n(t)$. Using Eqs. (17.72) and (17.75), we see that $D_n(t)$ satisfies

$$D_n(t) = \frac{P(t, T_\ell)\Pi_{j=\ell}^{n-1} \frac{1}{1+\Delta_j L_j(t)}}{P(t, T_\ell)\Pi_{j=0}^{\ell-1}(1 + \Delta_j L_j(T_j))} \tag{17.76}$$

Interestingly we have $P(t, T_\ell)$ in both the numerator and the denominator and it would be vanished through this operation and we get

$$D_n(t) = \frac{\Pi_{j=\ell}^{n-1} \frac{1}{1+\Delta_j L_j(t)}}{\Pi_{j=0}^{\ell-1}(1+\Delta_j L_j(T_j))} \tag{17.77}$$

In LIBOR market models, we assume the following generic stochastic differential equation for evolution of LIBOR rates under the spot measure

$$dL_n(t) = \mu_n(t)L_n(t)dt + L_n(t)\sigma_n^\top(t)dW_t, \quad 0 \le t \le T_n, n = 1, \ldots, M \tag{17.78}$$

where $\mu_n(t)$ is drift and a scalar, and $\sigma_n(t)$ is volatility and a d-dimensional vector, and both are adapted processes and W_t is a d-dimensional Brownian motion where d is the number of factors. We know under no-arbitrage condition, $\mu_n(t)$ can be found as a function of $\sigma_n(t)$. The goal is to find the expression for $\mu_n(t)$.

From (17.77) we see that if we take the log of $D_n(t)$ we would have a simpler expression for the log

$$\log D_n(t) = \log\left(\Pi_{j=\ell}^{n-1}\frac{1}{1+\Delta_j L_j(t)}\right) - \log\left(\Pi_{j=0}^{\ell-1}(1+\Delta_j L_j(T_j))\right) \tag{17.79}$$

$$= -\sum_{j=\ell}^{n-1}\log(1+\Delta_j L_j(t)) - \sum_{j=0}^{\ell-1}\log(1+\Delta_j L_j(T_j)) \tag{17.80}$$

If we take the derivatives, the second term vanishes and we get

$$d\log D_n(t) = -\sum_{j=\ell}^{n-1} d\log(1+\Delta_j L_j(t)) + 0 \tag{17.81}$$

Applying Itô's lemma to get

$$d\log(1+\Delta_j L_j(t)) = 0 + \frac{\Delta_j}{1+\Delta_j L_j(t)}dL_j(t) - \frac{1}{2}\frac{\Delta_j^2}{(1+\Delta_j L_j(t))^2}(dL_j(t))^2 \tag{17.82}$$

Substituting (17.78) and we have

$$d\log(1+\Delta_j L_j(t)) = \frac{\Delta_j}{1+\Delta_j L_j(t)}(\mu_j(t)L_j(t)dt + L_j(t)\sigma_j^\top(t)dW_t) \tag{17.83}$$

$$-\frac{1}{2}\frac{\Delta_j^2}{(1+\Delta_j L_j(t))^2}L_j(t)\sigma_j^\top(t)\sigma_j(t)dt \tag{17.84}$$

By gathering term we get

$$d\log D_n(t) = -\sum_{j=\ell}^{n-1} d\log(1+\Delta_j L_j(t)) = \sum_{j=\ell}^{n-1}\left(\frac{\Delta_j^2 L_j^2(t)\sigma_j^\top(t)\sigma_j(t)}{2(1+\Delta_j L_j(t))^2} - \frac{\Delta_j \mu_j(t) L_j(t)}{1+\Delta_j L_j(t)}\right)dt - \left(\sum_{j=\ell}^{n-1}\frac{\Delta_j L_j(t)\sigma_j^\top(t)}{1+\Delta_j L_j(t)}\right)dW_t \tag{17.85}$$

Knowing that $D_n(t)$ is a martingale under spot measure implies that there exists an $\mathbb{R}^d$-valued process $v_n(t)$ such that

$$dD_n(t) = D_n(t)v_n^\top(t)dW_t \tag{17.86}$$

We can apply Itô's Lemma to $\log D_n(t)$. We see from (17.86) that

$$d\log D_n(t) = 0 + \frac{1}{D_n(t)}dD_n(t) - \frac{1}{2}\frac{1}{D_n^2(t)}(dD_n(t))^2 \tag{17.87}$$

Substituting (17.86) and we have

$$d \log D_n(t) = v_n^\top(t) dW_t - \frac{1}{2} v_n^\top(t) v_n(t) dt$$
$$= -\frac{1}{2} v_n^\top(t) v_n(t) dt + v_n^\top(t) dW_t \quad (17.88)$$

Comparing the volatility terms in (17.88) and (17.85) gives us

$$v_n(t) = -\sum_{j=\ell}^{n-1} \frac{\Delta_j L_j(t) \sigma_j(t)}{1 + \Delta_j L_j(t)} \quad (17.89)$$

We would now like to find an expression for the drift μ_j. By setting the drifts in Eqs. (17.85) and (17.88) equal, we get

$$\sum_{j=\ell}^{n-1} \left(\frac{\Delta_j^2 L_j^2(t) \sigma_j^\top(t) \sigma_j(t)}{2(1 + \Delta_j L_j(t))^2} - \frac{\Delta_j \mu_j(t) L_j(t)}{1 + \Delta_j L_j(t)} \right)$$
$$= -\frac{1}{2} v_n^\top(t) v_n(t) \quad (17.90)$$

For simplicity we define

$$\eta_j(t) = \frac{\Delta_j L_j(t)}{1 + \Delta_j L_j(t)} \quad (17.91)$$

and using (17.89) we can write

$$\frac{1}{2} \sum_{j=\ell}^{n-1} \eta_j^2(t) \sigma_j^\top(t) \sigma_j(t) - \sum_{j=\ell}^{n-1} \eta_j(t) \mu_j(t)$$
$$= -\frac{1}{2} \left(\sum_{j=\ell}^{n-1} \eta_j(t) \sigma_j^\top(t) \right) \left(\sum_{j=\ell}^{n-1} \eta_j(t) \sigma_j(t) \right) \quad (17.92)$$

Assuming $n = \ell + 1$ we have

$$\frac{1}{2} \eta_\ell^2(t) \sigma_\ell^\top(t) \sigma_\ell(t) - \eta_\ell(t) \mu_\ell(t) = -\frac{1}{2} \eta_\ell^2(t) \sigma_\ell^\top(t) \sigma_\ell(t) \quad (17.93)$$

$$\eta_\ell(t) \mu_\ell(t) = \frac{1}{2} \eta_\ell^2(t) \sigma_\ell^\top(t) \sigma_\ell(t) + \frac{1}{2} \eta_\ell^2(t) \sigma_\ell^\top(t) \sigma_\ell(t) \quad (17.94)$$
$$= \eta_\ell^2(t) \sigma_\ell^\top(t) \sigma_\ell(t) \quad (17.95)$$

therefore

$$\mu_\ell(t) = \eta_\ell(t) \sigma_\ell^\top(t) \sigma_\ell(t) \quad (17.96)$$

Assuming $n = \ell + 2$, we get

$$\eta_\ell(t) \mu_\ell(t) + \eta_{\ell+1}(t) \mu_{\ell+1}(t)$$
$$= \frac{1}{2} \eta_\ell^2(t) \sigma_\ell^\top(t) \sigma_\ell(t) + \frac{1}{2} \eta_{\ell+1}^2(t) \sigma_{\ell+1}^\top(t) \sigma_{\ell+1}(t)$$
$$+ \frac{1}{2} (\eta_\ell(t) \sigma_\ell^\top(t) + \eta_{\ell+1}(t) \sigma_{\ell+1}^\top(t))(\eta_\ell(t) \sigma_\ell(t)$$
$$+ \eta_{\ell+1}(t) \sigma_{\ell+1}(t)) \quad (17.97)$$

Expanding terms and using (17.96) and regrouping, we get

$$\eta_{\ell+1}(t) \mu_{\ell+1}(t) = \eta_{\ell+1}^2(t) \sigma_{\ell+1}^\top(t) \sigma_{\ell+1}(t)$$
$$+ \eta_{\ell+1}(t) \eta_\ell(t) \sigma_{\ell+1}^\top(t) \sigma_\ell(t) \quad (17.98)$$

Hence,

$$\mu_{\ell+1}(t) = \eta_{\ell+1}(t) \sigma_{\ell+1}^\top(t) \sigma_{\ell+1}(t)$$
$$+ \eta_\ell(t) \sigma_{\ell+1}^\top(t) \sigma_\ell(t) \quad (17.99)$$
$$= \sigma_{\ell+1}^\top(t) (\eta_\ell(t) \sigma_\ell(t)$$
$$+ \eta_{\ell+1}(t) \sigma_{\ell+1}(t)) \quad (17.100)$$
$$= \sigma_{\ell+1}^\top(t) \sum_{j=\ell}^{\ell+1} \eta_j(t) \sigma_j(t) \quad (17.101)$$
$$= \sigma_{\ell+1}^\top(t) v_{\ell+2}(t) \quad (17.102)$$

By induction we can show that

$$\mu_n(t) = \sigma_n^\top(t) \sum_{j=\ell}^{n} \eta_j(t) \sigma_j(t)$$
$$= \sigma_n^\top(t) v_{n+1}(t)$$
$$= \sigma_n^\top(t) \sum_{j=\ell}^{n} \frac{\Delta_j L_j(t) \sigma_j(t)}{1 + \Delta_j L_j(t)}$$
$$= \sum_{j=\ell}^{n} \frac{\Delta_j L_j(t) \sigma_n^\top(t) \sigma_j(t)}{1 + \Delta_j L_j(t)} \quad (17.103)$$

therefore we obtain the following arbitrage-free $\mathbb{P}^*$-dynamics for the forward LIBOR rates

$$dL_n(t) = \left(\sum_{j=\ell}^{n} \frac{\Delta_j L_j(t) \sigma_n^\top(t) \sigma_j(t)}{1 + \Delta_j L_j(t)} \right) L_n(t) dt + L_n(t) \sigma_n^\top(t) dW_t, \quad 0 \leq t \leq T_n, n = 1, \ldots, M \tag{17.104}$$

17.3.3 Some Implications

The procedure followed in the previous section chose a bond which matured at time $t = t_3$ in order to obtain a synthetic probability (measure) under which the martingale equalities turned out to be more convenient for pricing purposes. The choice of $B^s_{t_1}$ as the normalizing factor was dictated partly by this desire for convenience.

In fact, any other asset can be chosen as the normalizing variable. Yet, the fact that $B^s_{t_1}$ matured at time t_3 made the time t_3 value of this bond *constant*. The convenience of the conditional expectations obtained under π is the result of this very simple fact. It is this last property that makes of the u, d-dependent terms $L^{ij}_{t_2}$ or $C^{ij}_{t_2}$ constant relative to the information set available at time t_1 in equations such as (17.28). Because they were constants, these coefficients could be factored out of the expectation operators. This is an important result because it eliminated the need to calculate complex correlations between spot rates and future values of interest rate dependent prices. Also, due to this we avoided working with random discount factors.

But the choice of normalization was important for another reason as well. Under carefully chosen normalization, forward rates such as F_{t_1} become martingales, and were *unbiased* estimators for future values of spot rates such as L_{t_2}. This implies that one can, heuristically speaking, replace the future value of a spot rate by the corresponding forward rates to find the current arbitrage-free price of various interest rate dependent securities.

We close this section by applying what was said in this chapter to two pricing examples in a continuous time setting.

17.3.3.1 The FRA Contract

Suppose we have an FRA contract that pays the sum $(F_t - L_T)N$ at some future date $T + \Delta, t < T$, where N is a notional amount and F_t is the forward price of the random variable L_T. The F_t is observed at contract time t.

Because this is a cash flow that belongs to a future date, $T + \Delta$, the current value denoted by V_t of the cash flow will be given by the "usual" martingale equality, where the future cash flow is discounted by using the risk-free rate $r_s, t \leq s \leq T + \delta$. Under the risk-neutral measure we can write:

$$V_t = \mathbb{E}^{\mathbb{Q}} \left[e^{-\int_t^{t+\delta} r_u du} (F_t - L_T) N\delta \right] \tag{17.105}$$

Now, we know that forward contracts do not involve any exchange of cash at the time of initiation. Thus, at contract initiation[10] we have

$$V_t = 0 \tag{17.106}$$

How can this price be zero given the formula in (17.56)? Because Fl is chosen so that the right-hand side expectation vanishes. If the spot rates are assumed to be deterministic, this will be very easy to do. A value for F_t can be easily obtained by factoring out the discount factor,

$$V_t = e^{-\int_t^{t+\delta} r_u du} \mathbb{E}^{\mathbb{Q}} [F_t - L_T] N\delta \tag{17.107}$$

then setting the V_t equal to zero and canceling:

$$0 = \mathbb{E}^{\mathbb{Q}} [F_t - L_T] \tag{17.108}$$

The F_{t_1} that makes the current price of the forward contract zero is the one where

$$F_t = \mathbb{E}^{\mathbb{Q}} [L_T] \tag{17.109}$$

[10] Any margins that may be required are not cash exchanges, but are provided as a guarantee toward settlements in the future.

That is, when spot rates are deterministic, the forward price is equal to the "best" forecast of the future L_T under the risk-neutral measure Black–Scholes framework exploits the assumption of constant interest rates at various points in pricing stock options. But, the same assumption is not usable when one is dealing with interest-sensitive securities. The most important reason that such securities are traded is the need to hedge interest rate risk. Obviously, assuming deterministic spot rates would not be very appropriate here. But, if the assumption of deterministic r_t is dropped, then the discount factor does not factor out and we cannot use Eq. (17.59) under $\mathbb{Q}$.

The forward measure can provide a convenient solution. Using the arbitrage-free price of the discount bond $B(t, T+\delta)$, we can instead write the pricing equation under the *forward* measure:

$$V_t = \mathbb{E}^{\pi}\left[B(t, T+\delta)(F_t - L_T) N\delta\right] \tag{17.110}$$

where $\delta > 0$ is the tenor of L_T. Here $B(t,T)$ is a value observed at time t; hence, it factors out of the expectation operator:

$$V_t = B(t, T+\delta)\,\mathbb{E}^{\pi}\left[(F_t - L_T)\right] N\delta \tag{17.111}$$

Now, use the fact that $V_t = 0$:

$$F_t = \mathbb{E}^{\pi}\left[L_T\right] \tag{17.112}$$

This is an equation that one can exploit conveniently to find the arbitrage-free value of F_t. The critical point is to make sure that, in calculating this average, one uses the forward measure π and not the risk-neutral probability $\mathbb{Q}$.

17.3.3.2 A Caplet

As a second example to the power of the forward measure discussed above we consider pricing issues involving a caplet. Let C_t be the current price of a caplet written on some Libor rate L_t, with tenor δ and with cap rate K. Suppose the notional amount is $N = 1$ and that the caplet expires at time T. We let $\delta = 1$, except for the notation on L_t.

According to this, the buyer of the caplet will receive the payoff

$$C_T = \max\left[L_{T-\delta} - K, 0\right]$$

at time T. As mentioned earlier, this instrument will protect the buyer against increases in $L_{T-\delta}$ beyond the level K. Normally $0 < \delta < 1$ and in the above the right-hand side will be proportional to δ.

How does one price this caplet? Suppose we decide to use the risk-neutral probability $\mathbb{Q}$. We know that the arbitrage-free price will be given by:

$$C_t = \mathbb{E}^{\mathbb{Q}}\left[e^{-\int_t^T r_u du} \max\left[L_{T-\delta} - K, 0\right]\right] \tag{17.113}$$

We also know that at time $T-\delta$ the $F_{T-\delta}$ will coincide with $L_{T-\delta}$.[11] So we may decide to use F_t as the "underlying." After all, as time passes, this variable will eventually coincide with the future spot rate $L_{T-\delta}$. This process is called the *forward Libor process*.

This suggests that we model log-normal forward rate dynamics with a Wiener process W_t defined under the original probability $\mathbb{P}$,

$$dF_t = \mu F_t dt + \sigma F_t dW_t$$

as in a Black–Scholes environment, and then apply the Black–Scholes logic to determine the C_t.

If we proceed this way, the first step will be to switch to $\mathbb{Q}$ the risk-neutral probability. But this creates a problem. The F_t is not a martingale under $\mathbb{Q}$. So, as we switch probabilities and use the Wiener process, W_t^*, defined under $\mathbb{Q}$, the forward rate dynamics will become

$$dF_t = \mu^* F_t dt + \sigma F_t dW_t^*$$

where the μ^* is the new *risk-adjusted* drift implied by the Girsanov theorem. Under $\mathbb{Q}$ this drift is not known at the outset. So, unlike the Black–Scholes

[11] At any time, the forward rate for an immediate loan of tenor δ will be the same as the spot rate for that period.

case, where the drift of the underlying stock price is replaced by the *known* (and constant) spot rate r, we now end up with a difficult *unknown* to determine.

Consider what happens to the forward rate dynamics if we use the forward measure π instead. Under the forward measure obtained with $B(t, T)$-normalization, the forward rate F_t defined for time $T - \delta$ will be a martingale.[12] Hence we can write:

$$dF_t = \sigma F_t dW_t^{\pi}$$

where the W_t^{π} is a Wiener process under π. A very convenient property of this SDE is that the drift is equal to zero and the F_t is an unbiased estimator of $L_{T-\delta}$:

$$F_t = \mathbb{E}^{\pi}\left[L_{T-\delta}\right]$$

There is no additional difficulty of determining an unknown drift. We can go ahead with a Black–Scholes type argument and price this caplet in a straightforward fashion.[13]

17.3.3.3 Normalization as a Tool

Above we discussed the important implications of normalization and measure choice from the point of view of asset pricing, with particular emphasis on interest rate sensitive securities. Are there any implications for the mathematics of financial derivatives?

We see from the above discussion that the fundamental variables are in fact the state prices $\left\{\psi^{ij}\right\}$. When there are no arbitrage opportunities, these prices will exist, they will be positive, and will be unique. Once this is determined, the financial analyst has a great deal of flexibility concerning the martingale measure that he or she can choose. The synthetic probability can be selected as the classical risk-neutral measure $\mathbb{Q}$ or the forward measure π, depending on the instruments one is working with. Hence, the issue of which measure to work with becomes another tool for the analyst.

In fact, as suggested by Girsanov theorem, one can go back and forth between various probabilities depending on the requirements of the pricing problem. In fact, consider a normalization with respect to B_t^s and the corresponding measure π that we just used. Clearly, we could also have normalized with the longer maturity bond B_t and obtained a new probability, say $\widetilde{\pi}$, given by:

$$\widetilde{\pi}^{ij} = \frac{1}{B_t}\psi^{ij} \tag{17.114}$$

All prices that mature at time T would then be martingales once they are normalized by the Bl.

Note that the ratio

$$\frac{\pi^{ij}}{\widetilde{\pi}^{ij}} = \left[\frac{\frac{1}{B_t^s}}{\frac{1}{B_t}}\right] \tag{17.115}$$

can be used to write:

$$\pi^{ij} = \widetilde{\pi}^{ij}\left[\frac{\frac{1}{B_t^s}}{\frac{1}{B_t}}\right] \tag{17.116}$$

This way one can go from one measure to another.

Would such adjustments be any use to us in pricing interest rate sensitive securities? The answer is again yes. When we deal with an instrument that depends on *more* than one L_T with different tenors T, we can first start with one forward measure, but then by taking the derivative with respect to the other, we can obtain the proper "correction terms" that need to be introduced.

17.4 CONCLUSION

In this chapter we introduced the notions of normalization and forward measure. These tools

[12] It is important to realize that under a different normalization this particular forward rate will not be a martingale.

[13] A remaining difference is in the units used here. There is no need to discount the caplet payoff to the present if we use the forward rate dynamics. This is unlike the Black–Scholes environment, where the stock price dynamics dS_t are expressed in time t dollars.

play an important role in pricing derivative securities in a convenient fashion. More than just theoretical concepts, they should be regarded as important tools in pricing assets in real world markets. They are especially useful for any derivative whose settlement is done at a future date, in future dollars.

Some of the main results were the following. When we use the forward measure π_T obtained from a default-free discount bond $B(t, T)$, three things happen:

- The price of all assets considered here, once normalized by the arbitrage-free price of a zero-coupon bond of $B(t, T)$, becomes a martingale under π_T.
- The forward prices *that correspond to the same maturity* become martingales themselves, without any need for normalization.[14]
- The discount factors become deterministic and factor out of pricing equations for derivatives with expiration date T.

17.5 REFERENCES

The book by Musiela and Rutkowski (1997) is an excellent source for a reader with a strong quantitative background. Although it is much more demanding mathematically than the present text, the results are well worth the efforts. Another possible source is the last chapter in Pliska (1997). Pliska treats these notions in discrete time, but our treatment was also in discrete time.

17.6 EXERCISES

1. Suppose you are given the following information on the spot rate r_t:

   - The r_t follows:
   $$dr_t = \mu r_t dt + \sigma r_t dW_t$$
   - The annual drift is
   $$\mu = 0.01$$
   - The annual volatility is
   $$\sigma = 12\%$$
   - The current spot rate is assumed to be 6%.

   (a) Suppose instruments are to be priced over a year. Determine an appropriate time interval Δ, such that binomial trees have five steps.
   (b) What would be the implied u and d in this case?
   (c) Determine the tree for the spot rate r_t.
   (d) What are the "up" and "down" probabilities implied by the tree?

2. Suppose at time $t = 0$ you are given four default-free zero-coupon bond prices, $P(t, T)$, with maturities from 1 to 4 years:
   $$P(0,1) = 0.94, \quad P(0,2) = 0.92, \quad P(0,3) = 0.87, \quad P(0,4) = 0.80$$

   (a) How can you "fit" a spot-rate tree to these bond prices? Discuss.
   (b) Obtain a tree consistent with the term structure given above.
   (c) What are the differences, if any, between the tree approaches in questions (a) and (b)?

3. Select ten standard, normal random numbers using Mathematica, Maple, or Matlab. Suppose interest rates follow the SDE:
   $$dr_t = 0.02 r_t dt + 0.06 r_t dW_t$$
   Assume that the current spot rate is 6%.

   (a) Discretize the SDE given above.

[14] This is the case because the forward price, F_{t_1}, itself belongs to the same forward date, unlike, say, the C_t, which is a value expressed in time t dollars.

(b) Calculate an estimate for the following expectation, using a time interval $\Delta = 0.04$,

$$\mathbb{E}\left[e^{-\int_0^1 r_s ds} \max(r_1 - .06, 0)\right]$$

and the random numbers you selected. Assume that the expectation is taken with respect to the *true* probability.

(c) Calculate the sample average for

$$\mathbb{E}\left[e^{-\int_0^1 r_s ds}\right]$$

and then multiply this by the sample average for

$$\mathbb{E}\left[\max(r_1 - .06, 0)\right]$$

Do we obtain the same result?

(d) Which approach is correct?

(e) Can you use this result in calculating bond prices?

(f) In particular, how do we know that the interest rate dynamics displayed in the above SDE are arbitrage-free?

(g) What would happen to the above interest rate dynamics if we switched to risk-neutral measure $\mathbb{Q}$?

(h) Suppose you are given a series of arbitrage-free bond prices. How can you exploit this within the above framework in obtaining the arbitrage-free dynamics for r_t?

4. We observe that current price of a zero-coupon bond with one year maturity is \$0.97 (paying \$1 at $t = 1$). We also observe the current implied forward rate is 2.5%. Under the forward measure, price a six month caplet paying six month Libor. Assume the following dynamics under the forward measure: $dF_t = \sigma F_t dW_t$. Assume volatility is set at 9%, a strike rate of 3%.

5. Price a zero-coupon bond with maturity one year under the Vasicek model

$$dr_t = \alpha(\mu - r_t)dt + \sigma dW_t \tag{17.117}$$

with $r_0 = 0.01, \alpha = 0.2, \mu = 0.01, \sigma = 0.05$.

6. Show that the price of a *swap* is the same as the price of a *cap* minus the price of a *floor*.

7. Show that the price of a payer swaption minus the price of a receiver swaption is equal to the price of a forward swap.

8. Consider a payer forward start swap where the swap begins at time T_n and it matures at time T_N. Assume that the accrual period is of length δ. In swaps, payments are made in-arrears, therefore the first payment occurs at $T_{n+1} = T_n + \delta$.

(a) Assuming a notional principal of \$1, show that the total value of the payer swap at time $t \leq T_n$ is

$$\Pi_p(t) = P(t, T_n) - P(t, T_N) - S\delta \sum_{j=n+1}^{N} P(t, T_j)$$

where S is the fixed rate (annualized) specified in the contract.

(b) Find the expression for the *forward swap rate*, $R_{swap}(t)$, at time $t \leq T_n$ using the results in part (a). Here $R_{swap}(t)$ is the fixed rate S above which gives $\Pi_p(t) = 0$.

(c) Also show that the forward swap rate, $R_{swap}(t)$, can be written as weighted average of simple forward rates, that is

$$R_{swap}(t) = \sum_{j=n+1}^{N} w_j(t) F(t; T_{j-1}, T_j)$$

and find $w_j(t)$. $F(t; T_{j-1}, T_j)$ is the simple forward rate for $[T_{j-1}, T_j]$ prevailing at t.

9. Forward Par Swap Rate $y_{n,N}(t)$ is defined

$$y_{n,N}(t) = \frac{P(t, T_n) - P(t, T_N)}{\sum_{j=n+1}^{N} \delta P(t, T_j)} = \frac{P(t, T_n) - P(t, T_N)}{P_{n+1,N}(t)}$$

$P_{n+1,N}(t)$ is called the present value of a basis point (PVBP).

A swaption gives the holder the right not the obligation to enter into a particular swap contract. A swaption with option maturity T_n and swap maturity T_N is termed a $T_n \times T_N$-swaption. The total time-swap associated with the swaption is then T_n+T_N. A payer swaption gives the holder the right not the obligation to enter into a payer swap and can be seen as a call option on a swap rate. The option has the payoff at time T_n, the option maturity, of

$$\begin{aligned}\left[V_{n,N}^{Payer}(T_n)\right]^+ &= \left[\{1-P(T_n,T_N)\} - \kappa \sum_{j=n+1}^{N} \delta P(T_n,T_j)\right]^+ \\ &= \left[y_{n,N}(T_n)P_{n+1,N}(T_n) - \kappa P_{n+1,N}(T_n)\right]^+ \\ &= P_{n+1,N}(T_n)\left[y_{n,N}(T_n)-\kappa\right]^+\end{aligned}$$

where κ denotes the strike rate of the swaption. The second line follows directly from the definition of the forward swap rate. We let $B_t = \exp(\int_0^t r_s ds)$ be the money market account at time t. Assuming absence of arbitrage, the value of a payer swaption at time $t < T_n$ denoted by $\mathbf{PS}_t$ can be expressed by the following risk-neutral conditional expectation,

$$\frac{\mathbf{PS}_t}{B_t} = \mathbb{E}_t^{\mathbb{Q}}\left\{\frac{\left[V_{n,N}^{Payer}(T_n)\right]^+}{B_{T_n}}\right\}$$

$$\frac{\mathbf{PS}_t}{B_t} = \mathbb{E}_t^{\mathbb{Q}}\left\{\frac{P_{n+1,N}(T_n)}{B_{T_n}}\left[y_{n,N}(T_n)-K\right]^+\right\}$$

a. Use $P_{n+1,N}(t)$ as a numeraire to find a new probability measure, $\mathbb{P}^{n+1,N}$, that we call *swap measure*.

b. Under the swap measure show that

$$\mathbf{PS}_t = P_{n+1,N}(t)\mathbb{E}_t^{\mathbb{P}^{n+1,N}}\left\{\left[y_{n,N}(T_n)-K\right]^+\right\}$$

Note that under this swap measure the corresponding swap rate, $y_{n,N}(t)$, is a martingale. The change of numeraire shows explicitly why swaptions can be viewed as options on swap rates.

10. Use induction to establish that drifts, $\mu_1,\ldots,\mu_n$, must satisfy (17.103) under the no-arbitrage assumption. Show that $D_{n+1}(t)$ is a martingale if and only if $L_n(t)D_{n+1}(t)$ is a martingale and then apply Itô's lemma to obtain (17.103).

CHAPTER

18

Modeling Term Structure and Related Concepts

OUTLINE

18.1 INTRODUCTION

The previous chapter was important because it discussed the Fundamental Theorem of Finance when interest-sensitive securities are included in the picture. We obtained new results. The issue of normalization, the use of forward measures within Libor instruments, and ways of handling the simultaneous existence of bonds with differing maturities was introduced using a simple model. Now it is time to take some steps backward and discuss the basic concepts in more detail before we utilize the results obtained in Chapter 17.

In particular, we need to do two things. The new concepts from fixed income are much more fragile and somehow less intuitive than the straightforward notions used in the standard Black–Scholes world. These fixed income concepts need to be defined first, and carefully

Financial Derivatives, *Third Edition*.
http://dx.doi.org/10.1016/B978-0-12-384682-2.00018-9

motivated second. Otherwise, some of the reasoning behind the well-known bond pricing formulas may be difficult to grasp.

Next, at this point we need to introduce some important arbitrage relationships that are used repeatedly in pricing interest-sensitive securities. The next chapter will consider two fundamentally different methodologies used in pricing interest-sensitive securities. These are the so-called *classical* approach and the *Heath-Jarrow-Morton* approach, respectively. Our main purpose will be to show the basic reasoning behind these fundamentally different methodologies and highlight their similarities and differences. But to do this we must first introduce a number of new arbitrage relations that exist between the spot rates, bond prices, and forward rates.

The first arbitrage relation that we need to study is the one between investment in very short-term savings accounts and bonds. Suppose both of these are default-free. How would the long-term bond prices relate to depositing money in a short-term savings account and then rolling this over continuously?[1] It is clear that when one buys a longer term bond, the commitment is for more than one night, or one month. During this "long" period, several risky events may occur, and these may affect the price of the bond adversely. Yet, the overnight investment will be mostly immune to the risky events because the investor's money is returned the next "day," and hence can be reinvested at a higher overnight rate. Thus, it appears that long-term bonds should pay a premium relative to overnight money, in order to be held by risk-averse investors. In the Black–Scholes world the switch to the risk-neutral measure eliminated these risk premia and gave us a pricing equation. Can the same be done with interest-sensitive securities and random spot rates? We will see that the answer is yes. In fact, the classical approach to pricing interest-sensitive securities exploits this particular arbitrage relation extensively.

The second arbitrage relation is specific to fixed income. Fixed income markets provide many liquid instruments that are almost identical except for their maturity. For example, we have a spectrum of discount bonds that are differentiated only by their maturity. Similarly, we have forward rates of different maturities. It turns out that this multidimensional aspect of interest-sensitive instruments permits writing down complex arbitrage relations between a set of zero-coupon bonds and a set of forward rates. In fact, if we have a k-dimensional *vector* of bond prices, we can relate this to a *vector* of forward rates using arbitrage arguments. These arbitrage relations form the basis of the Heath-Jarrow-Morton approach to pricing interest-sensitive securities.

Thus, one way or another, the material in the present chapter should be regarded as a necessary background to discussing pricing of interest-sensitive securities.

18.2 MAIN CONCEPTS

We begin with some definitions, some of which were introduced earlier. The price of a *discount bond* maturing at time T observed at time $t < T$ will be represented by the symbol $B(t, T)$. The r_t will again denote the *instantaneous* spot rate on riskless borrowing. The spot rate is instantaneous in the sense that the loan is made at time t and is repaid after an infinitesimal period dt. The spot rate is also *riskless* in the sense that there is no *default risk*, and the return to this instantaneous investment is known with certainty.[2] These two definitions were seen earlier.

The first new concept that we now define is the continuously compounded *yield*, $R(t, T)$, of the

[1] In practice, the shortest-term investment will earn an overnight interest rate.

[2] But, as we saw earlier, the spot rate itself can be a random variable and the investment may well have a market risk if rolled over.

discount bond, $B(t,T)$. Given the current price of the bond $B(t,T)$ and with par value \$1, the $R(t,T)$ is defined by the equation:

$$B(t,T) = e^{-R(T,t)(T-t)} \tag{18.1}$$

It is the rate of return that corresponds to an investment of $B(t,T)$ dollars which returns \$1 after a period of length $[T-t]$. Here, the use of an exponential function justifies the term continuously compounded. Note that there is a one-to-one relationship between the bond price and the yield. Given one, we know the other. They are also indexed by the same indices, T and t.

Next, we need to define a continuously compounded *forward* rate, $F(t,T,U)$. This concept represents the interest rate on a loan that begins at time T and matures at time $U > T$. The rate is contracted at time t, although cash transactions will take place at future dates T and U. The fact that the rate is continuously compounded implies that the actual interest calculation will be made using the exponential function. In fact, if \$1 is loaned at time T, the money returned at time U will be given by:

$$e^{F(t,T,U)(U-T)} \tag{18.2}$$

Note that the $F(t,T,U)$ has three time indices whereas discount bond prices each came with two indices. This suggests that to obtain a relation between forward rates and bond prices, we may have to use two different bonds, $B(t,T)$ and $B(t,U)$, with maturities T and U, respectively. Between them, these two bond prices will have the same time indices (t,T,U).

18.2.1 Three Curves

The basic concepts defined in the previous section can be used to define three "curves" used routinely by market professionals. These are the *yield* curve, the *discount* curve, and the *credit-spread* curve. The so-called *swap* curve, which is perhaps the most widely used curve in fixed income markets, is omitted. This is due to the limited scope of this book. We do not consider instruments. We only deal with mathematical tools to study them. The *forward* curve, which consists of a spectrum of interest rates on forward loans contracted for various future dates, will be discussed later in the chapter.

18.2.1.1 The Yield Curve

The yield curve is obtained from the relationship between the yield $R(t,T)$ and the discount bond price $B(t,T)$. We have:

$$B(t,T) = e^{-R(T,t)(T-t)}, \quad t < T \tag{18.3}$$

where $B(t,T)$ is the arbitrage-free price of the T-maturity discount bond. Thus, to obtain the yield $R(t,T)$ of a bond, we first need to obtain its price. Then, Eq. (18.3) is used to get the continuously compounded yield:

$$\begin{aligned} R(T,t) &= \frac{\log(1) - \log B(t,T)}{T-t} \\ &= \frac{-\log B(t,T)}{T-t} \end{aligned} \tag{18.4}$$

Here, we have $0 < B(t,T) < 1$ as long as $t < T$. Thus $\log[B(t,T)]$ will be a negative number, and hence the $R(t,T)$ will be positive.

Now, assume that at time t there exist zero-coupon bonds with a full spectrum of maturities $T \in [t, T_{\max}]$, where $T_{\max}$ is the longest maturity available in the market. Let the price of these bonds be given by the set $\{B(t,T), T \in [t, T_{\max}]\}$. For each $B(t,T)$ in this set, we can use Eq. (18.4) and obtain the corresponding yield $R(t,T)$. Then we have the following definition.

Definition 22. The spectrum of yields $\{B(t,T), T \in [t, T_{\max}]\}$ is called the yield curve.

The yield curve is a correspondence between the yields of the bonds belonging to a certain risk class and their respective maturities.

The definition of the yield curve given above is an extension of the yield curve notion used by practitioners. Observed yield curves provide the spectrum of yields on, say, Treasuries, at a finite number of maturities. Here, we assume not only that time is continuous, but that at any time

t, there is a continuum of pure discount bonds. An investor can always buy and sell a liquid T-maturity bond, for any value of $T < T_{\max}$. These maturities extend from the immediate tenor,

$$T = t + dt \tag{18.5}$$

to the longest possible maturity $T = T_{\max}$, providing a continuous yield curve. According to this assumption, given an arbitrary $T < T_{\max}$, there will be no need to "interpolate" the corresponding yield because it will be directly observed in the markets.

18.2.1.2 The Discount Curve

In spite of the popularity of the term "yield curve," most market applications instead use the *discount* curve.

Definition 23. The spectrum of default-free zero-coupon bond prices $\{B(t,T), T \in \{t,T\}\}$, with a continuum of maturities that belong to the same risk class, is called the discount curve.

The discount curve is more convenient to use in valuing general cash flows. In fact, let the $\{cf_{T_1}, \ldots, cf_{T_n}\}$ represent a general cash flow to be received at arbitrary times $T_1 < T_2 < \cdots < T_n$. The present value, CF_t, of this general cash flow can be obtained by simply multiplying the amount to be received at time T_i by the corresponding $B(t, T_i)$. In fact, the discounted value can easily be obtained by using arbitrage-free zero-coupon bond prices with maturities falling to the corresponding T_i. This present value is

$$CF_i = \sum_{i=1}^{n} B(t, T_i)\, cf_{T_i} \tag{18.6}$$

The reason why this works is simple. The price $B(t, T_i)$ is simply the *current* arbitrage-free value of \$1 to be paid at time T_i. The discount is directly *quoted* by the market. Hence, the discount curve will play an essential role in the daily work of a market practitioner.

18.2.1.3 The Credit Spread Curve

Yield curves and the discount curves are obviously valid for bonds of a given risk class. When we look at the spectrum of bonds $\{B(t,T), T \in [t, T_{\max}]\}$, we implicitly assume that the default risk on these bonds is the same. Otherwise, the difference between yields would not just be due to differences in the corresponding maturities.

Hence, for each risk class we obtain a different yield (discount) curve. The difference between these yield (discount) curves will indicate the *credit spreads*, the supplemental amount riskier credits have to pay to borrow money at the same maturity. The coexistence of different yield curves that represent different risk classes leads to the so-called *credit spread curve*.

Definition 24. Given two yield curves $\{R(t,T), T \in [t, T_{\max}]\}$ that correspond to default-free bonds and the $\{\widetilde{R}_t^T, T \in [t, T_{\max}]\}$ that correspond to bonds with a given default probability, the spectrum of the spreads, $\{s(t,T) = \widetilde{R}_t^T - R(t,T), T \in [t, T_{\max}]\}$, is called the credit spread curve.

Indeed, some practitioners prefer to work with a correspondence between the credit spread and the maturity, instead of dealing with the yield curve itself. The use of the credit spread curve will be more practical if the traded instruments are written on the spreads rather than the underlying interest rates. In this book, we omit a discussion of credit instruments and assume throughout that there is no default risk. Hence, there is only one risk class and there the default risk is assumed to be zero.

18.2.2 Movements on the Yield Curve

Before we deal with more substantial issues we also would like to discuss the comparison between a shift in the yield curve and a movement along it. Given a yield curve, $\{R(t,T), T \in [t, T_{\max}]\}$ continuous in T, note that at time t we can consider two different incremental changes. First, at any instant t, we can ask what happens to a particular $R(t, T)$ as T changes by a small amount denoted by dT. Here, we are modifying the maturity of a particular bond under consideration, namely the one that has maturity T, by dT. In other words, we are moving along the same yield curve. According to this, if the yield curve is continuous and "smooth," we can obtain the derivative:

$$\frac{dR\,(t,T)}{dT} = g(T) \tag{18.7}$$

This is simply the slope of the yield curve $\{R\,(t,T)\,, T \in [t, T_{\max}]\}$. These quantities are shown in Figure 18.1. The $g(T)$ is the slope of the tangent to the continuous yield curve at maturity T. Figure 18.2 displays the corresponding situation with the discount curve.

Yield curves are generally classified as negatively sloped, positively sloped, and flat. They can also exhibit "humps." As the shape of the curve changes, the slope changes as well. It is important to realize that an incremental change in T would not involve any unknown random shocks. It is an experiment involving bonds with different maturities at the same instant t and, at time t, every $R(t, T)$ is known. Also, because there are no Wiener increments involved in these movements, the derivative can be taken in a standard fashion without having recourse to Ito's Lemma.

A second type of incremental change that we can contemplate is a variation in the time parameter t. The incremental change in the spectrum of yields $R(t, T)$, due to a change in time t, will involve random shocks. As t changes, time will pass, new Wiener increments are drawn, random shock(s) affect the spot rate, and the yield curve *shifts*. It is important to realize that as t increases by dt, the *entire* spectrum of yields will, in general, change. Thus, the dynamics of fixed income instruments are essentially the dynamics of a curve rather than the dynamics of a single stochastic process. The implied arbitrage restrictions will be much more complicated than the case of the Black–Scholes environment. After all, we need to make sure that the movements of an entire curve occur in a fashion that rules out arbitrage.

Also, note that for stochastic differentials such as $dR(t, T)$, Ito's Lemma needs to be used due to involved Wiener components. We are now ready to introduce a fundamental pricing equation that will be used throughout the second part of this book, namely the bond pricing equation.

18.3 A BOND PRICING EQUATION

In this section we start discussing the first substantial issue of this chapter. We derive an equation that gives the arbitrage-free price of a default-free zero-coupon bond $B(t, T)$, maturing at time T. We go in steps. We first begin with a simplified case where the instantaneous spot rate is constant, and then move to a stochastic risk-free rate. This way of proceeding makes it easier to understand the underlying arbitrage arguments.

18.3.1 Constant Spot Rate

Thus we first let the spot rate r_t be constant:

$$r_t = r \tag{18.8}$$

Then, the price of a default-free pure discount bond paying \$1 at time T will be given by:

$$B\,(t,T) = e^{-r(t,T)} \tag{18.9}$$

Consider the rationale behind this formula. The r is the continuously compounded instantaneous interest rate. The function plays the role of a discount factor at time t. At $t = T$ the exponential function equals 1, which is the same as the maturity value of the bond. At all other times, $t < T$, the exponential factor is less than 1. Hence, the right-hand side of Eq. (18.9) represents the present value of one time-T dollar, discounted to t at a constant, continuously compounded rate r.

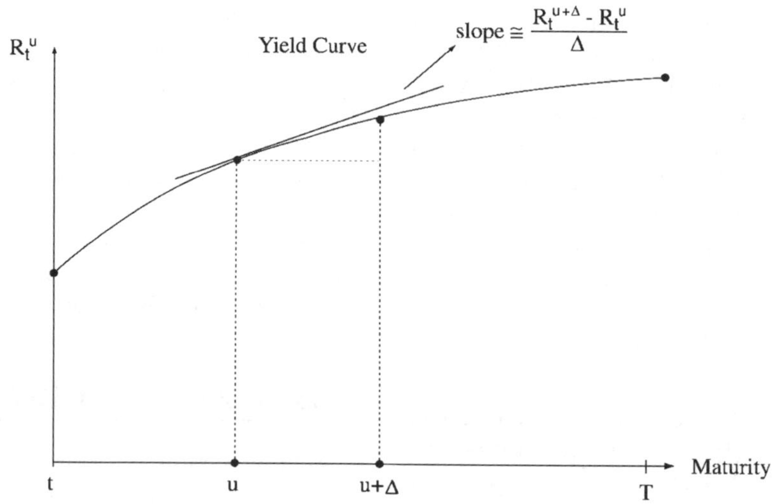

FIGURE 18.1 Slope of the yield curve.

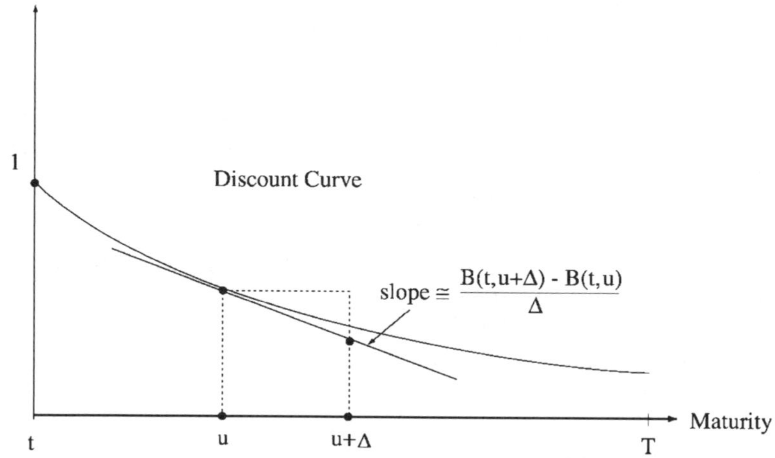

FIGURE 18.2 Slope of the discount curve.

Now, an investor who faces these instruments has the following choices. He or she can invest dollars in a risk-free savings account now, and at time T this will be worth \$1. Or, the investor can buy the T-maturity discount bond and pay $B(t,T)$ dollars now. This investment will also return \$1 at time T. Clearly we have two instruments, with no default risk, and with the same payoff at time T. There are no interim payouts either. If interest rates are constant and if there is no default risk, any bond that promises to pay \$1 at time T will have to have the same price as the initial investment of $e_t^{-r(t,T)}$ to risk-free lending. That is, we must have:

$$B(t,T) = e^{-r(t-T)} \tag{18.10}$$

Otherwise, there will be arbitrage opportunities. It is instructive to review the underlying arbitrage argument within this setting.

First, suppose $B(t,T) > e^{-r(t.T)}$. Then, one can short the bond during the period $[t,T]$ and invest of the proceeds to risk-free lending. At time T, the short (bond) position is worth $-\$1$. But the risk-free lending will return $+\$1$. Hence, at time T the net cash flow will be zero. But, at t the investor is still left with some cash in the pocket because

$$B(t,T) > e^{-r(t.T)}$$

The other possibility is $B(t,T) < e^{-r(t.T)}$. Then at time t one would borrow dollars, and buy a bond at a price of $B(t,T)$. When the maturity date arrives, the net cash flow will again be zero. The $+\$1$ received from the bond can be used to pay the loan off. But at time t, there will be a net gain:

$$e^{-r(t.T)} - B(t,T) > 0$$

The only condition that would eliminate such arbitrage opportunities is when the "bond pricing equation" holds:

$$B(t,T) = e^{-r(t.T)} \tag{18.11}$$

Hence, this relationship is *not* a definition, or an assumption. It is a restriction imposed on bond prices and savings accounts by the requirement that there are no arbitrage opportunities. Notice that in obtaining this equation we did not use the Fundamental Theorem of Finance, but doing this would have given exactly the same result.[3]

18.3.2 Stochastic Spot Rates

When the instantaneous spot rate r_t becomes stochastic, the pricing formula in (18.9) will have to change. Suppose r_t represents the risk-free rate earned during the infinitesimal interval $[t,t+dt]$. Thus, r_t is known at t, but its future values fluctuate randomly as time passes. The Fundamental Theorem of Finance can be applied to obtain an arbitrage relation between $B(t,T)$ and the stochastic spot rates, $r_t, t \in [t,T]$.

[3]See Exercise 2 at the end of the chapter.

We utilize the methodology introduced in Chapter 17. We take the current bond price $B(t,T)$ and normalize by the current value of the savings account, which is \$1. Next, we take the maturity value of the bond, which is \$1, and normalize it by the value of the savings account. This value is equal to $e^{\int_t^T r_s ds}$ because it is the return at time T to \$1 rolled-over at the instantaneous rate r_s for the entire period $s \in [t,T]$. Dividing \$1 by this value of the savings account, we get $e^{-\int_t^T r_s ds}$.

According to Chapter 17, this normalized bond price must be a martingale under the risk-neutral measure $\mathbb{Q}$. Thus, we must have

$$B(t,T) = \mathbb{E}^{\mathbb{Q}}\left[e^{-\int_t^T r_s\, ds}\right] \tag{18.12}$$

where the term $e^{-\int_t^T r_s ds}$ can also be interpreted as a random discount factor applied to the par value \$1.

Some further comments about this formula are in order. First, the bond price formula given in (18.12) has another important implication. Bond prices depend on the whole spectrum of future short rates $r_s, t < s < T$. In other words, we can look at it this way: the yield curve at time t contains all the relevant information concerning future short rates.[4]

Second, there is the issue of which probability measure is used to calculate these expectations. One may think that with the class of Treasury bonds being *risk-free* assets, there is no *risk premium* to eliminate, and hence, there is no need to use the equivalent martingale measure. This is, in general, incorrect. As interest rates become stochastic, prices of Treasury bonds will contain "market risk." They depend on the future behavior of spot rates and this behavior is stochastic. To eliminate the risk premium associated with such risks, we need to use equivalent martingale measures in evaluating expressions as in (18.12).

[4]Remember that conditional expectations provide the *optimal* forecasts in the sense of minimum mean square error given an information set I_t.

We now discuss this formula using discrete intervals of size $0 < \Delta$. This will show the passage to continuous time, and explain the mechanics of the bond pricing formula better.

18.3.2.1 Discrete Time

Consider the special case of a three-period bond in discrete time. If Δ represents some time interval less than one year and if t is the "present," then the price of a three-period discount bond will be given by:

$$B(t, t+3\Delta) = \mathbb{E}^{\mathbb{Q}}\left[\frac{1}{(1+r_t\Delta)(1+r_{t+\Delta}\Delta)\left(1+r_{t+2\Delta}\Delta\right)}\right] \tag{18.13}$$

where r_t is the known current spot rate on loans that begin at time t and end at time $t+\Delta$, and $r_{t+\Delta}, r_{t+2\Delta}$, are unknown spot rates for the two future time periods. Unlike the case of continuous time, these are simple interest rates and, by market convention, are multiplied by Δ.

According to Eq. (18.13), the bond's price is equal to the discounted value of the payoff at maturity. The discount factor is random and an (conditional) expectation operator needs to be used. The expectation is taken with respect to the risk-neutral probability $\mathbb{Q}$. We normalize the value of the bond at times t and $t+3\Delta$ using the "risk-free" saving and borrowing. As mentioned earlier, for time t we divide the bond price by 1, the amount invested in risk-free lending and borrowing. For time $t+3\Delta$ we divide the value of the bond at maturity, which is \$1, by the value of rolling the investment over at future spot rates $r_{t+\Delta}, r_{t+2\Delta}$. The expectation is conditional on the information set available at time t. This information set contains the current value of r_t.

18.3.3 Moving to Continuous Time

We now show the heuristics of moving to continuous time in the present setting. As we move from three periods to an n-period setting, the formula (18.13) becomes:

$$B(t, t+3\Delta) = \mathbb{E}^{\mathbb{Q}}\left[\frac{1}{(1+r_t\Delta)(1+r_{t+\Delta}\Delta)\cdots(1+r_{t+n\Delta}\Delta)}\right] \tag{18.14}$$

with the condition that the Δ is selected so that $T = t + n\Delta$. Now, recall the approximation that when r_j is small one can write:

$$\frac{1}{1+r_j\Delta} \approx e^{-r_j\Delta} \tag{18.15}$$

Next, apply this to each ratio on the right-hand side of (18.14) separately, to obtain the approximation

$$\frac{1}{(1+r_t\Delta)(1+r_{t+\Delta}\Delta)\cdots(1+r_{t+n\Delta}\Delta)} = e^{-r_t\Delta}e^{-r_{t+\Delta}\Delta}\cdots e^{-r_{t+\nu\Delta}\Delta} \tag{18.16}$$

$$= e^{-[r_t+r_{t+\Delta}+\cdots+r_{t+n\Delta}]\Delta} \tag{18.17}$$

or as $\Delta \to 0$:

$$e^{-\sum_{i=1}^{n}(r_{t+\Delta})\Delta} \to e^{-\int_t^T r_s\,ds} \tag{18.18}$$

given that all technical conditions are satisfied. Thus, as $\Delta \to 0$ we move from discrete-time discounting toward continuous-time discounting with variable spot rates. As a result, discrete-time discount factors get replaced by the exponential function. Because interest rates are continuously changing, an integral has to be used in the exponent. Thus we obtain the continuous-time bond pricing formula:

$$B(t,T) = \mathbb{E}^{\mathbb{Q}}\left[e^{-\int_t^T r_s\,ds}\right]$$

18.3.4 Yields and Spot Rates

We can also derive a relation between the yields $R(t, T)$ and the short rate r_t. We can relate future short rates to the yield curve of time t using the two equations in (18.12) and (18.3). Equating the right-hand sides:

$$e^{-R(t,T)(T-t)} = \mathbb{E}^{\mathbb{Q}}\left[e^{-\int_t^T r_s\, ds}\right] \tag{18.19}$$

Taking logarithms:

$$R(t,T) = \frac{-\log \mathbb{E}^{\mathbb{Q}}\left[e^{-\int_t^T r_s\, ds}\right]}{T-t} \tag{18.20}$$

We see that the yield of a bond can (roughly) be visualized as some sort of average spot rate that is expected to prevail during the life of the bond. In fact, in the special case of a constant spot rate,

$$r_s = r, \quad t \le s \le T \tag{18.21}$$

we obtain:

$$R(t,T) = \frac{-\log \mathbb{E}^{\mathbb{Q}}\left[e^{-\int_t^T r_s\, ds}\right]}{T-t} = \frac{-\log e^{-r(T-t)}}{T-t} \tag{18.22}$$

$$= r \tag{18.23}$$

Hence, the yield equals the spot rate, if the spot rate is indeed constant.

18.4 FORWARD RATES AND BOND PRICES

In this section we obtain another arbitrage relation that shows how *forward* rates relate to bond prices. It turns out that this relationship plays a crucial role in the modern theory of fixed income.

Let $F(t,T,U)$ be the current forward rate, contracted at time t, on a loan that begins at time T and matures at time $U > T$.

As mentioned earlier, to derive a relationship between the $B(t,T)$ and $F(t,T,U)$ we need a second bond, $B(t,U)$, that matures at time U. This is easy to see. The $F(t,T,U)$ is a market price that incorporates time t information concerning the (future) period *between* the times T and U. We expect the bond $B(t,T)$ to incorporate all relevant information up to time T. The longer maturity bond $B(t,U)$, on the other hand, is a price that will incorporate all information up to time U. Hence, we should in principle be able to extract from $B(t,T)$ and $B(t,U)$ all necessary information concerning the $F(t,T,U)$. As before, we obtain this relationship first in discrete time, using intervals of length Δ, and then take the continuous-time limit.

18.4.1 Discrete Time

To motivate the discussion, we begin with two periods. If Δ represents a small but noninfinitesimal time interval, and if t is the "present," then the price of a two-period bond will be given by $B(t,t+2\Delta)$. This bond will yield a cash flow of \$1 at maturity date $t+2\Delta$. Thus, one pays $B(t,t+2\Delta)$, at time t, and the investment pays off \$1 at time $t+2\Delta$.

Now suppose liquid markets are available in forward loans and consider the following alternative investment at time t. We make a forward loan that begins at time $t+\Delta$, which pays an interest of $F(t,t+\Delta,t+2\Delta)\Delta$ at $t+2\Delta$. Let the total amount loaned be such that, at time $t+2\Delta$, we receive \$1. Thus, the *amount* of the loan contracted for time $t+\Delta$ and denoted by $B^*_{t+\Delta}$ is[5]:

$$B^*_{t+\Delta} = \frac{1}{1+F(t,t+\Delta,t+2\Delta)\,\Delta} \tag{18.24}$$

Now this is an amount that belongs to time $t+\Delta$. We need to discount the to time t using the current spot rate. This gives the time t value of the forward loan, which we call B^*_t:

$$B^*_t = \frac{1}{1+r_t\Delta}\left[\frac{1}{1+F(t,t+\Delta,t+2\Delta)\,\Delta}\right] \tag{18.25}$$

[5]In the following expressions the reader may notice that the forward rates, say, $F(t,T,T+\Delta)$, are multiplied by Δ. This is needed because the $F()$ are assumed to be annual rates, whereas the Δ is supposedly a small arbitrary interval. By market convention, the forward interest earned during Δ is not $F(t,T,T+\Delta)$, but $F(t,T,T+\Delta)$ times Δ. For example, if the annual forward rate is 6%, and if one year is made of 360 days, then a three-month loan will earn six times 1/4%.

Finally, after recognizing that for period t, r_t is also the trivially defined forward rate $F(t, t, t+\Delta)\Delta$, the B_t^* becomes[6]:

$$B_t^* = \frac{1}{[1+F(t,t,t+\Delta)\Delta]} \frac{1}{[1+F(t,t+\Delta,t+2\Delta)\Delta]} \quad (18.26)$$

According to this, if, at time t, we invest the amount at a rate r_t, and then at $t+\Delta$ roll this investment at a predetermined rate $F(t, t+\Delta, t+2\Delta)$, we get a payoff of \$1 at time $t+2\Delta$. But this is exactly the same payoff given by the strategy of buying the bond $B(t, t+2\Delta)$. Thus, if the credit risk involved in the two strategies is the same, we must have

$$B(t,t+2\Delta) = B_t^*$$

or

$$B(t,t+2\Delta) = \frac{1}{[1+F(t,t,t+\Delta)\Delta]} \frac{1}{[1+F(t,t+\Delta,t+2\Delta)\Delta]} \quad (18.27)$$

Note that since all the quantities on the right-hand side of this equation are *known* at time t, there is no need to use any expectation operators in this formula. The relationship between the bond prices and the *current* forward rates is exact.

What happens when this arbitrage relation does not hold? One would simply short sell the expensive investment and buy the cheaper one. The payments and receipts of two positions will cancel each other at time $t+2\Delta$, while leaving some profit at time t. Hence, there will be an arbitrage opportunity.

18.4.2 Moving to Continuous Time

Suppose now we consider n discrete time periods, each of length Δ, so that $T = t + n\Delta$. The formula becomes

$$B(t,T) = \frac{1}{[1+F(t,t,t+\Delta)\Delta]\cdots[1+F(t,t+(n-1\Delta),t+n\Delta)\Delta]} \quad (18.28)$$

Now use the approximation that when the $F(t, T, U)$ and Δ are small, one has

$$\frac{1}{[1+F(t,T,U)\Delta]} \approx e^{-F(t,T,U)\Delta} \quad (18.29)$$

and write the $B(t, T)$ as

$$B(t,T) \approx \left[e^{-F(t,t,t+\Delta)\Delta}\right]\left[e^{-F(t,t+\Delta,t+2\Delta)\Delta}\right] \cdots \left[e^{-F(t,t+(n-1)\Delta,t+n\Delta)\Delta}\right] \quad (18.30)$$

But products of exponential terms can be simplified by adding the exponents. So

$$B(t,T) \approx e^{-F(t,t,t+\Delta)\Delta - F(t,t+\Delta,t+2\Delta)\Delta - \cdots - F(t,t+(n-1)\Delta,t+n\Delta)\Delta} \quad (18.31)$$

$$= e^{-\sum_{i=1}^{n} F(t,t+(i-1)\Delta,t+i\Delta)\Delta} \quad (18.32)$$

which means that we can let $\Delta \to 0$ and increase the number of intervals to obtain the continuous version of the relation between instantaneous forward rates and bond prices,

$$B(t,T) = e^{-\int_t^T F(t,s)\,ds} \quad (18.33)$$

given that the recurring technical conditions are all satisfied. The $F(t, s)$ is now the instantaneous forward rate contracted at time t, for a loan that begins at s and ends after an infinitesimal time interval ds. Thus, as $\Delta \to 0$, we move from discrete time toward continuous-time discounting. As a result, two things happen. First, the discrete-time discount factors need to be replaced by the exponential function. Second, instead of discrete forward rates we need to use instantaneous forward rates. Because instantaneous forward rates may be different, an integral has to be used in

[6] Any loan that begins now can be called a trivial forward loan.

the exponent. Again, note that there is no expectation operator in this equation because all $F(t,s)$ are quantities known at time t.

The formula:

$$B(t,T) = e^{-\int_t^T F(t,s)\,ds} \tag{18.34}$$

gives prices of default-free zero-coupon prices as a function of instantaneous forward rates.

We can also go in the opposite direction and write $F(t,T,U)$ as a function of bond prices. We prefer to do this for the maturities T and $U = T+\Delta$.[7] Thus, consider two bonds, $B(t,T)$ and $B(t,T+\Delta)$, whose maturities differ only by a small time interval $\Delta > 0$. Then writing the formula (18.32) twice:

$$B(t,T) = e^{-\int_t^T F(t,s)\,ds} \tag{18.35}$$

and

$$B(t,T+\Delta) = e^{-\int_t^{T+\Delta} F(t,s)\,ds} \tag{18.36}$$

Take logarithms of these equations and subtract:

$$\log B(t,T) - \log B(t,T+\Delta) = -\int_t^T F(t,s)\,ds + \int_t^{T+\Delta} F(t,s)\,ds \tag{18.37}$$

$$= \int_T^{T+\Delta} F(t,s)\,ds \tag{18.38}$$

Now, suppose Δ is small so that the $F(t,T)$ can be considered "constant" during the small time interval $[T, T+\Delta]$. We can write:

$$\log B(t,T) - \log B(t,T+\Delta) \approx F(t,T)\,\Delta \tag{18.39}$$

This equation becomes exact, after taking the limit:

$$F(t,T) = \lim_{\Delta\to 0} \frac{\log B(t,T) - \log B(t,T+\Delta)}{\Delta} \tag{18.40}$$

[7]This will facilitate the derivations of HJM arbitrage conditions later.

That is, the instantaneous forward rate $F(t,T)$ is closely related to the derivative of the logarithm of the discount curve.

By going through a similar argument, we can derive a similar expression for the noninstantaneous, but continuously compounded forward rate $F(t,T,U)$[8]:

$$F(t,T,U) = \frac{\log B(t,T) - \log B(t,U)}{U-T}$$

where $F(t,T,U)$ is the continuously compounded forward rate on a loan that begins at time $T < U$ and ends at time U. The contract is written at time t.

Clearly, by letting $T \to U$ we get the instantaneous forward rate $F(t,T)$:

$$F(t,T) = \lim_{T\to U} F(t,T,U) \tag{18.41}$$

It is obvious from these arguments that the existence of $F(t,T)$ assumes that the discount curve, that is, the continuum of bond prices, is differentiable with respect to T, the maturity date. Using Eq. (18.39) and assuming that some technical conditions are satisfied, we see that

$$F(t,t) = r \tag{18.42}$$

That is, the instantaneous forward rate for a loan that begins at the current time t is simply the spot rate r_t.

18.5 CONCLUSIONS: RELEVANCE OF THE RELATIONSHIPS

It is time to review what we have obtained so far. We have basically derived three relationships between the bond prices $B(t,T)$, the bond yields

[8]Earlier in this section when discussing the discrete time case, $F(t,T,U)$ was used as a symbol for simple forward rates. In moving to continuous time, and switching to the use of the exponential function, the same symbol now denotes continuously compounded rates. A more appropriate way of proceeding would perhaps be to use different symbols for the two concepts. But the notation of this chapter is already too complicated.

$R(t,T)$, the forward rates $F(t,T,U)$, and the spot rates r_t.

The first relation was simply definitional. Given the bond price, we defined the continuously compounded yield to maturity $R(t,T)$ as:

$$R(t,T) = \frac{-\log B(t,T)}{T-t} \tag{18.43}$$

The second relationship was the result of applying the same principle that was used in the first part of the book to bond prices; namely, that the expectation under the risk-neutral measure $\mathbb{Q}$ of payoffs of a financial derivative would equal the current arbitrage-free price of the instrument, once discounted by the instantaneous interest rate r_t. The bond $B(t,T)$ paid \$1 at maturity, and the discounted value of this was

$$\left[e^{-\int_t^T r_s\, ds}\right]$$

The spot rate r_t being random, we apply the (conditional) expectation operator under the risk-neutral measure to obtain the relation:

$$B(t,T) = \mathbb{E}^{\mathbb{Q}}\left[e^{-\int_t^T r_s\, ds}\right] \tag{18.44}$$

Thus, this second relationship is based on the no-arbitrage condition and as such is a *pricing* equation. That is, given a proper model for r_t, it can be used to obtain the "correct" market price for the bond $B(t,T)$.

The third relationship was derived in the previous section. Using again an arbitrage argument we saw that the (arbitrage-free) prices of the bonds $B(t,T), B(t,U)$ with $U > T$, and continuously compounded forward rate $F(t,T,U)$ were related according to:

$$F(t,T,U) = \frac{\log B(t,T) - \log B(t,U)}{U-T}, \quad t < T < U \tag{18.45}$$

This can also be used as a pricing equation, except that if we are given a $F(t,T,U)$ we will have one equation and two unknowns to determine here, namely the $B(t,T), B(t,U)$. Thus, before we can use this as a pricing equation we need to know at least one of the $B(t,T), B(t,U)$. The addition of other forward rates would not help much because each forward rate equation would come with an additional unknown bond price.[9]

To sum up, the first relation is simply a definition. It cannot be used for pricing. But the other two are based on arbitrage principles and would hold in liquid and well-functioning markets. They form the basis of the two broad approaches to pricing interest-sensitive instruments. The so-called *classical approach* uses the second relation, whereas the recent Heath-Jarrow-Morton, *HJM approach*, uses the third. We will study these in the next chapter.

18.6 REFERENCES

The reader can consult this excellent book for discrete-time fixed income models, Jarrow (1996). Rebonato (1998) is one source that contains a good and comprehensive review of interest rate models. The other good source is the publication by Risk (1996). For a survey of recent issues, see Jegadeesh and Tuckman (2000).

18.7 EXERCISES

1. Consider the SDE for the spot rate r_t

$$dr_t = \alpha\,(m - r_t)\,dt + \sigma dW_t \tag{18.47}$$

Suppose the parameters α, μ, σ are known, and that, as usual, W_t is a Wiener process.

(a) Show that

$$\mathbb{E}\left[r_s | r_t\right] = \mu + (r_t - \mu)\, e^{\alpha(s-t)}, \quad t < s \tag{18.48}$$

[9] Suppose we brought in another equation containing $B(t,U)$:

$$F(t,U,S) = \frac{\log B(t,U) - \log B(t,S)}{S-U}, \quad t < U < S \tag{18.46}$$

$$\mathbb{V}[r_s|r_t] = \frac{\sigma^2}{2\alpha}\left(1 - e^{2\alpha(s-t)}\right), \quad t < s \tag{18.49}$$

(b) What do these two equations imply for the conditional mean and variance of spot rate as $s \to \infty$?

(c) Suppose the market price of interest rate risk is constant at λ (i.e., the Girsanov transformation adjusts the drift by $\sigma\lambda$). Using the bond price function given in the text, show that the drift and diffusion parameters for a bond that matures at time s are given by

$$\mu^B = r_t + \frac{\sigma\lambda}{\alpha}\left(1 - e^{\alpha(s-t)}\right) \tag{18.50}$$

$$\sigma^B = \frac{\lambda}{\alpha}\left(1 - e^{\alpha(s-t)}\right) \tag{18.51}$$

(d) What happens to bond price volatility as maturity approaches? Is this expected?

(e) What happens to the drift coefficient as maturity approaches? Is this expected?

(f) Finally, what is the drift and diffusion parameter for a bond with very long maturity, $s \to \infty$?

2. Consider a world with two time periods and two possible states at each time $t = 0, 1, 2$. There are only two assets to invest. One is risk-free borrowing and lending at the risk-free rate $r_i, i = 0, 1$. The other is to buy a two period bond with current price B_0. The bond pays \$1 at time $t = 2$ when it matures.

(a) Set up a 2×4 system with state prices $\psi^{ij}, i, j = u, d$ that gives the arbitrage-free prices of a savings account and of the bond B.

(b) Show how one can get risk-neutral probabilities in this setting.

(c) Show that if one adopts a savings account normalization, the arbitrage-free price of the bond will be given by

$$B = \mathbb{E}_0^{\mathbb{Q}}\left[\frac{1}{(1+r_0)(1+r_1)}\right]$$

3. What is the expected value of a zero-coupon bond, that is:

$$B(0,t) = \mathbb{E}(e^{-\int_t^s r_u du}) \tag{18.52}$$

Under the following dynamics:

$$dr_t = \alpha(\mu - r_t)\,dt + \sigma\,dW_t \tag{18.53}$$

What distribution does the zero-coupon bond follow?

Hint: Consider the distribution of $\int_t^s r_u du \approx \sum_t^s r_u du$. In addition you are given that: $var(\int_t^s r_u du) = \frac{\sigma^2}{\alpha^2}\{s-t+\frac{1-e^{-2\alpha(s-t)}}{2\alpha}-2\frac{1-e^{-\alpha(s-t)}}{\alpha}\}$.

4. Consider an interest rate derivative which pays \$1 at expiration T if the short rate is greater than 0.01, and pays 0 otherwise. Price this derivative with maturity $T = 1$ year under the Vasicek model

$$dr_t = \alpha(\mu - r_t)dt + \sigma\,dW_t \tag{18.54}$$

with $r_0 = 0.01, \alpha = 0.2, \mu = 0.01, \sigma = 0.05$.

CHAPTER

19

Classical and HJM Approach to Fixed Income

OUTLINE

19.1 INTRODUCTION

Market practice in pricing interest-sensitive securities can proceed in two different ways depending on which of the two arbitrage relations developed in the previous chapter is taken as a starting point. In fact, Chapter 18 discussed in detail the bond pricing equation

$$B(t,T) = \mathbb{E}^{\mathbb{Q}}\left[e^{-\int_t^T r_s ds}\right]$$

which gave arbitrage-free prices of default-free discount bonds $B(t,T)$ under the risk-neutral measure $\mathbb{Q}$. This was a relation between spot rates

Financial Derivatives, Third Edition.
http://dx.doi.org/10.1016/B978-0-12-384682-2.00019-0

r_t and bond prices B (t, T) that held only when there were no-arbitrage possibilities.

The second arbitrage relation of Chapter 18 was between instantaneous forward rates $F(t, T)$ and bond prices:

$$B(t, T) = e^{-\int_t^T r_s ds}$$

Obviously, both relations can be exploited to calculate arbitrage-free prices of interest-sensitive securities.

The market practice is to start with a set of bond prices $\{B(t, T)\}$ that can reasonably be argued to be arbitrage-free. Then either one of the above relations can be used to go "backwards" and determine a model for r_t or for the set of forward rates $\{F(t, s), s \in [t, T]\}$. Because the two relations above hold under no-arbitrage conditions, the model that one obtains for r_t, or for the instantaneous forward rates, will also be "risk-adjusted." That is, they will be valid under the risk-neutral measure $\mathbb{Q}$.

The so-called classical approach uses the first arbitrage relation and tries to extract from the $\{B(t, T)\}$ a risk-adjusted model for the spot rate r_t. This will involve modeling the drift of the spot rate dynamics, as well as calibration to observed volatilities. An assumption on the Markovness of r_t is used along the way.

The Heath-Jarrow-Morton (HJM) approach, on the other hand, uses the second arbitrage condition and obtains arbitrage-free dynamics of k-dimensional instantaneous forward rates $F(t, T)$. It involves no drift modeling, but volatilities need to be calibrated. It is more general and, usually, less practical to use in practice. The HJM approach does not need spot-rate modeling. Yet, it also demonstrates that the spot rate r_t is in general not Markov.

In this chapter we provide a discussion of these methods used by practitioners in pricing interest-sensitive securities. Given our limited scope, numerical issues and details of the pricing computations will be omitted. Interested readers can consult several excellent texts on these. Our focus is on the understanding of these two fundamentally different approaches.

19.2 THE CLASSICAL APPROACH

The relationship between bond prices and instantaneous spot rates,

$$B(t, T) = \mathbb{E}^{\mathbb{Q}}\left[e^{-\int_t^T r_s ds}\right] \tag{19.1}$$

can be exploited in (at least) two different ways by market practitioners.

First, if an accurate and arbitrage-free discount curve $\{B(t, T)\}$ exists, one can use these in Eq. (19.1), go "backwards," and try to obtain an arbitrage-free model for the spot rate r_t. One can then exploit the arbitrage-free characteristic of this spot-rate model to price interest rate derivatives *other* than bonds.

Second, one may go the other way around. If there are no reliable data on the discount curve $B(t, T)$, one may first posit an appropriate arbitrage-free model for the spot rate r_t, estimate it using historical data on interest rates, and then use Eq. (19.1) in getting "fair" market prices for illiquid bonds and other interest-sensitive derivatives. Both of these will be called the *classical approach* to pricing interest rate derivatives. We will see that, one way or another, the classical approach is based on modeling the instantaneous interest rate r_t, in the first case by starting from a "reliable" set of bond prices $\{B(t, T)\}$, and in the second case, from data available on r_t process itself.

None of these are straightforward, so we start by looking at some simple examples.

19.2.1 Example 1

First, consider the case in which we prefer to model r_t directly.

Suppose, in an economy where discount bonds do not trade actively, we have reasons to believe that r_s is constant at r. That is,

$$r_s = r, \quad s \geq t$$

Using relation (19.1) we can write:

$$B(t, T) = \mathbb{E}^{\mathbb{Q}}\left[e^{-\int_t^T r ds}\right]$$

Because r is constant, we "take" the expectation trivially and obtain:

$$B(t,T) = e^{-r(T-t)}$$

Thus, starting from a posited model for r_t we obtained a bond pricing equation, namely a closed-form formula that depends on the known quantities T, t, and r.

Using this equation, we can price illiquid bonds. To give an example, suppose $r = 0.05$. We then have the following prices for 1, 2, 3, 4 year maturity discount bonds:

$$B(t,t+1) = 0.95$$
$$B(t,t+2) = 0.90$$
$$B(t,t+3) = 0.86$$
$$B(t,t+4) = 0.82$$

If our original assertion about the constancy of r_t is correct, these bond prices will be arbitrage-free.

However, note that if we had posited a non-deterministic model for r_t, the application of the same procedure would be problematic. In fact, this would require knowing the drift of the spot-rate process under the risk-neutral measure $\mathbb{Q}$. In the above, the r_t was constant, and hence its drift under $\mathbb{Q}$ was zero.

19.2.2 Example 2

Now suppose we do not know what type of stochastic process r_t follows in reality. In fact, suppose our purpose is to determine this process from observations on liquid bonds that trade in the market. In particular, suppose we observe the following discount curve:

$$B(t,t+1) = 0.95$$
$$B(t,t+2) = 0.90$$
$$B(t,t+3) = 0.86$$
$$B(t,t+4) = 0.82$$

We can then *infer* from these prices that the r_t process is in fact following the SDE

$$dr_t = a(r_t,t)\,dt + b(r_t,t)\,dW_t$$

with

$$a(r_t,t) = 0, \quad b(r_t,t) = 0$$

That is, the r_t process is in fact constant at r.[1]

Using this information, we can price interest-sensitive derivatives written on r_t or on $B(t,T)$. For example, a bond option will have an arbitrage-free price equal to zero because r_t is constant.

19.2.3 The General Case

Suppose one obtains a reasonably accurate observation on the discount curve $\{B(t,u), t < u \leq T\}$ that one can assume to be arbitrage-free. Then, Eq. (19.1) says that the same spot rate process r_s must satisfy the following set of equations:[2]

$$B(t,T_0) = \mathbb{E}^{\mathbb{Q}}\left[e^{-\int_t^{T_0} r_s ds}\right] \tag{19.2}$$

$$B(t,T_1) = \mathbb{E}^{\mathbb{Q}}\left[e^{-\int_t^{T_1} r_s ds}\right] \tag{19.3}$$

$$\cdots = \cdots \tag{19.4}$$

$$B(t,T_n) = \mathbb{E}^{\mathbb{Q}}\left[e^{-\int_t^{T_n} r_s ds}\right] \tag{19.5}$$

where

$$T_0 < T_1 < \cdots < T_n \tag{19.6}$$

are the $n+1$ maturities at which we have reasonably accurate and arbitrage-free bond prices.

Let us discuss these equations. Given the martingale measure that defines the expectation operator $\mathbb{E}^{\mathbb{Q}}[\cdot]$, the right-hand sides of *all* these equations depend on the *same* spot rate process r_t, albeit with different T_i. The *market* will determine the left-hand side of these $n+1$ equations. The problem faced by the practitioner is to determine *one* model for the interest rate process r_t such that all these equations are satisfied simultaneously. How can we guarantee that the specific model selected for the r_t will be consistent with these $n+1$ set of equations? This is indeed

[1] These prices are identical to those in Example 1.

[2] We are assuming that bonds have a par value of 1.

no straightforward task. Let us illustrate some of the difficulties involved.

In fact consider what this requires. First, one has to postulate a *spot rate model*:

$$dr_t = a(r_t, t)\, dt + b(r_t, t)\, dV_t \tag{19.7}$$

and second, one has to select the $a(r_t, t), b(r_t, t)$ *and* the probabilistic behavior of the driving process V_t, such that the system of equations shown in (19.5) are satisfied.[3] We consider two examples.

19.2.3.1 A Geometric SDE

To see that the system in (19.2)–(19.5) comes with "hidden" complications. Suppose we select for the spot rate r_t a geometric SDE driven by a Wiener process under the $\mathbb{Q}$:

$$dr_t = \mu r_t dt + \sigma r_t dW_t \tag{19.8}$$

Hence we have postulated that in (19.7) the drift and the diffusion coefficients are given by:

$$a(r_t, t) = \mu r_t, \sigma(r_t, t) = \sigma r_t, V_t = W_t \tag{19.9}$$

There will immediately be some headaches. A spot-rate process obeying this model would eventually go to plus or minus infinity depending on the sign of μ, as $t \to \infty$. Also, the percentage volatility of the r_t will be constant. Clearly, these do not seem to be ideal properties to represent the behavior of overnight rates observed in reality. First, interest rates do not have "trends." Second, in reality percentage interest rate volatility seems to be a complicated nonlinear function of the level of spot rate r_t, rather than being just a constant.[4]

[3]Note the implicit assumption here. The increment in the spot rate depends only on the current r_t, and hence the spot rate has a Markovian character. As we see below, this will be a special case in fixed-income markets where arbitrage conditions are satisfied.

[4]The selection of V_t as a Wiener process is already made. This also may not be appropriate because real-world spot-rate processes may contain jumps.

But putting these two difficulties aside, consider the problem mentioned above: namely, how to select the μ and σ such that all the equations in the system in (5) are satisfied simultaneously?

This is no simple task. In fact, given the reasonably accurate observations on the bond prices, $\{B(t, T_i), i = 0, \ldots, n\}$, in (19.2)–(19.5) we have $n+1$ equations with known left-hand sides. But the free parameters of the interest rate model that we can choose are only the μ and the σ. Hence we have to satisfy a system of $n+1$ equations by choosing two unknowns. This is not going to be possible unless there are strong interdependencies among observed bond prices $\{B(t)Ti)$, $i = 0, .., n\}$, so that $n-1$ of these equations are in fact redundant. Then, the system would in fact reduce to two equations in two unknowns and a set of μ, σ that fits the observed arbitrage-free discount curve $\{B(t, T)\}$ can be found.

But how attractive is it to postulate such strong dependencies among the $n+1$ bond prices that one observes in liquid markets? Obviously, the spot-rate process postulated in Eq. (19.8) is quite inadequate for practical pricing purposes. Other models must be sought.

19.2.3.2 A Mean-Reverting Model

The geometric SDE may be inappropriate for describing the dynamics of the spot rate, but from the above arguments we learned something. First, an appropriate SDE should be selected for r_t, and then the parameters of this model should be determined (calibrated) so that the spot-rate model "fits" the discount curve $\{B(t, Ti)\}$ given by liquid markets. If this can be done, and if the observed discount bond prices $\{B(t, T_i), i = 0, .., n\}$ are arbitrage-free, then the resulting model for the spot-rate process r_t would also be arbitrage-free. It could be used to price interest-sensitive derivatives.

Thus, one may ask if one can postulate an SDE more realistic than the geometric process discussed in the first example. In fact, consider the mean-reverting spot-rate process with variable "mean" θ_t and a square-root diffusion component:

$$dr_t = \lambda\left(\theta_t - r_t\right)dt + \sigma\sqrt{r_t}dW_t \quad (19.10)$$

Here, for each time period t, the parameter θ_t is allowed to assume a different known value. This augments the number of free parameters that one has in system (5). For example, in a discrete setting with:

$$t_0 < t_1 < \cdots < t_m \quad (19.11)$$

there will be $m + 3$ free parameters to select in the interest rate model, namely the

$$\left\{\theta_{t_0}, \theta_{t_0}, \ldots, \theta_{t_0}, \lambda, \sigma\right\} \quad (19.12)$$

This gives more flexibility in fitting the interest rate process to the observed discount curve, $\{B(t, T_i), i = 0, \ldots, n\}$.[5] In fact, we can not only fit the r_t process to bond prices, but fit it to bond volatilities as well.[6] See Hull and White (1990) for example.

Besides, unlike geometric processes, mean-reverting processes are known under the right conditions *not* to explode as $t \to \infty$. Also, given infinitesimal steps, the r_t process that will be generated by the mean-reverting model will not become negative given the diffusion component adopted here.

19.2.4 Using the Spot-Rate Model

Suppose one successfully completes the project to extract an arbitrage-free *model* for the spot rate r_t from the pricing equation:

$$B(t, T) = \mathbb{E}^{\mathbb{Q}}\left[e^{-\int_t^{T_1} r_s ds}\right]$$

How would this model be used?

[5]The parameters n and m need not be the same.

[6]That is, we can calibrate the free parameters of the SDE shown in (19.10) so that the volatilities of $B(t, T)$ obtained from Eq. (19.1) match the volatilities observed in liquid options markets on these bonds.

The answer to this question was briefly mentioned at the beginning of this chapter. The bond pricing equation is used to extract an arbitrage-free spot-rate model from the existing term structure because using this model one can then price other interest-sensitive securities and obtain arbitrage-free prices without having to look at the markets for these securities.[7]

To see the use of the spot-rate model, consider the following setup. A reliable term structure $B(t, T)$ is given and is exploited to extract the arbitrage-free model for r_t:

$$dr_t = \widetilde{a}\left(r_t, t\right)dt + b\left(r_t, t\right)dW_t$$

where the drift $\widetilde{a}(r_t, t)$ has a "tilde" because it is assumed to be adjusted for the interest rate risk and consequently the W_t is a Wiener process under the risk-neutral measure $\mathbb{Q}$. We consider two cases.

19.2.4.1 A One-Factor Model

Suppose we want to price a derivative instrument that is sensitive to r_t only. Its price is denoted by $C(r_t, t)$. The expiration date is T and the expiration payoff is given by the known function $G(r_T, T)$:

$$C\left(r_T, T\right) = G\left(r_T, T\right)$$

One could immediately use the pricing equation:

$$C\left(r_t, T\right) = \mathbb{E}^{\mathbb{Q}}\left[e^{-\int_t^T r_s ds} G\left(r_T, T\right)\right]$$

This expectation can be evaluated using Monte Carlo methods; or it can be solved for a closed-form solution, if one exists; or it can be converted

[7]We can mention at least three specific uses, but there are many others that we do not go into because of the limited scope of this book: (1) It may be that there is a traded instrument $C(r_t, t)$ that can be synthetically replicated using the traded bonds $B(t, T)$. (2) The $C(rt, t)$ may be a new instrument that does not yet trade. (3) There may be some suspicion that $C(r_t, t)$ is mispriced by the markets. Then, by using an arbitrage-free model for r_t one can calculate a "fair" price for the instrument and take proper hedging, arbitrage, or speculative positions. Or, one could simply use the price in investment banking operations.

into a PDE, as will be seen in Chapter 21; or it can be evaluated in a tree model. This will be possible because we would already have a dynamics for r_t under the:

$$dr_t = \widetilde{a}(r_t, t)\, dt + b(r_t, t)\, dW_t$$

The rest is just computation.

19.2.4.2 A Second Factor

Things can get somewhat more complicated if we want to price a derivative instrument that is sensitive to r_t and, say, to R_t, a long rate, which is not perfectly correlated with r_t. Suppose the price of this new instrument is denoted by $C(r_t, R_t, t)$. The expiration date is again T, and the expiration payoff is given by the known function $G(r_T, R_T, T)$:

$$C(r_t, R_t, t) = G(r_T, R_T, T)$$

One could again write the pricing equation:

$$C(r_t, R_t, t) = \mathbb{E}^{\mathbb{Q}}\left[e^{-\int_t^T r_s ds} G(r_T, R_T, T)\right]$$

But, the model would not be complete. In fact, we do not yet have an arbitrage-free model, given the "second factor," R_t. Before we can proceed and calculate the price, we need to obtain a risk-adjusted SDE for R_T as well. For these issues we refer the reader to Brennan and Schwarz (1979) and the related literature. It must be realized that the two processes, r_t and R_T, may have complex time-varying correlation properties and computationally the problem may get much more difficult than the case of a single factor. [8]

19.2.4.3 The Importance of Calibration

It is important to understand the process by which one obtains the spot-rate model in (2.4.1). If one used only econometric methods and estimated a continuous-time drift $a(r_t, t)$ and diffusion $\sigma(r_t, t)$, the resulting model written as

$$dr_t = a(r_t, t)\, dt + \sigma(r_t, t)\, dW_t^*$$

would *not* be called arbitrage-free. Econometric methods yield estimates for the *real-world parameters*, and the model would be valid under the real-world probability $\mathbb{P}$. The Wiener process W_t^* can be directly estimated from the data as continuous-time regression residuals.

It is the backward extraction of the r_t process using

$$B(t, T) = \mathbb{E}^{\mathbb{Q}}\left[e^{-\int_t^T r_s ds}\right]$$

that yields an arbitrage-free model because the probability used in this pricing equation is the $\mathbb{Q}$. Hence, arbitrage-free spot-rate modeling is more than just an estimation or calibration problem. It is also based on judicious choice of pricing models.

19.2.5 Comparison with the Black–Scholes World

We see that the classical approach to pricing interest-sensitive securities amounts, essentially, to spot-rate modeling. We also see that this calibration effort is not trivial, especially when discount bond prices are not perfectly related to each other across maturities.

More importantly, if one pursues the classical approach, arbitrage restrictions will be incorporated into the model *indirectly*, through fitting to the initial yield curve. One first starts with a set of discount bond prices, or the corresponding yields, and then one tries to find a model for r_t that "fits" the observed term structure so that

$$B(r_t, T) = \mathbb{E}^{\mathbb{Q}}\left[e^{-\int_t^T r_s ds}\right]$$

is satisfied for *every* T.

This is quite different from the philosophy used in the Black–Scholes world discussed in

[8] In the following chapters we will have a different notation for the risk-adjusted drift. As we develop new concepts that we can use, we will be able to write the risk-adjusted drift as $a(r_t, t) - \lambda_t b(r_t, t)$, where the λ_t is the Girsanov drift adjustment, or, in this case, the market price of interest rate risk.

the first part of this book. There, the arbitrage restrictions were *directly* and explicitly incorporated into the model by replacing the unknown drift of the underlying process by the known spot rate. There was no need to model the drift term of the stock price process. The latter was simply replaced by the (constant) spot rate r. As a result, Black–Scholes approach reduced the problem to one of volatility modeling. The assumption of a geometric process for the underlying process S_t simplified this further and percentage volatility was assumed to be constant.

Thus, in this sense, the spot-rate modeling that forms the basis of the *classical approach* appears to be a *fundamentally* different methodology from the arbitrage-free pricing as seen until now.

This leads to the following question: Is there another approach that one can use, which will be more in line with the philosophy of Black–Scholes? The answer is yes and it is the Heath-Jarrow-Morton (HJM) Model.

19.3 THE HJM APPROACH TO TERM STRUCTURE

The arbitrage restrictions that we have been studying are the result of common random processes that influence discount bonds that are identical except for their maturity. If the liquid bonds that determine the term structure $\{B(t,T)\}$ are all influenced by the same unpredictable Wiener process W_t, the respective prices must somehow be related to each other as suggested by the pricing relation:

$$B(t,T) = \mathbb{E}^{\mathbb{Q}}\left[e^{-\int_t^T r_s ds}\right]$$

The classical approach to pricing interest-sensitive securities is an attempt to extract these arbitrage relations from the $B(t,T)$ and then summarize them within an arbitrage-free spot-rate model:

$$dr_t = \widetilde{a}(r_t, t)\, dt + b(r_t, t)\, dW_t^*$$

This is indeed a complicated task of indirect accounting for a complex set of arbitrage relations between market prices. The Heath-Jarrow-Morton, or as known in the market, HJM, approach attacks these arbitrage restrictions directly by bringing the forward rates to the forefront.

The idea is based on the second arbitrage relation developed extensively in Chapter 18. As mentioned there, there are direct relations between discount bonds that are identical except for their maturity and forward rates. It is sufficient to review a simple case.

Let $B(t,T)$ and $B(t,U)$ be two default-free zero-coupon bonds that are identical except for their maturity $U > T$. Let $F(t,T,U)$ be the interest rate contracted at time t on a default-free forward loan that starts at T and ends at U. Here, $F(\cdot)$ is a percentage rate for period $U - T$. Thus, no days adjustment factor is needed. Then, the discussion in Chapter 18 permits writing the no-arbitrage condition:[9]

$$[1 + F(t,T,U)] = \frac{B(t,T)}{B(t,U)}$$

We thus have two bonds with different maturities in a single expression that contains $F(t,T,U)$. Now consider the joint dynamics of these variables. Because bonds *are* traded assets, in the corresponding SDEs we can replace the drift parameters by the risk-free rate r_t. Thus, up to this point everything is identical to Black–Scholes

[9] Let us repeat the arbitrage condition using somewhat different language. The $B(t,U)$ is the present value of a sure dollar to be received at a later date U. Its inverse is the time U value of \$1 that we have now. Dividing the inverse by $1 + F(t,T,U)$ brings a time U value to time T:

$$\frac{1}{[1 + F0t,T,U]B(t,U)}$$

Multiplying this by $B(t,T)$ should bring it back to \$1, the amount that we originally started with:

$$B(t,T)\frac{1}{[1 + F0t,T,U]B(t,U)} = 1$$

This is the case since $B(t,T)$ is the present value of \$1 to be received at time T.

derivation. But, note that according to the arbitrage relation above, the *ratio* of the two risk-neutral bond dynamics will be captured by the movements of a single forward rate $F(t,T,U)$. In other words, once risk-neutral dynamics of the bonds are written, the SDE for the forward rate $F(t,T,U)$ will be *determined*. There will be no need to calibrate and/or estimate any additional drift coefficients, or for that matter to adjust these coefficients for risk. All these will *automatically* be incorporated in the forward rate dynamics.

In other words, if we decided to model the forward rates $F(t,T,U)$ instead of the spot rate r_t, the arbitrage relations can be directly built into the forward rate dynamics similar to the case of Black–Scholes. The development of the HJM approach is based on this idea. Of course, in this framework we still have to calibrate the volatilities. Also, we need to select the exact forward rates that the pricing will be based on.

19.3.1 Which Forward Rate?

Here we have several options because the arbitrage relation can be written in several different ways.

The original approach used by HJM is to model the continuously compounded instantaneous forward rates $F(t,T)$—that is, use the relation developed in Chapter 19:

$$B(t,T) = e^{-\int_t^T F(t,s)ds}$$

where $F(t,s)$ is the rate on a forward loan that begins at time s and ends after an infinitesimal time period ds.

Writing the arbitrage relation as

$$\frac{B(t,T)}{B(t,U)} = e^{-\int_T^U F(t,s)ds}$$

we can obtain an arbitrage restriction on the dynamics for continuously compounded instantaneous rates $F(t,T)$, as will be done in the next section.

But this is only one way HJM models can proceed. Another option is to use forward rates for discrete, noninfinitesimal periods. That is, we can use models that are based on the $F(t,T,U)$. Letting $U = T + \Delta$, we can model arbitrage-free dynamics using the relationship:

$$[1 + F(t,t+\Delta,t+2\Delta)\,\Delta] = \frac{B(t,T)}{B(t,T+\Delta)}$$

Here, we can keep the $\Delta > 0$ fixed and consider the joint dynamics of the $B(t,T), B(t,T+\Delta)$ as t changes. The joint dynamics can be modeled with the risk-neutral measure or, depending on the instrument to be priced, with the forward measure introduced in Chapter 17. Proceeding this way leads to the so-called BGM models, after the work in Brace et al. (1997). The remaining part of this chapter will proceed along the lines of the original HJM approach by using the instantaneous forward rate $F(t,T)$.

19.3.2 Arbitrage-Free Dynamics in HJM

From the relationship between the default-free pure discount bond prices $B(t,T_i), T_i < T_{\max}$, with maturity T_i and forward rates $F(t,T)$ derived in Chapter 19, we have:

$$B(t,T) = e^{-\int_t^T F(t,u)du} \tag{19.13}$$

Recall that there is no expectation operator involved in this expression, because the $F(t,u)$ are all forward rates observed at time t. They are rates on forward loans that will begin at future dates $u > t$ and last an infinitesimal period du.

For the next section, adopt the notation $Bt = B(t,T)$ and assume that for a typical bond with maturity T we are given the following stochastic differential equation:

$$dB_t = \mu(t,T,B_t)\,B_t dt + \sigma(t,T,B_t)\,B_t dV_t^T \tag{19.14}$$

where the V_t^T is a Wiener process with respect to the real-world probability $\mathbb{Q}$. We need to emphasize three points concerning this SDE. First, the diffusion parameter is written in terms of percentage bond volatility, but is not necessarily of geometric form.[10] Second, the SDE is driven by a Wiener process indexed by T. This means that, in principle, every bond with different maturity is allowed to be influenced by some different shock. Later, we will see the single factor case where all the V_t^T will be required to be the same. And third, note the new way we write the diffusion parameter. The $\sigma(t,T,B_t)$ is explicitly made a function of the maturity T. This is needed in the derivation below, but will be abandoned in later chapters.

Now, bonds are traded assets. In a risk-neutral world with application of the Girsanov theorem, the drift coefficient can be modified as in the case of the Black–Scholes framework:

$$dB_t = r_t B_t dt + \sigma\,(t,T,B_t)\, B_t dW_t^T \qquad (19.15)$$

where r_t is the risk-free instantaneous spot rate, and W_t^T is the new Wiener process under the risk-neutral measure $\mathbb{Q}$. That is, by switching from $\mathbb{P}$ to $\mathbb{Q}$, we have eliminated the unknown drift in the bond dynamics.

Given these SDEs for bonds, we can get the dynamics of the $F(t,T)$ from Eq. (19.13). Begin with the arbitrage relation introduced in Chapter 19 and discussed above:

$$F(t,T,T+\Delta) = \frac{\log B(t,T) - \log B(t,T+\Delta)}{T+\Delta-T} \qquad (19.16)$$

where a noninfinitesimal interval $0 < \Delta$ is used to define the non-instantaneous forward rate, $F(t,T,T+\Delta)$, for a loan that begins at time T and ends at time $T+\Delta$. This is done by considering two bonds that are identical in all aspects, except for their maturity, which are Δ apart.

Now, to get the arbitrage-free dynamics of forward rates, apply Ito's Lemma to the right-hand side of (19.16) and use the risk-adjusted drifts whenever needed.[11] Apply Ito's Lemma first to $\log B(t,T)$ to get:

$$d[\log B(t,T)] = \frac{1}{B(t,T)}dB(t,T) - \frac{1}{2B(t,T)^2}\sigma \times (t,T,B_t)^2\, B(t,T)^2 dt \qquad (19.17)$$

Simplifying and then substituting from the SDE for the risk-adjusted bond dynamics in (19.15):

$$d[\log B(t,T)] = \left(r_t - \frac{1}{2}\sigma\,(t,T,B_t)^2\right)dt + \sigma\,(t,T,B_t)\,dW_t \qquad (19.18)$$

Now apply Ito's Lemma to $d\log B(t,T+\Delta)$ and get the equivalent expression with T replaced by $T+\Delta$:[12]

$$d[\log B(t,T+\Delta)] = \left(r_t - \frac{1}{2}\sigma\,(t,T+\Delta,B_t)^2\right)dt + \sigma\,(t,T+\Delta,B_t)\,dW_t \qquad (19.19)$$

It is important to realize that the first terms in drift of the SDEs for $B(t,T)$ and $B(T,T+\Delta)$ are the same because the dynamics under consideration are arbitrage-free. Under $\mathbb{Q}$, discount bonds with different maturities will have expected rates of returns that equal the risk-free rate r_t. This is essentially the same argument used in switching to the (constant) risk-free rate r in the drift of the SDE for a stock price S_t utilized in Black–Scholes derivation.

Now substitute the stochastic differentials (19.18) and (19.19) in the definition of $F(t,T,T+\Delta)$ given in (19.16) and cancel the *common* $r_t dt$

[10] A geometric SDE would have the diffusion parameter written as σB_t, with σ constant. Here we have $\sigma(t,T,B_t)$ depend on B_t as well. Hence, percentage bond volatility is not constant here.

[11] Here, applying Ito's Lemma means varying the t parameter. The reader may mistakenly think at this point that we are trying to take the limit as $\Delta \to 0$. This will be done, but for the time being the Δ is kept constant.

[12] After all, the two bonds are identical, except for their maturities.

terms:

$$\begin{aligned} dF(t,T,T+\Delta) &= \frac{1}{2\Delta}\Big[\sigma\,(t,T+\Delta,B\,(t,T+\Delta))^2 \\ &\qquad -\sigma\,(t,T,B\,(t,T))^2\Big]dt \\ &\quad +\frac{1}{2\Delta}\,[\sigma\,(t,T+\Delta,B\,(t,T+\Delta)) \\ &\qquad -\sigma\,(t,T,B\,(t,T))]\,dW_t \end{aligned} \tag{19.20}$$

This is the final result of applying Ito's Lemma to (19.16). This equation gives the arbitrage-free dynamics of a forward rate on a loan that begins at time T and ends Δ period later.

Now, we can let $\Delta \to 0$. This will give the dynamics of the instantaneous forward rate. To do this, note the way expression (19.20) is written. On the right-hand side, we have two terms that are of the form:

$$\frac{g\,(x+\Delta)-g(x)}{\Delta}$$

In expressions like these, letting $\Delta \to 0$ means taking the (standard) derivative of $g(\cdot)$ with respect to x. Writing these terms in brackets separately and then letting $\Delta \to 0$ amounts to taking the derivative of the two terms on the right-hand side with respect to T. Doing this gives

$$\begin{aligned} \lim_{\Delta\to 0}\frac{1}{2\Delta}&\Big[\sigma\,(t,T+\Delta,B\,(t,T+\Delta))^2 \\ &\quad -\sigma(t,T,B(t,T))^2\Big] \\ &= \sigma(t,T,B(t,T))\left[\frac{\partial\sigma(t,T,B(t,T))}{\partial T}\right] \end{aligned}$$

$$\begin{aligned} \lim_{\Delta\to 0}\frac{1}{\Delta}&\left[\sigma\left(t,T+\Delta,B(t,T+\Delta)\right)\right. \\ &\left. -\sigma\left(t,T,B(t,T)\right)\right] = \left[\frac{\partial\sigma(t,T,B(t,T))}{\partial T}\right] \end{aligned}$$

Putting these together in (19.20) we get the corresponding SDE for the instantaneous forward rate:

$$\lim_{\Delta\to 0} dF(t,T,T+\Delta) = dF(t,T)$$

or,

$$\begin{aligned} dF(t,T) &= \sigma(t,T,B(t,T))\left[\frac{\partial\sigma(t,T,B(t,T))}{\partial T}\right]dt \\ &\quad +\left[\frac{\partial\sigma(t,T,B(t,T))}{\partial T}\right]dW_t \end{aligned} \tag{19.21}$$

where the $\sigma(\cdot)$ are the bond price volatilities.

We have several comments to make on this result.

19.3.3 Interpretation

The HJM approach is based on imposing the no-arbitrage restrictions directly on the forward rates. First, a relation between forward rates and bond prices is obtained using an arbitrage argument. Then arbitrage-free dynamics are written for $B(t,T)$. Given the SDEs for bond prices, a SDE that an instantaneous forward rate should satisfy is obtained. To see the real meaning of this, suppose we postulate a general SDE for the instantaneous forward rate $F(t,T)$:

$$dF(t,T) = a(F(t,T),T)dt + b(F(t,T),T)dW_t \tag{19.22}$$

where the $a(F(t,T),T)$ and $b(F(t,T),T)$ are supposed to be the risk-adjusted drift and the diffusion parameters, and the W_t is the risk-neutral probability.

A reader may wonder how one would obtain these risk-adjusted parameters that are valid under the condition of no-arbitrage. Well, the previous section just established that under no-arbitrage, risk-adjusted drift can be *replaced* by:

$$a(F(t,T),t) = \sigma(t,T,B(t,T))\left[\frac{\partial\sigma(t,T,B(t,T))}{\partial T}\right] \tag{19.23}$$

The diffusion parameter will be given by:

$$b(F(t,T),t) = \left[\frac{\partial\sigma(t,T,B(t,T))}{\partial T}\right] \tag{19.24}$$

Hence, the previous section derived the *exact* no-arbitrage restrictions on the drift coefficient for instantaneous forward rate dynamics. This is similar to the Black–Scholes approach that was seen several times in the first part of the book. There, the drift term μ of the SDE for a stock price S_t was replaced by the risk-free interest rate r under the condition that there were no-arbitrage possibilities. Here, the drift is replaced not by r, but by a somewhat more complicated term that depends on the volatilities of the bonds under consideration. But, in principle, the drift is determined by arbitrage arguments and will hold only under the condition that there are no-arbitrage possibilities between the forward loan markets and bond prices. Throughout this process no "forward rate modeling" was done.

It is worth emphasizing that the risk-adjusted drift of instantaneous forward rates depends only on the *volatility* parameters. This is again similar to the Black–Scholes environment where there was no need to model the expected rate of return on the underlying stock, but modeling or calibrating the volatility was needed. It is in this sense that the HJM approach can be regarded as a true extension of the Black–Scholes methodology to fixed-income sector.

19.3.4 The r_t in the HJM Approach

Further, note that in the HJM approach there is no need to model any short-rate process. In particular, an exact model for the spot rate r_t is not needed. Yet, suppose there is a spot rate in the market. What would the SDEs obtained for the forward rates $F(t,T)$ imply for this spot rate? The question is relevant because the spot rate corresponds to the nearest infinitesimal forward loan, the one that starts at time t.

Thus, realizing that

$$r_t = F(t,t) \tag{19.25}$$

for all t, we can in fact derive an equation for the spot rate starting from the SDEs for forward rates. Before we start, we simplify the notation and write: $b(F(s,T),t) = b(s,t)$ in (19.24). Then, write the integral equation for $F(t,T)$ using the new $b(\cdot)$ notation:

$$F(t,T) = F(0,T) + \int_0^t b(s,T)\left[\int_s^T b(s,u)du\right]ds + \int_0^t b(s,T)dW_s$$

where we used (19.23) and (19.24) in (19.21). Next, select $T = t$ to get a representation for the spot rate r_t:

$$r_t = F(0,T) + \int_0^t b(s,T)\left[\int_s^T b(s,u)du\right]ds + \int_0^t b(s,T)dW_s \tag{19.26}$$

where the $b(s,t)$ is the volatility of the $F(s,t)$.

The first important result that we obtain from this equation is that the forward rates are *biased* estimators of the future spot rates under the risk-free measure. In fact, consider taking the conditional expectation of some future spot rate r_t with initial point $t < \tau$:

$$\begin{aligned}\mathbb{E}^{\mathbb{Q}}[r_\tau] = \mathbb{E}^{\mathbb{Q}}[F(t,\tau)] &+ \mathbb{E}^{\mathbb{Q}}\left[\int_t^\tau b(s,\tau)\left[\int_s^\tau b(s,u)du\right]ds\right] \\ &+ \mathbb{E}^{\mathbb{Q}}\left[\int_t^\tau b(s,\tau)dW_s\right]\end{aligned} \tag{19.27}$$

Here, the forward rate in the first expectation is known at time t; hence it comes out of the expectation sign. The third expectation on the right-hand side is zero because it is taken with respect to a Wiener process. But the second term is in general positive and does not vanish. Hence we have:

$$F(t,\tau) \neq \mathbb{E}^{\mathbb{Q}}[r_\tau] \tag{19.28}$$

The second major implication of the SDE for r_t has to do with the non-Markovness of the spot rate. To see this, note that the r_t given by Eq. (19.26) depends on the term:

$$\int_0^t b(s,t)\left[\int_s^t b(s,u)du\right]ds \tag{19.29}$$

that, in general, will be a complex function of all past forward rate volatilities. In particular, this term is not simply an "accumulation" of past changes the way a typical drift or diffusion term would lead to

$$\int_0^t \mu(r_s, \tau)\, ds \tag{19.30}$$

or

$$\int_0^t b(r_s, t)\, dW_s \tag{19.31}$$

In fact, the new term in the equation for r_t is more like a cross product. Hence, the similar term for an interest rate observed Δ period before the r_t would be

$$\left[\int_0^{t-\Delta} b(s, t-\Delta)\left[\int_s^{t-\Delta} b(s,u)du\right] ds\right] \tag{19.32}$$

and would not be captured by a state variable. The difference between (19.29) and (19.32) will depend on interest rates observed before $t - \Delta$. This would make the interest rate non-Markov in general.

Next, we see an example.

19.3.4.1 *Constant Forward Volatilities*

Suppose all forward rates $F(t, T)$ have volatilities that are constant at b. Then for each one of these forward rates the equation under no-arbitrage will be given by:

$$dF(t, T) = b^2(T - t)dt + bdW_t \tag{19.33}$$

The dynamics of the bond price will be

$$dB(t, T) = r_t B(t, T)dt + b(T - t)B(t, T)dW_t \tag{19.34}$$

From these we can derive the equation for the spot rate by taking the integrals in (19.26):

$$r_t = F(0, t) + \frac{1}{2}b^2t^2 + bW_t \tag{19.35}$$

which gives the SDE

$$dr_t = \left(F_t(0, t) + b^2t\right) dt + bdW_t \tag{19.36}$$

where the $F_t(0, t)$ is given by

$$F_t(0, t) = \frac{\partial F(0, t)}{\partial t} \tag{19.37}$$

Note that according to this model, the spot rate has a time-dependent drift and a constant volatility.

19.3.5 Another Advantage of the HJM Approach

The HJM approach exploited the arbitrage relation between forward rates and bond prices to impose restrictions on the dynamics of the instantaneous forward rates directly. By doing this it eliminated the need to model the expected rate of change of the spot rate.

But the approach has other advantages as well. As was seen in earlier chapters, a k-dimensional Markov process would in general yield non-Markov univariate models. Hence, within the HJM framework one could in principle impose Markovness on the behavior of a *set* of forward rates and in a multivariate sense this would be a reasonable approximation. Yet, in a univariate sense when we model the spot rate, the latter would still behave in a non-Markovian fashion.

This point is important because current empirical work indicates that spot-rate behavior in reality may fail to be Markovian. Hence, from this angle, the HJM approach provides an important flexibility to market practitioners.

19.3.6 Market Practice

The HJM approach is clearly the more appropriate philosophy to adopt from the point of view of arbitrage-free pricing. It incorporates arbitrage restrictions directly into the model and is more flexible.

However, it appears that market practice still prefers the classical approach and continues to use spot-rate modeling one way or another. How can we explain this discrepancy?

As discussed in Musiela and Rutkowski (1997), modeling the instantaneous spot rate has its own difficulties. When one imposes a Gaussian structure to SDEs that govern the dynamics of the $dF(t,T)$ and when one uses constant percentage volatilities, the processes under consideration explode in finite time. This is clearly not a very desirable property of a dynamic model. It can introduce major instabilities in the pricing effort.

It is also true that there are significant resources invested in spot-rate models, both financially and timewise. There is, again, a great deal of familiarity with the spot-rate models, and it may be that they provide good approximations to arbitrage-free prices anyway.

The recent models that exploit the *forward measure* seem to be an answer to problems of instantaneous forward rate modeling, and should be considered as a promising alternative.

19.4 HOW TO FIT r_t TO INITIAL TERM STRUCTURE

At several points in this chapter we discussed how a spot-rate model can be "fit" to an existing term structure known to be arbitrage-free. But, during this discussion, we never showed how this could be done in practice. This book tries to keep numerical issues to a minimum, but there are some cases where a discussion of practical pricing methods facilitates the understanding of the conceptual issues. Some simple examples of how an arbitrage-free spot-rate model can be obtained fall into this category. We discuss this briefly at the end of the chapter.

Suppose we are given an arbitrage-free family of n bond prices $B(t,T_i)$, $i = 1,\ldots,n$. Supppose also that we decided to use the classical approach to price interest-sensitive securities. Assuming a one factor model, we first need to fit a risk-adjusted spot-rate model

$$dr_t = a(r_t,t)\,dt + b(r_t,t)\,dW_t$$

to this term structure. How can this be done in practice?

Several methods are open to us. They all start by positing a class of plausible spot-rate models and then continue by discretizing it. Thus, we can let r_t follow the Vasicek model:

$$dr_t = \alpha(\kappa - r_t)\,dt + \sigma dW_t$$

and then discretize this using the straightforward Euler scheme:[13]

$$r_t = r_{t-\Delta} + \alpha(\kappa - r_t)\,\Delta + \sigma(W_t - W_{t-\Delta}) \quad (19.38)$$

where Δ is the discretization interval. The remaining part of the calibration exercise depends on the method adopted. We discuss some simple examples.

19.4.1 Monte Carlo

Suppose we know that increments $[W_t - W_{t-\Delta}]$ are independent and are normally distributed with mean zero and variance Δ. Suppose we have also calibrated the volatility parameter σ and the speed of mean reversion α. Hence, there is only one unknown parameter κ. Finally, we also have the initial spot rate r_0.

Consider the following exercise. Select M standard normal random variables using some random number generator. Multiply each random number by $\sqrt{\Delta}$. Start with a historical estimate of κ and obtain the first Monte Carlo trajectory for r_t^1, starting with r_0 and using Eq. (19.38) recursively.

Repeat this N times to obtain N such spot-rate trajectories

$$\left[\left\{r_t^1\right\},\left\{r_t^2\right\}\ldots\left\{r_t^N\right\}\right]$$

Then calculate the prices by using the sample equivalent of the bond pricing formula:

$$\widehat{B}(t,T_i) = \frac{1}{N}\sum_{j=1}^{N} e^{-\sum_{i=1}^{M} r_t^j \Delta}$$

[13] Euler scheme replaces differentials by first differences. It is a first-order approximation that may end up causing significant cumulative errors.

where M may be different for each bond, depending on the maturity. Now, because κ was selected arbitrarily, the $\widehat{B}(t,T_i)$ will not be arbitrage-free.

But, we also have the observed term structure, which is known to be arbitrage-free. So, we can try to adjust the κ in a way to minimize the distance:

$$\sum_{i=1}^{T_{\max}} \left|\widehat{B}(t,T_i) - B(t,T_i)\right|^2$$

This way we find a value for κ such that the calculated term structure is as close as possible to the observed term structure. Once such a κ is determined, the r_t dynamics becomes (approximately) arbitrage-free, in the sense that using the model parameters, and this new κ, one can obtain bond prices that come "close" to the observed term structure.

19.4.2 Tree Models

The previous approach used a single parameter κ to make calculated bond prices comes as close as possible to an observed term structure. The fit was not perfect because the distance between the two term structures was not reduced to zero, although it was minimized. By adopting a general tree approach one can "improve" the fit.

Once we consider a binomial model for movements in r_t we can choose the relevant parameters so that the tree trajectories "fit" the arbitrage-free term structure and the relevant volatilities. For example, we can assume that we have N arbitrage-free bond prices. Suppose we also know the volatilities σ_i of each bond $B(t,T_k)$. Let the up and down movements in r_i at stage i be denoted by u_i, d_i, such that

$$u_i d_i = 1$$

Given this restriction, the tree will be recombining and at every stage we will have i unknown parameters. The next task will be to determine these u_i, d_i by using the equality:

$$B(0,T_k) = \frac{1}{N}\sum_{j=1}^{N_k} e^{-\sum_{i=1}^{T_k} r_i^j \Delta}$$

where the are the ith element of the jth tree trajectory and N_k is the number of tree trajectories for a bond that matures after T_k steps. These trajectories depend on the u_i, d_i, hence these equations can be used to determine the latter. To do this we need to impose enough restrictions such that the total number of unknown parameters in the tree becomes equal to the number of equations. The tree parameters can then be obtained from these equations. The tree will fit the initial term structure exactly. An example to this way of proceeding is in Black et al. (1990).

19.4.3 Closed-Form Solutions

Suppose we can analytically calculate the expectation

$$B(t,T) = \mathbb{E}^{\mathbb{Q}}\left[e^{-\int_t^t r_s ds}\right]$$

and get a closed-form solution for the $B(t,T)$, as will be discussed in the next chapter. Suppose this results in the function:

$$B(t,T) = G(r_t, T, \kappa)$$

Then, we can minimize the distance between the closed-form solution and the observed arbitrage-free yield curve by choosing κ in some optimal sense:

$$\sum_{i=1}^{T_{\max}} |B(t,T_i) - G(r_t,T,\kappa)|^2$$

This is another example of obtaining an (approximately) arbitrage-free model for r_t.

19.5 CONCLUSION

This chapter has briefly summarized the two major approaches to pricing derivative securities that depend on interest rates. The classical approach was shown to be an effort in spot-rate modeling. The arbitrage restrictions were incorporated indirectly through a process of "fitting an initial curve." The HJM approach, on the other hand, was an extension of the Black–Scholes formula to interest-sensitive securities.

19.6 REFERENCES

The best source on these issues is Musiela and Rutkowski (1998). Of course, this source is quite technical, but we recommend that readers who are seriously interested in fixed-income sector put in the necessary effort and become more familiar with it. The excellent discrete time treatment, Jarrow (1996), should also be mentioned here.

19.7 EXERCISES

1. Consider the equation below that gives interest rate dynamics in a setting where the time axis $[0, T]$ is subdivided into n equal intervals, each of length Δ:

$$r_{t+\Delta} = r_t + \alpha r_t + \sigma_1 (W_{t+\Delta} - W_t) + \sigma_2 (W_t - W_{t-\Delta})$$

where the random error terms

$$\Delta W_t = (W_{t+\Delta} - W_t)$$

are distributed normally as

$$\Delta W_t \sim \mathcal{N}\left(0, \sqrt{\Delta}\right)$$

(a) Explain the structure of the error terms in this equation. In particular, do you find it plausible that $\Delta W_{t-\Delta}$ may enter the dynamics of observed interest rates?

(b) Can you write a stochastic differential equation that will be the analog of this in continuous time? What is the difficulty?

(c) Now suppose you know, in addition, that long-term interest rates, R_t, move according to a dynamic given by

$$R_{t+\Delta} = R_t + \beta r_t + \theta_1 (\widetilde{W}_{t+\Delta} - \widetilde{W}_t) + \theta_2 (\widetilde{W}_t - \widetilde{W}_{t-\Delta})$$

where we also know the covariance:

$$\mathbb{E}\left[\Delta \widetilde{W} \Delta \widetilde{W}\right] = \rho\Delta$$

Can you write a representation for the vector process

$$X_t = \begin{bmatrix} r_t \\ R_t \end{bmatrix}$$

such that X_t is a first-order Markov?

(d) Can you write a continuous-time equivalent of *this* system?

(e) Suppose short or long rates are individually non-Markov. Is it possible that they are jointly so?

2. Suppose the (vector) Markov process X_t,

$$X_t = \begin{bmatrix} r_t \\ R_t \end{bmatrix}$$

has the following dynamics,

$$\begin{bmatrix} r_{t+\Delta} \\ R_{t+\Delta} \end{bmatrix} = \begin{bmatrix} \alpha_{11} & \alpha_{12} \\ \alpha_{21} & \alpha_{22} \end{bmatrix} \begin{bmatrix} r_t \\ R_t \end{bmatrix} + \begin{bmatrix} \Delta W^1_{t+\Delta} \\ \Delta W^2_{t+\Delta} \end{bmatrix}$$

where the error term is jointly normal and serially uncorrelated. Suppose r_t is a short rate, while R_t is a long rate.

(a) Derive a univariate representation for the short rate r_t.

(b) According to this representation, is r_t a Markov process?

(c) Under what conditions, if any, would the univariate process r_t be Markov?

3. Suppose at time $t = 0$, we are given four zero-coupon bond prices $\{B_1, B_2, B_3, B_4\}$ that mature at times $t = 1, 2, 3, 4$. This forms the term structure of interest rates. We also have one-period forward rates $\{f_0, f_1, f_2, f_3\}$, where each f_i is the rate contracted at time $t = 0$ on a loan that begins at time $t = i$ and ends at time $t = i + 1$. In other words, if a borrower borrows \$$N$ at time $t = i$, he or she will pay back $N(1 + f_i)$ at time $t = i + 1$. The spot rate is denoted by r_i. By definition we have

$$r_0 = f_0$$

The $\{B_i\}$ and all forward loans are default-free. At each time period there are two possible states of the world, denoted by $\{u_i, d_i := 1, 2, 3, 4, \}$.

(a) Looked at from time $i = 03$ how many possible states of the world are there at time $i = 3$?

(b) Suppose

$$\{B_i = 0.9,\ B_2 = 0.87,\ B_3 = 0.82,\ B_4 = 0.75\}$$

and

$$\{f_0 = 8\%,\ f_1 = 9\%,\ f_2 = 10\%,\ f_3 = 18\%\}$$

Form three arbitrage portfolios that will guarantee a net positive return at times $i = 1, 2, 3$ with no risk.

(c) Form three arbitrage portfolios that will guarantee a net return at time $i = 0$ with no risk.

(d) Given a default-free zero-coupon bond, B_n, that matures at time $t = n$, and all the forward rates $\{f_0, \ldots, f_{n-1}\}$, obtain a formula that expresses B_n as a function of f_i.

(e) Now consider the Fundamental Theorem of Finance as applied to the system:

$$\begin{bmatrix} B_1 \\ B_2 \\ B_3 \\ B_4 \end{bmatrix} = \begin{bmatrix} 1 & 1 \\ B_2^u & B_2^d \\ B_3^u & B_3^d \\ B_4^u & B_4^d \end{bmatrix} \begin{bmatrix} \psi_1 \\ \psi_2 \end{bmatrix}$$

Can all B_i be determined independently?

(f) In the system above can all the $\{f_i\}$ be determined independently?

(g) Can we claim that all f_i are normally distributed? Prove your answer.

4. Consider again the setup of Question 1. Suppose we want to price three European-style call options written on one-period (spot) LIBOR rates L_i with $i = 0, 1, 2, 3$, as in the previous case. Let these option prices be denoted by C_i. Each option has the payoff:

$$C^i = N \max[L_i - K, 0]$$

where N is a notional amount that we set equal to one without loss of any generality.

(a) How can you price such an option?

(b) Suppose we assume the following:

i. Each f_i is a current observation on the future unknown value of L_i.

ii. Each f_i is normally distributed with mean zero and constant variance σ_i.

iii. We can use the Black formula to price the calls.

(c) Would these assumptions be appropriate under the risk-neutral measure obtained using money market normalization? Explain.

(d) How would the use of the forward measure that corresponds to each L_i improve the situation?

(e) In fact, can you obtain the forward measures for times $t = 1, 2$?

(f) Price the call option for time $t = 2$ using the forward measure.

5. Consider a payer forward start swap where the swap begins at some fixed time T_n in the future and expires at time $T_M > T_n$. We assume the accrual period is of length δ, measured in years. Since payments are made in-arrears, the first payment occurs at $T_{n+1} = T_n + \delta$ and the final payment at T_{M+1}. First, convince yourself that at time $t < T_n$ the value, SW_t, of this forward start swap is

$$SW_t = \mathbb{E}\left(\delta \sum_{j=n}^{M} \frac{B_t}{B_{T_{j+1}}} (L(T_j, T_j) - R)\right) \tag{19.39}$$

where R is the fixed rate (annualized) specified in the contract and $L(T_j, T_j)$ is the spot LIBOR rate applying to the interval $[T_j, T_{j+1}] = [T_j, T_j + \delta]$.

Note that LIBOR rates are annualized rates based on simple compounding. This means that one dollar invested at time T_j until time $T_j + \delta$ at the LIBOR rate $L(T_j, T_j)$ will be worth $1 + \delta L(T_j, T_j)$ dollars at time $T_j + \delta$.

Assuming a notional principal of \$1, show that

$$SW_{T_n} = 1 - Z_{T_n}^{T_{M+1}} - R\delta \sum_{j=n+1}^{M+1} Z_{T_n}^{T_j} \tag{19.40}$$

6. Considering the following SDE for the spot rate:

$$dr_t = \lambda(\theta - r_t)dt + \sigma r_t dW_t \tag{19.41}$$

Using the following set of parameters:

$$r_0 = 0.03 \tag{19.42}$$

$$\lambda = 1.5 \tag{19.43}$$

$$\theta = 0.06 \tag{19.44}$$

$$\sigma = 0.25 \tag{19.45}$$

$$T = 1 \tag{19.46}$$

write a program to generate a sequence of short rates via simulation. Using this sequence of short rates, calculate the expectation and variance of the bond price.

CHAPTER

20

Classical PDE Analysis for Interest Rate Derivatives

OUTLINE

20.1 INTRODUCTION

The reader is already familiar with various derivations of the Black–Scholes formula, one of which is the partial differential equations (PDE) method. In particular, Chapter 12 showed how risk-free borrowing and lending, the underlying instrument, and the corresponding options can be combined to obtain risk-free portfolios. Over time, these portfolios behaved in such a way that small random perturbations in the positions taken canceled each other, and the portfolio return became *deterministic*. As a result, with no default risk the portfolio had to yield the same return as the risk-free spot rate r, which was assumed to be constant. Otherwise, there would be arbitrage opportunities. The application of Ito's Lemma within this context resulted in the fundamental Black–Scholes PDE. The Black–Scholes PDE was of the form:

$$-rF + F_t + rS_tF_s + \frac{1}{2}\sigma^2 S_t^2 F_{ss} = 0 \qquad (20.1)$$

with the boundary condition:

$$F(S_T, T) = \max[S_T - K, 0] \qquad (20.2)$$

Financial Derivatives, Third Edition.
http://dx.doi.org/10.1016/B978-0-12-384682-2.00020-7

The r is the constant risk-free instantaneous spot rate, the S_t is the price of a stock that paid no dividends, the F is the time t price of a European call option written on the stock. The K and the T are the strike price and the expiration date of the call, respectively. In Chapter 15, it was also mentioned that the solution of this PDE corresponded to the conditional expectation

$$F(S_t, t) = \mathbb{E}_t^{\mathbb{Q}}\left[e^{-r(T-t)}F(S_T, T)\right] \tag{20.3}$$

calculated with the risk-neutral probability $\mathbb{Q}$.

Given that we are now dealing with derivatives written on interest-sensitive securities, we can now ask (at least) two questions:

- Do we get similar PDEs in the case of interest rate derivatives? For example, considering the simplest case, what type of a PDE would the price of a default-free discount bond satisfy?
- Given a PDE involving an interest rate derivative, can we obtain its solution as a conditional expectation similar to (20.3)?

These questions can be answered in two different ways. First, we can follow the same approach as in Chapter 12 and obtain a PDE for discount bond prices along the lines similar to the derivation of the Black–Scholes PDE. In particular, we can form a "risk-free"portfolio and equate its deterministic return to that of a risk-free instantaneous investment in a savings account. Application of Ito's Lemma should yield the desired PDE.[1]

The second way of obtaining PDEs for interest-sensitive securities is by exploiting the martingale equalities and the so-called Feynman–Kac results directly. In fact, when we investigate the relationship between a certain class of expectations and PDEs, we are led to an interesting mathematical regularity. It turns out that there is a very close connection between a representation such as:

$$B(t, t) = \mathbb{E}_t^{\mathbb{Q}}\left[e^{-r\int_t^T r_s ds}B(T, T)\right] \tag{20.4}$$

and a certain class of partial differential equations. In stochastic calculus, these topics come under the headings of "Generators for Ito Diffusions,""Kolmogorov Backward Equation,"and more importantly, "Feynman–Kac formula." Using these methods, given a conditional expectation such as in (20.4), we can directly obtain a PDE that corresponds to it and vice versa. Of course, this correspondence depends on some additional conditions concerning the underlying random variables, but is clearly a very convenient tool for the financial market practitioner. Yet, the discussion of these "modern"methods should wait until the next chapter.

In this chapter we show that prices of interest rate derivatives will satisfy PDEs similar to the fundamental Black–Scholes PDE using the "classical steps." But, this derivation will still be fundamentally different than the one followed in Chapter 12 because the underlying variable will now be the spot rate r_t. Spot rate is *not* an asset price, in contrast to the S_t, which represented the price of a traded asset in the Black–Scholes world.[2] Obviously, the difficulties associated with spot-rate modeling will be present here also.

The derivation of the fundamental PDE for interest-sensitive securities will follow steps similar to the classic paper by Vasicek (1977). The essential idea is to incorporate in the dynamics of the returns the arbitrage conditions implied by a single Wiener process[3] that determines the

[1]We remind the reader that risk-free portfolios are not self-financing, and as a result the method is not mathematically accurate in continuous time. Yet, one still obtains the "correct"PDE because the extra cash flow invested or withdrawn over time has an expected value of zero. This issue was discussed in Chapter 12 in more detail. We keep utilizing this heuristic method with the condition that the reader keeps in mind this important point.

[2]The r_t is more like a percentage return, a pure number.

[3]Or in case of two-factor models, two independent Wiener processes.

random movements observed in more than one asset. In the case of the Black–Scholes approach, we worked with two securities, the underlying stock and the call option written on it. An infinitesimal random movement in the price of the stock also affected the price of the option. Hence we had two prices driven essentially by the same source of randomness. These securities could be combined in a careful fashion with risk-free borrowing and lending so that the unpredictable random movements canceled each other and the resulting portfolio became "riskless."

The same idea can be extended to interest-sensitive securities. For example, except for their maturities, bonds are "similar" instruments. They are expected to be influenced by the similar infinitesimal random fluctuations. Hence, under some conditions, a portfolio formed using two (or more) bonds can be made risk-free if portfolio weights are chosen carefully.

Yet, there are differences when compared with the case of stocks. In the classical Black–Scholes derivation, the spot rate was assumed to be constant. This assumption did not appear to be very severe. In the case of interest-sensitive securities, the assumption of a constant interest rate cannot be maintained. On the contrary, the randomness that drives the system comes from infinitesimal Wiener increments that affect instantaneous spot rate r_t. But this latter is not an asset price, as mentioned earlier. The unknown drift of interest rate dynamics cannot be simply made equal to the risk-free rate by invoking arbitrage arguments. This introduces major complications in the derivation and numerical estimation of PDEs for interest rate derivatives. In fact, although the steps in the following derivation are mathematically straightforward, they are somewhat more convoluted than in the case of plain vanilla call options written on stocks.

Finally, we should reiterate that the "classical" approach adopted here is heuristic just like the derivation of the Black–Scholes PDE. A technically correct derivation would incorporate in the argument the condition that the risk-free portfolios are also self-financing. As discussed earlier, the approach below may not yield self-financing portfolios.

20.2 THE FRAMEWORK

The first step is to set the framework. We assume that we are provided two SDEs describing the dynamics of two default-free discount bond prices, $B(t, T_1)$ and $B(t, T_2)$, with maturities T_1, T_2 such that $T_1 < T_2$. The bond prices are driven by the same Wiener process, W_t. To simplify the notation, in this section we ignore the time subscript t and write:

$$B^1 = B(t, T_1) \tag{20.5}$$

$$B^2 = B(t, T_2) \tag{20.6}$$

These bond prices are postulated to have the following dynamics:

$$dB^1 = \mu\left(B^1, t\right) B^1 dt + \sigma_1\left(B^1, t\right) B^1 dW_t \tag{20.7}$$

$$dB^2 = \mu\left(B^2, t\right) B^2 dt + \sigma_2\left(B^2, t\right) B^2 dW_t \tag{20.8}$$

Note two points. First, the diffusion terms are a function of the same W_t, but depend on different diffusion parameters $\sigma_i, i = 1, 2$. Second, the volatility parameters are written in terms of percentage volatility, but the bond dynamics are not necessarily given by geometric processes because the drift and diffusion parameters are also allowed to depend on $B_i, i = 1, 2$, and are not constant, as would be required by a geometric SDE.

Because we are adopting a "classical" approach we now need to posit an interest rate model. We let the dynamics of r_t be given by:

$$dr_t = a(r_t, t)\, dt + b(r_t, t)\, dW_t \tag{20.9}$$

where the drift $a(r_t, t)$ and the diffusion $b(r_t, t)$ parameters are assumed to be known. They are either estimated from historical data or, as in the practical approaches, calibrated using market prices. It is also worth emphasizing that the W_t here is a Wiener process with respect to the real world probability $\mathbb{P}$.

Note the critical restriction imposed on this spot-rate dynamics; the parameters $a(r_t, t), b(r_t, t)$ are assumed to depend only on the latest observation r_t, so that previous $r_s, s < t$ do not affect the drift and volatility parameters. We already know from the previous chapter that this Markov property of r_t will be violated in a general term-structure model. Still, the classical approach proceeds, assuming that it is a reasonable approximation.

20.3 MARKET PRICE OF INTEREST RATE RISK

To derive a PDE for a discount bond's price, we first need to form a risk-free portfolio $\mathcal{P}$ made of the two bonds B_1, B_2 at time t.[4] In particular, without any loss of generality, it is assumed that θ_1 units of B_1 are purchased, and θ_2 units of B_2 are shorted, for a total portfolio value:

$$\mathcal{P} = \theta_1 B^1 - \theta_2 B^2 \tag{20.10}$$

Suppose the portfolio weights are chosen as:

$$\theta_1 = \frac{\sigma_1}{B^1(\sigma_2 - \sigma_1)}\mathcal{P} \tag{20.11}$$

$$\theta_2 = \frac{\sigma_2}{B^1(\sigma_2 - \sigma_1)}\mathcal{P} \tag{20.12}$$

where $\sigma_i, i = 1, 2$ are the volatility parameters $\sigma_1(B_1, t), \sigma_2(B_2, t)$ of the two bonds as described in Eqs. (20.7) and (20.8). As time passes, this portfolio's value will change. Acting as if the portfolio weights are constant, the implied infinitesimal changes will be given by:

$$d\mathcal{P} = \theta_1 dB^1 - \theta_2 dB^2 \tag{20.13}$$

[4]The time subscript is ignored for notational simplicity.

or after replacing from the SDEs that give the dynamics of dB^1, dB^2:

$$\begin{aligned} d\mathcal{P} = \theta_1 & \left[\mu\left(B^1, t\right) B^1 dt + \sigma_1\left(B^1, t\right) B^1 dW_t\right] \\ & -\theta_2\left[\mu\left(B^2, t\right) B^2 dt + \sigma_2\left(B^2, t\right) B^2 dW_t\right] \end{aligned} \tag{20.14}$$

Grouping the Wiener increment dW_t, we see that its coefficient becomes zero after replacing the values of σ_1 and σ_2:

$$\begin{aligned} \left(\theta_1\sigma_1 B^1 - \theta_2\sigma_2 B^2\right) = & \left(\frac{\sigma_2}{B^1(\sigma_2 - \sigma_1)}\sigma_1 B^2 - \right. \\ & \left. \frac{\sigma_2}{B^2(\sigma_2 - \sigma_1)}\sigma_2 B^2\right)\mathcal{P} = 0 \end{aligned} \tag{20.15}$$

This gives the incremental changes in the portfolio value:

$$d\mathcal{P} = \left(\theta_1\mu_1 B^1 - \theta_2\mu_2 B^2\right) dt \tag{20.16}$$

These increments do not have a Wiener component and are completely predictable.

These steps justify the particular values chosen for the portfolio weights σ_1, σ_2. These weights were selected so that the dW_t term drops from the SDE of the portfolio $\mathcal{P}$. This is similar to the derivation of the Black–Scholes PDE. Indeed, replacing the θ_i, dividing and multiplying by $\mathcal{P}$, and arranging the $d\mathcal{P}$ can be written as:

$$d\mathcal{P} = \frac{(\sigma_2\mu_1 - \sigma_1\mu_2)}{(\sigma_2 - \sigma_1)}\mathcal{P}dt \tag{20.17}$$

This SDE does not contain a diffusion term and the dynamic behavior of $d\mathcal{P}$ is riskless. Hence, we can now use the standard argument and claim that this portfolio should not present any arbitrage opportunities and its deterministic return should equal the

$$r_t dt = \frac{(\sigma_2\mu_1 - \sigma_1\mu_2)}{(\sigma_2 - \sigma_1)}\mathcal{P}dt \tag{20.18}$$

Simplifying the $\mathcal{P}dt$ and rearranging, we obtain:

$$\frac{\mu_1 - r_t}{\sigma_1} = \frac{\mu_2 - r_t}{\sigma_2} \tag{20.19}$$

That is, the risk premia offered by bonds of different maturities are equal once normalized by the corresponding volatility parameter. Risk premia of per unit volatility are the same across bonds. Bonds with higher volatility pay proportionately higher risk premia.[5] This result is not very unexpected because, at the end, these bonds have the same source of risk, given the common dW_t factor. Obviously, if one of the bonds was a function of an additional and different Wiener process, say W_t^*, then even under a no-arbitrage condition, risk premia per volatility unit could be different across bonds. Note, in passing, that these risk premia can very well be negative.

Now, during this derivation the maturities of the underlying bonds were selected arbitrarily. Thus, similar equalities should be true for all discount bonds, as long as their dynamics are driven by the same Wiener process, W_t. This gives a term $\lambda(r_t, t)$ that is relevant to all bond prices, $B(t, T_i)$:

$$\frac{\mu_i - r_t}{\sigma_i} = \lambda(r_t, t) \tag{20.20}$$

This term is called the *market price of interest rate risk*. As can be seen from the derivation, it is in general a function of r_t and t. But in the following section we will simply write it as λ_t while assuming that this dependence is kept in mind. Note again that λ_t is independent of the bond maturity.

It is worth mentioning that a similar market price of equity risk was present in the Black–Scholes framework but was not used explicitly. In contrast to the case of Black–Scholes PDE, with interest-sensitive securities we do have to use the λ_t explicitly in deriving the PDEs here.

[5]Another way of saying this is that the Sharpe Ratios of the bonds are equal.

20.4 DERIVATION OF THE PDE

The third step of the PDE derivation for bond prices is to use the previous results in Ito's expansion for $B(t, T)$. Remembering that $B(t, T)$ is also a function of r_t, and applying Ito's rule:

$$dB(t, T) = B_t dr_t + B_t dt + \frac{1}{2} B_{rr} b(r_t, T)^2 dt \tag{20.21}$$

Substituting for dr_t from

$$dr_t = a(r_t, t)\,dt + b(r_t, t)\,dW_t \tag{20.22}$$

we get:

$$dB(r_t, t) = \left(B_r a(r_t, t)\, a(r_t, t) + B_t + \frac{1}{2} B_{rr} b(r_t, t)^2 \right) dt + b(r_t, t)\,dW_t \tag{20.23}$$

where again the W_t is a Wiener process with respect to the real-world probability $\mathbb{P}$. This SDE must be identical to the original equation that drives the bond price dynamics. Simplifying the notation, this SDE is:

$$dB = \mu(B, t)\,B dt + \sigma(B, t)\,B dW_t \tag{20.24}$$

under the probability $\mathbb{P}$. This means that we can equate the drift and diffusion coefficients. Setting the two diffusion coefficients in (20.23) and (20.24) equal to each other, we obtain:

$$b(r_t, t)\,B_r = \sigma B \tag{20.25}$$

where $\sigma(B, t)$ is abbreviated as σ. Equating the drifts in (20.23) and (20.24) gives:

$$\mu(B, t)\,B = B_r a(r_t, t)\, a(r_t, t) + B_t + \frac{1}{2} B_{rr} b(r_t, t)^2 \tag{20.26}$$

Here we have two Eqs. (20.25) and (20.26) that we can exploit in obtaining the PDE for bond prices. In fact, this last Eq. (20.26) is already a PDE, except for the fact that it contains the unknown $\mu(B, t)$. Also, note that up to this point we did nothing that would incorporate the arbitrage restrictions that we must have in this system.[6]

[6]There *will* be arbitrage restrictions because we have assumed that all bond prices are driven by the same Wiener process, W_t.

It turns out that the way to eliminate the "unknown" drift $\mu(B,t)$ from (20.26) is by using arbitrage arguments. Recall that in the case of Black–Scholes PDE, one simply "replaces" the $\mu(B,t)$ by the constant spot rate r. But in the present case this is not possible because we keep using the spot-rate drift $a(r_t,t)$ in (20.26). If we replaced the $\mu(B,t)$ by r_t, this would require adjusting the spot-rate drift $a(r_t,t)$ in (20.26) to its risk-neutral equivalent as well. But the r_t is not the price of an asset and it is not clear how this adjustment can be done. This problem can be resolved by utilizing the market price of interest rate risk λ_t.

In fact, Eq. (20.20) gives the market price of risk λ_t as:

$$\frac{\mu(B,t)-r_t}{\sigma}=\lambda_t \tag{20.27}$$

or, using the equivalence of diffusion parameters shown in (20.25):

$$\frac{B(\mu(B,t)-r_t)}{b(r_t,t)B_r}=\lambda_t \tag{20.28}$$

This gives:

$$\mu(B,t)B=r_tB+b(r_t,t)\lambda_t \tag{20.29}$$

Now substitute the right-hand side of this for $B\mu(B,t)$ in (20.26) and rearrange:

$$B_ra(r_t,t)+B_t+\frac{1}{2}B_{rr}b(r_t,t)^2-r_tB-b(r_t,t)B_r\lambda_t=0 \tag{20.30}$$

Note that the "unknown" drift $\mu(B,t)$ is now eliminated. This can finally be written as:

$$B_r\left(a(r_t,t)-b(r_t,t)\lambda_t\right)+B_t+\frac{1}{2}B_{rr}b(r_t,t)^2-r_tB=0 \tag{20.31}$$

This is a PDE for the price of a default-free pure discount bond $B(t,T)$. The associated boundary condition is simpler than the case of Black–Scholes. The bond is default-free and at maturity is guaranteed to have a value of 1, regardless of the level of spot rates at that time:

$$B(T,T)=1 \tag{20.32}$$

If one had an interest rate model with known drift $a(r_t,t)$ and diffusion coefficient $b(r_t,t)$, to use this PDE in practice, one would *still* need an estimate for the λ_t. Otherwise the equation is not usable. Also, it is worth realizing that in this PDE the coefficient of B_r is equivalent to a risk-adjusted drift of the spot-rate dynamics.

In fact, it is as if we are using the drift from the spot-rate dynamics, written under the risk-neutral measure $\mathbb{Q}$. Invoking the Girsanov theorem for Eq. (20.9), and switching from the Wiener process W_t defined under $\mathbb{P}$ to the Wiener process defined under, we obtain a new SDE for r_t:

$$dr_t=\left(a(r_t,t)-b(r_t,t)\lambda_t\right)dt+b(r_t,t)\,d\widetilde{W}_t \tag{20.33}$$

The drift of this SDE is now adjusted for "interest rate risk." Whenever the bond price drifts are switched from $\mu(\cdot)$ to r_t, one needs to switch the spot-rate dynamics from $a(r_t,t)$ to $(a(r_t,t)-b(r_t,t)\lambda_t)$.

We now summarize the major aspects of this derivation and compare it with the approach taken in the case of Black–Scholes PDE.

20.4.1 A Comparison

The general strategy in deriving the PDE was similar to the case of Black–Scholes. The main difference arises from the fact that the driving process in the present case is not S_t, the price of an asset, but is the spot rate r_t, which is a pure number. Hence, the no-arbitrage conditions have to be introduced in a different way than just making the unknown drift coefficient equal to the risk-free rate.

The approach was to modify the drift of the bond dynamics using the market price of risk for r_t. The reader should realize that letting as was done in Eq. (20.29) introduces the no-arbitrage condition in the equation implicitly.

$$\mu(B,t)B=r_tB+\left(b(r_t,t)B_r\right)\lambda_t \tag{20.34}$$

However, notice a rather important difference. In the case of the Black–Scholes derivation, by using the no-arbitrage condition we succeeded in *completely* eliminating the need to model and calibrate the drift of the stock price process S_t. In fact, in the Black–Scholes derivation, expected change in S_t did not matter at all. The option price depended on the relevant *volatilities* only.

In case of the spot-rate approach to pricing interest-sensitive securities, the use of no-arbitrage conditions will again introduce the spot rate r_t in the PDE. Yet, along with the r_t, *two* new parameters enter, namely the spot-rate drift $a(r_t, t)$ and the λ, market price of interest rate risk. These parameters need to be estimated or calibrated if the PDE is to be used in real-world pricing. As mentioned in the previous chapter, this is a departure from the practicality of the Black–Scholes approach, which required the modeling of volatilities only. But it is also a change in philosophy because, in a sense, a complete modeling of the r_t process is now needed.

A second fundamental point of the above derivation is the assumption of a single driving process r_t. Remember that the dynamics of all bond prices were assumed to be driven by the *same* univariate Wiener process W_t. Because the same Wiener process is present in the SDE for the spot rate r_t, this assumption enabled us to obtain a convenient no-arbitrage condition that was a function of a *single* market price of risk λ_t. Clearly, this may not be the case. Making a single stock price a function of a single random process, W_t, may be an acceptable approximation; doing the same thing for a *set* of discount-free bonds ranging from very short to very long maturities may be more questionable.

Nevertheless, our purpose in this book is to display the relevant tools rather than obtaining satisfactory pricing methods for actual markets. The assumption of a single factor is useful to this end.[7]

[7] It should be remembered that this assumption is often made in actual pricing projects as well.

20.5 CLOSED-FORM SOLUTIONS OF THE PDE

The fundamental PDE for bond prices can sometimes be solved for a closed-form solution. This way, an explicit formula that ties $B(t, T)$ to the maturity T, the "current"spot rate r_t, and the relevant parameters $a(r_t, t), b(r_t, t)$ and λ_t can be obtained.

The analogy is with the fundamental PDE of Black–Scholes and the Black–Scholes formula. Given enough assumptions on the S_t process and the constancy of the interest rates, one was able to solve that PDE to get the Black–Scholes formula. In the present framework, given enough assumptions about the interest rate process r_t, one can do the same for the bond price PDE. We discuss some simple examples.

20.5.1 Case 1: A Deterministic r_t

We begin with an extreme case. Suppose the spot rate is constant at $r_t = r$ for all t. Then the SDE for r_t:

$$dr_t = a(r_t, t)\, dt + b(r_t, t)\, dW_t$$

will have the following (trivial) parameters:

$$a(r_t, t) = b(r_t, t) = 0$$

Further, because there is no interest rate risk, no risk premia should be paid for it:

$$\lambda = 0$$

Thus, the fundamental PDE for a typical $B(t, T)$, which originally is given by

$$B_r\left(a - b\lambda\right) + B_t + \frac{1}{2}B_{rr}b^2 - r_tB = 0 \qquad (20.35)$$

will reduce to

$$B_t + rB = 0$$

with the boundary condition

$$B(T, T) = 1$$

But this is nothing other than the ordinary differential equation

$$\frac{dB(t,T)}{dt} + rB(t,T) = 0$$

with terminal condition $B(T) = 1$. Its solution will be given by

$$B(t,T) = e^{-r(T-t)}$$

This bond pricing function will satisfy the boundary condition and the fundamental PDE. It is the usual discount at a constant instantaneous rate r.

20.5.2 Case 2: A Mean-Reverting r_t

Suppose now the market price of risk is constant:

$$\lambda(r_t,t) = \lambda \tag{20.36}$$

but that the spot rate follows the mean-reverting SDE given by:

$$dr_t = \alpha(\kappa - r_t)\,dt + b\,dW_t \tag{20.37}$$

where W_t is a Wiener process under the real-world probability. Note that the volatility structure is restricted to be a constant *absolute* volatility denoted by b. Suppose further that the parameters α, κ, b, and λ are known exactly. Then, the fundamental PDE for a typical $B(t,T)$ will reduce to:

$$B_r\left(\alpha(\kappa - r_t)\,dt + b\,dW_t\right) \tag{20.38}$$

This setup is known as the Vasicek model, after the seminal work of Vasicek (1977).

It can be shown that the solution of this PDE is the closed-form expression given by the bond pricing formula $B(t,T)$, for time $t = 0$,

$$B(0,T) = e^{\frac{1}{\alpha}\left(1-e^{-\alpha T}\right)(R-r)-TR-\frac{b^2}{4\alpha^3}\left(1-e^{-\alpha T}\right)^2} \tag{20.39}$$

where

$$R = \kappa - \frac{b\lambda}{\alpha} - \frac{b^2}{\alpha^2} \tag{20.40}$$

and r is the current observation on the spot rate. Given some plausible estimates for the unknown parameters, we can then plot this function.

20.5.2.1 Example

For example, consider an economy where the long-run mean of the spot rate is 5% and where the spot rate is pulled toward the long-run mean at a rate of 0.25. We thus have

$$\alpha = 0.25 \quad \kappa = 0.05 \tag{20.41}$$

Further, suppose the absolute interest rate volatility is .015 during one year:

$$b = 0.015 \tag{20.42}$$

To apply the formula, we need the market price of interest rate risk. Assume that we have

$$\mu - r_t = -0.1\sigma \tag{20.43}$$

where the μ, σ are the unknown bond drift and volatility parameters. Then, we know that

$$\lambda = -0.10 \tag{20.44}$$

Using these parameters, we can calculate the bond pricing function $B(t,T)$ that will depend on the initial interest rate r and on the maturity parameter T. This is the so-called "discount curve" discussed earlier.

The graph of the $\{B(t,T), T \in [0, T_{\max}]\}$ with $\{\lambda = -0.10, b = 0.015, \alpha = 0.25, \kappa = 0.05\}$ at three different levels for the spot rate $r = 0.5\%, r = 5\%, r = 15\%$ are shown in Figure 20.1. Because these are discount bond prices, the short maturities have values close to 1, whereas longer maturities get progressively cheaper.

The corresponding *yield curve* is obtained by taking (minus) the logarithm of the discount curve and then dividing by the maturity. The yield curve is shown in Figure 20.2 for the same set of initial spot rates.

Note that the mean-reverting aspect of the interest rate SDE determines that the yield curve can have upward- or downward-sloping curves, as well as flat ones. This is because if the spot rate

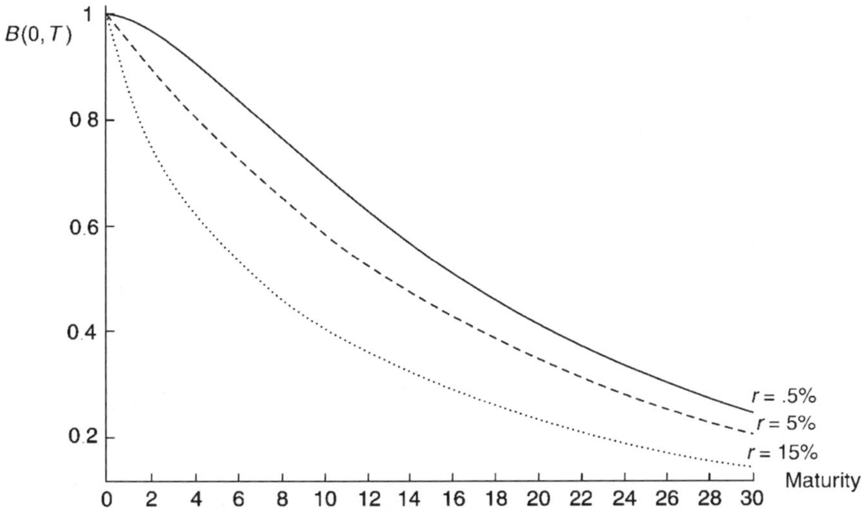

FIGURE 20.1 The graph of the zero-coupon bond at three different levels for the spot rate $r = 0.5\%$, $r = 5\%$, $r = 15\%$.

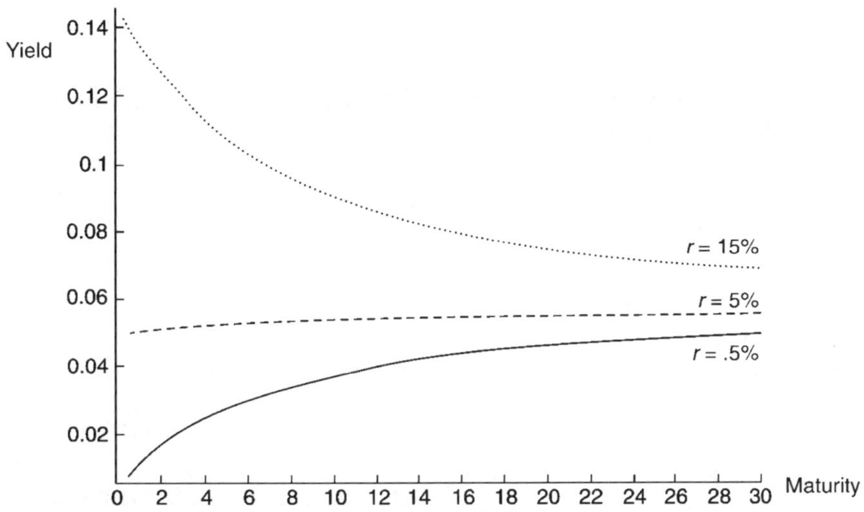

FIGURE 20.2 The yield curves for the same set of initial spot rates.

is 15% currently, the model assumes that it will go back toward its mean 5% as we consider the long bonds. Thus long bonds would automatically be priced by using rates on the average around 5%, whereas short bonds will be priced by using short rates closer to 15%. The case of a current short rate below the long-run mean is the reverse and gives an upward-sloping yield curve.

Figure 20.3 shows the effect of changing the value of market price of risk λ on the discount curve, assuming $r = 5\%$.

20.5.3 Case 3: More Complex Forms

There are several other models that result in closed-form solutions for bond prices.

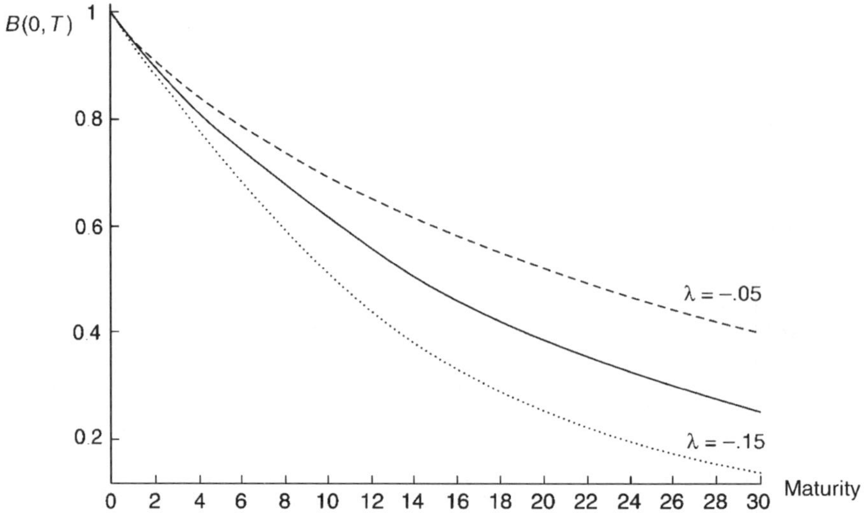

FIGURE 20.3 The effect of changing the value of market price of risk λ on the discount curve, assuming spot rate of $r = 5\%$.

For example, in the case of Cox-Ingersoll-Ross, the fundamental spot rate r_t is assumed to obey the slightly different SDE

$$dr_t = \alpha(\kappa - r_t)\,dt + br^{\frac{1}{2}}dW_t \tag{20.45}$$

which is known as the square-root specification for interest rate volatility.

The PDE that will correspond to this case will be given by:

$$\left(\alpha(\kappa - r_t) - b^2 r\lambda\right)B_r + B_t + \frac{1}{2}B_{rr}b^2 r - rB = 0 \tag{20.46}$$

with boundary condition

$$B(T,T) = 1 \tag{20.47}$$

This PDE can again be solved for a closed-form bond-pricing equation. The resulting expression is somewhat more complex than the case of Vasicek. It is given by

$$B(t,T) = A(t,T)\,e^{-C(t,T)r} \tag{20.48}$$

where the functions $A(t,T), C(t,T)$ are given by

$$A(t,T) = \left(2\frac{\gamma e^{1/2(\alpha+\lambda+\gamma)T}}{(\alpha+\lambda+\gamma)\left(e^{\gamma T}-1\right)+2\gamma}\right)^{2\frac{\alpha\kappa}{b^2}} \tag{20.49}$$

$$C(t,T) = 2\frac{e^{\gamma T}-1}{(\alpha+\lambda+\gamma)\left(e^{\gamma T}-1\right)+2\gamma} \tag{20.50}$$

and the γ is given by

$$\gamma = \sqrt{(\alpha+\lambda)^2 + 2b^2} \tag{20.51}$$

One can act in a similar fashion and plot the yield curves for this case.

20.6 CONCLUSION

This chapter dealt with the classical approach to deriving PDEs for interest-sensitive securities. We see that although the major steps are similar to the case of Black–Scholes, there are some major differences in terms of practical applications and the underlying philosophy between the two cases. The classical approach to pricing interest rate-sensitive securities rests on modeling the drifts of the underlying stochastic processes, whereas the Black–Scholes approach was one where only the volatilities needed to be modeled and calibrated.

20.7 REFERENCES

The PDE solution for bond prices can be found in all major sources. The reader may, however, prefer to read first the original paper by Vasicek that can be found in "Vasicek and Beyond." Two other good sources are Cox-Ingersoll-Ross (1985) and Hull and White (1990).

20.8 EXERCISES

1. Suppose you are given the following SDE for the instantaneous spot rate:

$$dr_t = \sigma r_t dW_t \tag{20.52}$$

where the W_t is a Wiener process under the real-world probability and the σ is a constant volatility. The initial spot rate r_0 is known to be 5%.

   (a) What does this spot-rate dynamics imply?
   (b) Obtain a PDE for a default-free discount bond price $B(t, T)$ under these conditions.
   (c) Can you determine the solution to this PDE?
   (d) What is the market price of interest rate risk? Can you interpret its sign?

2. You are given the spot-rate model:

$$dr_t = \alpha(\kappa - r_t)\,dt + b\,dW_t \tag{20.53}$$

where the W_t is a Wiener process under the real-world probability. Under this spot-rate model, the solution to the PDE that corresponds to a default-free pure discount bond $B(t, T)$ gives the closed-form bond pricing formula $B(t, T)$:

$$B(t, T) = e^{\frac{1}{\alpha}\left(1-e^{-\alpha(T-t)}\right)(R-r)-TR-\frac{b^2}{4\alpha^3}\left(1-e^{-\alpha(T-t)}\right)^2} \tag{20.54}$$

where

$$R = \kappa - \frac{b\lambda}{\alpha} - \frac{b^2}{\alpha^2} \tag{20.55}$$

Now consider the following questions that deal with properties of discount bonds whose prices can be represented by this formula.

   (a) Apply Ito's Lemma to the bond formula that gives $B(t, T)$ above and obtain the SDE that gives bond dynamics.
   (b) What are the drift and diffusion components of bond dynamics? Derive these expressions explicitly and show that the drift μ is given by:

$$\mu = r_t - \frac{b\lambda}{\alpha}\left(1 - e^{-\alpha(T-t)}\right)$$

   and that the diffusion parameter equals:

$$\frac{b}{\alpha}\left(1 - e^{-\alpha(T-t)}\right)$$

   (c) Is it expected that the diffusion parameter is independent of market price of risk λ?
   (d) What is the relationship between the maturity of a discount bond and its volatility?
   (e) Is the risk premium, that is, the return, in excess of risk-free rate proportional to volatility? To market price of risk? Is this important?
   (f) Suppose $T \to \infty$, what happens to the drift and diffusion parameters?
   (g) What does the R represent?

3. The canonical two-factor Vasicek model for the short rate, $r(t)$, assumes

$$dY_1(t) = -\lambda_1 Y_1(t)dt + dW_1(t) \tag{20.56}$$

$$dY_2(t) = -\lambda_{21}Y_1(t)dt - \lambda_2 Y_2(t)dt + dW_2(t) \tag{20.57}$$

$$r(t) = \delta_0 + \delta_1 Y_1(t) + \delta_2 Y_2(t) \tag{20.58}$$

where $W_1(t)$ and $W_2(t)$ are $\mathbb{Q}$-independent Brownian motions with case account takes as numeraire. Zero-coupon bond prices are then given by

$$Z_t^T = \mathbb{E}(e^{-\int_t^T r(s)ds}), \quad 0 \le t \le T \tag{20.59}$$

Because the solution to (20.1) and (20.2) is Markov, there must be a function $f(t, y_1, y_2)$ such that $Z_t^T = f(t, Y_1(t), Y_2(t))$. Show that $f(t, y_1, y_2)$ satisfies the following PDE

$$-(\delta_0 + \delta_1 y_1 + \delta_2 y_2)f(t, y_1, y_2) + f_t(t, y_1, y_2) \tag{20.60}$$

$$-\lambda_1 y_1 f_{y_1}(t, y_1, y_2) - \lambda_{21} y_1 f_{y_2}(t, y_1, y_2)$$
$$-\lambda_2 y_2 f_{y_2}(t, y_1, y_2) \tag{20.61}$$

$$+\frac{1}{2}f_{y_1y_1}(t, y_1, y_2) + \frac{1}{2}f_{y_2y_2}(t, y_1, y_2) = 0 \tag{20.62}$$

4. Consider the following SDE for the spot rate: $dr_t = \sigma r_t dW_t$.

(a) Write a program that simulates a path for r_t by simulating dW_t. Assume $r_0 = 0.01, T = 2.0$ and $\Delta = 0.05$. Calculate the expectation of the terminal spot rate, r_T.

(b) Solve the given SDE and repeat the above simulation by simulating a single W_t along each path. As above, calculate the expectation of the terminal spot rate, r_T.

(c) Show empirically that these two simulation methods converge as the number of draws increases and comment on their relative efficiency.

CHAPTER

21

Relating Conditional Expectations to PDEs

OUTLINE

21.1 INTRODUCTION

Throughout this book we keep alternating between mathematical tools for two major pricing methods. Using the Fundamental Theorem of Finance and normalizing by the money market account, we often used the representation

$$F(S_t, t) = \mathbb{E}_t^{\mathbb{Q}}\left[e^{-\int_t^T r_s ds} F(S_T, T)\right] \tag{21.1}$$

to price a derivative with expiration payoff $F(S_T, T)$, written on S_t. According to this, the conditional expectation under the risk-neutral measure, $\mathbb{Q}$, of future payoffs would equal the current arbitrage-free price $F(S_t, t)$,

Financial Derivatives, *Third Edition*.
http://dx.doi.org/10.1016/B978-0-12-384682-2.00021-9

once discounted by the random discount factor $e^{-\int_t^T r_s ds}$. When the r_t was constant, as was the case under the Black–Scholes assumptions, this formula simplified to:

$$F(S_t, t) = e^{-r(T-t)}\mathbb{E}_t^{\mathbb{Q}}[F(S_T, T)] \tag{21.2}$$

At other times, the pricing was discussed using PDE methods. For example, in the previous chapter, using the method of risk-free portfolios we derived the PDE that a default-free discount-bond price $B(t, T)$ must satisfy under the condition of no arbitrage:

$$B_r\left(a(r_t, t) - \lambda_t b(r_t, t)\right) + B_t + \frac{1}{2}B_{rr}b(r_t, T)^2 - r_t B = 0 \tag{21.3}$$

with the boundary condition

$$B(T, T) = 1 \tag{21.4}$$

Similarly, under the Black–Scholes assumptions with constant spot rate r, we earlier obtained the fundamental Black–Scholes PDE for a call option with strike price K and expiration T, written on S_t:

$$S_t F_s r_t + F_t + \frac{1}{2}F_{ss}\sigma_t^2 - rF = 0 \tag{21.5}$$

The boundary condition was

$$F(S_T, T) = \max[S_T - K, 0] \tag{21.6}$$

Thus, the pricing effort went back and forth between PDE approaches and approaches that used conditional expectations. Yet, both of these methods are supposed to give the same arbitrage-free price $F(S_t, t)$. This suggests that there may be some deeper correspondence between conditional expectations, such as in (21.1) or (21.2), and the PDEs that are shown in (21.3) or (21.5), respectively.

In fact, suppose we showed that when a function $F(S_t, t)$ is given by

$$F(S_t, t) = \mathbb{E}_t^{\mathbb{Q}}\left[e^{-\int_t^T r_s ds}F(S_T, T)\right] \tag{21.7}$$

where $F(S_t, t)$ is twice differentiable, the same $F(S_t, t)$ would automatically satisfy a specific PDE. And suppose we derived the general form of this PDE. This would be very convenient. We discuss some examples.

All interest rate derivatives have to assume that instantaneous spot rates are random. At the same time, the Fundamental Theorem of Finance would always permit one to write the derivatives' price $F(S_t, t)$ as

$$F(S_t, t) = \mathbb{E}_t^{\mathbb{Q}}\left[e^{-\int_t^T r_s ds}F(S_T, T)\right] \tag{21.8}$$

under the risk-neutral measure. As a result, such conditional expectations arise naturally in derivative pricing. This is especially the case for interest rate derivatives, where the spot rate cannot be assumed constant and, hence, the discount factors will have to be random.

But these conditional expectations are not always easy to evaluate. The stochastic behavior of r_t can make this a very complex task indeed. Often, there is no closed-form solution and numerical methods need to be used. Even when such expectations can be evaluated numerically, speed and accuracy considerations may necessitate alternative methods. Thus, it may be quite useful to have an alternative representation that avoids the (direct) evaluation of conditional expectations in calculating the arbitrage-free price $F(S_t, t)$. In particular, if we can obtain a PDE that corresponds to the conditional expectations (21.1) or (21.2), we can use numerical schemes to calculate $F(S_t, t)$. If one could establish a PDE that corresponds to such expectations, this could give a faster, more accurate, or simply a more practical numerical method for obtaining the fair market price $F(S_t, t)$ of a financial derivative written on S_t.[1]

Alternatively, a market practitioner can be given a PDE that he or she does not know how

[1] For example, in dealing with American-style derivatives, it will in general be more convenient to work with numerical PDE methods, instead of evaluating the conditional expectations through Monte Carlo.

to solve. If the conditional expectation in (21.8) is shown to be a solution for this PDE, then this may yield a convenient way of "solving"for $F(S_t, t)$. Again, the correspondence will be very useful.

In this chapter we discuss the mechanics of obtaining such correspondences and the tools that are associated with them.

21.2 FROM CONDITIONAL EXPECTATIONS TO PDEs

In this section we establish a correspondence between a class of conditional expectations and PDEs. Using simple examples, we illustrate that starting with a function defined via a certain class of conditional expectations, we can always obtain a corresponding PDE satisfied by this function, as long as some nontrivial conditions are satisfied. The main condition necessary for such a correspondence to exist is Markovness of the processes under consideration.

Our discussion will begin with a simple example that is not directly useful to a market participant. But this will facilitate the understanding of the derivations. Also, we gradually complicate these examples and show how the methods discussed here can be utilized in practical derivatives pricing as well.

21.2.1 Case 1: Constant Discount Factors

Consider the function $F(x_t)$ of a random process $x_t \in [0, \infty)$, defined by the conditional expectation:

$$F(x_t) = \mathbb{E}_t^{\mathbb{P}}\left[\int_t^{\infty} e^{-\beta s} g(x_s)\, ds\right] \tag{21.9}$$

where $\beta > 0$ represents a constant instantaneous discount rate, and $g(\cdot)$ is some continuous payout that depends on the value assumed by the random process, $x_t.IE_t^{IP}[.]$ which is the expectation under the probability $\mathbb{P}$ and conditional on the information set I_t, both of which are left unspecified at this point. The process x_t obeys the SDE:

$$dx_t = \mu\, dt + \sigma\, dW_t \tag{21.10}$$

where μ, σ are known constants.

This $F(x_t)$ can be interpreted as the expected value of some discounted future cash flow $g(x_s)$ that depends on an I_s-measurable random variable x_s. The discount factor $0 < \Delta$ is deterministic.

Clearly, the cash flows of interest in financial markets will, in general, be discounted by *random* discount factors. This is especially the case for interest rate derivatives, but we will leave this aside at the moment. All we want to accomplish at this point is to obtain a PDE that "corresponds" to the expectation in (21.9). We intend to study in detail the steps that will lead to this PDE. Once we learn how to do this, random discount factors can easily be introduced.

We now obtain a PDE that corresponds to expectation (21.9) in several steps. These steps are general and can be applied to more complicated expectations than the one in (21.9). We proceed in a mechanical way to illustrate the derivation. To simplify the notation we assume that the initial point is given by $t = 0$.

First, consider a small time interval $0 < \Delta$ and split the period $[0, \Delta)$ in two. One being the immediate future, represented by the interval $[0, \Delta]$, and the other represented by $[\Delta, \infty)$,

$$F(x_0) = \mathbb{E}_0^{\mathbb{P}}\left[\int_0^{\Delta} e^{-\beta s} g(x_s)\, ds + \int_{\Delta}^{\infty} e^{-\beta s} g(x_s)\, ds\right] \tag{21.11}$$

The second step involves some elementary transformations that are intended to introduce a future value of $F(\cdot)$ to the right-hand side of this expression. In fact, note that the second term in the brackets can be rewritten after multiplying and dividing by $e^{-\beta\Delta}$ as:

$$\mathbb{E}_0^{\mathbb{P}}\left[\int_{\Delta}^{\infty} e^{-\beta s} g(x_s)\, ds\right] = \mathbb{E}_0^{\mathbb{P}}\left[e^{-\beta\Delta}\int_{\Delta}^{\infty} e^{-\beta(s-\Delta)} g(x_s)\, ds\right] \tag{21.12}$$

The third step will apply the recursive property of conditional expectations. As seen earlier, when conditional expectations are nested, it is the expectation with respect to the *smaller* information set that matters. Thus, if we have $I_t \subseteq I_s$, we can write:

$$\mathbb{E}_t^{\mathbb{P}}\left[\mathbb{E}_s^{\mathbb{P}}[\cdot]\right] = \mathbb{E}_t^{\mathbb{P}}[\cdot] \tag{21.13}$$

This permits replacing the operator $\mathbb{E}_0^{\mathbb{P}}[\cdot]$ in (21.12) by the operator $\mathbb{E}_0^{\mathbb{P}}\left[\mathbb{E}_\Delta^{\mathbb{P}}[\cdot]\right]$.[2] Thus, we get:

$$\mathbb{E}_0^{\mathbb{P}}\left[\int_\Delta^\infty e^{-\beta s} g(x_s)\, ds\right] = \mathbb{E}_0^{\mathbb{P}}\left[e^{-\beta\Delta}\mathbb{E}_\Delta^{\mathbb{P}}\left[\int_\Delta^\infty e^{-\beta(s-\Delta)} g(x_s)\, ds\right]\right] \tag{21.14}$$

But we can recognize the term inside the inner brackets on the right-hand side as the $F(x_\Delta)$ and write[3]:

$$\mathbb{E}_0^{\mathbb{P}}\left[\int_\Delta^\infty e^{-\beta s} g(x_s)\, ds\right] = \mathbb{E}_0^{\mathbb{P}}\left[e^{-\beta\Delta}F(x_\Delta)\right] \tag{21.15}$$

This last expression can now be utilized in (21.11):

$$F(x_0) = \mathbb{E}_0^{\mathbb{P}}\left[\int_0^\Delta e^{-\beta s} g(x_s)\, ds + e^{\beta\Delta}F(x_\Delta)\right] \tag{21.16}$$

Grouping all terms on the right-hand side *and* moving them inside the expectation operator, we obtain:

$$\mathbb{E}_0^{\mathbb{P}}\left[\int_0^\Delta e^{-\beta s} g(x_s)\, ds + e^{\beta\Delta}F(x_\Delta) - F(x_0)\right] = 0 \tag{21.17}$$

[2]Recall that at time $t = \Delta$, we will have more information than at time $t = 0$.

[3]Here the $F(x_0)$ is the value of $F(\cdot)$ observed at time $t = 0$. It is conditional on x_0. The $F(x_\Delta)$, on the other hand, is the value that will be observed after a time interval of length Δ at $t = \Delta$. It will be conditional on x_Δ.

As the fourth step, we add and subtract $F(x_\Delta)$, divide all terms by Δ, and rearrange:

$$\frac{1}{\Delta}\mathbb{E}_0^{\mathbb{P}}\left[\int_0^\Delta e^{-\beta s} g(x_s)\, ds + \left(e^{\beta\Delta} - 1\right)F(x_\Delta) + \left[F(x_\Delta) - F(x_0)\right]\right] = 0 \tag{21.18}$$

As the last step, we take the limit as $\Delta \to 0$ of each term on the left-hand side. The second term is, in fact, a standard derivative of $e^{\beta x}$ evaluated at $x = 0$:

$$\lim_{\Delta\to 0}\left(e^{\beta\Delta} - 1\right) = -\beta \tag{21.19}$$

The first term is the derivative with respect to the upper limit of a Riemann integral:

$$\lim_{\Delta\to 0}\frac{1}{\Delta}\int_0^\Delta e^{-\beta s} g(x_s)\, ds = g(x_0) \tag{21.20}$$

The third term, on the other hand, involves the expectation of a stochastic differential and hence requires the application of Ito's Lemma. First, we approximate using Taylor series and write:

$$\frac{1}{\Delta}\mathbb{E}_0^{\mathbb{P}}\left[F(x_\Delta) - F(x_0)\right] \approx \frac{1}{\Delta}\mathbb{E}_0^{\mathbb{P}}\left[F_x\left[x_\Delta - x_0\right] + \frac{1}{2}F_{xx}\sigma_x^2\Delta\right] \tag{21.21}$$

Then let $\Delta \to 0$ and take the expectation to obtain:

$$\lim_{\Delta\to 0}\frac{1}{\Delta}\mathbb{E}_0^{\mathbb{P}}\left[F_x\left[x_\Delta - x_0\right] + \frac{1}{2}F_{xx}\sigma_x^2\Delta\right] = F_x\mu + \frac{1}{2}F_{xx}\sigma^2 \tag{21.22}$$

where μ is the drift of the random process x that enters the formula as a result of applying the expectation operator to $(x_\Delta - x_0)$.

Replacing the limits obtained in (21.19)–(21.22) in expression (21.18), we reach the desired PDE:

$$F_x\mu + \frac{1}{2}F_{xx}\sigma^2 - \beta F + g = 0 \tag{21.23}$$

where the F_x, F_{xx}, F, and g are all functions of x.

One may wonder what causes this correspondence between the conditional expectation (21.9) and this PDE. After all, these two concepts seemed to be quite unrelated at the outset. A heuristic answer to this question is the following.

The PDE corresponds to the expectation of the "present value"of cash flow stream $\{g(x_s)\}$. If this present value $F(\cdot)$ is given by the conditional expectation shown above, then it cannot be an arbitrary function of x_0 and its behavior over time must satisfy some constraints due to the expected future behavior of x. These constraints lead to the PDE.

More precisely, the function $F(x_0)$ is the result of an optimal forecast. This optimal forecast requires projecting ways in which $F(x_t)$ may change over time. Expected changes in the random variable x_t, deterministic changes in the time variable t, payouts $g(x_t)$, and the second-order Ito correction all cause various predictable changes in $F(\cdot)$. The optimal prediction should take these changes into account. The PDE that corresponds to the conditional expectation operators are obtained in such a way that the expected value of the prediction error is set equal to zero, and its variance is minimized, once these predictable changes are taken into consideration.[4]

21.2.2 Case 2: Bond Pricing

We now see a more relevant example to the correspondence between a class of conditional expectations and PDEs. In fact, we now apply the same derivation to obtain a PDE for default-free pure discount-bond prices.

Consider the price $B(t,T)$ of a default-free pure discount bond with maturity T in a no-arbitrage setting. Assume that the instantaneous spot rate r_t is a Markov process and write the price of the bond with par value \$1, using the familiar formula:

$$B(t,T) = \mathbb{E}^{\mathbb{Q}}\left[e^{-\int_t^T r_s ds}\right] \tag{21.24}$$

with

$$B(T,T) = 1$$

Here the expectation is taken with respect to the risk-neutral measure and with respect to the conditioning set available at time t, namely the I_t. This is assumed to include the current observation on the spot rate r_t. If r_t is a Markov process, $B(t,T)$ will depend only on the latest observation of r_t. Because we are in the risk-neutral world, as dictated by the use of the r_t will follow the dynamics given by the SDE

$$dr_t = \left[a(r_t,t) - \lambda_t b(r_t,t)\right]dt + b(r_t,t)\,dW_t \tag{21.25}$$

where W_t is a Wiener process under the risk-neutral measure $\mathbb{Q}$. The λ_t is the market price of interest rate risk defined by

$$\lambda_t = \frac{\mu - r_t}{\sigma} \tag{21.26}$$

with μ, σ being the shorthand notation for the drift and diffusion components of the bond price dynamics:

$$dB = \mu(B,t)\,B\,dt + \sigma(B,t)\,B\,dW_t$$

Thus, we again have a conditional expectation and a process that is driving it, just as in the previous case. This means that we can apply the same steps used there and obtain a PDE that corresponds to $B(t,T)$. Yet, in the present case, this PDE may also have some practical use in pricing bonds. It can be solved numerically or, if a closed-form solution exists, analytically.

The same steps will be applied in a mechanical way, without discussing the details. First, split the interval $[t,T]$ into two parts to write:

$$B(t,T) = \mathbb{E}^{\mathbb{Q}}\left[\left(e^{-\int_t^{t+\Delta} r_s ds}\right)\left(e^{-\int_{t+\Delta}^{T} r_s ds}\right)\right] \tag{21.27}$$

[4]In fact, note that in obtaining the PDE we replaced the Wiener component of the x_t with zero.

Second, try to introduce the future price of the bond, $B(t+\Delta,T)$, in this expression. In fact, the second exponential on the right-hand side can easily be recognized as $B(t+\Delta,T)$ once we use the recursive property of conditional expectations. Using

$$\mathbb{E}_t^{\mathbb{P}}\left[\mathbb{E}_{t+\Delta}^{\mathbb{P}}[\cdot]\right]=\mathbb{E}_t^{\mathbb{P}}[\cdot] \tag{21.28}$$

we can write

$$B(t,T)=\mathbb{E}^{\mathbb{Q}}\left[\left(e^{-\int_t^{t+\Delta} r_s ds}\right)B(t+\Delta,T)\right] \tag{21.29}$$

because

$$B(t+\Delta,T)=\mathbb{E}^{\mathbb{Q}}\left[e^{-\int_{t+\Delta}^{T} r_s ds}\right] \tag{21.30}$$

In the third step, group all terms inside the expectation sign, add and subtract $B(t+\Delta,T)$, and divide by Δ:

$$\frac{1}{\Delta}\mathbb{E}^{\mathbb{Q}}\left[\left(e^{-\int_t^{t+\Delta} r_s ds}-1\right)B(t+\Delta,T)\right.$$
$$\left.+\left[B(t+\Delta,T)-B(t,T)\right]\right]=0 \tag{21.31}$$

Note that this introduces the increment $[B(t+\Delta,T)-B(t,T)]$ to the left-hand side. This will be used for applying Ito's Lemma.

Fourth, take the limit as $\Delta\to 0$ of the first term in this equation[5]:

$$\lim_{\Delta\to 0}\frac{1}{\Delta}\mathbb{E}^{\mathbb{Q}}\left[\left(e^{-\int_t^{t+\Delta} r_s ds}-1\right)B(t+\Delta,T)\right]$$
$$=-r_t B(t,T) \tag{21.32}$$

Then, apply Ito's Lemma to the second term in (21.31) and take the expectation:

$$\lim_{\Delta\to 0}\frac{1}{\Delta}\mathbb{E}^{\mathbb{Q}}\left[B(t+\Delta,T)-B(t,T)\right]$$
$$=B_t+B_r\left[a(r_t,t)-\lambda_t b(r_t,t)\right]$$
$$+\frac{1}{2}B_{rr}b(r_t,t)^2 \tag{21.33}$$

where the drift and the diffusion of the spot-rate process $a(r_t,t), b(r_t,t)$ are used.[6]

In the final step, replace these limits in expression (21.31) to obtain the PDE that corresponds to the conditional expectation (21.24):

$$-r_t B+B_t+B_r\left[a(r_t,t)-\lambda_t b(r_t,t)\right]$$
$$+\frac{1}{2}B_{rr}b(r_t,t)^2=0 \tag{21.34}$$

with, of course, the usual boundary condition:

$$B(T,T)=1 \tag{21.35}$$

This is a PDE that must be satisfied by an arbitrage-free price of a pure discount bond with no-default risk. In Chapter 20, the same PDE was obtained using the method of risk-free portfolios.

21.2.3 Case 3: A Generalization

We have seen in detail two cases where the existence of a certain type of conditional expectation led to a corresponding PDE. In the first case there was a random cash flow stream depending on an underlying process x_t but the discount rate was constant. In the second case, the instrument paid a single, fixed cash flow at maturity, yet the discount factor was random.

Clearly, one can combine these two basic examples to obtain the PDE that corresponds to instruments that make spot-rate dependent payouts $g(r_t)$ and that need to be discounted by random discount factors:

$$F(r_t,t)=\mathbb{E}_t^{\mathbb{Q}}\left[\int_t^T\left(e^{-\int_t^u r_s ds}\right)g(r_u)\,du\right] \tag{21.36}$$

This $F(\cdot)$ would represent the price of an instrument that makes interest rate dependent payments at times $u\in[t,T]$, and hence needs to be evaluated using the random discount factor

[5]Here we are assuming that the technical conditions permitting the interchange of limit and expectation operators are satisfied.

[6]Unlike the previous example, here the $B(t,T)$ function depends on t as well as on r_t. Hence, there will be an additional B_t term that did not exist before.

D_u at each u:

$$D_u = e^{-\int_t^u r_s ds} \tag{21.37}$$

It is interesting to note that the expectation of this D_u is nothing other than the time t price of a default-free pure discount bond that pays \$1 at time u.[7]

Various instruments and interest rate derivatives, such as coupon bonds, financial futures that are marked to market, and index-linked derivatives fall into this category, where the arbitrage-free price will be given by conditional expectations such as in (21.36). Thus, the methods that were discussed in the last two sections can be applied to find the implied PDE if the process(es) that drive these expectations are Markov. The corresponding PDEs may be exploited for real-life pricing of these complex instruments.

21.2.4 Some Clarifications

We need to comment on some issues that may be confusing at the first reading.

21.2.4.1 The Importance of Markovness

The derivation used here in obtaining the PDE that corresponds to the class of conditional expectations is valid only if the underlying stochastic processes are Markov. It may be worthwhile to see exactly where this assumption of Markovness was used in the preceding discussion.

[7]Here we cannot directly apply the operator to D_u because the $g(r_u)$ will be correlated with the D_u. If such correlation did not exist, and if $g(\cdot)$ depended on an independent random variable, say x_u only, then we could take expectations separately and simply multiply the payout by the corresponding discount-bond price B_u to discount it:

$$\mathbb{E}_t^{\mathbb{Q}}\left[\int_t^T e^{-\int_t^T r_s ds} g(x_u)\, du\right] = \int_t^T B_u \mathbb{E}_t^{\mathbb{Q}}\left[g(x_u)\right] du \tag{21.38}$$

assuming that the necessary interchange of the operators is allowed. On the other hand, Eq. (21.38) can always be applied if we used the forward measure as discussed in Chapter 17.

During the derivation of the PDE, we used the conditional expectation operators $\mathbb{E}^{\mathbb{Q}}[\cdot]$ that we now express in the expanded form, showing the conditioning information set explicitly:

$$\mathbb{E}^{\mathbb{Q}}\left[\int_t^T e^{-\int_t^T r_s ds} g(x_u)\, du \middle| I_t\right]$$

$$= \mathbb{E}_t^{\mathbb{Q}}\left[\int_t^T B_u\left[g(x_u)\right] du \middle| r_t\right] \tag{21.39}$$

$$= F(r_t, t) \tag{21.40}$$

These operations are valid only when the r_t process is Markov. If this assumption is not true, then the conditional expectations that we considered would depend on more than just the r_t. In fact, past spot rates $\{r_s, s < t\}$ would also be determining factors of the price of the instrument. In other words, the latter price could no longer be written as $F(t, r_t)$, a function that depended on r_t and t only. The rest of the derivation would not follow in general.

Hence, we see that the assumption of Markovness plays a central role in the choice of pricing methods that one uses for interest rate derivatives.

21.2.5 Which Drift?

One may also wonder which parameter should be used as the drift of the random process in such PDE derivations. The answer is straightforward, but it may be worthwhile to repeat it.

The conditional expectations under study are obtained with respect to some (conditional) probability distribution. For example, when we write the arbitrage-free price of a bond as:

$$B(t, T) = \mathbb{E}_t^{\mathbb{Q}}\left[e^{-\int_t^T r_s ds}\right] \tag{21.41}$$

we take the expectation with respect to $\mathbb{Q}$, the risk-neutral probability. Given that the random process under consideration is r_t, this choice of risk-neutral probability requires that we use the risk-adjusted drift for rt and write the corresponding SDE as

$$dr_t = \left(a(r_t, t) - \lambda_t b(r_t, t)\right) dt + b(r_t, t)\, dW_t \tag{21.42}$$

instead of the "real world" SDE:

$$dr_t = a(r_t, t)\, dt + b(r_t, t)\, dW_t^* \tag{21.43}$$

where the W_t^* is a Wiener process with respect to real-world probability $\mathbb{P}$.

Hence, within the present context, while using Ito's Lemma, whenever a drift substitution for dr_t is needed we have to use $(a(r_t, t) - \lambda_t b(r_t, t))$ and not $a(r_t, t)$. This was the case, for example, in obtaining the limit in (21.33).

Will the nonadjusted drift ever be used? The question is interesting because it teaches us something about pricing approaches that use other than the risk-neutral measure; formulas that, in principle, should give the same answer, but may nevertheless not be very practical. In other words, the question will show the power of the martingale approach.

Indeed, during the same derivation, instead of using the risk-adjusted drift, we can indeed use the original drift of the spot-rate process. But this requires that the conditional expectation under consideration be evaluated using the real-world probability $\mathbb{P}$, instead of the risk-neutral probability. However, we know that an expression such as

$$B(t, T) = \mathbb{E}_t^{\mathbb{P}}\left[e^{-\int_t^T r_s ds}\right] \tag{21.44}$$

cannot hold in general if the $B(t, T)$ is arbitrage-free, and if the expectation is taken with respect to real-world probability $\mathbb{P}$. If one insists on using the real-world probability, then the formula for the arbitrage-free price will instead be given by:

$$B(t, T) = \mathbb{E}_t^{\mathbb{P}}\left[e^{-\int_t^T r_s ds} e^{\int_t^T \left[\lambda_s(r_s, s) dW_t^* - \frac{1}{2}\lambda(r_s, s)^2 ds\right]}\right] \tag{21.45}$$

where all symbols are as in (21.42) and (21.43).

One can in fact obtain the same PDE as in (21.34) by departing from this conditional expectation and using exactly the same steps as before. The only major difference will be at the stage when one calculates the limit corresponding to (21.33). There, one would substitute the real-world drift $a(r_t, t)$ instead of the risk-adjusted drift.

21.2.6 Another Bond Price Formula

The main focus of this chapter is the correspondence between PDEs and conditional expectations. But, in passing, it may be appropriate to discuss an application of equivalent martingale measures to bond pricing.

The preceding section considered *two* bond pricing formulas. One used the martingale measure and gave the compact expression:

$$B(t, T) = \mathbb{E}_t^{\mathbb{Q}}\left[e^{-\int_t^T r_s ds}\right] \tag{21.46}$$

The other used the real-world probability $\mathbb{P}$ and resulted in

$$B(t, T) = \mathbb{E}_t^{\mathbb{P}}\left[e^{-\int_t^T r_s ds} e^{\int_t^T \left[\lambda(r_s, s) dW_u^* - \frac{1}{2}\lambda(r_s, s)^2 ds\right]}\right] \tag{21.47}$$

Of course, the two $B(t, T)$ would be identical, except for the way they are characterized and calculated.

The question that we touch on briefly here is how to go from one bond price formula to the other. This provides a good example of the use of Girsanov theorem. First, we remind the reader that within the context of Chapter 15, two probabilities $\mathbb{Q}$ and $\mathbb{P}$ are equivalent if they are related by

$$d\mathbb{Q}_t = \xi_t\, d\mathbb{P}_t \tag{21.48}$$

where the Radon–Nikodym derivative ξ_t was given by

$$\xi_t = e^{\int_0^t \left[\lambda_u dW_u^* - \frac{1}{2}\lambda_u^2 du\right]} \tag{21.49}$$

where ξ_t is an I_t-measurable process.[8]

[8] In this particular case, λ_t will be the market price of spot interest rate risk.

We now show how to get pricing formula (21.47) starting from (21.46), assuming that all technical conditions of Girsanov theorem are satisfied.

Start with the bond pricing equation:

$$B(t,T) = \mathbb{E}_t^{\mathbb{Q}}\left[e^{-\int_t^T r_s ds}\right] \tag{21.50}$$

Write the same expression using the definition of the conditional expectation operator $\mathbb{E}_t^{\mathbb{Q}}[\cdot]$:

$$\mathbb{E}_t^{\mathbb{Q}}\left[e^{-\int_t^T r_s ds}\right] = \int_\Omega \left(e^{-\int_t^T r_s ds}\right) d\mathbb{Q} \tag{21.51}$$

where the Ω is the relevant range at which future r_t will take values. Now, use the equivalence between $\mathbb{Q}$ and $\mathbb{P}$ shown in (21.48) to substitute for $d\mathbb{Q}$ in this equation:

$$\mathbb{E}_t^{\mathbb{Q}}\left[e^{-\int_t^T r_s ds}\right] = \int_\Omega \left(e^{-\int_t^T r_s ds}\right) \xi_T \, d\mathbb{P} \tag{21.52}$$

Substituting for ξ_T, we get the desired equivalence:

$$\mathbb{E}_t^{\mathbb{Q}}\left[e^{-\int_t^T r_s ds}\right]$$

$$= \int_\Omega \left(e^{-\int_t^T r_s ds} e^{\int_0^T \left[\lambda_u dW_u^* - \frac{1}{2}\lambda_u^2 du\right]}\right) d\mathbb{P} \tag{21.53}$$

$$= \mathbb{E}_t^{\mathbb{P}}\left[e^{-\int_t^T r_s ds} e^{\int_t^T \left[\lambda(r_s,s) dW_u^* - \frac{1}{2}\lambda(r_s,s)^2 ds\right]}\right] \tag{21.54}$$

This is, indeed, the bond pricing formula with real-world probability obtained earlier.

Thus, the connection between the two characterizations of default-free pure discount-bond prices becomes very simple once the Girsanov theorem is utilized. Of course, in the above derivation, we did not show that the term λ_t is the market price for interest rate risk. But it is clearly a drift adjustment to the interest rate stochastic differential equation.

21.2.7 Which Formula?

Expressions (21.46) and (21.54) give two different characterizations for $B(t,T)$. But the second formula, derived with respect to real-world probability, seems to be messier because it is a function of λ_t, whereas characterization (21.50) does not contain this variable. Hence, one may be tempted to conclude that if one is utilizing Monte Carlo approach to calculate bond prices, or the prices of related derivatives, the formula in (21.50) is the one that should be used. It does not require the knowledge of λ_t.

The appearances are unfortunately deceiving in this particular case. Whether one uses (21.46) or (21.54), as long as one stays within the boundaries of the classical approach, Monte Carlo pricing of bonds or other interest-sensitive securities would necessitate a calibration of λ_t. In the case of (21.54) this is obvious, the λ_t is in the pricing formula. In the case of (21.50), some numerical estimate of the λ_t will also be needed in generating random paths for the r_t through the corresponding SDE *under the martingale probability* $\mathbb{Q}$:

$$dr_t = \left(a(r_t,t) - \lambda_t b(r_t,t)\right) dt + b(r_t,t)\, dW_t \tag{21.55}$$

Obviously, this equation becomes usable only if some numerical estimate for λ_t is plugged in.

Thus, in one case, the integral contains the λ_t but not the SDE. In the other case, the λ_t is in the SDE but does not show up in the integral. But in Monte Carlo pricing, the market participant has to use *both* the integral and the SDE. That is why the approach outlined here is still the "classical" approach and requires, one way or another, modeling underlying drifts. The HJM approach avoids this difficulty.

21.3 FROM PDEs TO CONDITIONAL EXPECTATIONS

Up to this point we showed that if the underlying processes are Markov and if some technical conditions are satisfied, then the arbitrage-free prices characterized as conditional expectations with respect to some appropriate measure would satisfy a PDE. That is, given a class of conditional expectations, we obtain a corresponding PDE.

In this section we investigate going in the opposite direction. Suppose we are given a PDE satisfied by an asset price $F(S_t, t)$. Can we go from there to conditional expectations as a possible solution class?

We discuss this within a special case. We let the $F(W_t, t)$ be the price of a financial derivative that is written on the Wiener process W_t defined with respect to probability. The choice of a W_t as the driving process may not seem to be very realistic, but it can easily be generalized. Further, it permits the use of a known PDE called the heat equation in engineering literature.

Suppose this price, $F(W_t, t)$, of the derivative was known to satisfy the following PDE:

$$F_t + \frac{1}{2} F_{ww} = 0 \tag{21.56}$$

and that we have the following boundary condition at expiration, $t = T$:

$$F(W_T, T) = G(W_T)$$

for some known function $G(\cdot)$.

We show that the solution of this PDE can be represented as a conditional expectation. To do this, we first assume that all technical conditions are satisfied and start by applying Ito's Lemma to $F(W_t, t)$:

$$dF = \left[\frac{\partial F}{\partial t} + \frac{1}{2}\frac{\partial^2 F}{\partial W^2}\right] dt + \frac{\partial F}{\partial W} dW_t \tag{21.57}$$

$$= \left[F_t + \frac{1}{2} F_{ww}\right] dt + F_w \, dW_t \tag{21.58}$$

where we use the fact that the Wiener process has a drift parameter that equals zero and a diffusion parameter that equals one.

This stochastic differential equation shows how $F(W_t, t)$ evolves over time. The next step is integrating both sides of this equality from t to T:

$$F(W_T, T) - F(W_t, t) = \int_t^T F_w \, dW_s$$

$$+ \int_t^T \left[F_t + \frac{1}{2} F_{ww}\right] ds \tag{21.59}$$

Recall that the partial derivatives F_t and F_{ww} are themselves functions of W_s and s.

Now we know something about the integrals on the right-hand side. As a matter of fact, using the PDE in (21.56), we know that the second integral equals zero:

$$\int_t^T \left[F_t + \frac{1}{2} F_{ww}\right] ds \tag{21.60}$$

Using this and taking the expectation with respect to of the two sides of Eq. (21.59), we can write:

$$\mathbb{E}_t^{\mathbb{Q}}\left[F(W_T, T)\right] = F(W_t, t) + \mathbb{E}_t^{\mathbb{Q}}\left[\int_t^T F_w \, dW_s\right] \tag{21.61}$$

Now, $F(W_T, T)$ is the value of $F(\cdot)$ at the boundary $t = T$, so we can replace it by the known function $G(W_T)$. Doing this and rearranging:

$$F(W_t, t) = \mathbb{E}_t^{\mathbb{Q}}\left[G(W_T)\right] - \mathbb{E}_t^{\mathbb{Q}}\left[\int_t^T F_w \, dW_s\right] \tag{21.62}$$

Thus, if we can show that the second expectation on the right-hand side is zero, then the (unknown) function $F(\cdot)$ can be determined by taking the expectation of the known function $G(\cdot)$. But this requires that:

$$\mathbb{E}_t^{\mathbb{Q}}\left[\int_t^T \frac{\partial F}{\partial W} dW_s\right] = 0 \tag{21.63}$$

To show that this is the case, we invoke an important property of Ito integrals with respect to Wiener processes. From Chapter 10 we know that if $h(W_t)$ is a nonanticipative function with respect to an information set I_t, and with respect to the probability $\mathbb{P}$, then the expectation of integrals with respect to W_t will vanish:

$$\mathbb{E}_t^{\mathbb{Q}}\left[\int_t^T h(W_s) \, dW_s\right] = 0 \tag{21.64}$$

Let us repeat why this is so. The W_t is a Wiener process. Its increments, dW_t, do not depend on the past, including the immediate past. But if $h(W_t)$ is nonanticipative, then $h(W_t)$ will not depend on the "future" either. So, in (21.56) we have the expectation of a product where the individual terms are independent of one another. Also, one of these, namely the dW_t, has mean zero.

Going back to equality (21.62), we see that the term we equate to zero, namely the

$$\mathbb{E}_t^{\mathbb{Q}}\left[\int_t^T \frac{\partial F}{\partial W} dW_s\right] \tag{21.65}$$

is exactly of this type. It is an integral of a nonanticipative function with respect to the Wiener process. This means that its expectation is zero, given that $F(\cdot)$ satisfies some technical conditions:

$$\mathbb{E}_t^{\mathbb{Q}}\left[\int_t^T \frac{\partial F}{\partial W} dW_s\right] = 0 \tag{21.66}$$

Thus we obtained:

$$F(W_t, t) = \mathbb{E}_t^{\mathbb{Q}}\left[G(W_T)\right] \tag{21.67}$$

which is a characterization of the price $F(W_t, t)$, as a conditional expectation of the boundary condition $G(W_T)$ and the probability. This function is also the solution of the heat equation. In fact, beginning with a PDE involving an unknown function $F(t, W_t)$, we determined the solution as an expectation of a known function with respect to a probability, with respect to which W_t is a Wiener process.

21.4 GENERATORS, FEYNMAN–KAC FORMULA, AND OTHER TOOLS

Given the importance of the issues discussed above, it is not very surprising that the theory of stochastic processes developed some systematic tools and concepts to facilitate the treatment of similar problems. Many of these tools simplify the notation and make the derivations mechanical. This is the case with the notion of a *generator*, which is the formal equivalent of obtaining limits such as in (21.33), and the Feynman–Kac theorem, which gives the probabilistic solution for a class of PDEs. We complete this chapter by formalizing these concepts utilized implicitly during the earlier discussion.

21.4.1 Ito Diffusions

A continuous stochastic process S_t that has finite first- and second-order moments was shown to follow the general SDE:

$$dS_t = a(S_t, t)\,dt + \sigma(S_t, t)\,dW_t, \quad t \in [0, \infty) \tag{21.68}$$

We now assume that the drift and diffusion parameters depend on S_t only.[9]

The SDE can be written as:

$$dS_t = a(S_t)\,dt + \sigma(S_t)\,dW_t, \quad t \in [0, \infty) \tag{21.69}$$

where the $a(\cdot)$ and $\sigma(\cdot)$ are the drift and diffusion parameters. Processes that have this characteristic are called time-homogenous Ito diffusions. The results below apply to those processes whose instantaneous drift and diffusion are not dependent on t directly. Usual conditions apply to $a(\cdot)$ and $\sigma(\cdot)$, in that they are not supposed to vary "too fast."

We can discuss two properties of Ito diffusions.

[9] In almost all cases of interest where there are no jumps involved, the SDEs utilized in practice are either of geometric or of mean reverting type. The latter is especially popular with interest rate derivatives because the short rate is widely believed to have a mean reverting character. Under these conditions, the drift and diffusion parameters would be a function of S_t only. However, often dependence on time is allowed to match the initial term structure.

21.4.2 Markov Property

This property was seen before. Let S_t be an Ito diffusion satisfying the SDE:

$$dS_t = a(S_t)\,dt + \sigma(S_t)\,dW_t, \quad t \in [0,\infty) \tag{21.70}$$

Let $f(\cdot)$ be any bounded function, and suppose that the information set I_t contains all $S_u, u \leq t$ until time t. Then we say that S_t satisfies the Markov property if:

$$\mathbb{E}\left[f(S_{t+h})\middle| I_t\right] = \mathbb{E}\left[f(S_{t+h})\middle| S_t\right], \quad h > 0 \tag{21.71}$$

That is, future movements in S_t, given what we observed until time t, are likely to be the same as starting the process at time t. In other words, the observations on S_t from the distant past do not help to improve forecasts, given the S_t.

21.4.3 Generator of an Ito Diffusion

Let S_t be the Ito diffusion given in (21.70). Let $f(S_t)$ be a twice differentiable function of S_t, and suppose the process S_t has reached a particular value s_t as of time t.

We may wonder how $f(S_t)$ may move, starting from the current state s_t. We define an *operator* to represent this movement. We let the operator A be defined as the expected rate of change for $f(S_t)$ as:

$$Af(s_t) = \lim_{\Delta \to 0} \frac{\mathbb{E}\left[f(S_{t+\Delta})\middle| f(s_t)\right] - f(s_t)}{\Delta} \tag{21.72}$$

Here, the small case letter s_t indicates an already observed value for S_t.The numerator of the expression on the right-hand side measures expected change in $f(S_t)$. As we divide this by Δ, the A operator becomes a *rate* of change. In the theory of stochastic processes, A is called the *generator* of the Ito diffusion S_t.

Some readers may wonder how we can define a rate of change for $f(S_t)$, which indirectly is a function of a Wiener process. A rate of change is like a derivative and we have shown that Wiener processes are not differentiable. So, how can we justify the existence of an operator such as A, one may ask.

The answer to this question is simple. A does not deal with the *actual* rate of change in $f(S_t)$. Instead, A represents an *expected* rate of change. Although the Wiener process may be too erratic and nondifferentiable, note that expected changes in $f(S_t)$ will be a smoother function and, under some conditions, a limit *can* be defined.[10]

21.4.4 A Representation for A

First note that A is an expected rate of change in the *limit*. That is, we consider the immediate future with an infinitesimal change of time. Then, it is obvious that such a change would relate directly to Ito's Lemma. In fact, in the present case where S_t is a univariate stochastic process:

$$dS_t = a(S_t)\,dt + \sigma(S_t)\,dW_t, \quad t \in [0,\infty) \tag{21.73}$$

the operator A is given by:

$$Af = a_t \frac{\partial f}{\partial S} + \frac{1}{2}\sigma_t^2 \frac{\partial^2 f}{\partial S^2} \tag{21.74}$$

It is worthwhile to compare this with what Ito's Lemma would give. Applying Ito's Lemma to $f(S_t)$, with S_t given by (21.73):

$$df(S_t) = \left[a_t \frac{\partial f}{\partial S} + \frac{1}{2}\sigma_t^2 \frac{\partial^2 f}{\partial S^2}\right] dt + \sigma_t \frac{\partial f}{\partial S} dW_t \tag{21.75}$$

Hence, the difference between the operator A and the application of Ito's Lemma is at two points:

1. The dW_t term in Ito's formula is replaced by its drift, which is zero.
2. Next, the remaining part of Ito's formula is divided by dt.

These two differences are consistent with the definition of A. As mentioned above, A calculates

[10]Every expectation represents an average. By definition, averages are smoother than particular values.

an expected rate of change starting from the immediate state s_t.

21.4.4.1 Multivariate Case

For completion, we should provide the multivariate case for A.

Let X_t be a k-dimensional Ito diffusion given by the (vector) SDE:

$$\begin{bmatrix} dX_{t_1} \\ \vdots \\ dX_{t_1} \end{bmatrix} = \begin{bmatrix} a_{1,t} \\ \vdots \\ a_{k,t} \end{bmatrix} dt + \begin{bmatrix} \sigma_t^{11} & \vdots & \sigma_t^{1k} \\ \cdots & \cdots & \cdots \\ \sigma_t^{k1} & \vdots & \sigma_t^{kk} \end{bmatrix} \begin{bmatrix} dW_{1t} \\ \vdots \\ dW_{1t} \end{bmatrix} \tag{21.76}$$

where the a_{it} are the diffusion coefficients depending on X_t and the σ_t^{ij} are the diffusion coefficients possibly depending on X_t as well. This equation is written in the symbolic form:

$$dX_t = a_t\, dt + \sigma_t\, dW_t \tag{21.77}$$

where $a(\cdot)$ is a $k \times 1$ vector and the σ_t is a $k \times k$ matrix.

The corresponding A operator will then be given by

$$Af = \sum_{i=1}^{k} a_{it} \frac{\partial f}{\partial X_i} + \sum_{i=1}^{k} \sum_{j=1}^{k} \frac{1}{2} \left(\sigma \sigma^T\right)^{ij} \frac{\partial^2 f}{\partial X_i \partial X_j} \tag{21.78}$$

where the term $\left(\sigma\sigma^T\right)^{ij}$ represents the ijth element of the matrix $\left(\sigma\sigma^T\right)$.

The difference between the univariate case and this multivariate formula is the existence of cross-product terms. Otherwise, the extension is immediate.

In most advanced books on stochastic calculus, it is this multivariate form of A that is introduced. The expression in (21.78) is known as the *infinitesimal generator* of $f(\cdot)$.

21.4.5 Kolmogorov's Backward Equation

Suppose we are given the Ito diffusion S_t. Also, assume that we have a function of S_t denoted by $f(S_t)$. Consider the expectation:

$$\widehat{f}\left(S^-, t\right) = \mathbb{E}\left[f\left(S_t\right) \middle| S^-\right] \tag{21.79}$$

where $\widehat{f}\left(S^-, t\right)$ represents the forecasted value and S^- is the latest value observed before time t. Heuristically speaking, S^- is the immediate past. Then, using the A operator, we can characterize how the $\widehat{f}\left(S^-, t\right)$ may change over time. This evolution of the forecast is given by Kolmogorov's backward equation:

$$\frac{\partial \widehat{f}}{\partial t} = A\widehat{f} \tag{21.80}$$

Remembering the definition of A:

$$A\widehat{f} = a_t \frac{\partial \widehat{f}}{\partial S} + \frac{1}{2}\sigma^2 \frac{\partial^2 \widehat{f}}{\partial S^2} \tag{21.81}$$

It is easy to see that the equality in (21.81) is none other than the PDE:

$$\widehat{f_t} = a_t \widehat{f_s} + \frac{1}{2}\sigma_t^2 \widehat{f_{ss}} \tag{21.82}$$

Thus, we again see the important correspondence between conditional expectations such as

$$\widehat{f}\left(S^-, t\right) = \mathbb{E}\left[f\left(S_t\right) \middle| S^-\right] \tag{21.83}$$

and the PDE in (21.81). As before, this correspondence can be stated in two different ways:

- The satisfies the PDE in Eq. (21.81).
- Given the PDE in Eq. (21.81), we can find an $\widehat{f}$ such that the PDE is satisfied.

This result means that $\widehat{f}\left(S^-, t\right)$ is a solution for the PDE in (21.81). Hence, Kolmogorov's backward equation is an example of the correspondence between an expectation of a stochastic process and PDEs seen earlier in this chapter.

21.4.5.1 Example

Consider the function:

$$p\left(S_t, S_0, t\right) = \frac{1}{\sqrt{2\pi t}} e^{-\frac{(S_t - S_0)^2}{2t}} \tag{21.84}$$

An inspection shows that this is the conditional density function of a Wiener process that starts from S_0 at time $t = 0$ and moves over time with zero drift and variance t.

If we were to write down a stochastic differential equation for this process, we would choose the drift parameter as zero and the diffusion parameter as one. The dS_t would satisfy:

$$dS_t = dW_t \tag{21.85}$$

We apply Kolmogorov's formula to this density. We know that a twice-differentiable function of S_t would satisfy Kolmogorov's backward equation:

$$\widehat{f}_t = a_t \widehat{f}_s + \frac{1}{2}\sigma_t^2 \widehat{f}_{ss} \tag{21.86}$$

But according to (21.85), in this particular case, we have:

$$a_t = 0 \tag{21.87}$$

and

$$\sigma_t^2 = 1 \tag{21.88}$$

Substituting these, Kolmogorov's backward equation becomes:

$$\widehat{f}_t = \frac{1}{2}\widehat{f}_{ss} \tag{21.89}$$

It turns out that the conditional density $p(S_t, S_0, t)$ is one such function. To see this, take the first partial derivative with respect to t and the second partial with respect to S_t and substitute in (21.89). The equation will be satisfied.

According to this result, the conditional density function of a (generalized) Wiener process satisfies Kolmogorov's backward equation. This PDE tells us how the probability associated with a particular value of S_t will evolve as time passes, given the initial point S_0.

21.5 FEYNMAN–KAC FORMULA

The Feynman–Kac formula is an extension of Kolmogorov's backward equation as well as being a formalization of the issues discussed earlier in this chapter. The formula provides a probabilistic solution $\widehat{f}$ that corresponds to a given PDE.

Feynman–Kac Formula: Given

$$\widehat{f}(t, r_t) = \mathbb{E}^{\mathbb{P}}\left[e^{-\int_t^u q(r_s)ds} f(r_u)\right] \tag{21.90}$$

we have

$$\frac{\partial \widehat{f}}{\partial t} = A\widehat{f} - q(r_t)\widehat{f} \tag{21.91}$$

where the operator A is given by:

$$A\widehat{f} = a_t \frac{\partial \widehat{f}}{\partial r_t} + \frac{1}{2}\sigma_t^2 \frac{\partial^2 \widehat{f}}{\partial r_t^2} \tag{21.92}$$

Hence, the Feynman–Kac formula provides conditional expectations as a solution that corresponds to a certain class of PDEs.

21.6 CONCLUSIONS

The correspondence between PDEs and some conditional expectations is very useful in practical asset pricing. Given an instrument with special characteristics, a market practitioner can use this correspondence and derive the implied PDEs. These can then be numerically evaluated.

21.7 REFERENCES

Several interesting cases using this correspondence are found in Kushner (2000). This source also gives practical ways of calculating the implied PDEs.

21.8 EXERCISES

1. Suppose the bond price $B(t, T)$ satisfies the following PDE:

$$-r_t B + B_t + B_r\left(\mu - \lambda\sigma^B\right) + \frac{1}{2}B_{rr}\sigma^2 = 0 \tag{21.93}$$

$$B(T, T) = 1 \tag{21.94}$$

Define the variable $V(u)$ as

$$V(u) = e^{-\int_t^u r_s ds} e^{\int_t^u \left[\lambda(r_s,s) W_s - \frac{1}{2}\lambda(r_s,s)^2 ds\right]} \tag{21.95}$$

where λ_s is the market price of interest rate risk.

(a) Let $B(t,T)$ be the bond price. Calculate the $d(BV)$.

(b) Use the PDE in (21.93) to get an expression for $dB(t,T)$.

(c) Integrate this expression from t to T and take expectations with respect to martingale equality to obtain the bond pricing formula:

$$B(t,T) = \mathbb{E}_t^{\mathbb{P}}\left[e^{-\int_t^T r_s ds} e^{\int_t^T \left[\lambda(r_s,s) dW_s - \frac{1}{2}\lambda(r_s,s)^2 ds\right]}\right] \tag{21.96}$$

where the expectation is conditional on the current rt, which is assumed to be known.

2. Let r_t denote the short rate at time t and suppose we have an affine term structure model so that

$$Z_t^T = e^{A(t,T)+B(t,T)r_t} \tag{21.97}$$

where $A(t,T)$ and $B(t,T)$ are deterministic functions of time t and maturity T.

(a) Let $f(t;s,s+\delta)$ denote the forward rate at time t for lending between times s and $s+\delta$, where $t < s$ and $\delta > 0$. Compute an expression for $f(t;s,s+\delta)$ in terms of r_t and the functions $A(.,.)$ and $B(.,.)$.

(b) Let $r_s(s+\delta)$ denote the spot rate at time s for lending or borrowing out to time $s+\delta$. Give an expression for $r_s(s+\delta)$ in terms of r_s and the function $A(.,.)$ and $B(.,.)$.

3. Consider the following spot-rate dynamics:

$$dr_t = \alpha(\kappa - r_t)dt + \sigma r_t^{\beta} dW_t \tag{21.98}$$

Write a program that uses simulation to compute the conditional expectation of:

$$P = \mathbb{E}(e^{-\int_0^T r_s ds} \max(r_T - K, 0)) \tag{21.99}$$

Assume $K = 0.05, r_0 = 0.052, T = 1$, and $\Delta = 0.01$. Additionally, assume $\alpha = 0.9, \kappa = 0.07$, $\sigma = 0.2$, and $\beta = 0.75$.

CHAPTER

22

Pricing Derivatives via Fourier Transform Technique

OUTLINE

Derivatives pricing begins with the assumption that the evolution of the underlying asset (which could be a stock, commodity, an interest rate, or an exchange rate) follows some stochastic process. In order to price a European call/put option, utilizing the conditional density of the underlier we integrate the discounted payoff against the conditional density. Without loss of generality we can just focus on call options. We begin with the following definitions. Let:

S_0 be today's spot price of the underlying security,
S_T be T-time price of the underlying security,
$f(S_T) \equiv f(S_T|S_0)$ be the conditional probability density function of S_T under risk-neutral measure,
$q(s_T) \equiv q(s_T|s_0)$ be density of log of the underlying security $s_T = \ln(S_T)$,
$k = \ln(K)$ be the log of the strike price,
$C(K) \equiv \widehat{C}(K, S_0, T)$ be the price of a T-maturity call with strike K,
$\widehat{C}(k)$ be the price of a T-maturity call with log strike $k = \ln(K)$.

Assuming a risk-free interest rate of r, we have for the call option premium

$$C(K) = \int_0^{\infty} e^{-rT}(S_T - K)^+ f(S_T|S_0)dS_T \quad (22.1)$$

Financial Derivatives, Third Edition.
http://dx.doi.org/10.1016/B978-0-12-384682-2.00022-0

$$= \int_K^{\infty} e^{-rT}(S_T - K)f(S_T)dS_T \qquad (22.2)$$

By change of variable $s = \ln S$ and $k = \ln K$, we get

$$\widehat{C}(k) = \int_k^{\infty} e^{-rT}(e^{s_T} - e^k)q(s_T|s_0)ds_T \quad (22.3)$$

$$= e^{-rT}\int_k^{\infty}(e^{s_T} - e^k)q(s_T)ds_T \qquad (22.4)$$

For pricing, we need to evaluate this integral numerically. One can apply various numerical integration routines such as the trapezoidal rule or the Simpson rule. Knowing that $q(s_T|s_0)$ approaches zero as s_T approaches infinity, we can use an upper bound $b = \ln B$ and use n equidistant intervals to set up a uniform grid $x_j = k + j\Delta x$ for $j = 0, 1, \ldots, n$, where $\Delta x = \frac{b-k}{n}$

$$\widehat{C}(k) = e^{-rT}\int_k^{\infty}(e^s - e^k)q(s)ds \qquad (22.5)$$

$$\approx e^{-rT}\int_k^{b}(e^s - e^k)q(s)ds \qquad (22.6)$$

$$= e^{-rT}\sum_{i=0}^{n-1}\int_{x_j}^{x_{j+1}}(e^s - e^k)q(s)ds \qquad (22.7)$$

In the trapezoidal rule, we replace the function $(e^s - e^k)q(s)$ in each subinterval by a line passing through end points of that subinterval and integrate. This is exactly the same as approximating the area underneath of the integral $\int_{x_j}^{x_{j+1}}(e^s - e^k)q(s)ds$ by the area of the trapezoid that is

$$\int_{x_j}^{x_{j+1}}(e^s - e^k)q(s)ds \approx \frac{x_{j+1} - x_j}{2}$$

$$\left[(e^{x_{j+1}} - e^k)q(x_{j+1}) + (e^{x_j} - e^k)q(x_j)\right] \quad (22.8)$$

$$= \frac{\Delta x}{2}\left[(e^{x_{j+1}} - e^k)q(x_{j+1}) + (e^{x_j} - e^k)q(x_j)\right] \quad (22.9)$$

inserting it back and we obtain the following expression for the trapezoidal rule

$$\widehat{C}(k) \approx e^{-rT}\frac{\Delta x}{2}$$

$$\left[1(e^{x_0} - e^k)q(x_0) + 2(e^{x_1} - e^k)q(x_1) + \cdots \right. \quad (22.10)$$

$$\left. + 2(e^{x_{n-1}} - e^k)q(x_{n-1}) + 1(e^{x_n} - e^k)q(x_n)\right] \quad (22.11)$$

In the Simpson rule, we consider two adjacent subintervals[1] at a time and find a quadratic function that passes through those three points and replace the integrand by the quadratic function and integrate that gives us

$$\int_{x_j}^{x_{j+2}}(e^s - e^k)q(s)ds \qquad (22.12)$$

$$\approx \frac{x_{j+2} - x_j}{6}\left[(e^{x_j} - e^k)q(x_j) + 4(e^{x_{j+1}} - e^k)q(x_{j+1})\right.$$

$$\left. + (e^{x_{j+2}} - e^k)q(x_{j+2})\right] \qquad (22.13)$$

$$= \frac{\Delta x}{3}\left[(e^{x_j} - e^k)q(x_j) + 4(e^{x_{j+1}} - e^k)q(x_{j+1})\right.$$

$$\left. + (e^{x_{j+2}} - e^k)q(x_{j+2})\right] \qquad (22.14)$$

inserting it back and we obtain the following expression for the Simpson rule

$$\widehat{C}(k) \approx e^{-rT}\frac{\Delta x}{3}\left[1(e^{x_0} - e^k)q(x_0) + 4(e^{x_1} - e^k)q(x_1)\right.$$

$$+ 2(e^{x_2} - e^k)q(x_2) + \cdots \qquad (22.15)$$

$$+ 2(e^{x_{n-2}} - e^k)q(x_{n-2}) + 4(e^{x_{n-1}} - e^k)q(x_{n-1})$$

$$\left. + 1(e^{x_n} - e^k)q(x_n)\right] \qquad (22.16)$$

Both schemes of order $O(n)$. Note that this formulation is model-free.

To see how well these schemes work, we present an example of a model that a European call premium will be made available.

[1]This implies $\frac{n}{2}$ cases and naturally the assumption is that n is even.

Example 1. Geometric Brownian motion/ Black Scholes Model

Assuming a stock price process evolves according to geometric Brownian motion that is

$$dS_t = (r - q)S_t\,dt + \sigma S_t\,dW_t \tag{22.17}$$

where r is risk-free interest rate, q is continuous dividend rate, and σ is instantaneous volatility. As shown before, the solution to this stochastic differential equation via Itô Lemma is given by

$$S_t = S_0 e^{(r-q-\frac{\sigma^2}{2})t+\sigma W_t} \tag{22.18}$$

$$= S_0 e^{(r-q-\frac{\sigma^2}{2})t+\sigma\sqrt{t}Z} \tag{22.19}$$

$$= e^{\ln S0+(r-q-\frac{\sigma^2}{2})t+\sigma\sqrt{t}Z} \tag{22.20}$$

where Z is a standard normal random variable, which implies S_t follows a log-normal distribution with mean $\ln S_0 + (r - q - \frac{\sigma^2}{2})t$ and standard deviation $\sigma\sqrt{t}$ that is

$$f(S_T|S_0) = \frac{e^{-\frac{1}{2}\left(\frac{\ln S_T - \ln S_0 - (r-q-\sigma^2/2)T}{\sigma\sqrt{T}}\right)^2}}{\sigma S_T\sqrt{2\pi T}} \tag{22.21}$$

Therefore when pricing a call we have

$$C(K) = \int_K^\infty e^{-rT}(S_T - K)f(S_T|S_0)dS_T \tag{22.22}$$

$$= \int_K^\infty e^{-rT}(S_T - K) \times \frac{e^{-\frac{1}{2}\left(\frac{\ln S_T - \ln S_0 - (r-q-\sigma^2/2)T}{\sigma\sqrt{T}}\right)^2}}{\sigma S_T\sqrt{2\pi T}} dS_T \tag{22.23}$$

$$= \int_k^\infty e^{-rT}(e^{s_T} - e^k) \times \frac{e^{-\frac{1}{2}\left(\frac{s_T - s_0 - (r-q-\sigma^2/2)T}{\sigma\sqrt{T}}\right)^2}}{\sigma\sqrt{2\pi T}} ds_T \tag{22.24}$$

$$= e^{-rT}\int_k^\infty (e^{s_T} - e^k)q(s_T|s_0)ds_T \tag{22.25}$$

$$= \widehat{C}(k).$$

In (22.24) we use change of variable $s_T = \ln S_T$ and $k = \ln K$.

Example 2. Numerical Pricing of a Black–Scholes European call

For the parameter set spot price $S_0 = 100$, strike price $K = 100$, maturity $T = 1$, volatility $\sigma = 30\%$, risk-free rate of $r = 2\%$, and dividend rate of $q = 1.25\%$, Black–Scholes premiums for this call option are 12.1040. Following Trapezoidal and Simpson schemes described earlier, we get the following numerical results tabulated in Table 22.1. As expected, results are quite sensitive to the level of $b = \ln B$. For large values of b, both schemes match very closely to exact solution almost independent of n.

There is a major issue with this approach. For many stochastic processes a closed form for condition probability distribution function is either not available or, if it is, could be relatively expensive to calculate. However, it happens that for many stochastic processes, their characteristic functions are known in closed form, even though their probability distribution function is not available.

First, we define the characteristic function of a random process and give an example of finding the characteristic function for geometric Brownian motion. We then show how one can price a European call option, having its characteristic function.

Definition 25. Fourier transform

The Fourier transform of a function $f(x)$ is defined as

$$\phi(\nu) = \int_{-\infty}^{\infty} e^{i\nu x}f(x)dx \tag{22.26}$$

Definition 26. Inverse Fourier transform

Knowing the Fourier transform of a function, $\phi(\nu)$, the function $f(x)$ can be recovered via the inverse Fourier transform as

$$f(x) = \frac{1}{2\pi}\int_{-\infty}^{\infty} e^{-i\nu x}\phi(\nu)d\nu \tag{22.27}$$

TABLE 22.1 Black–Scholes Premiums via Trapezoidal and Simpson Rules for Various Values of B and n (Black–Scholes Premium 12.1040)

		Numerical Integration Routine	
B	**N**	**Trapezoidal Rule**	**Simpson Rule**
200	2^4	11.1868	11.0992
	2^5	11.2100	11.1637
	2^6	11.2154	11.1917
	2^7	11.2167	11.2047
	2^8	11.2170	11.2110
	2^{10}	11.2171	11.2156
400	2^4	12.0111	12.1053
	2^5	12.0821	12.1037
	2^6	12.0984	12.1037
	2^7	12.1024	12.1037
	2^8	12.1033	12.1037
	2^{10}	12.1037	12.1037
800	2^4	11.1868	12.1141
	2^5	12.0554	12.1045
	2^6	12.0922	12.1040
	2^7	12.1011	12.1040
	2^8	12.1033	12.1040
	2^{10}	12.1040	12.1040

Definition 27. Characteristic function

If $f(x)$ is the probability density function of a random variable x that is $f(x) \geq 0$ and $\int f(x) = 1$, it's Fourier transform is called the *characteristic function*

$$\phi(v) = \int_{-\infty}^{\infty} e^{ivx} f(x)dx \tag{22.28}$$

$$= \mathbb{E}(e^{ivx}) \tag{22.29}$$

And obviously the probability density function $f(x)$ can be found from its characteristic function via the inverse Fourier transform. Note that characteristic function evaluated at $-iv$ is equivalent to moment-generating function. As in the case of moment-generating function, characteristic function can be used to find moments of a random variable.

Example 3. Characteristic function of the standard normal distribution

Most commonly used distribution in the mathematics of finance is the standard normal distribution. It is the main component of a diffusion process and is absolutely central to most of the models discussed here. If $Z \sim \mathcal{N}(0,1)$, then its characteristic function is calculated as

$$\phi_Z(v) = \mathbb{E}(e^{ivZ}) = \int_{-\infty}^{\infty} \frac{1}{\sqrt{2\pi}} \exp\left(ivz - \frac{1}{2}z^2\right) dz \tag{22.30}$$

We first consider the following integral

$$\mathbb{E}(e^{sZ}) = \int_{-\infty}^{\infty} \frac{1}{\sqrt{2\pi}} \exp\left(sz - \frac{1}{2}z^2\right) dz \tag{22.31}$$

Complete the square in the integrand and we get

$$\int_{-\infty}^{\infty} \frac{1}{\sqrt{2\pi}} \exp\left(sz - \frac{1}{2}z^2\right) dz$$

$$= \int_{-\infty}^{\infty} \frac{1}{\sqrt{2\pi}} \exp\left(-\frac{1}{2}(z^2 - 2sz)\right) dz \tag{22.32}$$

$$= \int_{-\infty}^{\infty} \frac{1}{\sqrt{2\pi}} \exp\left(-\frac{1}{2}(z-s)^2 + \frac{1}{2}s^2\right) dz \tag{22.33}$$

$$= \frac{1}{\sqrt{2\pi}} e^{\frac{1}{2}s^2} \int_{-\infty}^{\infty} \exp\left(-\frac{1}{2}(z-s)^2\right) dz \tag{22.34}$$

utilizing

$$\int_{-\infty}^{\infty} e^{-\frac{1}{2}u^2 du} = \sqrt{2\pi} \tag{22.35}$$

we get

$$\mathbb{E}(e^{sZ}) = \exp\left(\frac{s^2}{2}\right) \tag{22.36}$$

Following the argument in Grimmett and Stirzaker (1992), we can substitute iv for s to get

$$\phi_Z(v) = \mathbb{E}(e^{ivZ}) = e^{-\frac{v^2}{2}} \tag{22.37}$$

This is possible, by the theory of analytic continuation of functions of a complex variable.

Example 4. The Characteristic Function of a Normal Distribution

A normal random variable with a mean of μ and a standard deviation of σ can be constructed from a standard normal variable using $X = \mu + \sigma Z$, so that $X \sim \mathcal{N}(\mu, \sigma^2)$. Thus, its characteristic function can be calculated as

$$\phi_X(\nu) = \mathbb{E}(e^{i\nu X}) \tag{22.38}$$
$$= \mathbb{E}(e^{i\nu(\mu+\sigma Z)}) \tag{22.39}$$
$$= e^{i\nu\mu}\mathbb{E}(e^{i\nu\sigma Z}) \tag{22.40}$$
$$= e^{i\nu\mu}e^{-\frac{(\sigma\nu)^2}{2}} \tag{22.41}$$
$$= e^{i\mu\nu-\frac{\sigma^2\nu^2}{2}} \tag{22.42}$$

Brownian motion W_t is a key component in many models of asset prices. We know that

$$W_t - W_0 = W_t \sim \mathcal{N}(0, t) \tag{22.43}$$

Therefore if $X_t = W_t$ its characteristic function is

$$\mathbb{E}(e^{i\nu X_t}) = \mathbb{E}(e^{i\nu W_t}) = \mathbb{E}(e^{i\nu\sqrt{t}Z}) = e^{-\frac{\nu^2 t}{2}} \tag{22.44}$$

Example 5. Characteristic function of logarithmic of geometric Brownian motion

One of most commonly used models of asset prices in mathematics of finance is geometric Brownian motion. The model still is one of the pivotal moments in quantitative finance and is the standard by which most modern derivative pricing models are judged. Its invention helped to create an enormous and liquid market in options. This example will give a brief description of the derivation of the model, its stochastic differential equation, and its characteristic function.

When modeling asset prices, the underlying process assumes that the asset price S_t, at time t, will satisfy the following stochastic differential equation, known as the Black–Scholes stochastic differential equation:

$$dS_t = (r - q)S_t\,dt + \sigma S_t\,dW_t \tag{22.45}$$

This equation models the asset's log returns as growing at a constant rate of $r - q$ and having a volatility of σ.

By means of Itô's lemma, the solution to the stochastic differential equation is given by

$$S_T = S_0 e^{(r-q-\frac{\sigma^2}{2})T+\sigma W_T} \tag{22.46}$$
$$= S_0 e^{(r-q-\frac{\sigma^2}{2})T+\sigma\sqrt{T}Z} \tag{22.47}$$

Or equivalently

$$s_T = \ln S_T = s_0 + \left(r - q - \frac{\sigma^2}{2}\right)T + \sigma\sqrt{T}Z \tag{22.48}$$

where $s_0 = \ln S_0$, therefore

$$s_T \sim \mathcal{N}\left(s_0 + \left(r - q - \frac{\sigma^2}{2}\right)T, \sigma^2 T\right) \tag{22.49}$$

Using the characteristic function of a normal random variable, we can easily derive the characteristic function of the log of the asset price as

$$\phi_{s_T}(\nu) = e^{i(s_0+(r-q-\frac{\sigma^2}{2})T)\nu-\frac{\sigma^2\nu^2}{2}T} \tag{22.50}$$

22.1 DERIVATIVES PRICING VIA THE FOURIER TRANSFORM

In this chapter, we will discuss the use of *transform techniques* for pricing derivatives. The characteristic function of the distribution of asset prices is merely the Fourier transform of its probability distribution function. Thus its probability distribution function can be recovered from the characteristic function through Fourier inversion. This is particularly important for many classes of models which have a closed form only in their characteristic function representation. We will outline techniques for pricing derivatives under a variety of different models using transform methods, focusing on Fast Fourier Transform (FFT)-based techniques. The first major development in the pricing of derivatives using Fourier techniques was proposed by Carr and Madan (1999). While this method was a considerable breakthrough in numerical options pricing, like most methods, the Fast Fourier Transform (FFT) pricing method involves a number of trade-offs. This method is very useful as it allows us to efficiently price derivatives under any model with a known characteristic function, some of which are only expressible in this form. Also, this method is very fast when using FFT-based Fourier inversion, solving derivatives pricing problems in $O(n\ \ln(n))$ time. Furthermore, this method also allows us to compute not just the desired option price in $O(n\ \ln(n))$ time, but also the price for options at n different strikes. While there are some restrictions on which option prices are computed for *free*, we are able to extract more information from this method than many others, which is important for calibration.

This method cannot be used to price all of the derivatives. In particular, the method as originally presented is restricted to the pricing of derivatives with European payoffs which are completely path independent. Furthermore, the derivation of this method is very dependent on the payoff type, with only two payoffs presented in the original paper. Thus we are restricted to a small, but important, subset of derivative payoffs. In order to make this method work we need to define a damping factor α, whose optimal value must be determined. Finally, this method degrades in accuracy when the option to be priced becomes very far out-of-the-money options.

22.1.1 Call Option Pricing via the Fourier Transform

Let $\phi(\nu)$ be the characteristic function of the log of the underlying security s_T that is

$$\phi(\nu) = \int_{-\infty}^{\infty} e^{i\nu s_T} q(s_T) ds_T \tag{22.51}$$

The European call option price $C(K)$ can be expressed as:

$$C(K) = e^{-rT}\mathbb{E}_t\left[(S_T - K)^+\right] \tag{22.52}$$

$$= e^{-rT}\int_K^{\infty} (S_T - K) f(S_T|S_0) dS_T \tag{22.53}$$

$$= e^{-rT}\int_k^{\infty} (e^{s_T} - e^k) q(s_T) ds_T \tag{22.54}$$

$$= e^{-rT}\int_k^{\infty} (e^s - e^k) q(s) ds \tag{22.55}$$

$$= \widehat{C}(k) \tag{22.56}$$

Note that, for simplicity, we drop the subscript T in the last integral equation. Now that we have expressed the option price $\widehat{C}(k)$ in terms of the log price density, we can use this representation to calculate Fourier transform of $\widehat{C}(k)$ to link it to the characteristic function. We define it as $\psi(\nu)$, which is

$$\psi(\nu) = \int_{-\infty}^{\infty} e^{i\nu k}\widehat{C}(k) dk \tag{22.57}$$

$$= \int_{-\infty}^{\infty} e^{i\nu k}\left(e^{-rT}\int_k^{\infty} (e^s - e^k) q(s) ds\right) dk \tag{22.58}$$

$$= e^{-rT}\int_{-\infty}^{\infty} e^{i\nu k}\left(\int_k^{\infty} (e^s - e^k) q(s) ds\right) dk \tag{22.59}$$

In order to integrate we need to change the order of the integral. To do that, we utilize Fubini's theorem, which allows the order of integration to be changed in iterated integrals.

$$\psi(\nu) = e^{-rT}\int_{-\infty}^{\infty}\int_{-\infty}^{s} e^{i\nu k}(e^s - e^k) q(s) dk\, ds \tag{22.60}$$

$$= e^{-rT}\int_{-\infty}^{\infty} q(s)\left(\int_{-\infty}^{s} e^{i\nu k}(e^s - e^k) dk\right) ds \tag{22.61}$$

Now we can evaluate the inner integral

$$\int_{-\infty}^{s} e^{i\nu k}(e^s - e^k) dk$$

$$= \int_{-\infty}^{s} e^{i\nu k} e^s dk - \int_{-\infty}^{s} e^{i\nu k} e^k dk$$

$$= e^s \frac{e^{i\nu k}}{i\nu}\Bigg|_{-\infty}^{s} - \frac{e^{(i\nu+1)k}}{i\nu + 1}\Bigg|_{-\infty}^{s} \tag{22.62}$$

We can now see that the first integral does not converge and the first term is undefined. To make it work, we should reformulate the problem Carr and Madan (1999) by defining

$$c(k) = e^{\alpha k}\widehat{C}(k) \tag{22.63}$$

the option premium multiplied by an exponential of the strike. This term becomes a damping component in the inner integral, which forces convergence and allows the Fourier transform to be calculable. We redefine $\psi(\nu)$ to be the characteristic function of the modified option price $c(k)$ and the derivation now becomes

$$\begin{aligned}\psi(\nu) &= \int_{-\infty}^{\infty} e^{i\nu k}c(k)dk \\ &= \int_{-\infty}^{\infty} e^{i\nu k}\left(e^{-rT}e^{\alpha k}\int_{k}^{\infty}(e^s - e^k)q(s)ds\right)dk \\ &= e^{-rT}\int_{-\infty}^{\infty}\int_{-\infty}^{s} e^{(\alpha+i\nu)k}(e^s - e^k)q(s)dk\,ds \\ &= e^{-rT}\int_{-\infty}^{\infty} q(s)\left(\int_{-\infty}^{s} e^{(\alpha+i\nu)k}(e^s - e^k)dk\right)ds\end{aligned} \tag{22.64}$$

Now having the damping factor, the inner integral converges

$$\begin{aligned}&\int_{-\infty}^{s} e^{\alpha k}e^{i\nu k}(e^s - e^k)dk \\ &\quad = \int_{-\infty}^{s} e^{(\alpha+i\nu)k}(e^s - e^k)dk \qquad (22.65) \\ &\quad = e^s\frac{e^{(\alpha+i\nu)k}}{(\alpha+i\nu)}\Bigg|_{-\infty}^{s} - \frac{e^{(\alpha+i\nu+1)k}}{(\alpha+i\nu+1)}\Bigg|_{-\infty}^{s} \qquad (22.66)\end{aligned}$$

Both terms now vanish at negative infinity for a positive α and so we have

$$\begin{aligned}&\int_{-\infty}^{s} e^{(\alpha+i\nu)k}(e^s - e^k)dk \\ &\quad = e^s\frac{e^{(\alpha+i\nu)s}}{(\alpha+i\nu)} - \frac{e^{(\alpha+i\nu+1)s}}{(\alpha+i\nu+1)} \\ &\quad = \frac{e^{(\alpha+i\nu+1)s}}{(\alpha+i\nu)(\alpha+i\nu+1)} \qquad (22.67)\end{aligned}$$

Now by substituting (22.67) into (22.64) we get the following quantity for the characteristic function of the modified option premium

$$\psi(\nu) = e^{-rT}\int_{-\infty}^{\infty} q(s)\frac{e^{(\alpha+i\nu+1)s}}{(\alpha+i\nu)(\alpha+i\nu+1)}ds \tag{22.68}$$

Factoring out those terms that are independent of s, we get

$$\psi(\nu) = \frac{e^{-rT}}{(\alpha+i\nu)(\alpha+i\nu+1)}\int_{-\infty}^{\infty} e^{(\alpha+i\nu+1)s}q(s)ds \tag{22.69}$$

Factoring out i, we can see that the integral is the characteristic of the log asset price evaluated at $(\nu - (\alpha+1)i)$ that is

$$\begin{aligned}\psi(\nu) &= \frac{e^{-rT}}{(\alpha+i\nu)(\alpha+i\nu+1)}\int_{-\infty}^{\infty} e^{i(\nu-(\alpha+1)i)s}q(s)ds \qquad (22.70) \\ &= \frac{e^{-rT}}{(\alpha+i\nu)(\alpha+i\nu+1)}\phi(\nu-(\alpha+1)i) \qquad (22.71)\end{aligned}$$

Now having the characteristic function of the log of an underlying security price $\phi(\nu)$, we can now calculate $\psi(\nu)$. The Fourier transform of the modified call,

$$\begin{aligned}\psi(\nu) &= \int_{-\infty}^{\infty} e^{i\nu k}c_T(k)dk \qquad (22.72) \\ &= \frac{e^{-rT}}{(\alpha+i\nu)(\alpha+i\nu+1)}\phi(\nu-(\alpha+1)i) \qquad (22.73)\end{aligned}$$

Since we now have the characteristic function of the modified call price

$$\begin{aligned}\psi(\nu) &= \int_{-\infty}^{\infty} e^{i\nu k}c_T(k)dk \qquad (22.74) \\ &= \int_{-\infty}^{\infty} e^{i\nu k}e^{\alpha k}C_T(k)dk \qquad (22.75)\end{aligned}$$

we can utilize the inverse Fourier transform to get

$$\widehat{C}(k) = \frac{e^{-\alpha k}}{2\pi} \int_{-\infty}^{\infty} e^{-i\nu k} \psi(\nu) d\nu \tag{22.76}$$

However, $\widehat{C}(k)$ is a real number, which implies that its Fourier transform, $\psi(\nu)$, is even in its real part and odd in its imaginary part. Therefore we are only concerned with the real part for the option price. So we can treat this as an even function and we get

$$\widehat{C}(k) = \frac{e^{-\alpha k}}{\pi} \int_{0}^{\infty} e^{-i\nu k} \psi(\nu) d\nu \tag{22.77}$$

the call option premium. Using equation (22.73) for the characteristic function and the inverse Fourier transform of $\psi(\nu)$, we can calculate $\widehat{C}(k)$.

$$\widehat{C}(k) = \frac{e^{-\alpha k}}{\pi} \int_{0}^{\infty} e^{-i\nu k} \psi(\nu) d\nu \tag{22.78}$$

where $\psi(\nu)$ is a known function that will be determined and some suitable parameter $\alpha > 0$.

22.1.2 Evaluating the Pricing Integral

The Fourier techniques presented so far give us a method for calculating option prices for models where a closed-form probability distribution function is not available, except when a closed-form characteristic function is presented. However, we need to perform the integral to solve for the option premium. It remains to be seen why we would use this method, as we still need to calculate the integral. Note that

$$\widehat{C}(k) = \frac{e^{-\alpha k}}{\pi} \int_{0}^{\infty} e^{-i\nu k} \psi(\nu) d\nu \tag{22.79}$$

This integral can be computed easily using simple numerical integration techniques. First, we approximate the integral by defining B as the upper bound for the integration. We can numerically integrate this truncated integral via a simple trapezoidal rule. We let n be the number of equidistant intervals, $\Delta\nu = \frac{b}{n} = \eta$ be the distance between the integration points, and $\nu_j = (j-1)\eta$ be the endpoints for the integration intervals for $j = 1, \ldots, n+1$. Applying the Trapezoidal rule, we get

$$\begin{aligned}
\widehat{C}(k) &= \frac{e^{-\alpha k}}{\pi} \int_{0}^{\infty} e^{-i\nu k} \psi(\nu) d\nu && (22.80)\\
&\approx \frac{e^{-\alpha k}}{\pi} \int_{0}^{B} e^{-i\nu k} \psi(\nu) d\nu && (22.81)\\
&\approx \frac{e^{-\alpha k}}{\pi} \Big(e^{-i\nu_1 k} \psi(\nu_1) + 2e^{-i\nu_2 k} \psi(\nu_2) \\
&\quad + \cdots + 2e^{-i\nu_N k} \psi(\nu_N) && (22.82)\\
&\quad + e^{-i\nu_{N+1} k} \psi(\nu_{N+1}) \Big) \frac{\eta}{2} && (22.83)
\end{aligned}$$

Since the terms are decaying exponentially, we can just discard the final term and we end up with

$$C(k) \approx \frac{e^{-\alpha k}}{\pi} \sum_{j=1}^{n} e^{-i\nu_j k} \psi(\nu_j) w_j \tag{22.84}$$

where $w_j = \frac{\eta}{2}(2 - \delta_{j-1})$. For a more accurate result we could also use the Simpson rule, which would yield

$$\widehat{C}(k) \approx \frac{e^{-\alpha k}}{\pi} \sum_{j=1}^{n} e^{-i\nu_j k} \psi(\nu_j) w_j \tag{22.85}$$

where $w_j = \frac{\eta}{3}(3 + (-1)^j - \delta_{j-n})$. Here

$$\delta_j = \begin{cases} 1 & j = 0 \\ 0 & \text{otherwise} \end{cases}$$

22.1.2.1 *Fast Fourier Transform*

While the direct integration is sufficient, it is not an efficient method of evaluating the pricing integral. The Fast Fourier transform (FFT) algorithm developed by Cooley and Tukey (1965) and later extended by many others provides a more efficient algorithm for calculating a set of discrete inverse Fourier transforms with sample points that are powers of two. These transforms take the following form

$$\omega(m) = \sum_{j=1}^{n} e^{-i\frac{2\pi}{n}(j-1)(m-1)} x(j) \text{ for } m = 1, \ldots, n \tag{22.86}$$

These equations would appear to take n multiplications per inverse transform for a total of n^2 multiplications, however the Cooley–Tukey FFT algorithm can reduce this to $n\log(n)$ multiplications by using a divide and conquer algorithm to break down Discrete Fourier Transforms (DFTs). This is crucial for approximating the Fourier integral, as this can greatly accelerate the speed at which we can compute option prices under the FFT method.

We can convert our option pricing formula into the FFT form by creating a range of strikes around the strike for which we wish to calculate an accurate option price. A typical case would be an at-the-money option for a particular underlier, and in this case we define the range of logarithmic strikes as

$$k_m = \beta + (m-1)\Delta k = \beta + (m-1)\lambda, \text{ for } m = 1, \ldots, n \tag{22.87}$$

with $\beta = \ln S_0 - \frac{\lambda n}{2}$, which will cause at-the-money strike to fall in the middle of our range of strikes. For $\widehat{C}(k_m)$, we now have

$$\widehat{C}(k_m) \approx \frac{e^{-\alpha k_m}}{\pi} \sum_{j=1}^{n} e^{-i\nu_j k_m} \psi(\nu_j) w_j \text{ for } m = 1, \ldots, n \tag{22.88}$$

$$= \frac{e^{-\alpha k_m}}{\pi} \sum_{j=1}^{n} e^{-i\nu_j(\beta+(m-1)\Delta k)} \psi(\nu_j) w_j \tag{22.89}$$

$$= \frac{e^{-\alpha k_m}}{\pi} \sum_{j=1}^{n} e^{-i(j-1)\eta(m-1)\lambda} e^{-i\beta\nu_j} \psi(\nu_j) w_j \tag{22.90}$$

$$= \frac{e^{-\alpha k_m}}{\pi} \sum_{j=1}^{n} e^{-i\lambda\eta(j-1)(m-1)} e^{-i\beta\nu_j} \psi(\nu_j) w_j \tag{22.91}$$

So, we can see by setting

$$\lambda\eta = \frac{2\pi}{n} \tag{22.92}$$

and

$$x(j) = e^{-i\beta\nu_j} \psi(\nu_j) w_j \tag{22.93}$$

we get back the original form of FFT (22.86). Now we can generate n option prices using only $O(n \log n)$ multiplications required by the FFT. This entire operation is slower than the $O(n)$ multiplications needed to get a simple option price using direct integration, however it is rare to only price a single option on an underlier by itself. Utilizing the FFT method gives us a clear advantage. The $O(n \log n)$ multiplications, when amortized over n options, are only $O(\log n)$ per option.

However, these n options are not likely to be exactly the n options needed for producing a sensible implied volatility surface with points at market traded strikes because the FFT method prices strikes are determined by

$$k_m = \beta + (m-1)\Delta k = \beta + (m-1)\lambda \tag{22.94}$$

We can modify $\eta = \frac{b}{n}$ by modifying n, which will in turn change the strikes for which you get option prices using the FFT method. This allows us to extract enough information to interpolate a volatility surface with very small errors in considerably less time than direct integration, which would take $O(n^2)$ for n strikes.

22.1.3 Implementation of Fast Fourier Transform

In brief, having characteristic function of the log of the underlying process X_t, is $\phi(\nu)$. Here we choose η and $n = 2^l$, and calculate $\lambda = \frac{2\pi}{n\eta}$, $\nu_j = (j-1)\eta$ and set α. Now form vector $\mathbf{x}$

$$\mathbf{x} = \begin{pmatrix} x_1 \\ x_2 \\ \vdots \\ x_n \end{pmatrix}$$

$$= \begin{pmatrix} \frac{\eta}{2}\frac{e^{-rT}}{(\alpha+i\nu_1)(\alpha+i\nu_1+1)}e^{-i(\ln S_0-\frac{\lambda n}{2})\nu_1}\phi\left(\nu_1-(\alpha+1)i\right) \\ \eta\frac{e^{-rT}}{(\alpha+i\nu_2)(\alpha+i\nu_2+1)}e^{-i(\ln S_0-\frac{\lambda n}{2})\nu_2}\phi\left(\nu_2-(\alpha+1)i\right) \\ \vdots \\ \eta\frac{e^{-rT}}{(\alpha+i\nu_n)(\alpha+i\nu_n+1)}e^{-i(\ln S_0-\frac{\lambda n}{2})\nu_n}\phi\left(\nu_n-(\alpha+1)i\right) \end{pmatrix} \tag{22.95}$$

Vector **x** is the input to the FFT routine, and its output is vector **y** of the same size (for example in MATLAB we simply achieve this by typing **y** = fft(**x**)), then call prices at strike k_m for $m = 1, \ldots, n$ are

$$\begin{pmatrix} \widehat{C}(k_1) \\ \widehat{C}(k_2) \\ \vdots \\ \widehat{C}(k_n) \end{pmatrix} = \begin{pmatrix} \frac{e^{-\alpha(\ln S_0-\frac{n}{2}\lambda)}}{\pi}\mathrm{Re}(y_1) \\ \frac{e^{-\alpha(\ln S_0-(\frac{n}{2}-1)\lambda)}}{\pi}\mathrm{Re}(y_2) \\ \vdots \\ \frac{e^{-\alpha(\ln S_0-(\frac{n}{2}-(n-1))\lambda)}}{\pi}\mathrm{Re}(y_n) \end{pmatrix} \tag{22.96}$$

where $\mathrm{Re}(y_j)$ is the real part of y_j.

22.2 FINDINGS AND OBSERVATIONS

Use of the damping factor α made it possible to price vanilla options via a Fourier transform. At a glance it seems that α does not come into the calculation of the integrand, as it is hidden in $\psi(\nu)$. We already know that α has to be positive for calls and it seems that for any positive value of α we should get roughly the same results. However, this is not the case. One can demonstrate how sensitive the outcome is to the choice of α and one can find a suitable range for its value and illustrate its dependence on the choice of stochastic model.

Also we see that the following relationship between λ and η

$$\lambda\eta = \frac{2\pi}{n} \tag{22.97}$$

with

$$\eta = \frac{b}{n} \tag{22.98}$$

which implies of the four parameters in consideration, n, b, η, and λ. Only two can be chosen independently, as η is determined by b and the last one is determined via the constraint (22.97). If we assume a fixed computational budget, we would dictate a fixed number of integral terms n. Considering these assumptions, we have only two free variables, b, the upper bound of the integral, and $\lambda = \Delta k$, the spacing of the $\log(K)$ grid on which we calculate solutions, and they are inversely proportional. So we have an inherent trade-off between the upper bound of the integral, which determines the accuracy of our integration, and the step size in strikes, which will determine if we get relevant pricing information at strikes which are close to traded market strikes. The choice of b will determine how accurate our integral approximation will be; however, if we assume that we want a fixed spacing between integration points η to ensure a given degree of accuracy in the integration, we have to impose a restriction on the λ, which will determine the spacing in the log strikes of the solutions we calculate.

22.3 CONCLUSIONS

For a stochastic process with a known probability density function, we can integrate the payoff via some numerical integration procedure and get its option price.

In most cases, we do not know probability density function analytically or in an integrated form. However, we can often find characteristic function of an underlying security price or rather the characteristic function of the log of the underlying security price analytically or semi-analytically. It is shown in Carr and Madan (1999) that if we have the characteristic function analytically, we then can efficiently obtain option premiums via the inverse Fourier transform. Following the work in Carr and Madan (1999), we begin by formulating the option pricing problem for a European call in terms of the density of the log asset price, which will allow us to use Fourier transforms to obtain the option premium.

Many derivative instruments, including vanilla options, caps, floors, and swaptions, can be expressed as a simple call or put option. For that reason, this setup can be used to price those instruments.

The FFT approach can be used for any model for which a characteristic function for the asset price distribution exists. It can be applied for strictly path independent European options and for a restricted set of terminal payoffs. The method is fast and for n option prices is $O(n\log(n))$ time and generates n option prices in a single run. The problem with the scheme is that it needs to be re-derived for any change in the payoff structure. It requires estimation of proper α and can be exceedingly inaccurate for highly out of the money options.

22.4 PROBLEMS

1. Show in order to price a European put via FFT we just need to choose $\alpha < -1$.
2. In FFT, we define the range for log of strikes as $k_m = \beta + (m-1)\Delta k = \beta + (m-1)\lambda$, for $m = 1, \ldots, N$ and some β. There are many choices for β. One of which is to set $\beta = \ln S_0 - \frac{\lambda N}{2}$. This choice for β would cause at-the-money strike to fall in the middle of our range of strikes where S_0 is today spot. For this choice of β, if we are interested in finding the premium for $k = \log(K)$ we would typically interpolate. (a) If we are interested in finding the premium for a specific strike, say $k_0 = \log(K_0)$ without any interpolation, what β would we choose? Find the index for this β. (b) What β would choose such that the first entry coincides to the premium for k_0?
3. As shown, the characteristic function of the log of stock price in the Black–Scholes framework is:

$$\phi(u) = \mathbb{E}(e^{iu\ln S_T}) \tag{22.99}$$

$$= \mathbb{E}(e^{ius_T}) \tag{22.100}$$

$$= \exp\left(i(s_0 + (r - q - \sigma^2/2)T)u - \frac{1}{2}\sigma^2 u^2 T\right) \tag{22.101}$$

Use the FFT method to price a European call option for the following set of parameters: spot price $S_0 = \$100$, strike price $K = 100$, risk-free interest rate $r = 1.25\%$, dividend rate $q = 1.0\%$, time to maturity $T = 2$ years, and volatility $\sigma = 20\%$, and compare it with the close-form call price. Use various values for B, N, and α. Compare the results with analytical solution and conclude on sensitivity to B, N, and α.

4. In the Heston stochastic volatility model, the stock price follows the following stochastic differential equation:

$$\begin{aligned} dS_t &= rS_t\,dt + \sqrt{v_t}S_t\,dW_t^{(1)} \\ dv_t &= \kappa(\theta - v_t)dt + \sigma\sqrt{v_t}\,dW_t^{(2)} \end{aligned} \tag{22.102}$$

where the two Brownian components, $W_t^{(1)}$ and $W_t^{(2)}$, might be correlated with correlation ρ. The variable v_t represents the mean reverting stochastic volatility, where θ is the long-term variance, κ is the speed of the mean reversion, and σ is the volatility of the variance. The presence of the $\sqrt{v_t}$ term in the diffusion component of this equation prevents the volatility from becoming negative by forcing the diffusion component to zero as the volatility approaches zero. Its characteristics function for the log of stock price process under Heston model Hirsa (2012) is given by

$$\phi(u) = \mathbb{E}(e^{iu\ln S_t}) \tag{22.103}$$

$$= \frac{\exp\{\frac{\kappa\theta t(\kappa - i\rho\sigma u)}{\sigma^2} + iutr + iu\ln S_0\}}{(\cosh\frac{\gamma t}{2} + \frac{\kappa - i\rho\sigma u}{\gamma}\sinh\frac{\gamma t}{2})^{\frac{2\kappa\theta}{\sigma^2}}} \times \exp\left\{\frac{-(u^2 + iu)v_0}{\gamma\coth\frac{\gamma t}{2} + \kappa - i\rho\sigma u}\right\} \tag{22.104}$$

where $\gamma = \sqrt{\sigma^2(u^2+iu)+(\kappa - i\rho\sigma u)^2}$. Use the FFT method to price a European put using the following parameters: spot price $S_0 = \$100$, strike price $K = 100$, maturity $T = 2$ years, risk-free rate $r = 1.25\%$, volatility $\sigma = 20\%$, $\kappa = 1$, $\theta = 0.025$, $\rho = -0.7$, $\nu_0 = 0.05$. Use various values for B, N, and α.

CHAPTER

23

Credit Spread and Credit Derivatives

OUTLINE

This chapter covers credit spread and standard credit contracts and summarizes modeling and pricing of some credit derivatives contracts which include default correlation, defaultable bonds, credit default swaps, and basket default swaps.

23.1 STANDARD CONTRACTS

23.1.1 Credit Default Swaps

Credit default swaps (CDS) are the most basic credit derivatives instruments. A credit default

Financial Derivatives, Third Edition.
http://dx.doi.org/10.1016/B978-0-12-384682-2.00023-2

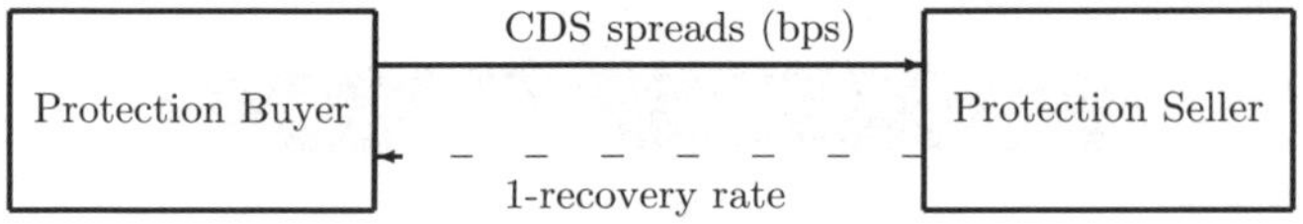

FIGURE 23.1 Credit default swap.

swap (CDS) is a kind of insurance against credit risk. It is a privately negotiated bilateral contract. The buyer of protection pays an initial upfront payment plus a fixed fee (or premium) to the seller of protection for a set period of time. If a certain prespecified *credit event* occurs, the protection seller pays compensation to the protection buyer. Figure 23.1 illustrates the payoff structure for a CDS contract: In this diagram, the counterparty X is buyer of (long) default protection which means he is short the reference credit. The counterparty Y is the seller of (short) default protection which means she is long the reference credit. X pays an up-front payments at the start of the deal and also some fee with some frequency conditional on the reference credit survival. In case of default, the counterparty Y pays $1 - R$ to counterpart X. R is the recovery rates for the reference credit. The recovery rate represents the cents per dollar of the face value of the bond that are recovered when default occurs. It is a number between zero and one. The reference bond is no guaranteed cashflows.

A *credit event* that triggers a payment from the protection seller can be a bankruptcy of a company, known as *reference entity*, or a default of a bond or other debt issued by the reference entity. If no credit event occurs during the term of the swap, the protection buyer continues to pay the premium until maturity. In contrast, should a credit event occur at some point before the contract's maturity, the protection seller owes a payment to the buyer of protection, thus insulating the buyer from a financial loss. Usually there is no exchange of money when two parties enter in the contract, but they make payments during the term of the contract, thus explaining the term credit default *swap*.

Another application of CDSs is a securitization product called synthetic collateralized debt obligations (CDO) as will be discussed later.

23.1.1.1 *What Does the Global CDS Market Look Like Today?*

The CDS market originally evolved from privately tailored agreements between banks and their customers. Perhaps because of its over-the-counter character, it is not clear exactly when the CDS market came into existence.

According to the Office of the Comptroller of the Currency (OCC), the total notional amount of interest rate and currency derivatives as of the end of the second quarter of 2013 stood at \$217.12 trillions, while the total notional amount of credit default swaps outstanding was \$11.824 trillion (as of September 27, 2013), or about 5% of the overall derivatives market. As recently as 2nd quarter 2013, credit derivatives (\$13.382 trillion) accounted for 5.72% of the derivatives market globally. As a further sign of growth, now the credit derivatives market has surpassed the size of the equity derivatives market, which stood at \$2.08 trillion at the end of 2nd quarter 2013 (source OCC).

The largest provider of CDSs are commercial banks. Traditionally, a bank's business has involved credit risk as it originates loans to corporations. The CDS market offers a bank an attractive way to transfer risk without removing assets from its balance sheet and without involving borrowers. Further, a bank may use CDSs to diversify its portfolio, which often is concentrated in certain industries or geographic areas. Banks are the net buyers of credit derivatives, and according to insurance companies are increasingly becoming dominant participants in the CDS market, primarily as sellers of protection, to enhance investment yields.

CDO total amount outstanding at the end of 2012 was $179.5 billion. As of October 1, 2013 that amount is $150.713 billion (Source SIFMA). The players are mostly insurance companies as well as financial guarantors and hedge funds.

There are no limits on the size or maturity of CDS contracts. However, most contracts fall between $10 million and $20 million in the notional amount. Maturity usually ranges from 1 to 10 years, with the 5-year maturity being the most common tenor. Mean size and median size of a single-name CDS are $18 million and $10 million respectively (Source NY Fed).

23.1.1.2 Why are Credit Default Swaps Attractive to Investors?

Credit default swaps can be a useful tool to hedge or manage one's exposure to credit risk. There are several important features that make CDS unique.

While the risk profile of a CDS is similar to a corporate bond of the reference entity, there are some important differences. A CDS does not require an initial funding, which allows leveraged positions. Moreover, a CDS transaction can be entered where a cash bond of the reference entity of a particular maturity is not available. Further, by entering a CDS as protection buyer, one can easily create a *short* position in the reference credit. With all these attributes, CDSs can be a great tool for diversifying or hedging one's portfolio.

While an investor can enter into a CDS transaction to get an exposure to single-name credit, trading based on indexes of CDSs has become more popular in recent years. The Dow Jones CDX Indices is the competing index family. The CDS indexes are linked to the 100–125 most liquid CDSs, equally weighted, and allow a quick diversification of CDS exposure.

23.1.1.3 What Exactly is a Credit Event?

Through its early stages of development, the credit default swap market has experienced many problems in the absence of widely accepted standardized documentation, since the terms and conditions of contracts were not precise enough, leaving many blind spots and technical loopholes. As credit events occurred, disputes often erupted between the buyers and the sellers over the specific terms and conditions of the CDS contract. The problem is that the protection buyer would want to interpret the scope of protection as widely as possible, while the seller would want to interpret it narrowly. This is understandable because a CDS is like an insurance policy, and the protection buyer, as the insured, would want to claim as much as possible for the insurance coverage, while the insurance company would always like to find the way to deny a claim and to pay as little as possible. The lack of the standardized documentation was so aggravating that it became an impediment to the growth of the CDS market. In 1999, a major breakthrough came when ISDA published its new Master Agreement designed for credit derivatives contracts, followed with a series of amendments to improve the documentation for credit derivatives. More recently, ISDA published its new 2003 ISDA credit derivatives definitions and 2002 Master Agreement, addressing the issues that had been raised earlier.

Market participants generally view the following three to be the most important triggers to a credit event

- Bankruptcy—The clearest concept of all, is the reference entity's insolvency or inability to repay its debt.
- Failure to Pay—It occurs when the reference entity, after a certain grace period, fails to make payment of principal or interest.
- Restructuring—It refers to a change in the terms of debt obligations that are adverse to the creditors.

Restructuring could be part of the credit event and could be problematic. In case of restructuring, there might be a limitation on the remaining maturity of deliverable securities in order to control the value of cheapest-to-deliver option. There are four options for treating the issue of restructuring according to the current ISDA agreement:

No Restructuring—This option excludes restructuring altogether from the contract, eliminating the possibility that the protection seller suffers a soft credit event that does not necessarily result in losses to the protection buyer.
Full Restructuring—This allows the protection buyer to deliver bonds of any maturity after restructuring of debt in any form occurs.
Modified Restructuring—Modified restructuring limits deliverable obligations to bonds with maturity of less than 30 months after a restructuring. This is popular in United States and Canada.
Modified-Modified Restructuring—This is a *modified* version of the modified restructuring option, which resulted from the criticism of the modified restructuring that it was too strict with respect to deliverable obligations. Under the modified-modified restructuring deliverable obligations can be maturing in up to 60 months after a restructuring. This restructuring is common and popular in Europe.

23.1.1.4 What Happens When a Credit Event Occurs?

The seller of protection has a choice of the type of the recovery value. Two very common settlement types are physical settlement and cash settlement. In physical settlement, the buyer of default protection delivers the bond from the credit default swap underlying pool. The buyer has a cheapest-to-deliver (CTD) option. Deliverable notional for zero-coupon bonds are adjusted for accreted value. In cash settlement, the buyer of the default protection pays a protocol defined recovery rate in cash.

23.1.1.5 What is a CDS Spread?

The premium paid by the protection buyer to the seller, often called *spread*, is quoted in basis points per annum of the contract's notional value and is typically paid quarterly. These spreads are not the same type of concept as yield spread of a corporate bond or a government bond. Rather, CDS spreads are the annual price of protection quoted in basis points of the notional value, and not based on any risk-free bond or any benchmark interest rates.

Periodic premium payments allow the protection buyer to deliver the defaulted bond at par or to receive the difference of par and the bond's recovery value. Therefore, a CDS works like a put option on a corporate bond and similar to the put option, the protection buyer is protected from losses incurred by a decline in the value of the bond as a result of a credit event. Accordingly, the CDS spread can be viewed as a premium on the put option, where payment of the premium is spread over the term of the contract.

As an example, 5-year CDS spread for Morgan Stanley was quoted 150 bps on April 22, 2013 according to Bloomberg. This implies if one is interested in buying 5-year protection for a \$20 million exposure to Morgan Stanley credit, one would pay $\frac{150}{4} = 37.5$ bps or $\frac{\$20{,}000{,}000 \times 37.5}{10{,}000} = \$75,000$, every quarter as an insurance premium for the protection she/he would receive.

23.1.1.6 What is the Connection Between the CDS and Bond Markets?

In theory, CDS spreads should be closely related to bond yield spreads, or excess yields to risk-free government bonds. To see this, consider, on one hand, a portfolio composed of:

(1) A short position (i.e., selling protection) in the CDS of a company.
(2) A long position in a risk-free bond.

On the other hand, consider an outright long position in the company's corporate bond, all with the same maturity and par and notional values of \$100. These two investments should provide identical returns, resulting in the CDS spread equaling the corporate bond spread.

If no default occurs, principal payoff at maturity of the portfolio of a CDS and a risk-free bond will be \$100, as no payment is made on the CDS short position and a risk-free bond pays \$100. The corporate bond will also pay \$100, if no default occurred. On the other hand, if a default occurs, the portfolio of a CDS and a risk-free bond will pay the amount equal to \$100 minus the contingent payment on the CDS upon default. This payment depends on the recovery rate of the defaulted corporate bond. If we assume, for example, the recovery rate of 40%, the protection seller must pay \$60, or 60%, on \$100 notional. Using the same recovery rate, the investment in the corporate bond would also result in a payoff value of \$60 upon default. These two investments have the identical payoff and risk profile. Accordingly, the CDS and the corporate bond should be traded at the same spread level.

23.1.1.7 *A Heuristic Link between Defaultable and Risk-free Bonds*

Heuristically we can find an expression for spread by looking at the link between a zero-coupon bond and a risky/defaultable bond. Recalling that price of a zero-coupon bond is given by

$$P(t,T) = \mathbb{E}_t\left(e^{-\int_t^T r_u du}\right) \tag{23.1}$$

assuming constant discount rate $r_u = r$ and we get

$$P(t,T) = e^{-r(T-t)} \tag{23.2}$$

Now, if we also assume constant hazard rate or credit spread then based on the above argument, the price of the risky/defaultable bond would be

$$\widetilde{P}(t,T) = e^{-(r+s)(T-t)} \tag{23.3}$$

The link between these two bonds is as follows: if we let p be the probability of default, then in case the risky bond defaults the bond holder receives R, recovery rate, at maturity with probability of p and if it does not default by maturity the holder receives the face value of the bond that is \$1 with probability of $1-p$ at maturity. Then the holder would receive $(1-p)\times \$1 + p\times R$ at maturity which is the expectation. Hence the price of the risky bond is equal to the expectation discounted back to current time (today) which means

$$\widetilde{P}(t,T) = P(t,T)((1-p)\times \$1 + p\times R) \tag{23.4}$$

Or equivalently

$$e^{-(r+s)(T-t)} = e^{-r(T-t)}((1-p)+pR) \tag{23.5}$$

$$e^{-s(T-t)} = 1-p(1-R) \tag{23.6}$$

From this equality we can obtain an expression for the credit spread by calculating it for s

$$s = \frac{-\ln(1-p(1-R))}{T} \tag{23.7}$$

There are two possible cases:

having probability of default p and solve for spread s
or having spread s and calculating probability of default p

Hence having s we can calculate p to obtain

$$p = \frac{1-e^{-s(T-t)}}{1-R} \tag{23.8}$$

Example 6. Having s calculating p.

Consider the case of a 5 year CDS spread on Morgan Stanley, whose quoted CDS spread is 150 bps[1] and assuming recovery rate is 50%. Using Eq. (23.8) we get

$$p = \frac{1-e^{-0.015(5)}}{1-0.5} \tag{23.9}$$

[1]Source Bloomberg.

$$p = 14.45\% \tag{23.10}$$

We can extend this heuristic link using floating rate notes (FRNs). Consider a Default free FRN that pays LIBOR and sells for $100 per $100 of par value. We can compare this to an issuer that pays LIBOR plus 150 bps on a 5-year FRN. In this case, the premium for 5-years of default protection should be 150 basis points (bps) per year. If default swap premium is less than 1.50% a dealer can buy the bond, finance it at LIBOR and buy default protection. The arbitrage profit would be the difference between 1.50% and the lower swap premium. If the rate is higher than 150 bps a dealer can short the bond and sell default protection and receive the difference between the higher swap rate and 150 bps.

23.1.2 Collateralized Debt Obligations

Collateralized debt obligations (CDOs) are a type of structured asset-backed security (ABS) with multiple tranches that are issued by special purpose entities and collateralized by debt obligations including bonds and loans. Each tranche offers a varying degree of risk and return so as to meet investor demand or appetite for risk. CDOs value and payments are derived from a portfolio of fixed-income underlying assets. CDO securities are split into different risk classes, or tranches, whereby senior tranches are considered the safest securities. Interest and principal payments are made in order of seniority, so that junior tranches offer higher coupon payments (and interest rates) or lower prices to compensate for additional default risk which are embedded in them.

23.1.2.1 Cash CDOs

Cash CDOs involve a portfolio of cash assets, such as loans, higher yield bonds like corporate bonds, asset-backed securities (ABS), or mortgage backed securities (MBS).

Furthermore, some CDO structures require active portfolio management by the servicer (portfolio manager) to maintain the asset base and reinvest maturing funds. For that reason, ownership of the assets is transferred to the legal entity that is known as a special purpose vehicle (SPV) which would issue the CDOs tranches. The risk of loss on the assets is divided among tranches in reverse order of seniority. Cash CDO issuance exceeded 800 billion in 2012.[2]

23.1.2.2 Synthetic CDOs

Synthetic CDOs do not own cash assets like bonds or loans. Instead, synthetic CDOs gain credit exposure to a portfolio of fixed-income assets without owning those assets through the use of credit default swaps, a derivatives instrument. Under such a swap, the credit protection seller, the CDO, receives periodic cash payments, called premiums, in exchange for agreeing to assume the risk of loss on a specific asset in the event that asset experiences a default or other credit event. Nonetheless similar to the cash CDO, the risk of loss on the CDOs portfolio is divided into tranches. Losses will first affect the equity tranche, next the junior tranches, and finally the senior tranche. Each tranche receives a periodic payment (the swap premium), with the junior tranches offering higher premiums.

A synthetic CDO tranche may be either funded or unfunded. Under the swap agreements, the CDO could have to pay up to a certain amount of money in the event of a credit event on the reference obligations in the CDOs reference portfolio. Some of this credit exposure is funded at the time of investment by the investors in funded tranches. Typically, the junior tranches that face the greatest risk of experiencing a loss have to fund at closing. Until a credit event occurs, the proceeds provided by the funded tranches are often invested in high-quality, liquid assets or placed in a guaranteed investment contract (GIC) account that offers a return that is a few basis points below LIBOR. The return from

[2]Source Bloomberg.

these investments plus the premium from the swap counterparty provide the cash flow stream to pay interest to the funded tranches. When a credit event occurs and a payout to the swap counterparty is required, the required payment is made from the GIC or reserve account that holds the liquid investments. In contrast, senior tranches are usually unfunded as the risk of loss is much lower. Unlike a cash CDO, investors in a senior tranche receive periodic payments but do not place any capital in the CDO when entering into the investment. Instead, the investors retain continuing funding exposure and may have to make a payment to the CDO in the event the portfolio's losses reach the senior tranche. Funded synthetic issuance exceeded 120 billion in 2012 (source Bloomberg). From an issuance perspective, synthetic CDOs take less time to create. Cash assets do not have to be purchased and managed, and the CDOs tranches can be precisely structured.

23.1.2.3 Characteristics of CDO Market

While cash CDOs involve a pool of corporate bonds or structured finance assets, such as residential mortgage backed securities (RMBS), commercial mortgage backed securities (CMBS), and asset-backed securities (ABS), synthetic CDOs are formed from a large pool (usually more than 100 names) of CDSs. Synthetic CDOs have become very popular in recent years, especially in Europe where over 90% of deals are synthetic. In United States, synthetic deals account for one third of all arbitrage CDOs. Synthetic CDOs allow more flexible structure than cash CDOs, due to the unique characteristics of CDS.

23.2 PRICING OF CREDIT DEFAULT SWAPS

There are two main approaches in pricing credit default swaps. The first one is called *structural approach* and second one is called *reduced-form approach*. From its name, in the structural approach, it is assumed that the default event is triggered by the market value of obligor[3] falling below its liabilities. In the reduced-form approach the default event is directly modeled as an stochastic (unexpected) arrival.

When pricing credit default swaps, the parameters to be modeled include

- The likelihood of default, probability of default.
- The recovery rate when/if default occurs.
- Some consideration for liquidity, regulatory, and market sentiment about the credit event.

In pricing credit default swaps, practitioners use LIBOR as a risk-free discount rate. Pricing theory shows that the price of a derivative is the cost of replicating the derivative in a risk-free portfolio using other securities. Considering that most market dealers in credit derivatives are banks which fund close to LIBOR, the cost of funding these other securities is also close to LIBOR. Therefore LIBOR rate is the effective risk-free rate for pricing CDS, as well as other credit derivatives.

Structural and reduced-form approaches in credit modeling are mutually exclusive. In this section we describe the general setup for CDS pricing, and then show how pricing is done using a reduced form approach. We show later, in section 23.4 some common structural approaches.

23.2.1 General Setup

In a typical CDS contract, we usually have two potential cash flow streams: a fixed leg and a contingent leg. On the fixed leg side, the buyer of protection makes a series of fixed, periodic payments of CDS premium until the maturity or default whichever happens sooner. On the contingent leg side, the protection seller makes one

[3]Obligor also called debtor could be an individual or a company that owes debt to another individual or company so called the creditor, as a result of borrowing or issuing bonds.

payment only if the reference credit defaults. The amount of a contingent payment is usually the notional amount multiplied by $(1-R)$, where R is the recovery rate, as a percentage of the notional.

Hence, the value of the CDS contract to the protection buyer at any given point of time is the difference between the present value of the contingent leg, which the protection buyer expects to receive, and that of the fixed (premium) leg, which he expects to pay, or,

$$\begin{aligned}&\text{Value of CDS (to the protection buyer)}\\&\quad = \text{PV [contingent leg]}\\&\qquad - \text{PV [fixed (premium) leg]}\end{aligned}$$

In order to calculate these present values, we need to know the default probability, that is credit curve of the reference credit, the recovery rate in the case of default, and risk-free discount factors, yield curve. A less obvious contributing factor is the counterparty risk. For simplicity, we assume that there is no counterparty risk and the notional value of the swap is \$10 million.

23.2.1.1 *Valuing the Premium Leg Periodic Payments*

First, let us look at the fixed leg. On each payment date, the periodic payment is calculated as the annual CDS premium, s, multiplied by Δ_t, the accrual days expressed in a fraction of 1 year between payment dates. For example, if the CDS premium is 200 basis points per annum and payments are made quarterly, the periodic payment will be: $\Delta_t s = \frac{200}{4} = 50$ bps. However, this payment is only going to be made when the reference credit has not defaulted by the payment date. Therefore we have to take into account the survival probability, or the probability that the reference credit has not defaulted on the payment date. For instance, if the survival probability of the reference credit in the first 3 months is 95%, the expected payment at t_1, or 3 months later, is, $q(t_1)\Delta_1 s = \frac{0.95\times 200}{4} = 47.5$ bps where $q(t)$ is the survival probability at time t as explained earlier. Then, using the discount factor for the particular payment date, $P(t_0, t_i)$, the present value for this payment is $P(t_0, t_i)q(t_i)\Delta_t s$. Adding present value of all these payments we obtain

$$\sum_{i=1}^{n} P(t_0, t_i)q(t_i)s\Delta_i \tag{23.11}$$

If we remove the spread s from the above equation, then we get the present value of a cashflow stream of one basis point on all payment dates, contingent on the reference entity not defaulting. This is an important quantity in CDS pricing, and is commonly known as *risky PV01* or *risky annuity*.

23.2.1.2 *Valuing the Premium Leg Accrual Payment*

To be more precise, we should also account for the accrued premium paid up to the date of default when default happens between the periodic payment dates in the fixed leg. The accrued payment can be approximated by assuming that default, if it happens, occurs in the middle of the interval between consecutive payment dates. Then, when the reference entity defaults between payment date t_{i-1} and payment date t_i, the accrued payment amount is $\frac{\Delta_i}{2}s$. This accrued payment has to be adjusted by the probability that the default actually occurs in this time interval. In other words, the reference credit survived through payment date t_{i-1}, but does not survive up to next payment date, t_i. This probability is given by $q(t_{i-1}) - q(t_i)$. Accordingly, for a particular interval, the expected accrued premium payment is $(q(t_{i-1}) - q(t_i))\frac{\Delta_i}{2}s$. Therefore, present value of all these expected accrued payments is given by

$$\sum_{i=1}^{n} P(t_0, t_i)(q(t_{i-1}) - q(t_i))s\frac{\Delta_i}{2} \tag{23.12}$$

Adding these two components of the fixed leg together, we get the present value of the fixed leg as follows

$$\sum_{i=1}^{n} P(t_0,t_i)q(t_i)s\Delta_i + \sum_{i=1}^{n} P(t_0,t_i)(q(t_{i-1}) - q(t_i))s\frac{\Delta_i}{2} \tag{23.13}$$

23.2.1.3 Valuing the Contingent Leg

Next, we compute the present value of the contingent leg. Assume the reference entity defaults between payment date t_{i-1} and payment date t_i. The protection buyer will receive the contingent payment of $(1 - R)$. This payment is made only if the reference credit defaults, and, therefore, it has to be adjusted by $q(t_{i-1}) - q(t_i)$, the probability that the default actually occurs in this time period. Discounting each expected payment and adding them over the term of a contract, the present value of the contingent leg would be

$$(1 - R)\sum_{i=1}^{n} P(t_0,t_i)(q(t_{i-1} - q(t_i)) \tag{23.14}$$

23.2.1.4 CDS Valuation Formula

Setting the present value of the fixed leg and contingent leg we arrive at a formula for calculating value of a CDS transaction.

$$\begin{aligned}\text{PV(CDS)} &= \sum_{i=1}^{n} P(t_0,t_i)q(t_i)s\Delta_i \\ &\quad + \sum_{i=1}^{n} P(t_0,t_i)(q(t_{i-1}) - q(t_i))s\frac{\Delta_i}{2} \\ &\quad -(1 - R) \times \sum_{i=1}^{n} P(t_0,t_i)(q(t_{i-1} - q(t_i))\end{aligned}$$

When two parties enter a CDS trade, the CDS spread is set so that the value of the swap transaction is zero (i.e., the value of the fixed leg equals that of the contingent leg). Hence, the following equality holds:

$$\sum_{i=1}^{n} P(t_0,t_i)q(t_i)s\Delta_i + \sum_{i=1}^{n} P(t_0,t_i)(q(t_{i-1}) - q(t_i))s\frac{\Delta_i}{2} = (1 - R)\sum_{i=1}^{n} P(t_0,t_i)(q(t_{i-1} - q(t_i)) \tag{23.15}$$

Given all the parameters, s, the annual premium payment is set as:

$$s = \frac{(1 - R)\sum_{i=1}^{n} P(t_0,t_i)(q(t_{i-1} - q(t_i))}{\sum_{i=1}^{n} P(t_0,t_i)q(t_i)\Delta_i + \sum_{i=1}^{n} P(t_0,t_i)(q(t_{i-1}) - q(t_i))\frac{\Delta_i}{2}} \tag{23.16}$$

Example 7. Valuing a hypothetical credit default swap (CDS) trade.

Consider a 5-year CDS with quarterly premium payments. For this example we assume credit spread is 180 basis points, recovery rate of 50% and total notional of 10 million. The discount curve and the survival probability for each payment date are shown in Table 23.1.

From the entries in Table 23.1 we can calculate present value of payments on fixed leg for each period. Having the survival probabilities, we can calculate default probabilities for each tenor by simply doing $q(t_i) - q(t_{i-1})$ and use it to compute present value of accrued payment. Also having those default probabilities we can compute present value of contingent payment for each period.

The present value of all expected fixed payments is obtained by multiplying each period's fixed payment by the corresponding survival probability, discounted at the risk-free rate and summed over the term of the credit default swap. For example, at the 9th period that corresponds to 27 month the discount factor is 0.99067, survival probability is 96.5% therefore the present value of fixed leg payment would be

$$P(t_0,t_i)q(t_i)\frac{s}{4}N = 0.99067 \times \frac{96.5}{100} \times \frac{\frac{200}{4}}{10{,}000} \times \$10{,}000{,}000 \tag{23.17}$$

$$= \$47{,}800 \tag{23.18}$$

TABLE 23.1 Discount Curve and Survival Probability

Time (in Months)	Discount Curve $P(t_0, t_i)$	Survival Probability up to t_i (%)
0	1.00000	100.0
3	0.99938	99.9
6	0.99875	99.7
9	0.99813	99.5
12	0.99750	99.2
15	0.99626	98.9
18	0.99501	98.4
21	0.99377	97.9
24	0.99253	97.2
27	0.99067	96.5
30	0.98881	95.5
33	0.98696	94.5
36	0.98511	93.0
39	0.98265	91.5
42	0.98020	89.5
45	0.97775	87.5
48	0.97531	85.5
51	0.97166	82.5
54	0.96802	79.5
57	0.96440	76.5
60	0.96079	73.5

as shown in Table 23.2. The amount \$905,447 at the bottom row of Table 23.2 is the sum of all present value of these payments for \$10 million notional amount. In the case of default between two payment dates, t_{i-1} and t_i, we assume it happens in the middle of time interval, the value of the accrued premium payment if a default occurs is a half of $\frac{200}{4} = 50$ bps, or 25 bps. Then, the expected value of the accrued payment for that period is 25 basis points multiplied by the probability of default for that period, that is $q(t_{i-1}) - q(t_i)$ and discount back to obtain its present value.

$$\begin{aligned} &P(t_0, t_i)(q(t_{i-1}) - q(t_i))\frac{\frac{s}{4}}{2}N \\ &= 0.99067 \times \frac{97.2 - 96.5}{100} \\ &\times \frac{\frac{200}{4}}{2 \times 10,000} \times \$10,000,000 \quad (23.19) \\ &= 0.99067 \times \frac{0.7}{100} \end{aligned}$$

TABLE 23.2 Present Value of Cash Flows for Both Fixed Leg and Contingent Leg

Time (in Months)	Discount Factor $P(t_0, t_i)$	Survival Probability up to t_i(%)	PV of Fixed Payment	Default Probability for Δ_i (%)	PV of Accrued Payment	PV of Contingent Payment
0	1.00000	100		0	0	0
3	0.99938	99.9	49,919	0.1	25	4997
6	0.99875	99.7	49,788	0.2	50	9988
9	0.99813	99.5	49,657	0.2	50	9981
12	0.99750	99.2	49,476	0.3	75	14,963
15	0.99626	98.9	49,265	0.3	75	14,944
18	0.99501	98.4	48,955	0.5	124	24,875
21	0.99377	97.9	48,645	0.5	124	24,844
24	0.99253	97.2	48,237	0.7	174	34,738
27	0.99067	96.5	47,800	0.7	173	34,673
30	0.98881	95.5	47,216	1.0	247	49,441
33	0.98696	94.5	46,634	1.0	247	49,348
36	0.98511	93.0	45,808	1.5	369	73,883
39	0.98265	91.5	44,956	1.5	368	73,699
42	0.98020	89.5	43,864	2.0	490	98,020
45	0.97775	87.5	42,777	2.0	489	97,775
48	0.97531	85.5	41,694	2.0	488	97,531
51	0.97166	82.5	40,081	3.0	729	145,749
54	0.96802	79.5	38,479	3.0	726	145,203
57	0.96440	76.5	36,888	3.0	723	144,660
60	0.96079	73.5	35,309	3.0	721	144,118
			$905,447		$6467	$1,293,431

$$\times \frac{25}{10,000} \times \$10,000,000 \tag{23.20}$$

$$= \$173 \tag{23.21}$$

Discount these values for all periods and adding them over the term of the swap and we get the present value of expected accrued fixed payments. The amount $6,467 at the bottom row of Table 23.2 is the sum of all present value of these payments. As expected it is a pretty small number because these are products of the default probability for each period and the accrued payment if a default occurs is 25 bps, which are both small numbers. From above, we can see that the present value of the fixed leg, or the present value of the expected payments by the protection buyer over the 5-year term, is for the notional value of $10 million.

The expected value of the contingent payment if a default occurs during each period is $(1 - R)$ multiplied by the probability of default for that period. For recovery rate of 50%, the expected contingent payment is 0.50 multiplied by the each period's default probability and as before we discount it back to get its present value.

$$
\begin{aligned}
&P(t_0, t_i)(q(t_{i-1}) - q(t_i))(1 - R)N \\
&= 0.99067 \times \frac{97.2 - 96.5}{100} \\
&\quad \times (1 - 0.50) \times \$10,000,000 \quad (23.22) \\
&= 0.99067 \times \frac{0.7}{100} \\
&\quad \times 0.50 \times \$10,000,000 \quad (23.23) \\
&= \$34,673 \quad (23.24)
\end{aligned}
$$

Add these values over the entire term of the swap to get the present value of expected contingent payments that is \$1,293,431 as shown in the bottom row of Table 23.2.

Hence, we can find the value of the 5-year CDS to the protection buyer (or the fixed payer) for spread is 200 bps per annum as

$$
\begin{aligned}
&\text{Value of CDS} \\
&= \text{PV of expected contingent payment} \\
&\quad - \text{PV of fixed leg} \quad (23.25) \\
&= \$1,293,431 - (\$905,447 + \$6467) \quad (23.26) \\
&= \$381,517 \quad (23.27)
\end{aligned}
$$

Intuitively, the average default probability over the term of the CDS is 5.3% per year because the survival rate after 5 years is 73.5%, and with recovery rate of 50%, the average expected loss per year is $(1-0.5)5.3\% = 2.65\%$. The CDS spread is 200 bps per year, which means that in this example the protection buyer gets protection for credit risk with the expected loss of 265 bps for a premium of only 200 bps. Not at all surprisingly, this is a valuable transaction for the CDS buyer, with a positive CDS value to the protection buyer, calculated above to be \$381,517 or 381 bps, for \$10 million notional.

23.2.1.5 *Valuation of a Credit Default Swap Position*

Assuming initiation time at zero, the t-time mark-to-market value of a credit default swap position is the difference between the value of the protection and the cost of the premium payments.

$$V(t) = \pm(Protection - Premium) \quad (23.28)$$

Assuming at time t the market implied spread is $s(t)$, then the value is the difference of the initial spread minus current spread multiply by risky PV01 that is

$$V(t) = (s(t) - s(0))\text{PV01} \quad (23.29)$$

t	q (t)
1	0.97
2	0.94
3	0.91
4	0.89
5	0.86

Here, PV01 is the value of a 1 basis point annuity paid over all payments dates specified in the CDS contract.

Under the assumption that the interest rate and credit curve are flat, the above calculation it is fairly adequate despite of its simplicity.

Example 8. An investor buys 25 m of 5-year protection at 100 bp. The contract pays its premium annually. A year later the credit is deteriorated and is traded at 200 bp. Assuming $r = 0.25\%$ and the following annual survival probabilities: the value of the CDS position can then be computed as:

$$V(t) = N(s(t) - s(0))\ \text{PV01} \quad (23.30)$$

$$V(t) = N\frac{(200 - 100)}{10,000}\ \text{PV01} \quad (23.31)$$

$$\text{PV01} = \sum_{i=1}^{n} P(t_0, t_i)q(t_i)\Delta_i \quad (23.32)$$

$$\text{PV01} = 4.5403 \quad (23.33)$$

$$V(t) = N\frac{(200 - 100)}{10,000}4.5403 = \$1.1351 \text{ m} \quad (23.34)$$

23.2.2 Reduced-form Approach—Hazard Rate Approach

The preceding section showed how to value CDS given survival and default probabilities at all payments dates. In this section we describe a reduced-form approach that relies on hazard rates in order to model these probabilities. Before going over the hazard rate approach, we first define hazard rate and will show how to calculate it.

23.2.2.1 Definition and Calculation of Hazard Rate

Define $q(t)\Delta t$ as the probability of default between times t and $t+\Delta t$ as seen at time zero. The hazard rate, $h(t)$ is defined so that $h(t)\Delta t$ is the probability of default between times t and $t+\Delta t$ as seen at time t assuming obviously no default has happened between time zero and time t. The relationship between $q(t)$ and $h(t)$ is given by

$$q(t) = h(t)e^{-\int_0^t h(\tau)d\tau} \tag{23.35}$$

To show it we take the following steps. Let $F(t)$ be the probability of default until time t, that is

$$F(t) = \int_0^t q(u)du \tag{23.36}$$

Then intuitively we can say that defaults between time zero and $t+\Delta t$ is equivalent to defaults between time zero and t plus surviving until time t and then defaulting between t and $t+\Delta t$, or mathematically

$$F(t+\Delta t) = F(t) + (1-F(t))h(t)\Delta t \tag{23.37}$$

This is equivalent to say that $h(t)$ is given by

$$h(t) = \frac{1}{1-F(t)}\frac{F(t+\Delta t)-F(t)}{\Delta t} \tag{23.38}$$

$$= \frac{1}{1-F(t)}\frac{\Delta F}{\Delta t} \tag{23.39}$$

Taking the limit as Δt approaches zero, we obtain

$$h(t) = \frac{1}{1-F(t)}\frac{dF}{dt} \tag{23.40}$$

Since survival probability is the complement to default probability we can say $S(t) = 1 - F(t)$. Substituting it into Eq. (23.40) we can see that

$$\frac{dS(t)}{dt} = -S(t)h(t) \tag{23.41}$$

with the initial condition $S(0) = 1$. We can solve this ordinary differential equation to get

$$\ln S(t) = -\int_0^t h(u)du + c \tag{23.42}$$

using the initial condition we get $c = 0$ and can write it as

$$S(t) = \exp(-\int_0^t h(u)du) \tag{23.43}$$

Subtracting survival probability at time t from survival probability at time $t+dt$ gives the probability of default between t and $t+dt$ as seen at time zero, that is

$$S(t+dt) - S(t) = q(t)dt \tag{23.44}$$

Or

$$q(t) = \frac{S(t+dt)-S(t)}{dt} = \frac{dS}{dt} \tag{23.45}$$

We can calculate $\frac{dS}{dt}$ be differentiating (23.43) and setting it equal to $q(t)$ which we get

$$q(t) = h(t)e^{-\int_0^t h(\tau)d\tau} \tag{23.46}$$

The hazard rate approach is the most widely used reduced-form approach. It is originally proposed by Jarrow and Turnbull in pricing derivatives on financial securities subject to credit risk (Jarrow and Turnbull, 1995). They characterize the credit event as the first event of a Poisson counting process that occurs at time t with a probability defined as

$$P(\tau \le t+dt|\tau > t) = h(t)dt \tag{23.47}$$

$h(t)$ is known as hazard rate. One can show that probability of survival is given by

$$\mathbb{E}^{\mathbb{Q}}\left(\exp\left(-\int_0^T h(u)du\right)\right) \tag{23.48}$$

and we are done. In this setup, it is clear that the hazard rate is deterministic and is independent of interest rate and recovery rate.[4]

Define $\lambda_t = \mathbb{1}_{\tau \le t}$. It can be shown that $\lambda_t - \int_0^t (1-\lambda_u)h_u du$ is a martingale. There are two possibilities for λ_t either 0 or 1. If $\lambda_t = 0$ that implies $\lambda_u = 0$ for all $u < t$ and therefore $\lambda_t - \int_0^t (1 - \lambda_u)h_u du = -\int_0^t h_u du$. If $\lambda_t = 1$ then there exists t^* between 0 and t such that for u in $(0, t^*)$ we have $\lambda_u = 0$. Therefore $\lambda_t - \int_0^t (1 - \lambda_u)h_u du = 1 - \int_0^{t^*} h_u du$.

The following stochastic exponential

$$\mathcal{E}\left(-\lambda_t + \int_0^t (1 - \lambda_u)h_u du\right) \tag{23.49}$$

is a martingale. Therefore

$$\mathcal{E}\left(-\lambda_t + \int_0^t (1 - \lambda_u)h_u du\right) = e^{\int_0^t (1-\lambda_u)h_u du}\Pi(1 - \Delta\lambda) \tag{23.50}$$

Hence

$$\mathbb{E}\left(e^{\int_0^t (1-\lambda_u)h_u du}(1 - \lambda_t)\right) = 1 \tag{23.51}$$

Here $(1 - \lambda_u)$ is redundant and we get

$$\mathbb{E}(1 - \lambda_t) = e^{-\int_0^t h_u du} \tag{23.52}$$

Iterative procedures can be used to extract default probabilities from credit spreads. Within the (Jarrow and Thurnbull, 1995) equivalent recovery model in which the recovery rate is taken to be an exogenous constant, one can assume that both the riskless interest rate $r(t)$ and the spread $s(t)$ evolve under the Longstaff & Schwartz framework (Longstaff and Schwartz, 1992) as stated in Hatgioannides and Petropoulos (2007).

Furthermore, for a small time dt, the short-term credit spreads are directly related to local default probabilities $q(t)$. Local default probabilities are the probability of default between t and $t + dt$, conditional on no default prior to time t as explained earlier. In addition, one can relate the probability of default to the intensity of the default process as in done (Arvanitis et al., 1999).

Now using the spread discount factors obtained under the spread-based model to derive the local default probabilities and, subsequently, the conditional default probabilities. Following the determination of the local default probabilities, the cumulative survival probabilities can be determined and taking into account the two possibilities at each time point: default and no-default. Using the ratings model of Jarrow, Lando and Turnbull one can demonstrate how the risk premia as estimated using the Longstaff & Schwartz (1992) model can be used to derive the implied ratings transition matrix.

23.2.2.2 Constant Hazard Rate Model

It would be natural to think that the breakeven spread in a credit default swap should be the spread that makes the present value of premium the same as the present value of the protection. Speaking of the present value means the timing of the credit event plays a very important role in pricing. This can be solved by conditioning on defaulting in each time interval $[t, t + dt]$ which means we have survived up to t and default $[t, t+dt]$. Probability of surviving up to t is given by $q(t)$ and therefore it is $q(t)h(t)dt$, after default we pay $(1 - R)$ and should discount this back to the present time using risk-free rate. Assuming a constant risk-free rate, we can then write the

[4]Credit portfolio models often assume that recovery rates are independent of default probabilities. John Frye (2003) presents empirical evidence showing that such assumptions are incorrect based on historical default data. We will address this later.

present value for the protection leg as

$$(1-R)\int_0^t h\exp(-(r+h)u)du = \frac{h(1-R)(1-e^{-(r+h)t})}{r+h} \tag{23.53}$$

The value of the premium leg is the present value of the payments which are made quarterly to maturity until default happens. Assuming constant spread then the present value of the premium leg is

$$s\int_0^T e^{-(r+h)u}du = \frac{s(1-e^{-(r+h)T})}{r+h} \tag{23.54}$$

Now by making the two equations equal and solving for s we will get

$$s = h(1-R) \tag{23.55}$$

The above relationship is known as the credit triangle considering that knowing two of them the third one can be calculated. The interpretation of the formula is that within any small time interval, the spread that the investor pays for protection compensates him/her from the risk of default.

Example 9. Credit triangle formula.

For a company with a spread curve at 200 bps, if we assume that recovery is 50% then from the formula we get implied hazard rate of 4% ($h = \frac{0.02}{0.5} = 0.04$). Using Eq. (23.55), we can find the survival probability curve going out 5 years.

23.2.2.3 Bootstrapping Hazard Rates

Under the assumption that the interest rate and credit curve are flat, the above calculation is fairly adequate despite of its simplicity. We can relax the assumption of a constant hazard rate and construct the term structure of hazard rate by *bootstrapping*. Bootstrapping is done through the following procedure:

We begin with the shortest maturity CDS and find the hazard rate that equates the PV of the two legs of the CDS trade. We then proceed to the next shortest maturity CDS and assume that the hazard rate is piecewise constant. That is, we assume that until the first maturity the hazard rate is h_1, and between time t_1 and t_2 we assume that the hazard rate is h_2. Keeping h_1 fixed, we then find the hazard rate that equates the PV of the second shortest CDS. We then continue this process until we have reached the maturity of the longest quoted CDS.

Assuming payment being paid every 6 months starting from the first CDS with spread of s_1 we have

$$(1-q_1)\frac{s_1}{2}Ne^{-L_1t_1} = q_1N(1-R)e^{-L_1t_1} \tag{23.56}$$

solving for q_1 we get

$$q_1 = \frac{\frac{1}{2}s_1 df_1}{\left((1-R)+\frac{s_1}{2}\right)df_1} \tag{23.57}$$

For the next CDS with spread of s_2 we have

$$(1-q_1)\frac{s_2}{2}Ne^{-L_1t_1} + (1-q_2)\frac{s_2}{2}Ne^{-L_2t_2} = q_1N(1-R)e^{-L_1t_1} + (q_2-q_1)N(1-R)e^{-L_2t_2} \tag{23.58}$$

solving for q_2 we get

$$q_2 = \frac{\frac{1}{2}s_2 df_2 + (1-q_1)\frac{s_2}{2}df_1 - (1-R)q_1(df_1-df_2)}{\left((1-R)+\frac{1}{2}s_2\right)df_2} \tag{23.59}$$

and by induction we can show that

$$q_k = \frac{\frac{1}{2}s_k df_k + \sum_{i=1}^{k-1}\left((1-q_i)\frac{s_k}{2}df_i - (1-R)q_i(df_i - df_{i+1})\right)}{\left((1-R)+\frac{1}{2}s_k\right)df_k} \tag{23.60}$$

In pricing of credit derivatives, the recovery rate is the expected recovery rate under risk-neutral following by a credit event which is only available from price information. However, given the price, it is difficult to find recovery rate because it is entangled with the probability of default. The common approach for practitioners is to use rating agency default studies to estimate the recovery rate. It could be via regression and proxy. It might be ad hoc considering that it is proxy and needs adjustment depending on US versus non-US entities data availability. A drawback with obtaining recovery rate this way is that these results would be backward looking and only include default and bankruptcy and not restructuring. Also, studies show that default and recovery rate are negatively correlated and to incorporate this effect researchers have been using models with stochastic recovery rates. The standard approach is to model the recovery rate via a beta distribution.

Real-world default probabilities are much lower than those calculated from pricing. This is similar to implied volatility from options versus realized historical volatility from prices. This is because of market risk aversion expressed through a risk premium additionally supply-and-demand imbalances.

23.3 PRICING MULTI-NAME CREDIT PRODUCTS

Many credit derivatives, such as synthetic CDO's rely on modeling the default dynamics of a portfolio of reference entities, rather than just a single entity. As a result, for these products we must not only model the default dynamics of each reference entity but also the default correlation of all assets in the portfolio. As we see in the following section, this is traditionally done through the use of structural models with a copula used to specify default co-dependence.

23.3.1 Modeling Default Correlation

This section describes how we can model joint defaults when pricing credit derivatives. In order to do so, one needs to extend structural and reduced-from approaches to account for multiple underlying entities. Principally the extension is doable, however, extension of structural models are more popular among practitioners due to ease of tractability in higher dimensions. Prior to looking at these extensions, we begin by reviewing the basic probabilistic concepts required for estimating joint defaults. There are various different models used to pricing these products.

For multivariate case one needs to extend structural and reduced-from approaches. Principally the extension is doable. However, extension of structural models is more popular among practitioners due to ease of tractability of the structural approach in higher dimension.

23.3.1.1 Estimation of Joint Default Probabilities

Consider two obligors A and B and a specified horizon T. Let

- p_i be the probability that obligor i defaults by time T.
- p_{ij} be joint default probability that i and j both default by time T.
- $p_{i|j}$ be the probability that i defaults by T, given that j has defaulted before T.

From probability theory we know that

$$p_{A|B} = \frac{p_{AB}}{p_B} \tag{23.61}$$

$$p_{B|A} = \frac{p_{AB}}{p_A} \tag{23.62}$$

$$\rho_{AB} = \text{linear correlation coefficient or default event correlation} \tag{23.63}$$

$$= \frac{p_{AB} - p_A p_B}{\sqrt{p_A(1-p_A)p_B(1-p_B)}} \tag{23.64}$$

Since default probabilities are very small, the correlation ρ_{AB} can have a much larger effect on the joint risk of a position.

It is important to notice that lower and upper limits are not -1 or 1 but are a function of the marginal probabilities themselves. Due to scarcity of default data estimation of default probabilities are very difficult. As a result practitioners have been searching for alternative ways to calibrate the frequency of joint defaults in modeling joint defaults, in order to avoid the need for a direct estimate of joint default probabilities like simulation approach or semi-analytical approach.

23.3.1.2 *Structural Approach to Joint Defaults*

There are various different ways proposed in the literature for modeling correlated defaults and pricing credit derivatives for multiasset. The model by Hull and White generates dependent default times by diffusion correlated assets and calibrating default thresholds to replicate a set of given marginal default probabilities. Most of these models for multiasset require simulation. Others like Finger and Gregory and Laurent exploit models that do not require simulation and the approach is semi-analytical based on a low-dimensional factor structure and conditional independence.

In simulation, we generate sample paths and knowing the default event get triggered we can generate defaults paths and the exact time of default on each path and this is done for each obligor in the basket which is drawn randomly from the joint default distribution known a priori. Knowing that like any other simulation would let us value any multiasset credit instrument for any specification.

The payoffs of multiasset credit default such as synthetic loss tranche or nth-to-default cannot be statically replicated by trading in a set of single asset credit default. Having said that market practitioners value correlation products using standard no-arbitrage arguments. It turns out that valuation of these instruments is to compute risk-neutral expectations over all possible default scenario.

Example 10. Joint default.

Considering that in the structural approach default happens when the asset value goes below some threshold then join defaults over a specified horizon must follow from the joint dynamics of asset values.

As an example, we consider the case where each asset follows the following dynamics:

$$\ln\left(S_t^{(i)} \big/ S_0^{(i)}\right) = \left(r - q - \frac{\sigma^2}{2}\right) t + \sigma W_t^{(i)} \quad (23.65)$$

$$= \left(r - q - \frac{\sigma^2}{2}\right) t + \sigma \sqrt{t} Z_i \quad (23.66)$$

where Z_i is standard norm random variable. Assume risk-free interest rate of 2% and zero-dividend and volatility of 30% for each obligor starting from $S_0 = 100$ with horizon of 1 year, $t = 1$. Assume also that $\ln(S_t/S_0) < L = -0.3$ is the default event. We consider five different correlation scenarios: 0%, 25%, 50%, and 75%, 90%. To be a fair comparison, we use \$10,000 samples draw for each obligor and through each correlation scenario we use the same samples that we draw (see Figure 23.2).

It is pretty easy to draw the following conclusion from Example 10 and Figure 23.3 that as asset return correlation increases so is the joint defaults. That implies asset correlation leads to default correlation.

23.3.1.3 *Gaussian Copula Model*

Gaussian copula model by David Li presents an inexpensive and simple approach for simulating correlated defaults. It is based on the assumption that multivariate distribution of default times and the multivariate distribution of underlying returns have the same correlation structure. In the paper, the assumption is that the correlation structure is Gaussian and by that assumption having the correlation structure the entire structure would be fully characterized.

Assume marginal distribution of jth defaultable instrument is F_j for $j = 1, \ldots, k$. Under copula model, we can join these marginal distributions with a correlation matrix Λ. Since asset returns are not observable directly, we use proxy asset correlations using equity correlations.

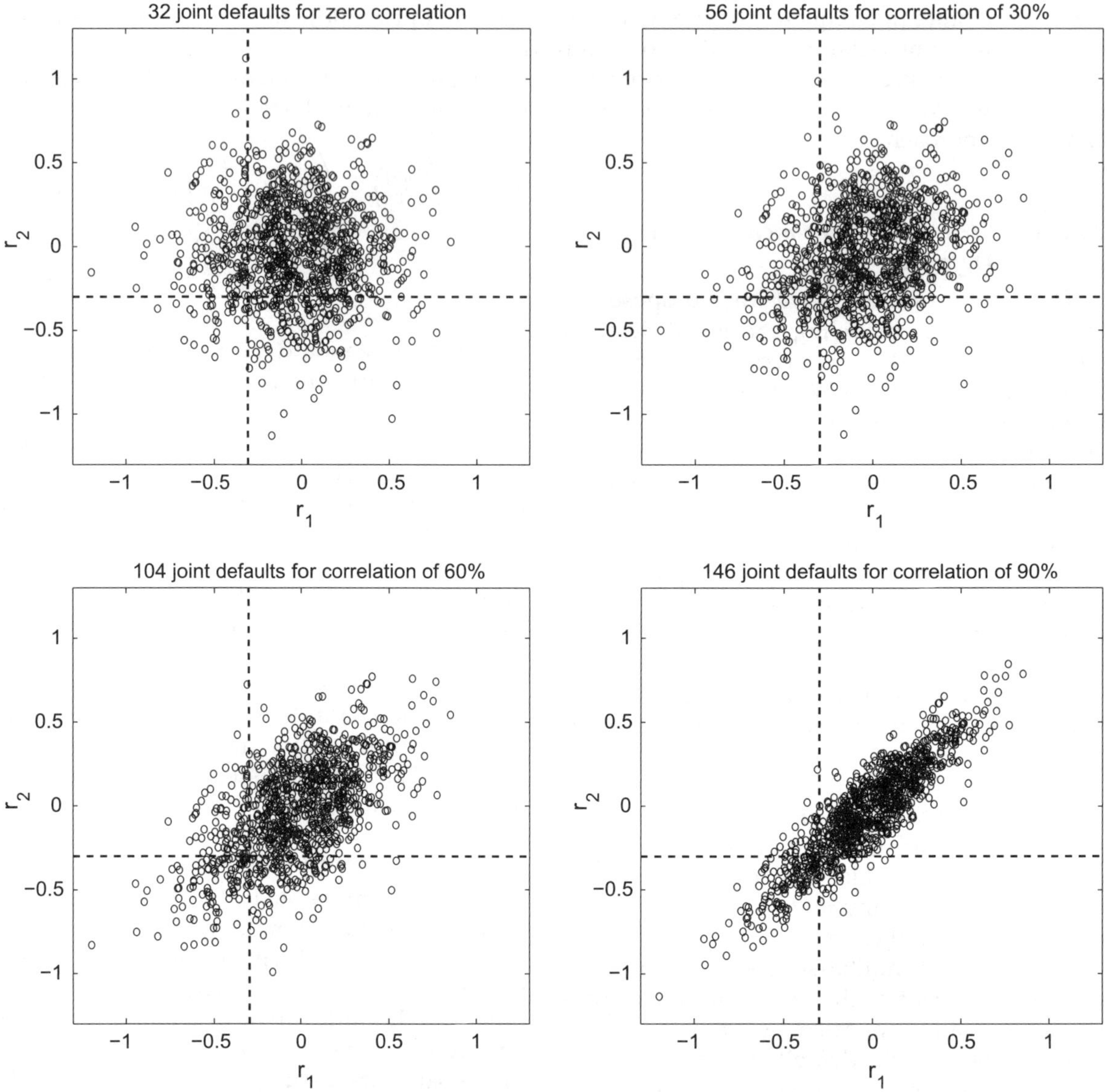

FIGURE 23.2 Default correlation for various correlation scenarios for two obligors.

In case of having two assets, we can write the joint distribution function of default times by

$$P(\tau_1 < x_1, \tau_2 < x_2) \\ = \Phi_{2,\Lambda}(\Phi^{-1}F_1(x_1), \Phi^{-1}(F_2(x_2))) \quad (23.67)$$

where $\Phi_{2,\Lambda}$ is the bivariate cumulative standard normal distribution with correlation Λ and Φ^{-1} is the inverse of a cumulative standard normal distribution. It is easy to see how one would extend it to k-dimensional case.

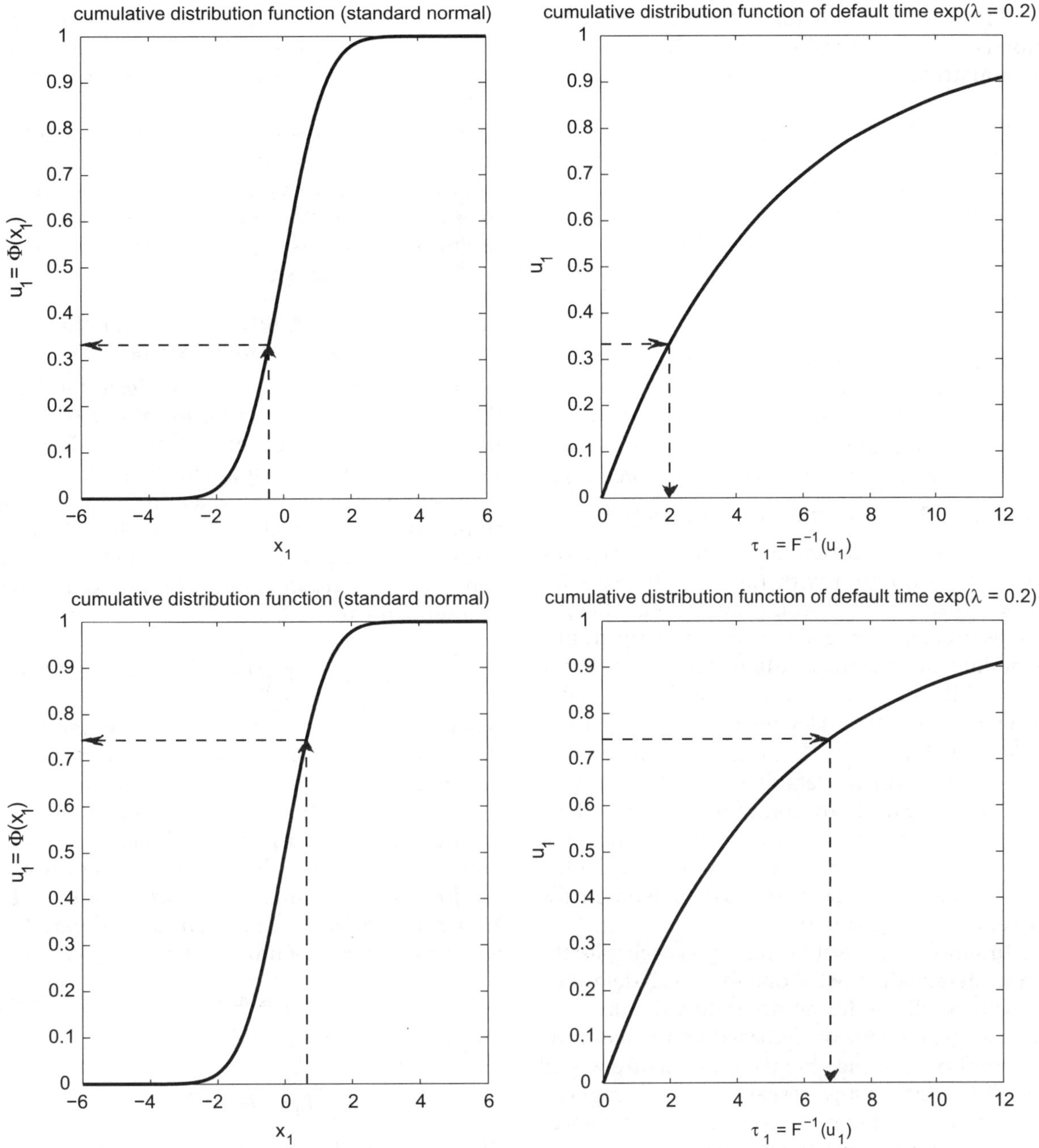

FIGURE 23.3 Mapping a standard normal random variable to a default time.

Simulating default times from this distribution is simple. Assuming k obligors and correlation matrix Λ we do it as follows:

using Cholesky decomposition of the correlation matrix Λ to simulate a multivariate normal random variable vector x that is k-dimensional vector.
form $U = (\Phi(x_1), \ldots, \Phi(x_k))$ that is a unit hypercube
$\tau = (\tau_1, \ldots, \tau_k) = \left(F_1^{-1}(u_1), \ldots, F_k^{-1}(u_k)\right)$

It is easy to verify that τ has the given marginals and a normal dependence structure fully characterized by the correlation matrix Λ. Here we show two examples: one is for a one-dimensional case and the other for a three-dimensional case

Example 11. One-dimensional example.

Assume default time distribution is an exponential distribution with rate $\lambda = 0.2$ or scale $\theta = \frac{1}{\lambda} = 5$. This assumption implies default times are, on average, 5 years. Drawing from a standard normal distribution, z and calculating $U = \Phi(z)$ and then finding $\tau = F^{-1}(u)$ where $F(x) = 1 - \exp(-\lambda x)$ which implies $F^{-1}(u) = -\frac{\ln(1-u)}{\lambda}$. Now that we know how to sample from the distribution of default times we can price a correlation trade by computing $\mathbb{E}(f(\tau)$, where $\tau = (\tau_1, \ldots, \tau_k)$ is the vector of default times as explained earlier and f is the function calculating the discounted cash flows associated with the asset under consideration.

In simulation considering that at each path the exact default time of the obligor which defaulted is known allows for an accurate valuation and risk measurements of credit with multi assets no matter how complex the payoff structure would be. There are many research papers suggesting modeling stochastic recoveries and considering negative correlation between recovery and default for a more realistic and better valuation of credit derivatives instruments.

Like all simulation method, there is a cost associated with simulation approach, especially when calculating sensitivity measures. Knowing that defaults are rare events the use of important sampling and low-discrepancy sequences could reduce the noise and variance.

A semi-analytical approach is intrigued by fast computation for sensitivity measure. In order to achieve it, certain steps should be taken. One is to assume one-factor model and also allow a set of finite set of times that default could occur. As far as accuracy this should suffice.

23.3.1.4 *A Semi-Analytical One-Factor Gaussian Copula Model*

Here we utilize a one-factor model to construct the risk-neutral loss distribution and show how to use a sequence of loss distributions at different time horizons to price synthetic loss trenches.

Assume that the jth issuer from today and some given time in the future is described by a standard normal random variable with mean zero and standard deviation of one and is of the following form

$$a_j = \rho_j z_m + \sqrt{1 - \rho_j^2} z_j \tag{23.68}$$

where z_j for $j = 1, \ldots, d$ are identically independent distributed $\mathcal{N}(0, 1)$. z_m is the common market factor due to systemic risk and z_j are specific risk or idiosyncratic risk of the j issuer.

The default event happens if the asset return falls below a certain threshold, L_j. That implies the jth issuer defaults in the case that $a_j < L_j$. Knowing that a_j is a standard normal variable, we can write the default event as

$$p_j = \Phi(L_j) \tag{23.69}$$

which implies

$$L_j = \Phi^{-1}(p_i) \tag{23.70}$$

This means calibration of the marginal probabilities is simply inverting the cumulative function of a standard normal variable.

In this setup conditioning on z_m, the asset returns are independent, and this makes it pretty easy to calculate conditional default probabilities. Now, conditioning on a realization of the common factor, say $z_m = u$ for some u being a real number, we can see that a default for the jth issuer will happen if

$$a_j = \rho_j z_m + \sqrt{1-\rho_j^2}\, z_j < L_j \tag{23.71}$$

solving for z_j and conditioning on $z_m = u$ we get

$$z_j \leq \frac{L_j - \rho_j u}{\sqrt{1-\rho_j^2}} \tag{23.72}$$

Therefore, the conditional default probability $p_j(z_j)$ of the individual jth issuer is given by

$$p_j(u) = \Phi\left(\frac{L_j - \rho_j u}{\sqrt{1-\rho_j^2}}\right) \tag{23.73}$$

If the assumption is that the loss on default for each of the issuer is of the same unit, then constructing the portfolio loss distribution could be done iteratively by adding assets to the portfolio. This is achieved according to the following steps:

Starting with one asset, there are two possibilities, either a loss of l with probability $p_1(u)$ or no loss with probability of $1 - p_1(u)$.

Now add a second asset to the portfolio and adjust each of the previous losses. No loss means that the new asset also survives and therefore has a probability of $(1 - p_1(u))(1 - p_2(u))$. A loss of l means the first one defaults and the second one survives or the first one has survived and the second one has defaulted which implies $p_1(u)(1-p_2(u))+p_2(u)(1-p_1(u))$ which is equal to $p_1(u) + p_2(u) - 2p_1(u)p_2(u)$. A loss of $2l$ means both have defaulted that is $p_1(u)p_2(u)$.

We keep doing this until the very last one is added to the portfolio.

We then repeat this procedure for a different value for z_m and integrate the loss distribution over all possible values for z_m which is obviously from a standard normal distribution.

We can see that in this approach there is no simulation and therefore no noise associated with it. We can also use other methods including fast Fourier techniques.

Both the times-to-default model and the one-factor model are Gaussian copula models that means as long as we use the same correlation matrix and marginals the results would be the same.

23.3.2 Valuation of Correlation Products

Now knowing about different methodologies for default correlation we can utilize them to price correlation products. The correlation products that we will be look at are basket default swaps and synthetic loss tranches.

The payoffs of these correlation products cannot be statically replicated by trading in a set of single asset credit default. Having said that market practitioners value correlation products using standard no-arbitrage arguments. It turns out that valuation of these instruments is to compute risk-neutral expectations over all possible default scenario.

23.3.2.1 Basket Default Swaps

A basket default swap is inherently an idiosyncratic product. This means that the identity of the defaulted asset is known. There are various ways of pricing default baskets. One of which is to use the times-to-default model.

The generation of default times are done using simulation. Given that it is known when each asset in the portfolio will default in each simulated path we do the following steps to compute the spread

First we sort the default times in an ascending order, then denote the kth time to default $\tau_k(j)$, which implies jth asset is defaulted. Here j is the label for the defaulted asset.

Then we calculate the present value of the one basis point coupon stream paid up to time that is the minimum of the default time $\tau_k(j)$ or the

maturity time T the maturity of the basket. Calculate the average over all paths which is called basket PV01.
It is clear that if $\tau_k(j) > T$ which means the jth asset survived which implies present value is equal to zero otherwise the present value is equal to $B(\tau_k(j))(1-R(j))$, where B is the LIBOR discount factor and $R(j)$ is the recovery rate for jth asset. Calculate the average over all paths.
To compute the fair-value spread we divided the protection leg by the basket PV01.

The advantage of this approach is that it is simple to implement and pretty fast. Gregory and Laurent (2005) introduces an analytical approach to build the basket n-to-default probabilities while retaining the identity of the default asset(s). Their approach results in a substantial reduction in computational time compared with Monte Carlo simulation techniques. Their approach is also parsimonious with respect to the number of parameters, therefore easing its calibration.

23.3.2.2 *Pricing of the Synthetic Loss Tranches*

To value a loss tranche, knowing when each asset in the portfolio would default in each simulation path, we proceed as follows:

For each path, we would calculate the present value of principal losses on the protection leg.
Now, compute the present value of one basis point (1 bp) on the premium leg.
Average the tranche PV01 over all paths, average present value of the protection leg over all paths, divide the protection present value by the tranche PV01 to find the so-called breakeven tranche spread.

This method is simple but could be computationally expensive or slow depending on the number of assets in the portfolio. An alternative to this approach is to utilize a semi-analytical method. The semi-analytical method is possible because a standard synthetic loss tranche can be priced from its loss distribution.

23.4 CREDIT SPREAD OBTAINED FROM OPTIONS MARKET

The motivation for this comes from the inherent connection between the equity and credit markets. To see this, we can consider the case of an investor believes that the future prospects for a company are bleak, and they are likely to default. The investor can express this view either through the credit market, through a CDS contract, or through the equity market, through a Put option. In order to express this view in the equity market in the cheapest possible manner, we would expect the investor to buy a deep out-of-the-money Put option.
There are two possibilities in default cases:

Anticipated default.
Unanticipated default.

In anticipated default, the underlying process diffuses to zero such as the arcsinh normal model. In unanticipated default, the underlying process jumps to zero, as explained the Merton model.
We are considering unanticipated default. Therefore we are assuming that the underlying stock is following a pure jump process with infinite activity near zero. Processes like these would reflect the unanticipated default and it assures us for short-term maturity bonds we can obtain spreads, unlike in case of diffusion processes.
Variance gamma (VG) is the underlying process for this case. There are many classes of these Lévy processes such as variance gamma with stochastic arrival (VGSA), CGMY, and the like.
In this section, we cover basic calibration of the Lévy processes to the option surface and utilizing the information in all traditional default models, namely, we consider the following models:

Merton default model.
Hazard rate approach.

Longstaff–Schwartz model.

Moreover, we review a simple approach proposed by Hirsa and Madan to estimate credit spread from option markets. For all four cases, we go through derivations and provide some examples.

23.4.1 The Merton Default Model Revised

Let V be the firm value, E be the value of equity, and F be the face value of the debt. Here we assume that the dynamics of firm value, V, is approximately the same as the dynamics of equity, E. The value of the risky bond at maturity is minimum of the face value of the debt and the firm value, that is,

$$\begin{aligned}&\text{bond value at maturity}\\ &\quad= \min(F, V) &(23.74)\\ &\quad= F - \max(F - V, 0) &(23.75)\end{aligned}$$

By discounting back, we get the today value of the risky/defaultable bond, we can see that the value of defaultable bond is given by

$$\begin{aligned}&\text{value of defaultable bond}\\ &\quad= Fe^{-rT} - \text{Put option value} \quad (23.76)\end{aligned}$$

which implies

$$\begin{aligned}Fe^{-(r+h)t} &= Fe^{-rt} - \text{Put}\\ Fe^{-rt}(1 - e^{-ht}) &= \text{Put}\\ e^{-ht} &= 1 - \frac{\text{Put}}{Fe^{-rt}}\end{aligned}$$

Here it begs these questions. How to bring information from option market? How to calculate the face value of debt and firm value?

We can simply assume that the firm value is following a pure jump process such as variance gamma and price the Put option.

$$\begin{aligned}e^{-ht} &= 1 - \frac{\text{Put}}{Fe^{-rt}}\\ &= 1 - \Phi(-d_2) + \frac{Ve^{rt}}{F}\Phi(-d_1)\\ &= 1 - \Phi(-d_2) + (\alpha + \beta S)\Phi(-d_1)\\ &= f(\alpha, \beta, \sigma, \nu, \theta, t)\end{aligned}$$

In calibration, the objective is to minimize

$$\sum_{t=1}^{n} |e^{-ht} - f(\alpha, \beta, \sigma, \nu, \theta, t)| \quad (23.77)$$

to get $\alpha, \beta, \sigma, \nu, \theta$ five underlying stock price process; function of volatility of stock; VG world: VG proper model of stock.

23.4.1.1 *How to Calculate the Value of the Put Option?*

$$\text{Put} = \mathbb{E}(e^{-rT}(F - V_T)^+)$$

where F as before is the face value of the debt, V is the value of the firm. Merton (1973) showed that stock could be considered as a call option on the firm with strike equal to the face value of the debt. Thus, at u years out, we have

$$S_u = C(V_u) = \mathbb{E}(e^{-r(T-u)}(V_T - F)^+) \quad (23.78)$$

By inverting the above equation numerically, we get

$$V_u = C^{-1}(S_u) \quad (23.79)$$

From put-call parity we have

$$\begin{aligned}P_u &= C(V_u) + Fe^{-r(T-u)} - V_u &(23.80)\\ &= S_u + Fe^{-r(T-u)} - V_u &(23.81)\\ &= S_u + Fe^{-r(T-u)} - C^{-1}(S_u) &(23.82)\end{aligned}$$

Assuming that the distribution of S_u is known, we have

$$\text{Put} = \mathbb{E}_0(P_u) \tag{23.83}$$

$$= \int_0^\infty e^{-ru} P_u f(S_u) dS_u \tag{23.84}$$

$$= \int_0^\infty e^{-ru} \left(S_u + Fe^{-r(T-u)} - C^{-1}(S_u)\right) f(S_u) dS_u \tag{23.85}$$

Therefore for the Merton default model T-forward spread is given by

$$h = -\frac{\ln\left(1 - \frac{\int_0^\infty \left(S_t + Fe^{-r(T-t)} - C^{-1}(S_t)\right) f(S_t) dS_t}{F}\right)}{T}$$

23.4.2 Option Implied Credit Spreads Equity Dependent Hazard (EDH) Rate Approach

Assuming that *hazard rate*[5] has the following form $h(u) = a - b\ln S_u$, where S_u follows a pure jump process.[6] The price of a defaultable bond is given by

$$\mathbb{E}\left\{e^{-\int_0^t (r(u)+h(u))du}\right\} \tag{23.86}$$

Assuming constant interest rate yields

$$e^{-rt}\mathbb{E}\left\{e^{-\int_0^t h(u)du}\right\} \tag{23.87}$$

To find a closed-form presentation of the above expression, we follow the work by Madan and Unal (2000) by assuming hazard rate $h(u)$ is linear in logarithmic of the stock price that is $h(u) = a - b\ln S_u$. As in Madan and Unal (2000), here we also assume S_u evolves according to a pure jump process and we can write $S_u = S_0 e^{(r-q+\omega)u + x*\mu}$ or equivalently

$$\ln S_u = \ln S_0 + (r - q + \omega)u + (x * \mu)_u \tag{23.88}$$

[5]This hazard rate approach is generalization of the two-factor hazard rate model for pricing risky debt and the term structure of credit spreads by Madan and Unal (2000).

[6]$S_u = S_0 e^{(r-q+\omega)u + x*\mu}$ where $x * \mu = \int_0^u \int_{-\infty}^\infty x\mu(dx, ds)du$.

where $x * \mu = \int_0^u \int_{-\infty}^\infty x\mu(dx, ds)$ where μ is an integer-valued random measure which counts the number of jumps of size x at time u. Substituting and integrating $h(u)$ from 0 to t yields

$$\int_0^t h(u)du = \int_0^t (a - b\ln(S_u))du \tag{23.89}$$

$$= \int_0^t (a - b(\ln S_0 + (r - q + \omega)u + (x * \mu)_u)du \tag{23.90}$$

$$= (a - b\ln S_0)t - (r - q + \omega)b\frac{t^2}{2} - b\int_0^t (x * \mu)_u du \tag{23.91}$$

To calculate the last term in Eq. (23.91), we first change the order of integration and integrate part of the expression and reduce the order and use Fubini's theorem and finally use the definition of $*$ that is

$$\int_0^t (x * \mu)_u du = \int_0^u \int_{-\infty}^\infty x\mu(dx, ds)du \tag{23.92}$$

$$= \int_{-\infty}^\infty \int_0^t \left(\int_s^t x du\right) \times \mu(dx, ds) \tag{23.93}$$

$$= \int_{-\infty}^\infty \int_0^t (t - s) \times x\mu(dx, ds) \tag{23.94}$$

$$= (t - s)x * \mu \tag{23.95}$$

Substituting it into Eq. (23.87) and taking the expectation to get

$$\mathbb{E}\left\{e^{-\int_0^t h(u)du}\right\} = e^{(a-b\ln S_0)t + (r-q+\omega)\frac{t^2}{2}}\mathbb{E}\left[\exp\left(b(t - s)x * \mu\right)\right] \tag{23.96}$$

Now we need to calculate the expectation. In order to do that we use compensator of a jump process. The following process is martingale

$$\left(e^{bx(t-s)} - 1\right) * (\mu - \nu) \tag{23.97}$$

if ν is the martingale compensator. Thus [7]

$$\mathcal{E}\left[(e^{b(t-s)x}-1)*(\mu-\nu)\right] \tag{23.98}$$

is also a martingale and therefore

$$\mathbb{E}\left\{\mathcal{E}\left[(e^{b(t-s)x}-1)*(\mu-\nu)\right]\right\}=1$$

Or equivalently

$$\mathbb{E}\left[e^{-\int_0^t\int_{-\infty}^{+\infty}(e^{b(t-s)x}-1)\nu(dx,ds)}e^{bx(t-s)*\mu}\right]=1$$

Therefore

$$\mathbb{E}\left[e^{bx(t-s)*\mu}\right]=e^{\int_0^t\int_{-\infty}^{+\infty}(e^{b(t-s)x}-1)k(x)dxds}$$

where $k(x)$ is the Lévy density of the underlying process. In case of having the variance gamma model we have the following expression for its Lévy density

$$k(y)=\frac{e^{-\lambda_p y}}{\nu y}\mathbf{1}_{y>0}+\frac{e^{-\lambda_n|y|}}{\nu|y|}\mathbf{1}_{y<0}$$

$$\lambda_p=\left(\frac{\theta^2}{\sigma^4}+\frac{2}{\sigma^2\nu}\right)^{\frac{1}{2}}-\frac{\theta}{\sigma^2}$$

$$\lambda_n=\left(\frac{\theta^2}{\sigma^4}+\frac{2}{\sigma^2\nu}\right)^{\frac{1}{2}}+\frac{\theta}{\sigma^2}.$$

Algebraic calculations yield the following expression for the expectation

$$e^{\int_0^t\int_{-\infty}^{+\infty}(e^{b(t-s)x}-1)k(x)dxds}$$
$$=e^{\frac{2t}{\nu}}\left(1-\frac{b}{\lambda_p}t\right)^{-\frac{1}{\nu}\left(1-\frac{b}{\lambda_p}t\right)}\left(1+\frac{b}{\lambda_n}t\right)^{-\frac{1}{\nu}\left(1+\frac{b}{\lambda_n}t\right)} \tag{23.99}$$

Therefore

$$\text{Price of defaultable bond}=e^{-rt}\mathbb{E}\left\{e^{-\int_0^t h(u)du}\right\} \tag{23.100}$$

$$=e^{-\left(r+(a-b\ln S_0)-\frac{2}{\nu}\right)t+(r-q+\omega)\frac{t^2}{2}}$$
$$\times\left(1-\frac{b}{\lambda_p}t\right)^{-\frac{1}{\nu}\left(1-\frac{b}{\lambda_p}t\right)}\left(1+\frac{b}{\lambda_n}t\right)^{-\frac{1}{\nu}\left(1+\frac{b}{\lambda_n}t\right)} \tag{23.101}$$

We observe that for the variance gamma model, the integral in the exponent is explicit. For each name and each time to maturity, we obtain VG parameters—σ,ν,θ—through calibration and construct an explicit term structure of credit spreads. We can show the impact of skew and kurtosis in options market on default spreads.

23.4.3 Longstaff–Schwartz Model

In this approach, we can compute defaultable claims as defined by Longstaff and Schwartz in their work in Longstaff and Schwartz (1995) using first passage time for Lévy processes. In finding the first passage time, we first define

$$G(s,t,T)=\mathbb{E}_t^{\mathbb{Q}}\left[\mathbf{1}_{(s(u)<H,0\le u\le T)}\right] \tag{23.102}$$

which is the conditional expectation of a random variable, here the stock level for the entity going below a prespecified threshold H which would constitute bankruptcy or default on the entity's bonds. By construction $G(s,t,T)$ is a martingale because it is a conditional expectation of a terminal random variable and hence $G(s,t,T)$ is the solution of the following PIDE:

$$G_t+(r+\omega)sG_s$$
$$+\int_{-\infty}^{\infty}\left(G(se^y,t,T)-G(s,t,T)\right)k(y)dy=0$$

subject to the following boundary conditions:

$$G(s,t,T)=1 \text{ if } s<H, \quad \text{for all } t$$
$$G(s,T,T)=0 \text{ if } s\ge H$$

[7]Here $\mathcal{E}$ is stochastic exponential. See *Limit Theorems for Stochastic Processes* by Jacod and Shirayev for more details.

For the variance gamma, CGMY, natural inverse Gaussian models we have Lévy density $K(y)$ and can solve the PIDE and get the probability of default for the entity.

23.4.4 Credit Spread Implied Explicitly from Option Premiums—A Simple Model

We start by assuming the density of the logarithmic price relative $z = \log(S_t/S_0)$ follows a stochastic process[8] is given by $h(z)$. Let p be the default probability for time horizon t that is the entity default prior to time t with probability p. Then $1-p$ is the survival probability for the entity. So long as the entity/company survives by time t, the logarithmic price relative has distribution $h(z)$, and in case it defaults before or at time t would be zero with probability of one. Define

$$f(z;p) = \begin{cases} h(z) & \text{with probability } 1-p \\ 1 & \text{with probability } p \end{cases} \tag{23.103}$$

where p is t-time default probability. Under this assumption, we have an alternative way of pricing call and put option prices. Call options can be calculated as

$$\tilde{C}(S_0, K, t) = e^{-rt} \int (S_t - K)^+ f(z;p)dz \tag{23.104}$$

$$= e^{-rt}(1-p) \int (S_t - K)^+ h(z)dz + p \int (0 - K)^+ dz \tag{23.105}$$

$$= (1-p)e^{-rt} \times \int (S_t - K)^+ h(z)dz \tag{23.106}$$

$$= (1-p)C(S_0, K, t) \tag{23.107}$$

[8] This includes the variance gamma (VG), constant normal vol. (Cox-Ross), CIR, CEV, logstable, Merton jump diffusion, Kou jump diffusion models.

Using the same approach, put options can be calculated as

$$\tilde{P}(S_0, K, t) = e^{-rt} \int (K - S_t)^+ f(z;p)dz \tag{23.108}$$

$$= e^{-rt} \left((1-p) \int (K - S_t)^+ h(z)dz + p \int (K - 0)^+ dz \right) \tag{23.109}$$

$$= (1-p)e^{-rt} \int (S_t - K)^+ h(z)dz + pKe^{-rt} \tag{23.110}$$

$$= (1-p)P(S_0, K, t) + pKe^{-rt} \tag{23.111}$$

Using Equations (23.107) or (23.111) p can be obtained from calibration to market prices (puts or calls) and having calculated p one can calculate credit spread s.

This way, we get credit spread directly from option prices. In the case of variance gamma model, we call the process *VG jump to ruin*. For each maturity t, we obtain p and the variance gamma model parameters σ, ν, and θ by calibrating the volatility surface using FFT techniques.

23.4.4.1 *Why Choose the Variance Gamma Process?*

An important component of the above model specification is the freedom in choosing the underlying process $h(z)$ that the asset follows in the absence of default. Variance gamma is a natural choice for this because it allows calibration of not only the mean and variance of the distribution, but also the skewness and kurtosis. As we are concerned with extracting a default probability, it is critical that we choose a distribution that accurately represents these higher moments of the distribution, as they may contain important information regarding market implied default probabilities.

23.4.5 Summary

In this section we described the following techniques for assimilating information from the options market in order to estimate default probabilities:

- Computing defaultable claims as defined in Longstaff and Schwartz using first passage time for Lévy processes.
- Basic calibration of the Lévy processes to the option surface and utilizing the information in all three traditional default models (Merton, Longstaff-Schwartz, hazard rate).

23.5 PROBLEMS

1. Following the derivation of price of defaultable bond for hazard rate approach, find the expression for defaultable bond assuming that underlying process follows the CGMY process. The Lévy density for CGMY is given by

$$k(y) = \frac{e^{-\lambda_p y}}{\nu y^{1+Y}} \mathbf{1}_{y>0} + \frac{e^{-\lambda_n |y|}}{\nu |y|^{1+Y}} \mathbf{1}_{y<0}$$

$$\lambda_p = \left(\frac{\theta^2}{\sigma^4} + \frac{2}{\sigma^2 \nu} \right)^{\frac{1}{2}} - \frac{\theta}{\sigma^2}$$

$$\lambda_n = \left(\frac{\theta^2}{\sigma^4} + \frac{2}{\sigma^2 \nu} \right)^{\frac{1}{2}} + \frac{\theta}{\sigma^2}$$

2. Problem on bootstrapping.
3. In Example 7 find the breakeven spread that is the spread that makes present value of the fixed leg the same of present value of the contingent leg.
4. Using the Hirsa-Madan model proposed in Section 23.4.4 to calibrate the following table of out-of-the-money WMT put option premiums. [9] What is the option markets belief of the firm's default probability? Assuming the recovery for WMT is 40%, use the credit triangle formula to estimate an approximate credit spread. Suppose you knew the traded credit spread was 50 bps. Comment on the accuracy of the option implied credit spread and the consistency of expectations in the equity and credit markets.

K	bid	ask	t	r
30	0.04	0.07	0.46175	0.0001
32.5	0.06	0.09	0.46175	0.0001
35	0.08	0.12	0.46175	0.0001
37.5	0.11	0.14	0.46175	0.0001
40	0.15	0.19	0.46175	0.0001
42.5	0.21	0.24	0.46175	0.0001
45	0.28	0.32	0.46175	0.0001
47.5	0.41	0.44	0.46175	0.0001
50	0.61	0.63	0.46175	0.0001
52.5	0.91	0.94	0.46175	0.0001
55	1.38	1.42	0.46175	0.0001
57.5	2.1	2.13	0.46175	0.0001

5. Consider a first-to-default basket (FTD) that pays $(1 - R)$ upon the default of the first credit in a basket of 3 credits. Protection on the FTD lasts 5 years and payments are made quarterly. The annualized contract spread is 800 bps. The three credits have constant hazard rates of 4%, 6%, and 10% respectively. Each credit has a correlation of $\rho = 0.2$ and a recovery of 0.5. Using a Gaussian copula, simulate default times and compute the present value of the contract. Assume a constant interest rate of 3%.

   Try again with $\rho = 0.5$. Comment on the relationship between default correlation and value of an FTD.

[9] Source OptionMetrics.

CHAPTER

24

Stopping Times and American-Type Securities

OUTLINE

24.1 INTRODUCTION

Options considered in this book can be divided into two categories. The first group was characterized using a pricing equation that depended on the current value of the underlying assets S_t and on the time t. For example, the price of a plain-vanilla call option at time t was written as:

$$C_t = F(S_t, t) \tag{24.1}$$

Given the observed value of S_t and the time t, the option price was determined by the function. Plain-vanilla European options, where the S_t was a geometric process, fell into this category.[1]

The second category of options, although not dealt with extensively in this book, were those that were classified as *path-dependent*. The price of these options at time t depended not only on the current S_t, but possibly on some or on all other values of S_t observed before time t as well. An option's payoff at expiration time T could

[1]Additional assumptions concerning no dividend payments and constant interest rates were also assumed to get a closed-form formula for $F(\cdot)$.

Financial Derivatives, Third Edition.
http://dx.doi.org/10.1016/B978-0-12-384682-2.00024-4

depend, for example, on the average of the last N values observed at discrete times:

$$t < t_1 < t_2 < \cdots < t_N = T \tag{24.2}$$

At expiration, a call option holder could for example be paid:

$$C_T = \max\left[\frac{S_{t_1} + S_{t_2} + \cdots + S_{t_N}}{N} - K, 0\right] \tag{24.3}$$

where K is some strike price.

Under these conditions the time T price of this call option could be written using:

$$C_T = F\left(S_{t_1}, S_{t_2}, \ldots, S_{t_N}\right) \tag{24.4}$$

Clearly, this expression will look somewhat more complicated for time $t, t < T$.[2] Yet, pricing this sort of exotic option is not necessarily more difficult than the case of plain-vanilla exotic options. In fact, according to what was said, the payoff of this option occurs at expiration date T, and in this sense the option is still European. The only complications are the additional S_{t_i} terms that show up in the expression. Thus, although the option is path-dependent, and the payoff depends on how one gets to an expiration value of the underlying asset, a Monte Carlo–type approach can give a reasonable approximation to C_t once the dynamics of S_t are correctly postulated.

Notice that for neither of these two categories of options the investor has to make another decision once the option is purchased. In both cases, one waits until expiration and exercises the right to buy if it is profitable to do so. Alternatively, the option holder can close the position and sell the option to somebody else. But no other decision has to be taken. Hence, no other variable enters the formula.[3]

Now consider an *American-style* option. These securities can be exercised at or *before* the expiration date T. Once the investor buys an American-style option he or she will have an *additional* decision to make. The time to exercise the option must now be chosen. The investor cannot just sit and wait until expiration. At some critical time denoted by θ, where $\theta \in [0, T]$, it may be more profitable to exercise the (call) option and realize the gain,

$$S_\theta - K \tag{24.5}$$

than hold onto the call until expiration to get

$$\max\left[S_T - K, 0\right] \tag{24.6}$$

In fact, at some critical time θ, the expectation under the martingale measure of the future payoff $\max[S_T - K, 0]$ may be less than what one may get if one exercised the option and received $S_\theta - K$. That is, with constant spot rates, we may have:

$$S_\theta - K > \mathbb{E}^{\mathbb{Q}}\left[e^{(T-\theta)r} \max\left[S_T - K, 0\right]\Big| I_\theta\right] \tag{24.7}$$

This means that the discounted value of the expected payoff may be less than what one gains by simply exercising the option at time θ.

From this it should be clear that with American-style securities the decision to exercise the option is equivalent to finding such critical time periods θ. Note that under these conditions, the pricing formula for the option may depend on the procedure used to select the θ's, as well as on the previously discussed variables.

Such θ's are called stopping times. When the date to exercise is chosen in some optimal fashion, they are called optimal stopping times and play a crucial role in pricing American-style securities.

24.2 WHY STUDY STOPPING TIMES?

Even if the notion of stopping times was limited to the class of American-style securities, it would still be necessary to study stopping

[2] In fact, no closed-form formula may exist.

[3] We always assume that the interest rates and volatility are constant.

times. It is true that most financial derivatives are American-style and stopping times are necessary to price them. But there is more to stopping times than just American-style derivatives. We need to study stopping times not just because they are theoretical notions useful in theoretical formulas, but also because there are some very specific numerical algorithms that one needs to use in determining dates of early exercise. That is, we study stopping times because of *numerical* considerations as well.

There are properties of optimal stopping times that make some approaches more convenient than others when it comes to pricing. By learning these properties we can reduce the time it takes to calculate whether, at a certain time t^*, an option should be exercised or not. Or, in terms of the θ, whether one has:

$$\theta = t^* \tag{24.8}$$

which means "exercise," or

$$\theta > t \tag{24.9}$$

which means "do not exercise." By doing these calculations faster or more accurately, one can reduce costs and capture arbitrage opportunities better. Hence, the properties of algorithms used to determine stopping times will be an important part of the pricing effort.

There are other reasons for studying stopping times. *Optimal* stopping times are in general obtained by using the so-called *dynamic programming* approach. Dynamic programming is a useful tool in its own right and should be learned whether one is interested in pricing derivatives or not. It just happens that the context of stopping times is a very natural setting for presenting the main ideas of dynamic programming.

24.2.1 American-Style Securities

American-type derivative securities contain implicit or explicit options, which can be exercised before the expiration date if desired. This causes significant complications both at a theoretical level, where one has to characterize the fair-market value of the security, and at a practical level, where one has to calculate this price.

Bermudan-style options are a mixture of American and European options. They can be exercised at some prespecified times other than the expiration date. Yet, they cannot be exercised at all times during $[0, T]$. At the date of issue the security specifies some specific dates $t_1 < t_2 < \cdots < t_n = T$ during which the option holder can exercise his or her option.

From the point of view of the "optimal stopping" perspective, the complications created by Bermudan options are very similar to American-style securities. The same introductory discussion of stopping times and the related tools will be sufficient for American as well as for Bermudan options. Hence, in the remainder of this chapter we work only with American options when dealing with stopping times.

24.3 STOPPING TIMES

Stopping times are special type random variables that assume as outcomes random time periods, t. For example, let τ be a stopping time. Then this means two things. First, that τ is random, and second, that the range of its possible values is $[0, T]$ for some $T > 0$. When an outcome is observed, it will be in the form:

$$\tau = t \tag{24.10}$$

That is, the outcome of the random variable is a particular time period.

Now consider an American-style option written on a bond. The option can be exercised at any time between the present $t = 0$ and the expiration date denoted by T. The option holder will exercise this option if he or she thinks that it is better to do so, rather than waiting until the expiration date of the contract.

Hence, we are dealing with a "random date," which is of great importance from the point of view of pricing the asset. In fact, the right to exercise early may have some additional value and pricing an American security must take this into account.

Thus, we let τ represent the early exercise date. It is obvious that given the information set, I_t, we will be able to tell whether the option has already been exercised or not. In other words, given I_t we can differentiate between the possibilities:

$$\tau \le t \tag{24.11}$$

which means that option has already been exercised, or

$$\tau > t \tag{24.12}$$

which means that the early exercise clause of the contract has not yet been utilized.

This property of τ is exactly what determines a stopping time.

Definition 28. A stopping time is an I_t-measurable nonnegative random variable such that:

1. Given I_t we can tell if

$$\tau \le t \tag{24.13}$$

2. We have

$$\Pr(\tau < \infty) = 1 \tag{24.14}$$

In case of derivative securities in general, we have a finite expiration period. So the options will either be exercised at a finite time or will expire unexercised. This means that the second requirement that τ be finite with probability one is always satisfied.

24.4 USES OF STOPPING TIMES

How can the stopping times, τ, be utilized in practice?

The most obvious use of τ is to let it denote the exercise date of an option. With European securities, there was no randomness in exercise dates. The security could only be exercised at expiration. Hence, we can write:

$$\Pr(\tau = T) = 1 \tag{24.15}$$

With American-type securities, τ is in general random.[4]

Consider an American-style call option $F(S_t, t)$ written on the underlying security S_t, where S_t follows an SDE:

$$dS_t = a(S_t, t)\,dt + \sigma(S_t, t)\,dW_t, \quad t \in [0, \infty) \tag{24.16}$$

with the drift and diffusion coefficients satisfying the usual regularity conditions.

The price of the derivative security can again be expressed using the equivalent martingale measure $\mathbb{Q}$. But this time there is an additional complication. The security holder does not have to wait until time T to exercise the option. He or she will exercise the option as soon as it is more profitable to do so than wait until expiration.

In other words, if one has to wait until expiration, the asset will be worth

$$F(S_t, t) = \mathbb{E}_t^{\mathbb{Q}}\left[e^{-r(T-t)} \max\{S_T - K, 0\}\right] \tag{24.17}$$

at time t. If the option can be exercised early, we can compare this with, say,

$$F\left(S_t, t^*\right) = \sup_{\tau \in \Phi_{t,T}} \left[\mathbb{E}_\tau^{\mathbb{Q}}\left[e^{-r(T-\tau)} F(S_\tau, t, \tau)\right]\right] \tag{24.18}$$

where the $\Phi_{t,T}$ is the set of all possible stopping opportunities[5] and the t^* is the optimal choice for τ. Here, τ represents a possible date where the option holder decides to exercise the call option.

Hence, at time t, we can calculate a spectrum of possible prices $F(S_\tau, t, \tau)$ indexed by τ using the possible values for the stopping time τ. To find the correct price, we then pick the supremum among all these $F(S_\tau, t, \tau)$.

[4]In some special cases it is never worth exercising the American-style option, and the corresponding τ will again equal T.

[5]That is, it is the set of possible outcomes for τ.

24.5 A SIMPLIFIED SETTING

We continue studying stopping times and the problem of optimal stopping using the simplified setting of a binomial model for pricing a plain-vanilla American-style call option. Yet, although the setting is "simple" and our main purpose is the understanding of tools related to stopping times, the actual pricing of American-style options often proceeds within frameworks similar to the one considered here. Thus, the discussion below is useful from the point of view of some simple numerical pricing calculations as well.

24.5.1 The Model

The model is a binomial setting for the price of an underlying asset S_t that behaves, in continuous time, as a geometric Wiener process:

$$dS_t = (r - \delta)\, S_t dt \,+\, \sigma S_t dW_t, \quad t \in [0, \infty) \tag{24.19}$$

where r is the constant instantaneous spot rate and the $0 < \delta$ is a known dividend rate. The W_t is a Weiner process with respect to risk-neutral measure $\mathbb{Q}$.

We let the C_t denote the price of an American-style call option, with strike K and expiration date $T, T < t$ that is written on S_t. Suppose we decide to price this call using a binomial tree approach.

The methodology was discussed earlier, but is summarized here for convenience. We first choose the grid parameter Δ and discretize the S_t in a standard way:[6]

$$S_i^u = S_{i-1} e^{\sigma\sqrt{\Delta}} \tag{24.22}$$

$$S_i^d = S_{i-1} e^{-\sigma\sqrt{\Delta}} \tag{24.23}$$

Here the up and down probabilities are assumed to be constant across n and across "states,"and are given by:

$$\Pr(u) = \frac{1}{2} + \frac{(r-\delta) - \frac{1}{2}\sigma^2}{2\sigma}\sqrt{\Delta} \tag{24.24}$$

$$\Pr(d) = 1 - \Pr(u) \tag{24.25}$$

That is, once the process reaches a point S_{i-1}, the next stage is either *up* S_i^u or *down* S_i^d, with a probability equal to $\Pr(u)$ or $\Pr(d)$.[7]

With this choice of discretization parameters, the discretized system converges to the geometric process as $\Delta \to 0$. That is, the drift and the diffusion parameters would be the same and the would follow the same trajectory as dictated for S_t in the SDE (24.19).

Note that the up and down parameters u, d are constant and are given by:

$$u = e^{\sigma\sqrt{\Delta}} \tag{24.27}$$

$$d = e^{-\sigma\sqrt{\Delta}} \tag{24.28}$$

Also, we have, as usual,

$$ud = 1 \tag{24.29}$$

That is, the tree is recombining.

The likely paths followed by the discretized S_t are shown in Figure 24.1. It is worthwhile to look at the structure of the tree. The horizontal movement represents the path taken by S_i over "time." The process begins from the initial point S_0 and then ends up at one of the six expiration states. During the times $i = 1, \ldots, 5$ the S_i can follow several trajectories. In fact, altogether there are $2n$ possible trajectories, where n is the number

[6]This is one possible choice for discretization. There are others. For example, we can let:

$$S_i^u = S_{i-1} e^{\left((r-\delta) - \frac{1}{2}\sigma^2\right)\Delta + \sigma\sqrt{\Delta}} \tag{24.20}$$

$$S_i^d = S_{i-1} e^{\left((r-\delta) - \frac{1}{2}\sigma^2\right)\Delta - \sigma\sqrt{\Delta}} \tag{24.21}$$

[7]As $\Delta \to 0$, the probabilities of up and down movements become equal and

$$\Pr(u) = \Pr(d) = \frac{1}{2} \tag{24.26}$$

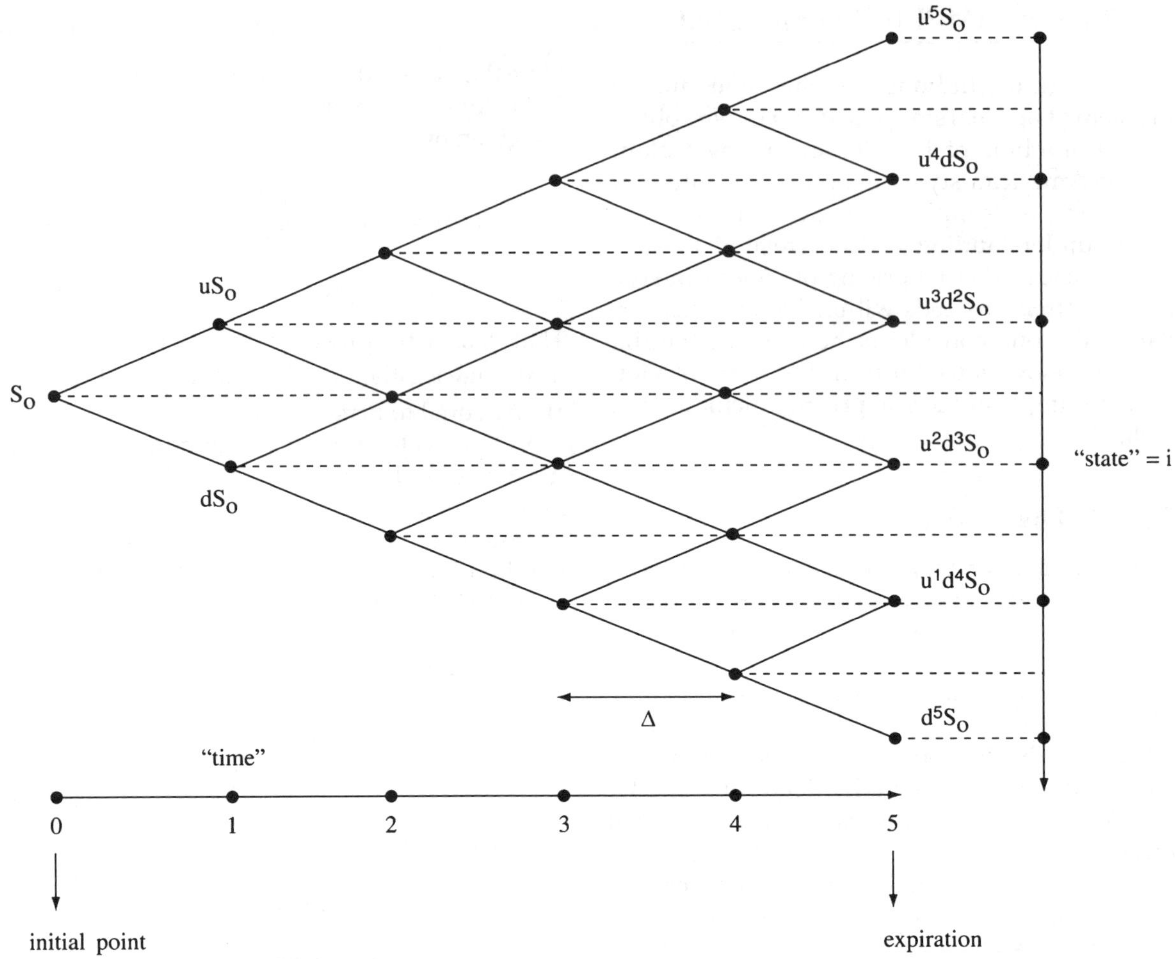

FIGURE 24.1 The standard way of looking at binomial trees which horizontally mimics the behavior of *stock* over time.

of "stages" given by $n = T\Delta$. In this particular case, this gives 32 possible trajectories that S_i can follow.

The call option's price will depend on the trajectory followed by S_i. For European options, this was discussed in earlier chapters. It turns out that for the American-style options, there is a completely different way of looking at the same tree and seeing the dependence between the S_i and C_n, the price of the call option.

The standard way of looking at binomial trees is shown in Figure 24.1, which horizontally mimics the behavior of S_i over "time." For analyzing stopping times and understanding the complications introduced by interim decisions on "stopping" or "continuing," it is worthwhile to look at the same tree from a different angle, literally. This may be a bit inconvenient at the beginning, but it greatly facilitates the understanding of some mathematical tools associated with stopping times.

Instead of looking at the tree over time as in Figure 24.1, consider now Figure 24.2, where on the horizontal axis we mark all possible values

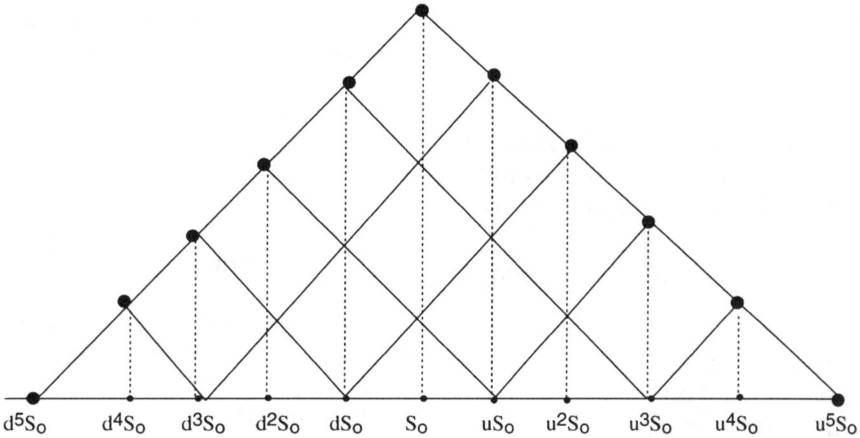

FIGURE 24.2 A different way of looking at the binomial tree where on the horizontal axis we mark all possible *stock values* during the move.

assumed by S_i during the move from $n = 0$ to $n = 5$, the expiration. During this period, S_i can assume 11 possible values. Denoting this set by E and using the condition $ud = 1$, we obtain:

$$E = \left\{u^5 S_0, u^4 S_0, u^3 S_0, u^2 S_0, u^1 S_0, S_0, d^1 S_0, d^2 S_0, d^3 S_0, d^4 S_0, d^5 S_0\right\} \tag{24.30}$$

Now, although binomial trees are normally visualized over time as in Figure 24.1, for stopping time problems, one gains additional insights when the tree is visualized as a function of the value assumed by S_i. These values are represented on the horizontal line shown in Figure 24.2. The line is none other than the representation of the set E. The figure is the same as in Figure 24.1, except we look at it "sideways" from right to left.

In the case of American options, we have the right to exercise early. And early exercise will naturally depend on the value of S_i at the point that we find ourselves. Now consider the way the process behaves on the set E, as represented by the horizontal line in Figure 24.2.

Initially, we are at point S_0 at the middle point. Next time $n = 1$ with probability $0.5, S_1$ will move either to the left, to S_0, or with probability 0.5 will move to the right, to uS_0. Once there, it can either come back to S_0, or move one step further to the right or to the left. Hence, at each point, the process can only move to adjacent states. The major exceptions are the two end points. Once the process gets there, it must stop by necessity because it takes exactly 5 time periods to get to those points and that is the expiration of the option.[8] This means that S_i can also represent the position of a Markov Chain at stage i.

Now, remember that we can at any stage "stop" the experiment if we desire to do so. If we do this, we receive the payoff:

$$S_i - K \tag{24.31}$$

In contrast, if we decide otherwise, and continue, then we will be in possession of a security that is valued at the price $C(S_i)$.

Let us now consider the way an optimal decision to stop can be made. We let the τ represent the random time period at which we decide to exercise the option before expiration. At the initial point $i = 0$ the τ is random because whether we stop or not depends on the random trajectory followed by S_i.

[8] In the terminology of Markov Chains the two end points are called absorbent. That is, once we get there, with probability one we stay there.

Suppose we consider stopping at stage i. That is, let:

$$\tau = i \tag{24.32}$$

Then, obviously this decision will be made by looking at the trajectory followed by S_i until the stage i. That is, we will have the observations:

$$\{S_0, S_1, \ldots, S_i\} \tag{24.33}$$

and the decision to stop will be a function of this history. According to this, the decision to exercise early does not depend on the knowledge of which states will occur *after* stage i. The "future" after i is still unknown. This is what we mean when we say τ is I_i-measurable.

Now, if we can determine a *strategy* to choose the τ then we may be able to obtain the probabilities associated with the random variable τ as well. But if such a strategy is not defined, then the properties of τ will not be known and the $C(\cdot)$ will not be a well-defined random variable. Thus the first task is to determine a strategy to choose the τ. How is this to be done?

Consider the following criterion, where the conditional expectation operator is written in the expanded form:

$$\begin{aligned} F(S_0) &= \max_{\tau} \mathbb{E}^{\mathbb{Q}}\left[C(S_\tau) \mid S_0\right] \\ &= \max_{\tau} \mathbb{E}^{\mathbb{Q}}\left[e^{-rT} \max\left[S_\tau - K, 0\right] \Big| S_0\right] \end{aligned} \tag{24.34}$$

According to this, we choose the τ so that when we stop, we stop optimally, in the sense that the value of the option now is maximized. [9]

Thus, we need to do two things. First, we need to obtain the variable:

$$\max_{\tau} \mathbb{E}^{\mathbb{Q}}\left[C(S_t)\right] \tag{24.35}$$

Second, we need to find a rule to determine the optimal stopping time τ^* such that:

$$\mathbb{E}^{\mathbb{Q}}\left[C(S_t)\right] \le C(S_{\tau^*}), \quad \text{for } t > \tau^* \tag{24.36}$$

When this is done, the optimal strategy will be of the form:

$$\tau^* = \min\left[k\text{: } S_k > B(k, S_k)\right] \tag{24.37}$$

where $B(_{k,} S_k)$ will be an optimal exercise boundary that depends on the k and on the current (and possibly past) values of S_k. This boundary is to be determined.[10]

24.6 A SIMPLE EXAMPLE

We now discuss a simple, yet important example in order to understand some deeper issues associated with stopping time problems, as the one above. Recall the following problem faced by the holder of an American-style call. At any instant during the life of the option, the option holder has the right to early exercise. Hence, at all $t \in [0, T]$ a decision should be made concerning whether to exercise early or not. But this decision is *much* more complicated than it looks.

It turns out that to make this choice, the investor has to calculate if he or she is likely to early exercise in the *future* as well. This means that before reaching a decision today, the investor must evaluate the odds of making the same decision *in the future*. It is only after analyzing possible *future* gains from the option that a decision to continue can be made. But a decision about early exercise possibilities in the future depends on the same assessment of the more distant future, and

[9] According to this setup, the $F(\cdot)$ is an objective function. If such $F(\cdot)$ is assumed to be bounded, all technical conditions will be satisfied for the following steps. In practice, pricing algorithms assume this boundedness implicitly.

[10] Let us see how we can interpret expectations such as $\mathbb{E}^{\mathbb{Q}}[S_\tau]$. Here one random variable is the τ. Hence possible values of the function S_τ need to be multiplied by the probability that $\tau = k$ among other things. There are other considerations, because even with τ fixed S_τ will still be random. On the other hand, in expectations such as $\mathbb{E}^{\mathbb{Q}}[S_\tau]$ one would multiply possible values of S_t by the probability that S_t will assume these values.

so on. At the end, the American-option holder is left with a complex decision that spans all time periods until expiration.

How should such decisions be made? Are there some mechanical rules that will help the decision maker to decide on the early exercise or not? Finally, how can we gain some insights into such interrelated complex decisions?

The simple example below is expected to shed some light on these questions. The reader will notice that the way the example is set is similar to the binomial tree model discussed in the previous section. In fact, we use the same notation.

Suppose we observe successive values of a random variable. We let $n = 5$ and assume that there are 11 possible values that S_i can take. These are given by the ordered set E:

$$E\,\{a_1, a_2, a_3, a_4, a_5, a_6, a_7, a_8, a_9, a_{10}\} \qquad (24.38)$$

The initial value S_0 is known to be a. The process is observed starting with $i = 1$, is Markovian, and behaves according to the following assumptions.

1. When the process assumes a particular value in E during stage i, in the next stage it can move either to the immediate left or to the immediate right. All other possibilities have zero probability. This means, for example, that if for $i = 3$ we have $S_3 = a_6$, then S_4 will assume either the value a or the value a_7.
2. Second, the states a_1 and a_{10} are *absorbent*. If the process reaches those states, it will stay there with probability one. These states can be reached only at "expiration."

The next stage in describing these types of models is to state explicitly the relevant transition probabilities.

According to the description above, the transition probabilities are given by:

$$\Pr\left(S_{i+1} = a_{j+1} \middle| S_i = a_j\right) = \frac{1}{2} \qquad (24.39)$$

$$\Pr\left(S_{i+1} = a_{j-1} \middle| S_i = a_j\right) = \frac{1}{2} \qquad (24.40)$$

$$\Pr\left(S_{i+1} = a_1 \middle| S_i = a_1\right) = \Pr\left(S_{i+1} = a_{10} \middle| S_i = a_{10}\right) = 1 \qquad (24.41)$$

All other transitions carry zero probability:

$$\Pr\left(S_{i+1} = a_m \middle| S_i = a_j\right) = 0, \quad |m| > j + 1 \qquad (24.42)$$

This implies that the process S_i cannot *jump* across states and that it has to move to an adjacent state. Finally, for the initial stage, we also have:

$$\Pr\left(S_1 = a_5 \middle| S_0 = a\right) = \frac{1}{2} \qquad (24.43)$$

$$\Pr\left(S_1 = a_6 \middle| S_i = a\right) = \frac{1}{2} \qquad (24.44)$$

This situation is shown in Figure 24.3. The horizontal axis represents the set E. The arrows on the horizontal line indicate the possible moves and the corresponding probabilities. Note that if the process reaches the two end points it stays there (gets absorbed).

We need to introduce one more component to discuss the optimal stopping decisions in this context. When the S_i visits a state, say, a_j in E, the decision maker is given an option to receive a payoff $F(a_j)$. If the decision maker accepts this payoff, then the game stops. If the payoff is not accepted, the game continues and the S_i moves to adjacent states. In Figure 24.3, the payoff associated with each state $a_j \in E$ is shown as the vertical line at the corresponding point.

The problem faced by the decision maker is the following. Successive values of S_i are observed and the corresponding $F(a_j)$ are revealed. The decision maker evaluates the payoff of stopping immediately, against the *expected* payoff of continuing and ending up with a better $F(a_j)$ in the future. How should this decision maker act? We discuss the optimal decision using Figure 24.3.

Note that at each stage we know where we are. In other words, we observe the current value of S_i. But, we do not know the future outcomes, even though we do know the possibilities. Consider these possibilities.

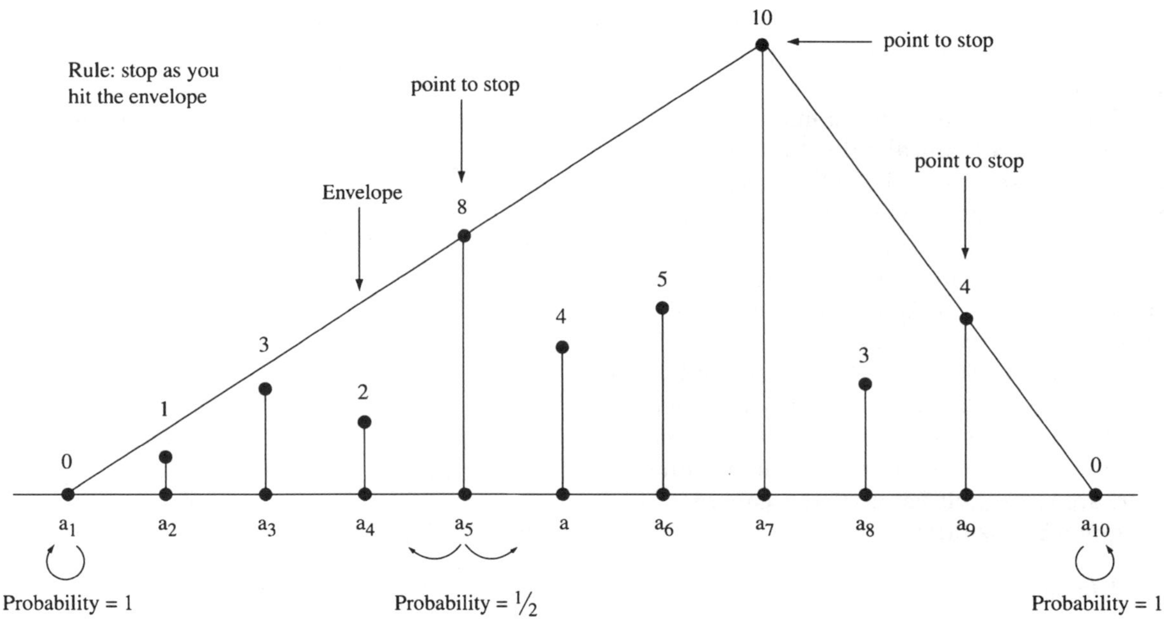

FIGURE 24.3 Illustration of the envelope – the points at which we end up stopping are those times when the payoff from the state is on a *boundary*.

Suppose at some stage we reach the two end points. Clearly we have no choice but to stop. The states are absorbent. We will never visit other states.

Next consider the state a_8. Should we stop once we have $S_i = a_8$, and accept the offered payoff $F(a_8)$? The answer is, obviously, no (unless, of course, $n = 5$ and we have to stop). It is clear that the next move of S_i will be to either a_7 or to a_9. As can be seen from Figure 24.3, either of these states has a payoff higher than $F(a_8)$. By continuing, we are guaranteed to do better. We should not stop.

Another obvious decision occurs at state a_7. This state is associated with the highest payoff ever, and we should clearly stop as soon as we reach it. We are not going to do any better by continuing.

Thus far, the decision to stop was not complicated at all. But now consider the two states a_3 and a_5. Here the decisions will be more complicated. Both of these states have the following property. The payoff is a *local maximum*. If we continue after reaching them, we go to adjacent states and these states have lower payoffs. The decision maker will be worse off at least in the immediate future. But by not stopping, one is also keeping the possibility of reaching a payoff such as $F(a_7)$ or $F(a_5)$ open. So which one is better? Should one accept the local maxima such as $F(a_3)$ or $F(a_5)$ and stop, or should one continue at these points and expect to stop at a future date when there is a higher payoff?

The answer is not obvious at the outset, and requires careful evaluation of future possibilities. In fact, the two states a_3 and a_5 will give different answers. It will be optimal to continue at a_3 and stop at a_5.

Begin with state a_5. Suppose at some i we have $S_i = a_5$. If we *stop*, we get $F(a_5) = 8$. How much do we expect to get if we *continue?*

It is easy to calculate the expected payoff of the immediate future:

$$\mathbb{E}\left[F\left(S_{i+1}\right)\middle|\, S_i = 5\right] = \frac{1}{2}F\left(a_4\right) + \frac{1}{2}F\left(a\right) \quad (24.45)$$

$$= 3 \quad (24.46)$$

This is clearly worse than the 8 we can guarantee by stopping now. But there is an additional point. The game does *not* end at the next stage and looking only at the immediate future would ignore the expected payoff that would result if we reached the state a_7.

What we would like to do is to obtain an optimal payoff function denoted by, say, $V(a_5)$ that represents the greater of two payoffs; namely, the current payoff if we stop, or the expected payoff if we decide to continue, *assuming that we continue in an optimal fashion*. That is, we want:

$$V(a_5) = \max\left[\mathbb{E}\left[\text{Payoff}\middle|\,\text{Stop}\right], \mathbb{E}\left[\text{Payoff}\middle|\,\text{Continue}\right]\right] \tag{24.47}$$

Here the expected payoff if we stop is known. It is the $F(a_5)$. On the other hand, the expected payoff if we continue is unknown. It should be calculated by using the *same* notion as $V(a_5)$ for future periods in an optimal fashion.

In other words, we need to write:

$$V(a_5) = \max\left[\mathbb{F}(a_5), \frac{1}{2}V(a_4) + \frac{1}{2}V(a)\right] \tag{24.48}$$

Note that this assumption assumes no discounting.

Thus, before we can calculate $V(a_5)$ we need to determine the $V(a_4)$ and the $V(a)$. But, there is the same problem with these. The V that corresponds to future states seems to be unknown.

Although the reasoning seems circular, it really is not. The problem is set up in a way that there are some stages where calculation of the $V(a_i)$ is immediate. For example, as we already know that we will stop when we reach a_1 or a_7:

$$V(a_7) = F(a_7) = 8 \tag{24.49}$$

Also, we know that

$$V(a_1) = F(a_1) = 0 \tag{24.50}$$

Thus, by substituting for "future" in (24.48), we can eventually get to's that are known to us. Then, the $V(a_5)$ can be evaluated.

So how can we take this into account in (24.48)? We do so by writing,

$$\mathbb{E}\left[F(S_{i+1})\middle|\, S_i = a_7, \text{ we stop optimally}\right] = \frac{1}{2}V(a_4) + \frac{1}{2}V(a) \tag{24.51}$$

instead of acting as we did in (24.46) and taking into account only the immediate future. In other words, we want to substitute the $V(a_4)$ and $V(a)$ in place of $F(a_4)$ and $F(a)$ in (24.46). This is the case because the latter does not take into account the possibility of reaching the higher future payoffs and then stopping there, whereas the $V(a_4)$ and $V(a)$ do. In other words the $V(a_4)$ and $V(a)$ incorporate the idea that the decision will be made optimally in the future. This way of reasoning is very similar to pricing derivatives by going "backwards" in a binomial tree. When this is done, it will be clear that we should continue at a_3 but stop at a_5. Note that the expectation on the left-hand side of (24.51) is different from (24.46) since it is now conditional on the fact that we stop optimally. Hence, it is in fact $V(a_5)$.

A final comment on the example. We should point out an interesting occurrence in Figure 24.3. The points at which we end up stopping are those times when the payoff from the state is on a *boundary* that we denote as the envelope in Figure 24.3. Thus, if a market practitioner is given this envelope, then the rule to pick optimal stopping times will be greatly simplified. All one has to do is to see whether the payoff is below the envelope or on it. One stops if the current payoff equals the value of the envelope at that point, and continues otherwise. Essentially, this is what is meant by Eq. (24.37), which gives an optimal stopping rule.

24.7 STOPPING TIMES AND MARTINGALES

We finish this chapter by looking at the role played by stopping times in the theory of martingales. It turns out that most of the results discussed in this book can be extended to stopping times. Below we simply give two such results without commenting extensively on them.

24.7.1 Martingales

Suppose M_t represents a continuous-time martingale with respect to a probability P, with

$$\mathbb{E}\left[M_{t+u}|\, I_t\right] = M_t \tag{24.52}$$

Would this martingale property be preserved if we consider randomly selected times as well?

The answer is yes under some conditions. Let τ_1 and τ_2 be two independent stopping times measurable with respect to I_t and satisfying:

$$P\left(\tau_1 < \tau_2\right) = 1 \tag{24.53}$$

Then, the martingale property will still hold:

$$\mathbb{E}\left[M_{\tau_2}\middle|\, I_{\tau_1}\right] = M_{\tau_1} \tag{24.54}$$

This property is clearly important in writing asset prices using equivalent martingale measures. The fact that the exercise date of a derivative is random does not preclude the use of equivalent martingale measures. With random τ, randomly stopped asset prices will still be martingales under the probability $\mathbb{Q}$.

24.7.2 Dynkin's Formula

Let B_t be a process satisfying:

$$dB_t = a\left(B_t\right)dt + \sigma\left(B_t\right)dW_t \tag{24.55}$$

Let $f(B_t)$ be a twice-differentiable bounded function of this process.

Now consider a stopping time τ such that:

$$\mathbb{E}^{\mathbb{Q}}\left[\tau\right] < \infty \tag{24.56}$$

Then we have

$$\mathbb{E}^{\mathbb{Q}}\left[f\left(B_\tau\right)\middle|\, B_0\right] = f\left(B_0\right) + \mathbb{E}^{\mathbb{Q}}\left[\int_0^\tau Af\left(B_s\right)ds\middle|\, B_0\right] \tag{24.57}$$

This expression is called Dynkin's formula. It gives a convenient representation for the expectation of a function that depends on a stopping time. The operator A is as usual the generator.

24.8 CONCLUSIONS

The chapter also introduced the notion of stopping times. This concept was useful in pricing American-style derivative products and in dynamic programming. We also illustrated the close relationship between binomial tree models and a certain class of Markov chains.

24.9 REFERENCES

There are three important topics in this chapter. First, there is the issue of stopping times. The early exercise is an optimal stopping problem. We can recommend Dynkin et al. (1999). This treatment is classic but still very intuitive. A reader interested in learning more about classical stopping time problems can read the book by Shiryayev (1978). The second major topic that we mentioned is dynamic programming, although this was a side issue for us. There are many excellent texts dealing with dynamic programming.

Finally, there is the issue of numerical calculation of stopping times. Here the reader can go to references given in Broadie and Glasserman (1998).

24.10 EXERCISES

1. A player confronts the following situation. A coin will be tossed at every time $t, t = 1, 2, 3, \ldots, T$ and the player will get a total reward W_t. He or she can either decide to stop or to continue to play. If he or she continues, a new coin will be tossed at time $t+1$, and so on. The question is, what is the best time to stop? We consider several

cases. We begin with the double-or-nothing game. The total reward received at time $t = T$ is given by:

$$W_T = \prod_{t=1}^{T} (z_t + 1)$$

where the z_t is a binomial random variable:

$$z_t = \begin{cases} 1 & \text{with probability } \frac{1}{2} \\ -1 & \text{with probability } \frac{1}{2} \end{cases}$$

Thus, according to this, the reward either doubles or becomes zero at every stage.

(a) Can you calculate the expected reward at time $T, \mathbb{E}[W_T]$, given this information?
(b) What is the best time to stop this game?
(c) Suppose now we sweeten the reward at every stage and we multiply the W_T by a number that increases and is greater than one. In fact, suppose the reward is now given by:

$$\overline{W}_T = \frac{2n}{n+1} \prod_{t=1}^{T} (z_t + 1)$$

with $T = 1, 2, 3, \ldots$ Show that the expected reward if we stop at some time T_k is given by:

$$\frac{2k}{k+1}$$

(Here, T_k is a stopping time such that one stops after the kth toss.)
(d) What is the maximum value this reward can reach?
(e) Is there an optimal stopping rule?

2. Consider the problem above again. Suppose we tossed a coin T times and the resulting z_t were *all* +1. The reward will be:

$$W_T = \frac{T\left(2^T + 1\right)}{T+1}$$

(a) Show that the conditional expected reward as we just play one more time is:

$$\mathbb{E}\left[\, W_{T+1} |\, W_T = w_T^* \right] = 2^{T+1} \frac{T+1}{T+2}$$

(b) How does this compare with W_T? Should the player then "stop?"
(c) But if the player never stops when he or she is in a winning streak, how long would the player continue playing the game?
(d) What is the probability that $z_t = -1$ at some point?
(e) How do you explain this puzzle?

3. Suppose you are given the following data: Risk-free interest rate is 6%, the stock price follows $dS_t = \mu S_t dt + \sigma S_t dW_t$, volatility is 12% a year and the stock pays no dividends, and the current stock price is 100. Using these data, you are asked to approximate the current value of an American call option on the stock. The option has a strike price of 100 and a maturity of 200 days.

(a) Determining an appropriate time interval Δ, such that the binomial tree has four steps. What would be the implied U and D?
(b) What is the implied "up" probability?
(c) Determine the tree for the stock price S_t.
(d) Determine the tree for the call premium C_t.
(e) Now the important question: would this option ever be exercised early?

4. Suppose the stock discussed above pays dividends. Assume all parameters are the same. Consider these three forms of dividends paid by the firm and in each case determine if the option will be exercised early.

(a) The stock pays a continuous, known stream of dividends at a rate of 4% per time.

(b) The stock pays 5% of the value of the stock at the third node. No other dividends are paid.
(c) The stock pays a \$5 dividend at the third node.

5. Consider a policy maker who uses an instrument k_t to control the path followed by some target variable Y_t. The policy maker has the following objective function

$$U = \sum_{i=1}^{4}\left[2\left(k_t - k_{t-1}\right)^2 + 100\left(Y_t\right)^2\right]$$

The environment imposes the following constraint on this policy maker:

$$Y_t = 0.2k_t + .6Y_{t-1}$$

The initial Y_0 is known to be 60.

(a) What is the best choice of k_t for period $t = 4$?
(b) What is the best choice of k_t for period $t = 3$?
(c) From these, can you iterate and find the best choice of k_t for $t = 1$?
(d) Determine the value function V_t that gives the optimal payoff for $t = 1, 2, 3, 4$.
(e) Plot the value function V_t and interpret it.

6. Prove that if $X_1, X_2, \ldots$ are independent and identically distributed random variables having finite expectations, and if N is a stopping time for $X_1, X_2, \ldots$ such that $\mathbb{E}(N) < \infty$, then

$$\mathbb{E}\left(\sum_{!}^{N} X_n\right) = \mathbb{E}(N)\mathbb{E}(X) \qquad (24.58)$$

7. Consider the following SDE that governs asset prices in our model:

$$dS_t = \mu S_t dt + \sigma S_t dW_t \qquad (24.59)$$

Assume that the risk-free interest rate is 3% and that there are no dividends in the model. Further, assume the current stock price is \$70. Write a program that prices a one-year American put option struck at \$55 in the above framework. Make sure that the optimal exercise decision is included in the tree.

CHAPTER

25

Overview of Calibration and Estimation Techniques

OUTLINE

Derivatives pricing begins with the assumption that the evolution of the underlying asset, which can be a stock, commodity, interest rate, or exchange rate, follows some stochastic process. For the model under consideration we develop a pricing routine to calculate the derivative contract premium. Pricing routines for derivative contracts take two mutually exclusive sets of parameters as input, which are (a) contractual/market parameters and (b) model parameters. Contractual/market parameters reflect features of the derivative which are specified in the contract, such as maturity of the contract, strike price, barrier level, and the like, which are clearly model independent. Model parameters, associate with the choice of the model for evolution of the underlying process; are subjective and utterly model dependent.

All pricing models require a parameter set to fully define the dynamics of each model. In order to make a model relevant to real markets and applicable for pricing, risk management, or trading, we must perform calibration. Calibration is the process of determining a parameter set such that model prices and market prices match very closely for a given set of liquidly traded instruments. These liquid instruments are called benchmark or calibration instruments and their

Financial Derivatives, Third Edition.
http://dx.doi.org/10.1016/B978-0-12-384682-2.00025-6

calibrated prices are typically recorded together in a market snapshot. The so-called *calibration* procedure delivers the optimal parameter set for the model based on these calibration instruments. For many applications, this set represents the majority, if not the entirety, of the set of liquidly traded derivatives in a given market. The calibration procedure is typically done very frequently, for example two or three times per day, in order to keep model derived prices close to their real world equivalents. This vital step in derivative pricing can be used to price other less liquid derivatives utilizing the prices of their more liquid counterparts. Lastly, they can also spot arbitrage opportunities among liquidly traded derivatives, among other applications.

In case of using a model in which the set of its parameters is larger than the set of prices for the calibration instruments, then the solution obviously would not be unique. This type of problem is over-parameterized. This is often the case in markets with small numbers of derivatives and complex dynamics. In cases of over-parametrization, no set of model parameters can be found which forces the model prices to exactly match the market prices of the calibration instruments; so in practice, an approximate solution is determined by solving an appropriately constructed optimization problem. On the contrary, there could be a case that the model class is too narrow to reproduce a full set of prices for the set of calibration instruments; then the solution does not exist. Moreover, this type of problem is under-parameterized. The best case for it is option pricing in the Black–Scholes framework, where there is only one free parameter, volatility, yet many liquidly traded options. In the case of under-parametrization, the model is typically calibrated in such a way that smaller subsets of calibration instruments have their market prices matched with the model prices under different model parameters. The most obvious example is the classic Black–Scholes volatility surface, where a calibrated volatility exists for every liquidly traded option. That is why, in general, the model calibration is not a well-posed problem (Cont, 2010).

25.1 CALIBRATION FORMULATION

Assuming a model for the evolution of the underling process and having a pricing routine for the derivative contract, we can now set the objective function in order to perform calibration. Objective function is typically set as the square root mean error. We then apply an optimization routine to minimize the objective function. In most generic cases, one can write the optimization problem as follows:

$$\inf_{\Theta\in\mathbb{O}} \sum_{i=1}^{I} H\left(C_i^{\Theta} - C_i\right) \tag{25.1}$$

where $\mathbb{O}$ is the space for all possible parameter sets for the model and H is an objective function applied to the discrepancy $C_i^{\Theta} - C_i$ between market and model prices. Once we have formulated this calibration problem, an optimization algorithm is then applied to compute a solution and determine the calibrated model parameters. General formulation for the objective function H could be as follows:

$$H(C_i^{\Theta} - C_i) = w_i|C_i^{\Theta} - C_i|^p$$

or

$$H\left(C_i^{\Theta} - C_i\right) = w_i|\frac{C_i^{\Theta} - C_i}{C_i}|^p$$

or

$$H\left(C_i^{\Theta} - C_i\right) = w_i|\ln\ C_i^{\Theta} - \ln\ C_i|^p$$

where $p \geq 1$ and w_i is a positive weight often chosen inversely proportional to the squared of bid-offer spread of C_i. The results may vary with the choice of the objective function. The most common formulation of the calibration problem is the least-squares formulation:

$$\inf_{\Theta\in\mathbb{O}} \sum_{i=1}^{I} w_i\left(C_i^{\Theta} - C_i\right)^2 \tag{25.2}$$

25.2 UNDERLYING MODELS

We will over three models, namely the Black–Scholes model, the local volatility model, and the variance gamma model.

25.2.1 Geometric Brownian Motion—Black–Scholes Model

The Black–Scholes model was described in Section 12. The Black–Scholes PDE gives us the price of derivative securities depending on the terminal and boundary conditions we apply to it. Consider the Black–Scholes PDE

$$\frac{\partial V}{\partial t} + \frac{1}{2}\sigma^2 S^2 \frac{\partial^2 V}{\partial S^2} + (r-q)S\frac{\partial V}{\partial S} = rV \qquad (25.3)$$

Assuming the terminal boundary condition is the payoff of a put option at maturity T

$$V(S,T) = \max(K-S,0) \qquad (25.4)$$

The analytical solution to this PDE at time $t < T$ is the Black–Scholes option pricing formula for a European put, which is given by

$$V(S,t) = Ke^{-r(T-t)}\Phi(-d_2) - Se^{-q(T-t)}\Phi(-d_1) \qquad (25.5)$$

where

$$d_1 = \frac{\ln(S/K) + (r-q+\sigma^2/2)(T-t)}{\sigma\sqrt{T-t}} \qquad (25.6)$$

$$d_2 = d_1 - \sigma\sqrt{T-t} \qquad (25.7)$$

The only non-observable model parameter in this model is the volatility of the underlying asset. It is the only variable which affects the value of the option. To find out the minimum and maximum possible value for the option premium, we calculate the case that volatility gets large (hypothetically infinity) and the case that volatility dies out (goes to zero).

Case 1—volatility approaches infinity ($\sigma \to \infty$). From Eqs. (25.6) and (25.7) we can see that $d_1 \to \infty$ and $d_2 \to -\infty$. Therefore $\Phi(-d_1) = \Phi(-\infty) = 0$ and $\Phi(-d_2) = \Phi(\infty) = 1$ which implies $V(S,t) = Ke^{-r(T-t)}$

Case 2—volatility approaches zero ($\sigma \to 0$). For this case it depends on the location of S versus the strike K.

(a) if $S > K$, it is easy to see from Eqs. (25.6) and (25.7) that d_1 and d_2 are both approaching infinity which means $\Phi(-d_1) = \Phi(-d_2) = \Phi(-\infty) = 0$ and that implies $V(S,t) = 0$

(b) if $S < K$, it is easy to see from Eqs. (25.6) and (25.7) that d_1 and d_2 are both approaching negative infinity which means $\Phi(-d_1) = \Phi(-d_2) = \Phi(\infty) = 1$ and that implies $V(S,t) = Ke^{-r(T-t)} - Se^{-q(T-t)}$.

It happens that what we just perform holds no matter what the underlying process is and, therefore, it is model-free. One can utilize the bisection method, the Newton–Raphson approach, or the like to find a volatility that the model price matches the given premium as long as the put premium is between zero and $V(S,t) = Ke^{-r(T-t)}$. That obtained volatility is called *implied volatility*.

In reality there are more than one option premiums available in the market and it is not realistic assuming that one volatility can describe the behavior of the surface, therefore using Black–Scholes is not feasible. Having said that, market practitioners quote option premiums based on Black–Scholes implied volatilities.

The parameter set, which we need to determine during calibration, is simply $\Theta = \{\sigma\}$, the implied volatility of the traded option. Using call and put premiums provided in Tables 25.1 and 25.2, we find the implied volatility for each option and form a surface for call and put premiums separately. The implied volatility surfaces are depicted in Figures 25.1(a) and 25.1(b) for calls and puts respectively.

25.2.2 Local Volatility Model

The Black–Scholes model is considered to be the simplest formulation for derivative pricing and is yet used for many other simpler derivative contracts; however, the need for a volatility

TABLE 25.1 S&P 500 Call Option Premiums on March 27, 2012

Maturity (in Days)	Strike	Bid	Ask	r	q	Forward Price
24	1405	21.7	23	0.2324	1.5314	1410.55
	1410	19.1	20	0.2324	1.5314	1410.80
	1415	15.9	17.1	0.2324	1.5314	1410.55
	1420	13.6	14.6	0.2324	1.5314	1410.70
	1425	11	12.2	0.2324	1.5314	1410.55
52	1405	30.8	32.6	0.3240	2.1087	1407.85
	1410	28.2	29.6	0.3240	2.1087	1408.00
	1415	25.5	26.8	0.3240	2.1087	1408.00
	1420	22.8	24.1	0.3240	2.1087	1407.95
	1425	20.6	21.6	0.3240	2.1087	1408.09
80	1405	39.5	41.4	0.4331	2.0990	1406.30
	1410	36.8	38.3	0.4331	2.0990	1406.50
	1415	33.8	35.6	0.4331	2.0990	1406.29
	1420	31.3	32.7	0.4331	2.0990	1406.54
	1425	28.4	30.1	0.4331	2.0990	1406.38
93	1375	61.8	66.3	0.4826	2.0819	1405.74
	1400	47	48.4	0.4826	2.0819	1405.81
	1425	32.8	34.1	0.4826	2.0819	1405.83
	1450	21.2	22.5	0.4826	2.0819	1405.04
	1475	12.6	13.9	0.4826	2.0819	1405.66
115	1375	67.4	69.7	0.5572	1.9816	1404.05
	1400	51.4	53.6	0.5572	1.9816	1404.41
	1425	37.3	39.2	0.5572	1.9816	1404.21
	1450	25.5	27.2	0.5572	1.9816	1404.27
	1475	16.2	17.7	0.5572	1.9816	1404.12
178	1375	80.7	83.2	0.7316	2.0899	1400.04
	1400	65.2	67.3	0.7316	2.0899	1400.45
	1425	50.7	53.1	0.7316	2.0899	1400.31
	1450	39	40.5	0.7316	2.0899	1400.22
	1475	28	29.7	0.7316	2.0899	1400.33
184	1375	82.2	84.5	0.7436	2.0980	1399.14
	1400	66.6	68.8	0.7436	2.0980	1399.15
	1425	52.5	54.5	0.7436	2.0980	1399.00
	1450	40.1	41.9	0.7436	2.0980	1399.86
	1475	29.5	31	0.7436	2.0980	1399.71
269	1375	95.5	98.5	0.8831	2.1546	1393.82
	1400	80.4	83.4	0.8831	2.1546	1393.31
	1425	66.5	69.1	0.8831	2.1546	1393.59
	1450	53.7	56.4	0.8831	2.1546	1393.63
	1475	42.4	44.8	0.8831	2.1546	1393.52

TABLE 25.2 S&P 500 Put Option Premiums on March 27, 2012

Maturity (in Days)	Strike	Bid	Ask	r	q	Forward Price
24	1405	16.2	17.4	0.2324	1.5314	1410.55
	1410	18.1	19.4	0.2324	1.5314	1410.80
	1415	20.3	21.6	0.2324	1.5314	1410.55
	1420	22.7	24.1	0.2324	1.5314	1410.70
	1425	25.3	26.8	0.2324	1.5314	1410.55
52	1405	27.9	29.8	0.3240	2.1087	1407.85
	1410	29.9	31.9	0.3240	2.1087	1408.00
	1415	32.1	34.2	0.3240	2.1087	1408.00
	1420	34.4	36.6	0.3240	2.1087	1407.94
	1425	36.9	39.1	0.3240	2.1087	1408.09
80	1405	38.1	40.2	0.4331	2.0990	1406.30
	1410	40.2	41.9	0.4331	2.0990	1406.50
	1415	42.3	44.5	0.4331	2.0990	1406.29
	1420	44.6	46.3	0.4331	2.0990	1406.54
	1425	47	48.7	0.4331	2.0990	1406.38
93	1375	32.6	34.1	0.4826	2.0819	1405.74
	1400	41	42.8	0.4826	2.0819	1405.81
	1425	51.6	53.6	0.4826	2.0819	1405.83
	1450	64.8	68.7	0.4826	2.0819	1405.04
	1475	80.1	84.9	0.4826	2.0819	1405.66
115	1375	38.5	40.6	0.5572	1.9816	1404.05
	1400	47.2	49	0.5572	1.9816	1404.41
	1425	57.8	60.2	0.5572	1.9816	1404.21
	1450	70.6	73.4	0.5572	1.9816	1404.27
	1475	86.2	89.2	0.5572	1.9816	1404.12
178	1375	55.7	58.3	0.7316	2.0899	1400.04
	1400	64.8	66.8	0.7316	2.0899	1400.45
	1425	75.4	77.6	0.7316	2.0899	1400.31
	1450	87.7	91	0.7316	2.0899	1400.22
	1475	102	104.5	0.7316	2.0899	1400.33
184	1375	57.9	60.7	0.7436	2.0980	1399.14
	1400	67	70.1	0.7436	2.0980	1399.15
	1425	77.7	81.1	0.7436	2.0980	1399.00
	1450	88.5	93.4	0.7436	2.0980	1399.86
	1475	102.3	108.2	0.7436	2.0980	1399.71
269	1375	77.1	79.5	0.8831	2.1546	1393.82
	1400	86.8	90.3	0.8831	2.1546	1393.31
	1425	97.6	100.4	0.8831	2.1546	1393.59
	1450	109.5	112.6	0.8831	2.1546	1393.63
	1475	122.9	126.2	0.8831	2.1546	1393.52

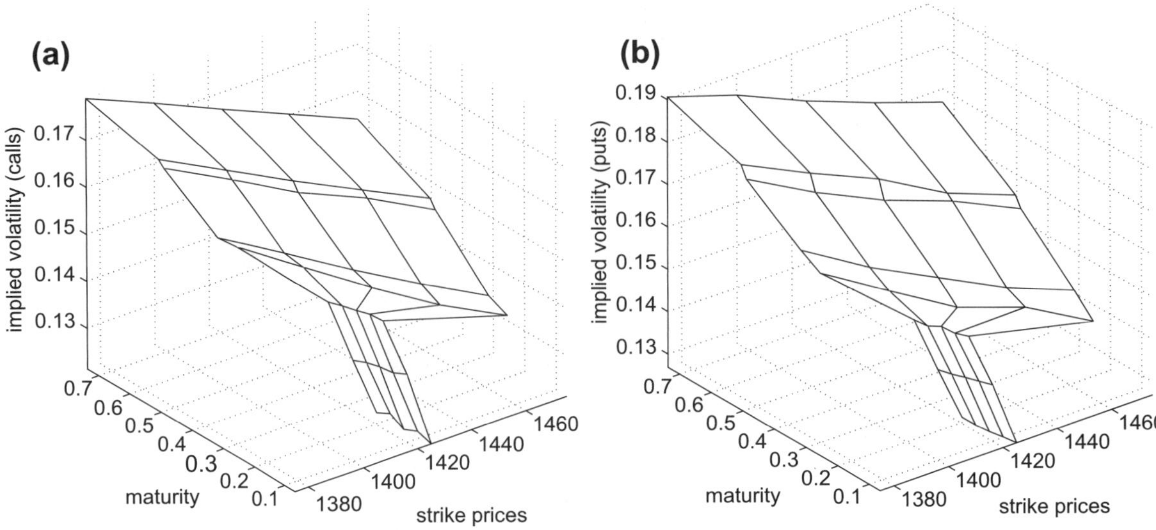

FIGURE 25.1 Implied volatility surface example: (a) calls, (b) puts.

surface, which implies different underlying parameters for every quoted option is needed and the model's inability to correctly replicate the evolution of the underlying asset limits its usefulness in pricing and hedging more exotic derivative contracts and thus extensions were developed.

To overcome the inadequacies and limitations of the Black–Scholes model, Derman and Kani (1994) proposed a local volatility model which parameterized underlier volatility in terms of the current underlier price and the calendar time. Its invention made the pricing and hedging derivatives in a manner consistent with the volatility smile which has been a major research area for over a decade and the development of the local volatility model was a significant step in improving performance in this area.

Extending the parameterization of the standard Black–Scholes model to include both underlier price and time which will allow us to simultaneously calibrate to many or all liquid vanilla option contracts. In addition, at the same time allowing this model to be more realistically applied to exotic derivatives. The local volatility model follows the following stochastic differential equation:

$$dS_t = (r(t) - q(t))S_t dt + \sigma(S_t, t)S_t dW_t \qquad (25.8)$$

Here they assume both interest rate and dividend rate have a deterministic term structure. Prices of options under this model will satisfy a Black–Scholes type equation which we call the generalized Black–Scholes PDE, which gives the pricing of derivatives securities, depending on the terminal condition and boundary conditions we apply to it.

$$\frac{\partial V}{\partial t} + \frac{1}{2}\sigma^2(S,t)S^2\frac{\partial^2 V}{\partial S^2} + (r(t) - q(t))S\frac{\partial V}{\partial S} = r(t)V(S,t) \qquad (25.9)$$

The attractiveness of the local volatility model is due to its simplicity. Arriving at solutions and derivative prices under the local volatility model requires only very few simple modifications of the Black–Scholes model. This, along with its additional flexibility in consistently pricing a full set of options, is why most traders and firms actively utilize this model. Marking and pricing of derivatives is simple and calculating hedge ratios is straightforward. Moreover, as explained in Hirsa et al. (2001), local volatility models re-engineer pure jump models for vanilla options. However, for path-dependent options, the local volatility model and pure jump models could behave very differently (Hirsa et al., 2003).

The set of calibration parameters under the local model is $\Theta = \{\sigma(S_t, t)\}$ and the set of contract parameters and deterministic market parameters is $\Lambda = \{S_0, K, T, r(t), q(t)\}$.

Thus, the objective of the calibration procedure for local volatility models is to find the local volatility surface $\sigma(S, t)$ so that model prices will closely match market prices for options at available strikes and maturities. One way to accomplish this is to formulate the calibration problem as follows:

$$\arg\min_{\Theta} \frac{1}{M}\frac{1}{N}\sum_{i=1}^{N}\sum_{j=1}^{M} \|V_{ij} - \widehat{V}_{ij}\| \tag{25.10}$$

where V_{ij} and $\widehat{V}_{ij}$ are market and model prices for strike K_i and maturity T_j, respectively. Note that for each set of strike and maturity we assume the same volatility surface $\sigma(S, t)$, yet we have to solve the generalized Black–Scholes PDE for each set of maturity and strike separately, in order to generate the market prices for any and all available options. For a fixed strike price, the option payoff is the same, but it is paid at a different time (maturity), which would result in a different set of option prices. For the fixed maturity, the option has a different payoff depending on the strike price, which would again result in a different set of option prices.

25.2.3 Forward Partial Differential Equations for European Options

As noted in the last section, the basic construction of the calibration problem for local volatility surfaces, when utilizing the generalized Black–Scholes PDE, requires us to solve the PDE, again for every calibrated derivative price and for every iteration of the optimization routine. This involves many PDE solutions and, if possible, we would like to be able to solve for every option price in the strike and maturity grid simultaneously. This could reduce our computation time for the calibration procedure significantly.

A breakthrough occurred in the mid-nineties with the recognition that in certain models European option values satisfied forward evolution equations in which the independent variables are the options strike and maturity. Specifically, Dupire (1994) showed that under deterministic carrying costs and a diffusion process for the underlying price, no arbitrage implies that European option prices satisfy a certain partial differential equation (PDE), now called the Dupire equation. If we assume that one can observe European option prices at all strikes and maturities, then this forward PDE can be used to explicitly determine the underlying's instantaneous volatility as a function of the underlying's price and time.

Once this volatility function is known, the value function for European, American and many exotic options can be determined by a wide array of standard methods. Because this value function relates to theoretical prices of these instruments to the underlying's price and time, it can also be used to determine many hedge parameters (Greeks) of interest as well. In addition to their usefulness in determining the volatility function, forward equations also serve another useful purpose. Once the volatility function is known, either by an explicit specification or by a prior calibration, the forward PDE can be solved via finite differences to efficiently value a collection of European options of different strikes and maturities, all written on the same underlying asset. Assuming a known local volatility surface, this will allow us to solve for every option price in strike and maturity space simultaneously on the same grid and, as pointed out in Andreasen (1998), all the Greeks of interest satisfy the same forward PDE and hence can also be efficiently determined in the same way. This will allow us to solve the implied volatility calibration problem in a much more computationally efficient manner.

The Dupire PDE that gives European call prices for all strikes and maturities is given by

$$-\frac{\partial C}{\partial T} + \frac{1}{2}\sigma^2(K,T)K^2\frac{\partial^2 C}{\partial K^2} - \big(r(T) - q(T)\big)K\frac{\partial C}{\partial K} = q(T)C \tag{25.11}$$

having the call surface would be straightforward to find the local volatility surface. For this PDE it is easy to see that if one assumes market quotes of option prices $C(K_i, T_j)$ are given/available, we can then calculate the local volatility surface from market prices explicitly using the following inversion formula:

$$\sigma(K,T) = \left(\frac{\frac{\partial C}{\partial T} + \big(r(T) - q(T)\big)K\frac{\partial C}{\partial K} + q(T)C}{\frac{1}{2}K^2\frac{\partial^2 C}{\partial K^2}}\right)^{1/2} \tag{25.12}$$

To apply the local volatility inversion formula from the Dupire equation, we must work under the assumption that we have a very smooth surface for the call price premiums in terms of strike and maturity, $C(K,T)$. In addition, we must also be able to calculate the calendar spread, $\frac{\partial C}{\partial T}$, butterfly spread, $\frac{\partial C}{\partial K}$ and second partial derivative with respect to strike price, $\frac{\partial^2 C}{\partial K^2}$, at arbitrary points on the surface of call price premiums.

In practice, this is the key difficulty in implementing this method. Option prices are only available in the market on a finite grid of strike prices and maturities, where interpolation schemes must be invoked to infer prices for the intermediate strike prices and maturities. Here interpolations used, may or may not be consistent with the requirements of the absence of at least static arbitrage across the strike price and maturity spectrum. Even when this is accomplished, the interpolation schemes can introduce non-differentiability at various levels, leading to local volatility functions that are erratic and inspire little confidence. The task of properly interpolating the surface of option prices, consistent with observed market prices, is essentially the task of formulating and estimating a market with a consistent option pricing model. As illustrated in Cont (2010), the local volatility can be very sensitive to small changes in inputs. In the coming section on calibration of local volatility surfaces, we will explore a number of different methods proposed to construct local volatility surfaces. It remains to be seen if they generate a smooth, stable, and consistent local volatility surface.

However, there are not enough option prices to construct a smooth call price surface in order to use the inversion formula, then what to do? The very first thing that comes to mind is to either use interpolation extrapolation to build a smooth surface (for example, bi-cubic spline interpolation) on call premiums or implied volatilities and invert them, or, assuming some function form for the local volatility (five or six variables), find those parameters via optimization. Note that in the first approach there is no need for an optimizer or an objective function.

Using call premiums provided in Table 25.1, we perform bi-cubic interpolation to generate call prices at finer mesh from the existing coarser mesh. Then, using Eq. (25.12), obtain the local volatility surface. The obtained local volatility surface is illustrated in Figure 25.2. As we see in that figure, the surface is pretty non-smooth. We assume the following functional from the local volatility surface which resembles sine hyperbolic function.

$$\sigma(K,T) = \frac{1}{2}\big(c_1\exp(K/c_2) + c_3\exp(-K/c_2)\big)\exp(-\beta T) \tag{25.13}$$

Through calibration we find the following parameters for the local volatility function

$$c_1 = -0.33 \tag{25.14}$$

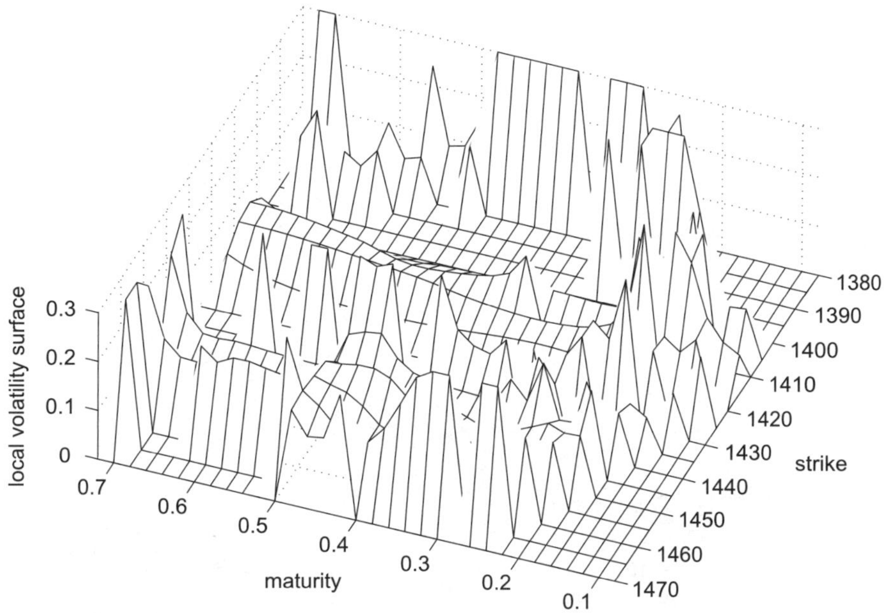

FIGURE 25.2 Local volatility surface obtained from interpolated call prices.

$$c_2 = 1404.27 \tag{25.15}$$

$$c_3 = 2.19 \tag{25.16}$$

$$c_4 = 2222.02 \tag{25.17}$$

$$\beta = -1.13 \tag{25.18}$$

In Figure 25.3 we illustrate the local volatility surface by substituting the parameters into the function in (25.13). We did not put any restriction on the signs of beta and we see that, unlike what we thought, we see that volatility increases with maturity. We plot the call price surface obtained from the surface in Figure 25.4 versus the market call prices, as illustrated we see a tight fit, the average discrepancies is 64.5 cents. In comparison with the bid-ask spread is pretty tight. That does not imply the local volatility obtained from it is admissible. For other models, such as the Heston stochastic volatility model or the variance gamma model, we need a pricing model for the instruments used for calibration purposes and simply setting up the objective function and using the optimizer to find the optimal parameter set minimizes the objective function.

In Hirsa et al. (2003), the authors propose inferring a local volatility surface from the calibrated parameters by simply calculating call prices for a range of strike prices and maturities for the model under consideration and substituting those premiums in Eq. (25.12) to obtain the local volatility surface. Any stochastic volatility model, such as the Heston stochastic volatility model, can be successfully calibrated to market European option prices across all strikes and maturities. The obvious result being those calibrated models can deliver a smooth surface for call/put prices in both dimensions. For a stochastic volatility model in a pure jump framework, they have shown the scheme generates a pretty smooth local volatility surface. The obtained local volatility surface is fully dependent on the model originally used to interpolate European option prices in strike and maturity space. Obviously it is not going to be unique. They illustrate the constructed local volatility

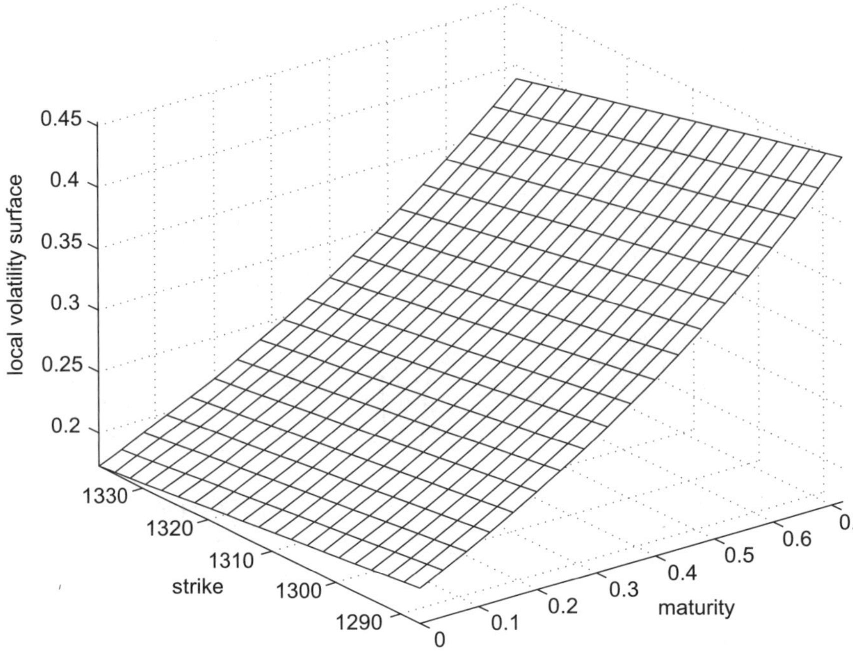

FIGURE 25.3 Local volatility surface obtained using a functional form for the local volatility surface.

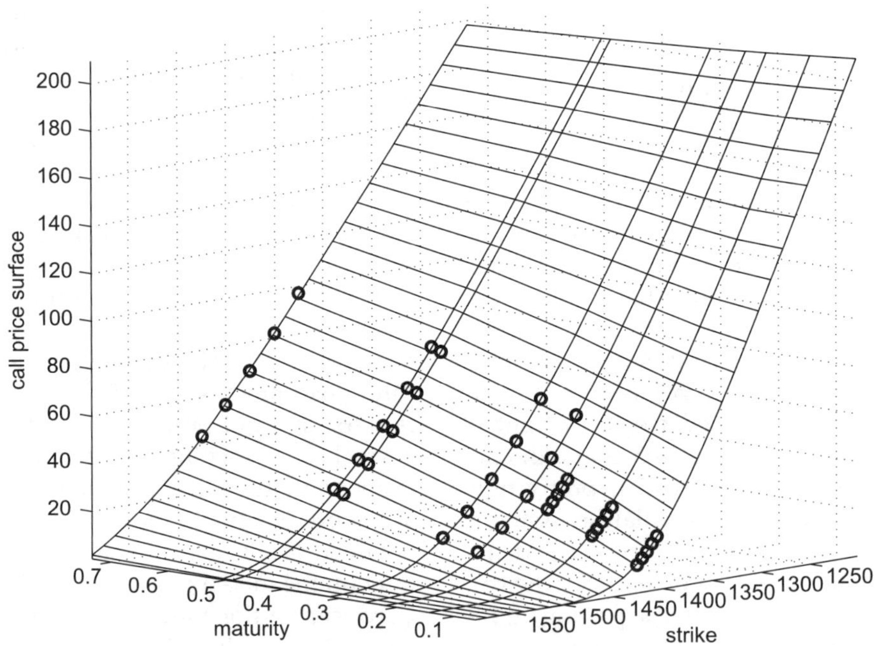

FIGURE 25.4 Call price surface versus market prices.

surface for the S&P 500 stock index as of using this approach for the VGSA and Heston stochastic volatility models, respectively.

25.2.4 Variance Gamma Model

The presence of sometimes very large volatility smiles for option prices, especially for short durations, can lead to very large changes in modeled volatilities under diffusion models. This will allow for non-constant volatility, such as the local volatility, constant elasticity of volatility, and Heston stochastic volatility models. This issue has led many researchers to concentrate their efforts on modeling underlying asset prices with jump models, which allow for discrete jumps in asset prices which may more readily explain the volatility smile. One such model is the variance gamma model which was described in Section 11.6.2. As explained, it is a three-parameter model with the calibration parameters being $\Theta = \{\sigma, \theta, \nu\}$, volatility, skewness, and kurtosis respectively.

Results indicate that for a fixed maturity the model does an adequate job fitting across various strike prices. This would suggest that for each maturity we should do a separate calibration and as a result would have a separate set of parameters. In calibration, we typically observe that the volatility parameter of the variance gamma model reduces from shorter to longer maturities; however, overall volatility stays in a tight range. Kurtosis increases for longer maturity; skewness, on the other hand, reduces. For equity options we typically observe negative skew.

Here we present a couple of calibration cases for the variance gamma model. In the first case, we look into the limiting behavior of the variance gamma model. We show that if out-of-the-money call and put prices from a Black–Scholes model with a constant volatility are provided as market prices, the variance gamma model can detect premiums are coming from Black–Scholes with a constant volatility and that indicates that variance gamma can recover a pure diffusion model under the special case of $\nu = 0$ and $\theta = 0$.

Parameters used for this example are spot price \$100, volatility $\sigma = 0.40$, maturity $T = 0.5$ year, risk-free rate $r = 2\%$, dividend rate $q = 1.0\%$, strike prices ranging from $80, 85, 90, \ldots, 125$. Initial parameters used for the variance gamma model are $\sigma = 0.10, \nu = 0.20$, and $\theta = -0.20$. Obtained parameters from calibration using Nelder–Mead optimization routine are $\sigma = 0.403928$, $\nu = 0.021429$, $\theta = -0.049330$. In Figure 25.5 we plot variance gamma calibrated prices versus Black–Scholes prices. Obviously, they match very closely and it is very difficult to distinguish them from each other.

In the second calibration case, we obtain variance gamma parameters from calibration to S&P 500 options. Table 25.3 displays the variance gamma parameters obtained from calibration of S&P 500 out-of-the-money European option prices (calls/puts) on March 27, 2012, across various different maturities. Initial guess for the variance gamma parameter set is $\sigma = 0.20$, $\nu = 0.20$, $\theta = -0.2$. S&P 500 spot price on March 27, 2012 was \$1412.52.

We use the option data provided in Tables 25.1 and 25.2 to obtain variance gamma models parameters via calibration to S&P 500 options for each maturity separately. Using the calibrated variance gamma model, we generate call and put option premiums and plot them against the market data. We note negative values for θ that overall decline with maturity. This is a reflection of negative skewness induced by risk aversion that declines in the implied annualized risk-neutral density. The parameter ν is constantly rising with time to maturity, which reflects an increase in excess kurtosis for the implied annualized risk-neutral density. The volatility parameter is increasing with time to maturity but in a relatively tight range. For illustrative purposes, we illustrate variance gamma premiums versus market premiums (mid prices) for the S&P 500 on March 27, 2012 for maturities of 24 and 93 days in Figures 25.6 and 25.7, respectively. On those

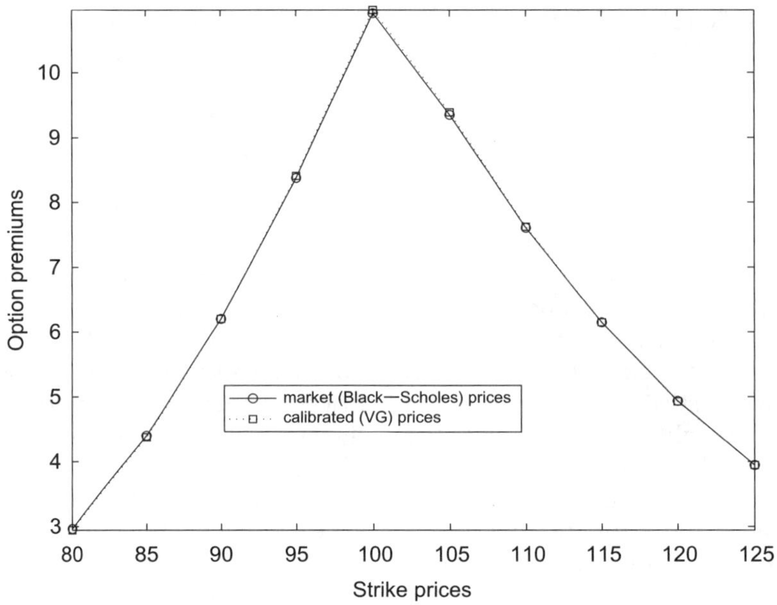

FIGURE 25.5 VG calibrated to Black–Scholes.

TABLE 25.3 VG Parameters Obtained from Calibration of the S&P 500 on March 27, 2012

Time to Maturity	σ	ν	θ	r	q	Spot
0.06575	0.1230	0.0425	−0.3398	0.00232	0.01531	1412.52
0.14247	0.1494	0.1567	−0.1512	0.00324	0.02109	1412.52
0.21918	0.1432	0.2306	−0.1952	0.00433	0.02099	1412.52
0.25479	0.1438	0.2333	−0.2082	0.00483	0.02082	1412.52
0.31507	0.1513	0.3385	−0.1727	0.00557	0.01982	1412.52
0.48767	0.1649	0.6021	−0.1460	0.00732	0.0209	1412.52
0.50411	0.1659	0.6344	−0.1487	0.00744	0.02098	1412.52
0.73699	0.1678	0.8594	−0.1430	0.00883	0.02155	1412.52

figures, the plot on the right is for call option premiums and the left plot is for put option premiums. We also show the error bars for bids and asks as well. In calibration we utilize Nelder–Mead optimization routine. What happens if instead we would use a grid search approach? Assume we lay down a grid having three parameters as

$$\bar{D} = \left\{ \begin{array}{l} \sigma_i = a + (i-1)\Delta\sigma; \ \Delta\sigma = \frac{b-a}{M}; \ i = 1, \ldots, M+1 \\ \nu_j = c + (j-1)\Delta\nu; \ \Delta\nu = \frac{d-c}{N}; \ j = 1, \ldots, N+1 \\ \theta_k = e + (k-1)\Delta\theta; \ \Delta\theta = \frac{f-e}{K}; \ k = 1, \ldots, K+1 \end{array} \right\} \tag{25.19}$$

and consider the following lower and upper limits: for volatility $a = 0.05$ and $b = 0.40$, for kurtosis $c = 0.01$ and $d = 0.90$, for skewness $e = -0.40$ and $f = 0.20$, and the number of grid points

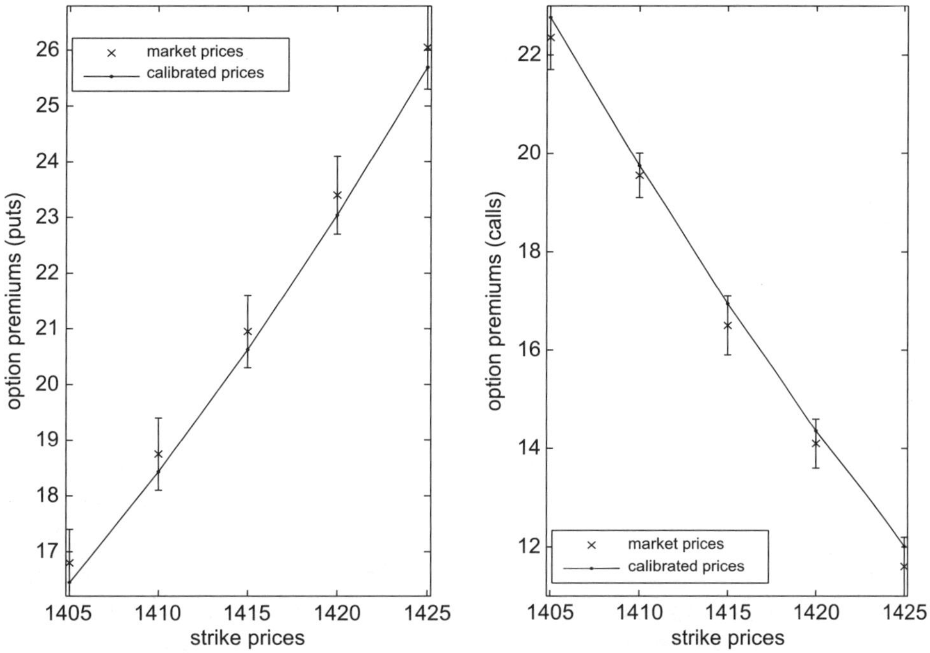

FIGURE 25.6 Variance gamma versus market premiums for the S&P 500 on March 27, 2012 for time to maturity of 24 days.

$M=15$, $N=89$, and $K=60$. Then for each maturity having 10 options (5 calls and 5 puts) we would have needed 878,400 option evaluations, which would be pretty expensive.

25.3 OVERVIEW OF FILTERING AND ESTIMATION

In the calibration procedure, the time series of prices do not come into process and as discussed we typically use the market snapshot of prices. It could be a case that we are interested in estimating parameters of a model by finding the parameter set that would best fit the history of prices as opposed to just the snapshots. We start by a simple example of estimation the parameter of geometric Brownian motion (the Black–Scholes model). Consider the following stochastic differential equation

$$dS_t = \mu S_t dt + \sigma S_t dW_t \qquad (25.20)$$

The exact solution to this stochastic differential equation is given by

$$S_t = S_0 \exp\left(\mu - \frac{\sigma^2}{2}\right) t + \sigma W_t \qquad (25.21)$$

$$= S_0 \exp\left(\mu - \frac{\sigma^2}{2}\right) t + \sigma\sqrt{t}Z \qquad (25.22)$$

where Z is a sample from a standard normal distribution, $\mathcal{N}(0,1)$. Assuming drift of 5%, $\mu = 0.05$, volatility of 20% and, $\sigma = 0.25$, we simulate a sample path of thousand daily observations as illustrated in Figure 25.8. Needless to say for simulation we start from S_0 and do it as follows

$$S_{t+1} = S_t \exp\left(\mu - \frac{\sigma^2}{2}\right) \Delta t + \sigma\sqrt{\Delta t}Z \qquad (25.23)$$

There are various ways of maximizing the likelihood. Here we achieve that by (a) grid search

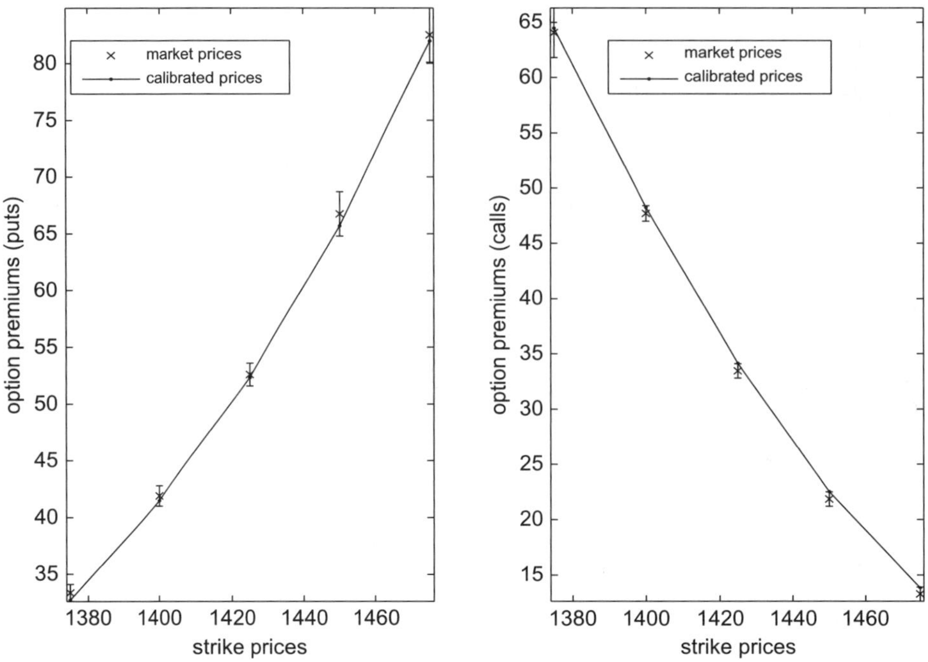

FIGURE 25.7 Variance gamma versus market premiums for the S&P 500 on March 27, 2012 for time to maturity of 93 days.

(brute-force) routine, (b) simplex (Nelder–Mead) algorithm. First, we use a grid search (brute-force) routine to maximize the likelihood. Setting up a grid as

$$\bar{D} = \left\{ \begin{array}{l} \mu_j = a + (j-1)\Delta\mu;\ \Delta\mu = \frac{b-a}{N};\ j = 1,\ldots,N+1 \\ \sigma_k = c + (k-1)\Delta\sigma;\ \Delta\sigma = \frac{d-c}{M};\ k = 1,\ldots,M+1 \end{array} \right\} \tag{25.24}$$

where in our case for the lower and upper bounds for the drift we choose $a = 0.01$ and $b = 0.20$, for the lower and upper bounds for volatility we pick $c = 0.05$ and $d = 0.50$, and assuming $N = 190$ and $M = 45$ steps in drift and volatility space respectively. That implies step sizes are $\Delta\mu = 0.001$ and $\Delta\sigma = 0.01$. For the geometric Brownian motion the likelihood at time t is given by

$$f(S_{t+1}, \mu, \sigma | S_t) = \frac{1}{\sqrt{2\pi}} \frac{1}{S_{t+1}\sigma\sqrt{\Delta t}} \times \exp\left(- \frac{\left(\log S_{t+1} - \log S_t - \left(\mu - \frac{\sigma^2}{2}\right)\Delta t\right)^2}{2\sigma^2 \Delta t} \right) \tag{25.25}$$

Therefore, the total (logarithmic of) likelihood would be

$$\sum_{i=1}^{N} \log f(S_i, \mu, \sigma | S_{i-1}) = \sum_{i=1}^{N} -\log\left(S_i \sigma \sqrt{\Delta t} - \frac{\left(\log S_i - \log S_{i-1} - \left(\mu - \frac{\sigma^2}{2}\right)\Delta t\right)^2}{2\sigma^2 \Delta t} \right) \tag{25.26}$$

By the grid search we obtain the maximum which occurs at $\mu = 0.045$ and $\sigma = 0.21$, which is pretty close to the true values of μ and σ.

Now using Nelder–Mead optimizer starting from initial parameters $[\hat{\mu} \quad \hat{\sigma}] = [0.01 \quad 0.01]$

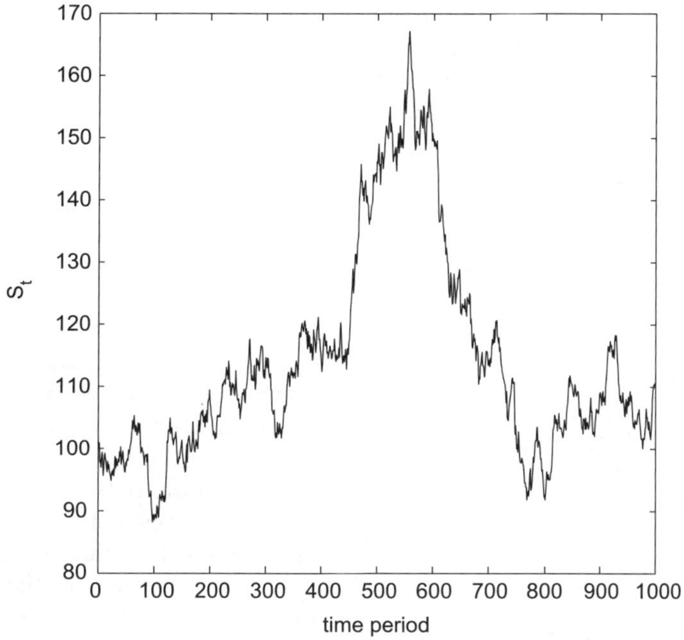

FIGURE 25.8 Example grid.

and maximizing the logarithmic of the likelihood (minimizing negative of the logarithmic of likelihood) we obtain $\mu = 0.0446$ and $\sigma = 0.2065$, which is pretty close to the results of grid search and the true values. In general, estimating volatility is much easier than the drift. See the exercise at the end of the chapter. The other item to know is that typically larger step size works better than shorter ones but that might not be feasible.

This above example on estimation is based on maximum likelihood function. In this example, the likelihood is available in an analytical form. We now consider an example where the likelihood is not known analytically. Assume a state evolves according to the following simple linear model:

$$x_{t+1} = ax_t + \lambda z_{t+1} \tag{25.27}$$

where $z_{t+1} \sim \mathcal{N}(0,1)$. Call y_{t+1} the market price (observation) at time $t+1$ and assume it is linked to the model price via the following relationship:

$$y_{t+1} = h\big(x_{t+1};\Theta\big) + \sigma u_{t+1} \tag{25.28}$$

where $u_{t+1} \sim \mathcal{N}(0,1)$ and h is the pricing model, with Θ being the parameter set that also includes a, λ, and σ. Having the time series of observations, y_{t+1} for $t = 1, \ldots, T$, we cannot calculate the likelihood in a closed-form, this is due to the fact that x_{t+1} is a variable and not known. Conditionally on the state x_t one can find that the (conditional) likelihood; however, there are many different values for x_{t+1} that we can conditional on. The state x_{t+1} is called hidden state and they are not directly observable.

There are two possible scenarios. The first one is the case that the parameter set Θ is already known/given and we are interested in finding the best estimate of x_{t+1} observing y_{t+1}. That implies starting at $t = 0$; we assign an arbitrary value to x_0 and given y_1 we find the best estimate for x_1 and at every time period we repeat this. By doing this, we construct a time series of the hidden state. We note that the obtained estimate of the state from the previous time step along with the observation at the current time step can be used to calculate the best estimate of the state at the current time. Therefore it is accepted to assume factors at time t, namely x_t, are given. Now, given an observation at time $t+1$ we want to calculate the best estimate of x_{t+1}, namely $\hat{x}_{t+1}$, that is,

$$\hat{x}_{t+1} = \mathbb{E}\left(x_{t+1} \mid y_{t+1}\right)$$

where y_{t+1} is the observation at time $t+1$. Assume the model price is given by $h(x_{t+1}; \Theta)$ where, as earlier stated, Θ is the parameter set. The assumption in filtering is that the market price (observation) at time $t+1$, y_{t+1}, is linked to the model price via the following relationship:

$$y_{t+1} = h\big(x_{t+1}; \Theta\big) + \sigma u_{t+1} \tag{25.29}$$

where $u_{t+1} \sim \mathcal{N}(0, 1)$. Both λ and σ are part of the parameter set Θ and therefore are already estimated and known.

Knowing the evolution equation for the hidden state x_{t+1}, we can first generate m samples for x_{t+1}.

$$x_{t+1}^{(i)} = ax_t + \lambda\mathcal{N}(0, 1) \tag{25.30}$$

for $i = 1, \ldots, m$. Having m samples of $x_{t+1}^{(i)}$, we now can calculate m samples for the model price, that is $h(x_{t+1}^{(i)}; \Theta)$. Now we can generate m samples for u_{t+1}, having observed the market price at time $t+1$:

$$u_{t+1}^{(i)} = y_{t+1} - h\left(x_{t+1}^{(i)}; \Theta\right) \tag{25.31}$$

We define $\mathcal{L}^{(i)}$ as the (conditional) likelihood function

$$\mathcal{L}^{(i)} \equiv \text{Likelihood}\left(u_{t+1}^{(i)} \mid x_{t+1}^{(i)}\right) \tag{25.32}$$

Hence $\mathcal{L}^{(i)}$ simply is

$$\mathcal{L}^{(i)} = \frac{e^{-\frac{\left(u_{t+1}^{(i)}\right)^2}{2\sigma^2}}}{\sqrt{2\pi}\sigma} \tag{25.33}$$

and therefore

$$\hat{x}_{t+1} = \mathbb{E}\left(x_{t+1} \mid z_{t+1}\right) \tag{25.34}$$

$$= \frac{\sum_{i=1}^{M} \mathcal{L}^{(i)} \times x_{t+1}^{(i)}}{\sum_{i=1}^{M} \mathcal{L}^{(i)}} \tag{25.35}$$

This is the best estimate of x_{t+1}. For the best estimate of the next time step, what we just obtained is used as the best estimate of the current step and proceed sequentially. The first finding is that during filtering we keep the parameter set fixed. As an example consider the case that the model price is given by $h\big(x_t; \Theta\big) = bx_t$. We consider the following parameter set: $a = 0.97, b = 1.01, \lambda = 0.1$, and $\sigma = 0.25$. In Figures 25.9(a) and 25.9(b) we illustrate a sample path for the hidden state and the observation respectively. In Figure 25.10 we display the hidden state obtained by filtering technique described versus the original hidden state. We use 100 samples, $m = 100$. The second scenario is the case that we do not have the parameter set and we seek to estimate it given the time series of observations. There are various ways of doing this, one of which is to start with an initial guess for the parameter set. Given the initial guess, we can do what we perform in the former case and predict the observation for the next time step, calculate the distance between the true observation and its prediction, calculate the sum of all those distances and regard it as a proxy for the likelihood. Then our criterion is to find the parameter set that minimizes the proxy and it would be the optimal parameter set. Some quick remarks are: (i) given any new guess for the parameter set, we have to conduct the filtering step and (ii) the time series of the hidden state is the byproduct of this procedure.

25.3.1 Kalman Filter

A more formal way to estimate is done through Kalman filter for linear cases and extended versions of Kalman filter such as extended Kalman filter, unscented Kalman filter, and particle filter for nonlinear cases.

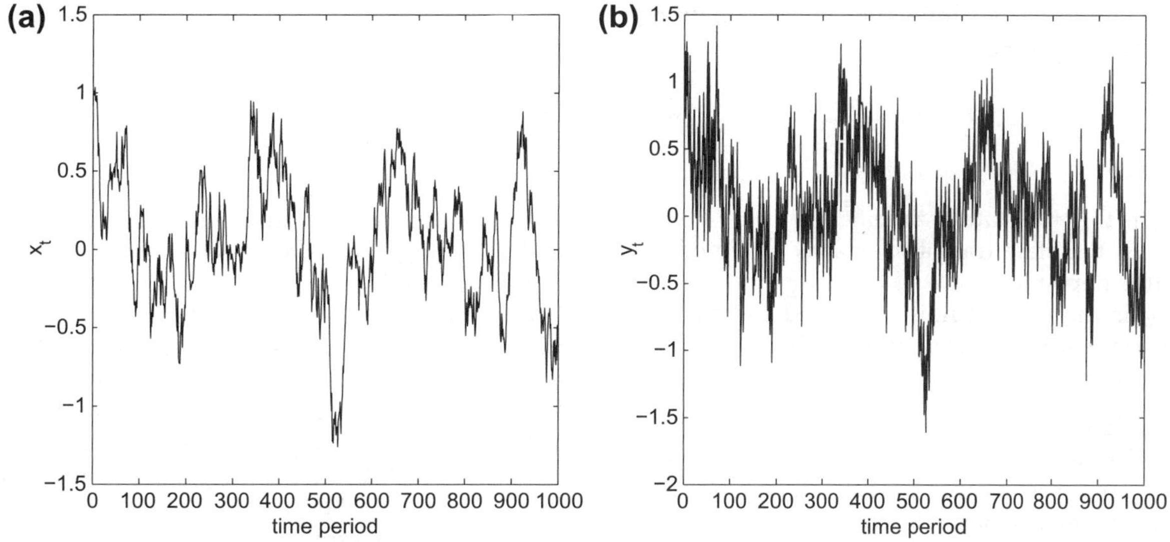

FIGURE 25.9 Filter example: (a) hidden state, (b) observation.

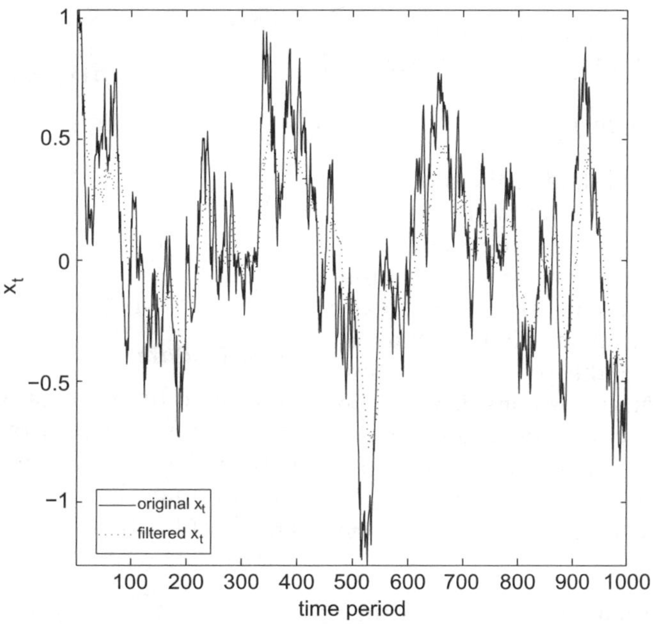

FIGURE 25.10 Original hidden state versus the filtered hidden state.

Assuming the following linear equation for evolution of the state

$$x_k = F_k x_{k-1} + w_k \tag{25.36}$$

where

x_k is the true state at time t_k.
F_k is the state transition matrix assumed to be time dependent.
w_k is the process noise assumed to be a multivariate normal distribution with zero mean and covariance Q_k, that is, $w_k \sim \mathcal{N}(0, Q_k)$.

Here x_k is an $n \times 1$ vector of real numbers and F_k is an $m \times n$ matrix. At time k an observation or measurement y_k of state x_k is made which is assumed to have the following measurement equation for its evolution:

$$y_k = H_k x_k + v_k \tag{25.37}$$

where H_k is the observation matrix and v_k is the observation noise assumed to be a white noise (Gaussian) with zero mean and covariance R_k, that is,

$$v_k \sim \mathcal{N}(0, R_k) \tag{25.38}$$

We assume that the initial state x_0, the state noise vectors $w_1, \ldots, w_k$, and measurement noise vectors $v_1, \ldots, v_k$ to be mutually independent. In reality, it is uncommon to presume a dynamical system follows the above framework. Having said that, considering the Kalman filter is designed to operate in the presence of noise makes it a useful filter and is considered to be a good start. There are variations and extensions on the Kalman filter that would allow more sophisticated models.

We characterize the state of the filter by the following symbols:

$\hat{x}_{k|k-1}$ presents the estimate of the state at time k given observations up to and including time $k-1$.
$\hat{x}_{k|k}$ presents the estimate of the state at time k given observations up to and including time k.
$P_{k|k-1}$ presents the error covariance matrix at time k given observations up to and including time $k-1$.
$P_{k|k}$ presents the error covariance matrix at time k given observations up to and including time k.

Error covariance matrix is a measure of the estimated accuracy of the state estimate. Formally filtering is done in two steps: (a) *time update or prediction* and (b) *measurement update or filtering*. In the prediction phase, we use the state estimate from the previous time step in order to produce an estimate of the state at the current time step. In the filtering phase, we use the measurement at the current time step to refine this prediction to obtain a new and (hopefully) a more accurate state estimate for the current time step.

Before we go through these steps, we need to define some invariants. If the model is precise and the values for $\hat{x}_{0|0}$ and $P_{0|0}$ accurately reflect the distribution of the initial state values, then the following invariants must be preserved: (a) all estimates have mean error zero, that is

$$\mathbb{E}(x_k - \hat{x}_{k|k}) = 0 \tag{25.39}$$

$$\mathbb{E}(x_k - \hat{x}_{k|k-1}) = 0 \tag{25.40}$$

and (b) covariance matrices accurately reflect the covariance of estimates, that is

$$P_{k|k} = \mathrm{cov}(x_k - \hat{x}_{k|k}) \tag{25.41}$$

$$P_{k|k-1} = \mathrm{cov}(x_k - \hat{x}_{k|k-1}) \tag{25.42}$$

Time Update

Starting at time k, what we have is the estimate of the state at the previous time, that is, $\hat{x}_{k-1|k-1}$. Having that, one can obtain the prediction for the state at time k by simply substituting $\hat{x}_{k-1|k-1}$ in Eq. (25.36) without the noise. Therefore the predicted state estimate is given by

$$\boxed{\hat{x}_{k|k-1} = F_k \hat{x}_{k-1|k-1}} \tag{25.43}$$

For the predicated state estimate, we are now interested in calculating the predicted estimate covariance that is $P_{k|k-1} = \mathrm{cov}(x_k - \hat{x}_{k|k-1})$.

$$
\begin{aligned}
P_{k|k-1} &= \text{cov}(x_k - \hat{x}_{k|k-1}) && \text{definition of } P_{k|k-1} \\
&= \text{cov}(x_k - F_k\hat{x}_{k-1|k-1}) && \text{substituting (25.43) for } \hat{x}_{k|k-1} \\
&= \text{cov}(F_k x_{k-1} + w_k - F_k\hat{x}_{k-1|k-1}) && \text{substituting (25.36) for } x_k \\
&= \text{cov}(F_k(x_{k-1} - \hat{x}_{k-1|k-1}) + w_k) && \text{gathering terms} \\
&= \text{cov}(F_k(x_{k-1} - \hat{x}_{k-1|k-1})) + \text{cov}(w_k) && \text{assuming the terms are independent} \\
&= F_k\text{cov}((x_{k-1} - \hat{x}_{k-1|k-1}))F_k^\top + Q_k && \text{using the fact that cov}(Ax) = A\text{cov}(x)A^\top \\
&= F_k P_{k-1|k-1} F_k^\top + Q_k && \text{definition of } P_{k-1|k-1}
\end{aligned}
$$

---

and therefore predicted estimate covariance is

$$
\boxed{P_{k|k-1} = F_k P_{k-1|k-1} F_k^\top + Q_k} \tag{25.44}
$$

Measurement Update

Having the predicted state estimate, $\hat{x}_{k|k-1}$, we can substitute it into the measurement equation (25.37) and calculate the measurement residual, that is,

$$
\boxed{\hat{\delta}_k = y_k - H_k\hat{x}_{k|k-1}} \tag{25.45}
$$

Calling S_k the residual covariance, one can calculate as follows

$$
\begin{aligned}
S_k &= \text{cov}(\hat{\delta}_k) && \text{definition of } S_k \\
&= \text{cov}(y_k - H_k\hat{x}_{k|k-1}) && \text{substituting (25.45)} \\
&= \text{cov}(H_k x_k + v_k - H_k\hat{x}_{k|k-1}) && \text{substituting (25.37)} \\
&= \text{cov}(H_k(x_k - \hat{x}_{k|k-1}) + v_k) && \text{regrouping terms} \\
&= \text{cov}(H_k(x_k - \hat{x}_{k|k-1})) + \text{cov}(v_k) && \text{assuming terms are independent} \\
&= H_k\text{cov}(x_k - \hat{x}_{k|k-1})H_k^\top + R_k && \text{using the fact that cov}(Ax) = A\text{cov}(x)A^\top \\
&= H_k P_{k|k-1} H_k^\top + R_k && \text{definition of } P_{k|k-1}
\end{aligned}
$$

Therefore the residual covariance is

$$
\boxed{S_k = H_k P_{k|k-1} H_k^\top + R_k} \tag{25.46}
$$

To update the state estimate, it seems trivial to think we add the residual $\hat{\delta}_k$ to the predicted state estimate $\hat{x}_{k|k-1}$, or to be more generic we write the updated state estimate as

$$
\boxed{\hat{x}_{k|k} = \hat{x}_{k|k-1} + K_k\hat{\delta}_k} \tag{25.47}
$$

where K_k is called the Kalman gain. At this point, we do not make any assumption on its optimality. Now, we are interested in updated estimate covariance matrix, so-called posterior estimate covariance matrix, that is, $P_{k|k}$. We can find an expression for the posterior estimate covariance matrix as follows

---

$$
\begin{aligned}
P_{k|k} &= \text{cov}(x_k - \hat{x}_{k|k}) && \text{definition of } S_k \\
&= \text{cov}(x_k - (\hat{x}_{k|k-1} + K_k\hat{\delta}_k)) && \text{substituting the updated state from Eq. (25.47)} \\
&= \text{cov}(x_k - (\hat{x}_{k|k-1} + K_k(y_k - H_k\hat{x}_{k|k-1}))) && \text{substituting for } \hat{\delta}_k \text{ using (25.45)} \\
&= \text{cov}(x_k - (\hat{x}_{k|k-1} + K_k(H_k x_k + v_k - H_k\hat{x}_{k|k-1}))) && \text{substituting for } y_k \text{ using (25.37)} \\
&= \text{cov}((I - K_kH_k)(x_k - \hat{x}_{k|k-1}) - K_kv_k) && \text{rewriting by gathering terms}
\end{aligned}
$$

The assumption is that measurement error v_k is uncorrelated with other terms and therefore we arrive at

$$
\begin{aligned}
P_{k|k} &= \text{cov}((I - K_kH_k)(x_k - \hat{x}_{k|k-1})) \\
&\quad + \text{cov}(K_kv_k) && (25.48) \\
&= (I - K_kH_k)\text{cov}(x_k - \hat{x}_{k|k-1}) \\
&\quad (I - K_kH_k)^\top + K_k\text{cov}(v_k)K_k^\top && (25.49)
\end{aligned}
$$

Now using the definition of $P_{k|k-1}$ and R_k the updated estimate covariance becomes

$$
\boxed{P_{k|k} = (I - K_kH_k)P_{k|k-1}(I - K_kH_k)^\top + K_kR_kK_k^\top} \tag{25.50}
$$

and it holds for any arbitrary value for K_k. So far we have not made any assumption on K_k. We can reduce this expression further for the optimal Kalman gain.

25.3.2 Optimal Kalman Gain, its Interpretation and Posterior Estimate Covariance

The Kalman filter minimizes posterior state estimation, which is equivalent to saying it minimizes mean-square error estimator. The error in the posterior state estimation is

$$
\epsilon_k \equiv x_k - \hat{x}_{k|k} \tag{25.51}
$$

The goal is to minimize the expected value of the square of ϵ_k

$$
\mathbb{E}(|x_k - \hat{x}_{k|k}|^2) \tag{25.52}
$$

which is the same as minimizing the trace of the posterior estimate covariance matrix $P_{k|k}$. The trace of an $n \times n$ square matrix A is defined to be

$$
\text{Tr}(A) = \sum_{i=1}^{n} a_{ii} \tag{25.53}
$$

that is the sum of the diagonal elements. We get the following after expanding the terms in $P_{k|k}$:

$$
\begin{aligned}
P_{k|k} &= P_{k|k-1} - K_kH_kP_{k|k-1} - P_{k|k-1}H_k^\top K_k^\top \\
&\quad + K_k(H_kP_{k|k-1}H_k^\top + R_k)K_k^\top && (25.54) \\
&= P_{k|k-1} - K_kH_kP_{k|k-1} \\
&\quad - P_{k|k-1}H_k^\top K_k^\top + K_kS_kK_k^\top && (25.55)
\end{aligned}
$$

Now taking the first derivative of the trace of $P_{k|k}$ with respect to K_k and setting it equal to zero, we find the optimal k_k:

$$
\frac{\partial \text{Tr}(P_{k|k})}{\partial K_k} = -2(H_kP_{k|k-1})^\top + 2K_kS_k = 0 \tag{25.56}
$$

Solving it for K_k yields

$$
K_kS_k = (H_kP_{k|k-1})^\top = P_{k|k-1}H_k^\top \tag{25.57}
$$

or

$$
K_k = P_{k|k-1}H_k^\top S_k^{-1} \tag{25.58}
$$

This Kalman gain minimizes the mean-square error estimate and is the optimal Kalman gain:

$$
K_kS_kK_k^\top = P_{k|k-1}H_k^\top K_k^\top \tag{25.59}
$$

Substituting it back into $P_{k|k}$ yields

$$
P_{k|k} = P_{k|k-1} - K_kH_kP_{k|k-1} - P_{k|k-1}H_k^\top K_k^\top + K_kS_kK_k^\top \tag{25.60}
$$

Notice that the last two terms cancel out, which gives us

$$P_{k|k} = P_{k|k-1} - K_k H_k P_{k|k-1} \qquad (25.61)$$
$$= (I - K_k H_k) P_{k|k-1} \qquad (25.62)$$

An interpretation of the Kalman filter could be based on linear regression. In the case of having time series of $\{\mathbf{x}_k\}$ and $\{\mathbf{y}_k\}$ the linear regression yields

$$\mathbf{y}_k = \alpha + \beta \mathbf{x}_k + \epsilon_k \qquad (25.63)$$

with α the intercept, β the slope, and ϵ_k the residual. Under linear regression we have

$$\beta = P_{k|k-1} H_k^\top S_k^{-1} \qquad (25.64)$$

which is the expression for the Kalman gain.

In the Kalman filter, the log likelihood for each time step is $\log p(z_t|z_{1:t-1})$, which is obtained by evaluating the log of the probability density function of a multivariate Gaussian density with mean zero and covariance of S_k evaluated at the values in $\hat{\delta}_k$, that is, the log likelihood of innovation

$$\log p(y_t|y_{1:t-1}) = \log\left(\frac{1}{(2\pi)^{d/2}|S_k|^{1/2}} \exp\left(-\frac{1}{2}(\hat{\delta}_k - 0)^\top S_k^{-1}(\hat{\delta}_k - 0)\right)\right) \qquad (25.65)$$
$$= -\frac{d}{2}\log(2\pi) - \frac{1}{2}\log|S_k| - \frac{1}{2}(\hat{\delta}_k - 0)^\top S_k^{-1}(\hat{\delta}_k - 0) \qquad (25.66)$$
$$= -\frac{d}{2}\log(2\pi) - \frac{1}{2}\log|S_k| - \frac{1}{2}(z_k - H_k\hat{x}_{k|k-1})^\top S_k^{-1} (z_k - H_k\hat{x}_{k|k-1}) \qquad (25.67)$$

Calling this log likelihood $\ln(l_k)$, then the log likelihood for the entire time series is $\ln(L_{1:T}) = \sum_{k=1}^{T}\ln(l_k)$. Having the log likelihood, we can either use an optimization routine (for example, Nelder–Mead approach) to maximize (minimize the negative) log likelihood for parameter estimation. It is important to notice that the Kalman gain does not come into the parameter estimation; it only enters into the filtering step.

25.4 EXERCISES

1. Knowing the log likelihood equation in the Kalman filter given by Eq. (25.67), use it to estimate the parameter of the linear model from the example done in the chapter.
2. To see the impact of the choice of the objective function on parameter set obtained from calibration to the same data set, we redo the example done in this chapter under various different objective function. Use the option data given in Tables 25.1 and 25.2 to calibrate the variance gamma parameters for each maturity using out-of-the-money options for the following objective functions

   a. $\sum_{i=1}^{n}\left(C_i^\Theta - C_i\right)^2$
   b. $\sum_{i=1}^{n}|C_i^\Theta - C_i|$
   c. $\sum_{i=1}^{n}\left(\frac{C_i^\Theta - C_i}{C_i}\right)^2$
   d. $\sum_{i=1}^{n}\left|\frac{C_i^\Theta - C_i}{C_i}\right|^2$
   e. $\sum_{i=1}^{n} w_i\left(C_i^\Theta - C_i\right)^2$ where $w_i = \frac{1}{\left(C_i^{\text{ask}} - C_i^{\text{bid}}\right)^2}$
   f. $\sum_{i=1}^{n} w_i\left(C_i^\Theta - C_i\right)^2$ and w_i as before

   Compare and conclude on your results.
3. Use data in Tables 25.1 and 25.2, and Eq. (25.12) to construct the local volatility surface. In order to do it, set up a uniform grid, then use bi-cubic spline interpolation/extrapolation to find the values at those grid points and then apply Eq. (25.12) to evaluate local volatility value at each grid point. Some hints on how to evaluate the partial derivatives will be provided.

References

Andreasen, Jesper, 1998. Implied Modelling, Stable Implementation, Hedging, and Duality. Manuscript, University of Aarhus.

Arvanitis, A., Gregory, J., Laurent, J.P., 1999. Building models for credit spreads. Journal of Derivatives,.

Bertoin, Jean, 1996. Lévy Processes. Cambridge University Press, Cambridge. Tracts in Mathematics 121.

Carr, Peter, Madan, Dilip B., 1999. Option valuation using the fast Fourier transform. Journal of Computational Finance 2 (4), 61–73.

Carr, Peter, Madan, Dilip B., 2009. Saddlepoint methods for option pricing. Journal of Computational Finance 13 (1) Fall.

Carr, Peter, Geman, Hélyette, Madan, Dilip B., Yor, Marc, 2002. The fine structure of asset returns: an empirical investigation. Journal of Business 75 (2), 305–332.

Carr, Peter, Geman, Hélyette, Madan, Dilip B., Yor, Marc, 2003. Stochastic volatility for Lévy processes. Mathematical Finance 13 (3), 345–382.

Cont, Rama, 2010. Model Calibration. In: Cont, Rama (Ed.). Encyclopedia of Quantitative Finance, vol. 3. John Wiley and Sons Ltd., Southern Gate, Chichester, West Sussex, England, pp. 1210–1218.

Cont, Rama, Tankov, Peter, 2003. Financial Modelling with Jump Processes. Chapman & Hall/CRC Financial Mathematics Series.

Cooley, James W., Tukey, John W., 1965. An algorithm for the machine calculation of complex Fourier series. Mathematics of Computation 19 (90), 297–301.

Derman, Emanuel, Kani, Iraj, 1994. The volatility smile and its implied tree. Risk 7 (2), 32–39.

Dupire, Bruno, 1994. Pricing with a smile. Risk 7 (1), 18–20.

Gregory, Jon, Laurent, Jean-Paul, 2005. Basket Default Swaps, CDOs and Factor Copulas. Review of Risk 7 (4), 103–122.

Grimmett, Geoffrey R., Stirzaker, David R. , 1992. Probability and Random Processes, second ed. Oxford University Press Inc., New York, USA.

Hirsa, Ali, 2012. Computational Methods in Finance. Chapman & Hall/CRC Financial Mathematics Series.

Hatgioannides, John, Petropoulos, George, 2007. On credit spreads, credit spread options and implied probabilities of default. Faculty of Finance, Cass Business School, UK and Eurobank EFG, Athens, Greece (January).

Hirsa, Ali, Courtadon, Georges, Madan, Dilip B., March 2001. Local Volatility Reengineers Semi-martingle Models. Morgan Stanley.

Hirsa, Ali, Courtadon, Georges, Madan, Dilip B., 2003. The effect of model risk on the valuation of barrier options. Journal of Risk Finance 4 (2).

Jacod, Jean, Shiryaev, Albert N. , 1987. Limit Theorems for Stochastic Processes. Springer-Verlag, Berlin, New York.

Jarrow, Robert A., Thurnbull, Stuart M., 1995. Pricing of options on financial securities subject to credit risk. Journal of Finance, 53–85.

Jarrow, Robert A., Turnbull, Stuart M., 1995. Pricing derivatives on financial securities subject to credit risk. Journal of Finance 50 (1), 53–85.

Longstaff, Francis A., Schwartz, Eduardo S., 1992. Interest rate volatility and the term structure: a two-factor general equilibrium model. Journal of Finance 47, 1259–1282.

Longstaff, Francis A., Schwartz, Eduardo S., 1995. A simple approach to valuing risky fixed and floating rate debt. Journal of Finance 50 (3), 789–819.

Madan, Dilip B., Carr, Peter, Chang, Eric C., 1998. The variance gamma process and option pricing. European Finance Review 2, 79–105.

Madan, Dilip B., Unal, Haluk, 2000. A two-factor hazard rate model for pricing risky debt and the term structure of credit spreads. Journal of Financial and Quantitative Analysis 35 (1), 43–65.

Merton, Robert C., 1976. Option pricing when underlying stock returns are discontinuous. Journal of Financial Economics 3, 125–144.

Øksendal, Bernt , 2000. Stochastic Differential Equations. An Introduction with Applications, fifth ed. Springer-Verlag, New York.

Papapantoleon, Antonis, 2008. An Introduction to Lévy Processes with Applications in Finance. Lecture notes taught

Financial Derivatives, Third Edition.
http://dx.doi.org/10.1016/B978-0-12-384682-2.00031-1

at the University of Piraeus, University of Leipzig, and at the Technical University of Athens.

Protter, Philip , 1990. Stochastic Integration and Differential Equations. Springer-Verlag, New York.

Sato, Ken-iti, 2001. Basic results on Lévy processes. In: Barndorff-Nielsen, Ole E., Mikosch, Thomas, Resnick, Sidney I. (Eds.), Lévy Processes: theory and Applications. Birkhauser, Boston, MA, USA, pp. 3–37.

Index

C

D

E

F

G

H

I